GEOPHYSICAL DATA ANALYSIS AND INVERSE THEORY WITH MATLAB® AND PYTHON

GEOPHYSICAL DATA ANALYSIS AND INVERSE THEORY WITH MATLAB® AND PYTHON

FIFTH EDITION

WILLIAM MENKE

Department of Earth and Environmental Sciences, Columbia University, New York, NY, United States

ELSEVIER

ACADEMIC PRESS
An imprint of Elsevier

Academic Press is an imprint of Elsevier
125 London Wall, London EC2Y 5AS, United Kingdom
525 B Street, Suite 1650, San Diego, CA 92101, United States
50 Hampshire Street, 5th Floor, Cambridge, MA 02139, United States

Notices

Knowledge and best practice in this field are constantly changing. As new research and experience broaden our understanding, changes in research methods, professional practices, or medical treatment may become necessary.

Practitioners and researchers must always rely on their own experience and knowledge in evaluating and using any information, methods, compounds, or experiments described herein. In using such information or methods they should be mindful of their own safety and the safety of others, including parties for whom they have a professional responsibility.

To the fullest extent of the law, neither the Publisher nor the authors, contributors, or editors, assume any liability for any injury and/or damage to persons or property as a matter of products liability, negligence or otherwise, or from any use or operation of any methods, products, instructions, or ideas contained in the material herein.

ISBN: 978-0-443-13794-5

For information on all Academic Press publications
visit our website at https://www.elsevier.com/books-and-journals

Cover Credit: William Menke
Publisher: Joseph Hayton
Acquisitions Editor: Jennette McClain
Editorial Project Manager: Teddy A. Lewis
Production Project Manager: Kumar Anbazhagan
Cover Designer: Greg Harris

Typeset by STRAIVE, India

Working together
to grow libraries in
developing countries

www.elsevier.com • www.bookaid.org

Dedication

To my parents
William Henry Menke
1925–1998

and

Lillian Alice Otto Menke
1919–2019

who taught me how to learn

Contents

6. Solution of the linear, Normal inverse problem, viewpoint 3: Maximum likelihood methods

7. Data assimilation methods including Gaussian process regression and Kalman filtering

8. Nonuniqueness and localized averages

9. Applications of vector spaces

10. Linear inverse problems with non-Normal statistics

11. Nonlinear inverse problems

This book has a companion website hosting complementary materials. Visit this URL to access it: https://www.elsevier.com/books-and-journals/book-companion/9780443137945

Preface

*For now we see in a mirror, dimly, but then … **Paul of Tarsus (53)***

Every researcher in the applied sciences who has analyzed data has practiced inverse theory. Inverse theory is simply the set of methods used to extract useful inferences about the world from physical measurements. The fitting of a straight line to data involves a simple application of inverse theory. Tomography, popularized by the physician's CT and MRI scanners, uses it on a more sophisticated level.

The study of inverse theory, however, is more than the cataloging of methods of data analysis. It is an attempt to organize these techniques, to bring out their underlying similarities and pin down their differences, and to deal with the fundamental question of the limits of information that can be gleaned from any given data set.

Physical properties fall into two general classes: those that can be described by discrete parameters (e.g., the mass of the Earth or the position of the atoms in a protein molecule) and those that must be described by continuous functions (e.g., temperature over the face of the Earth or electric field intensity in a capacitor). Inverse theory employs different mathematical techniques for these two classes of parameters: the theory of matrix equations for discrete parameters and the theory of integral equations for continuous functions.

Being introductory in nature, this book deals mainly with "discrete inverse theory," that is, the part of the theory concerned with parameters that either are truly discrete or can be adequately approximated as discrete. By adhering to these limitations, inverse theory can be presented on a level that is accessible to most first-year graduate students and many college seniors in the applied sciences. The only mathematics that is presumed is a working knowledge of the calculus and linear algebra and some familiarity with general concepts from probability theory and statistics.

Nevertheless, the treatment in this book is in no sense simplified. Realistic examples, drawn from the scientific literature, are used to illustrate the various techniques. Since in practice the solutions to most inverse problems require substantial computational effort, attention is given to how realistic problems can be solved.

The treatment of inverse theory in this book is divided into four parts. Chapter 1 is a brief primer in MATLAB® and Python that concentrates on subjects essential to data analysis, especially matrix manipulation and graphics. Chapters 2 and 3 provide a general background, explaining what inverse problems are and what constitutes their solution as well as reviewing some of the basic concepts from probability theory that will be applied throughout the text. Chapters 4–9 discuss the solution of the canonical inverse problem: the linear problem with Normal statistics. This is the best understood of all inverse problems because the theory relating unknowns to data takes the form of a simple matrix equation. It is here that the fundamental notions of uncertainty, uniqueness, and resolution can be most clearly developed. Chapters 10–12 extend the discussion to problems that are non-Normal and nonlinear.

Chapter 13 develops factor analysis and empirical function analysis, which provide ways of searching for patterns within a data set. They differ from the inverse theory methods discussed in earlier chapters in being undirected, that is, requiring no detailed theory to provide an explanation for the patterns.

Chapter 14 devotes special attention to the so called adjoint method, a mathematical technique that, over the past two decades, has become an increasingly important tool for solving inverse problems in seismology and climate science. Chapters 15 and 16 provide examples of the use of inverse theory and a discussion of the steps that must be taken to solve inverse problems on a computer. Chapter 17 develops several specialized mathematical methods and provides a series of step-by-step method summaries that pull together important results from throughout the book.

MATLAB® and Python scripts are used throughout the book as a means of communicating how the formulas of inverse theory can be used in computer-based data processing scenarios. MATLAB® is a commercial software product of The MathWorks, Inc. Python and its many add-ons, such as the NumPy linear algebra package, the SciPy scientific function package, and the Jupyter Notebook/JupyterLab environment, are freeware that offer an alternative to MATLAB® and other commercial products with license fees. Both are widely used in university settings as an environment for scientific computing.

All of the book's examples, its recommended homework problems, and the case studies in Chapter 13 use MATLAB® and Python extensively. Further, all MATLAB® and Python scripts used in the book are made available to readers through the book's website. The book is self-contained; it can be read straight through, and profitably, even by someone with no access to MATLAB® or Python. But it is meant to be used in a setting where students are actively using either MATLAB® or Python as an aid to studying (that is, by reproducing the examples and case studies described in the book) and as a tool for completing the recommended homework problems.

Many people helped me write this book. I am very grateful to my students at Columbia University and at Oregon State University for the helpful comments they gave me during the courses I have taught on inverse theory. Mike West, of the Alaska Volcano Observatory, did much to inspire recent revisions of the book by inviting me to teach a mini-course on the subject in the fall of 2009. The use of MATLAB® in this book parallels the usage in Environmental Data Analysis with MATLAB® (Menke and Menke, 2011), a data analysis textbook that I wrote with my son Joshua Menke in 2011. The many hours we spent working together on its tutorials taught us both a tremendous amount about how to use that software in a pedagogical setting. I thank my Summer Interns Emily Carrero Mustelier, Thomas Harper, Anna Ledeczi, Michael Pirrie, and Peter Skryzalin for assistance in the data analysis that underpins the cover image of the Northern Appalachian Anomaly (Menke et al. 2016), which I created using the GeoMapApp mapping software (https://www.geomapapp.org; Ryan et al., 2009). Finally, I thank the many hundreds of scientists and mathematicians whose ideas I drew upon in writing this book.

References

Menke, W., Menke, J., 2011. Environmental Data Analysis With MATLAB. Elsevier, Amsterdam, The Netherlands, 259 pp.

Menke, W., Skryzalin, P., Levin, V., Harper, T., Darbyshire, F., Dong, T., 2016. The Northern Appalachian Anomaly: a modern asthenospheric upwelling. Geophys. Res. Lett. 43, 10173–10179. https://doi.org/10.1002/2016GL070918.

Paul of Tarsus, 53. First Letter to the Corinthians 13:12 (translated from the Greek), (NASB®) New American Standard Bible®, Copyright © 1960 by The Lockman Foundation. Used by permission. All rights reserved. lockman.org.

Ryan, W.B.F., Carbotte, S.M., Coplan, J., O'Hara, S., Melkonian, A., Arko, R., Weissel, R.A., Ferrini, V., Goodwillie, A., Nitsche, F., Bonczkowski, J., Zemsky, R., 2009. Global multi-resolution topography (GMRT) synthesis data set. Geochem. Geophys. Geosyst. 10, Q03014. https://doi.org/10.1029/2008GC002332. Data https://doi.org/10.1594/IEDA.0001000.

1

Getting started with MATLAB® or Python

The practice of inverse theory requires computer-based computation. A person can learn many of the concepts of inverse theory by working through short pencil-and-paper examples and by examining precomputed figures and graphs. But beginners cannot become proficient in the practice of inverse theory that way because it requires skills that can only be obtained through the experience of working with large data sets. Three goals are paramount: to develop the judgment needed to select the best solution method among many alternatives, to build confidence that the solution can be obtained even though it requires many steps, and to strengthen the critical faculties needed to assess the quality of the results. This book devotes considerable space to case studies and homework problems that provide the practical problem-solving experience needed to gain proficiency in inverse theory.

Computer-based computation requires software. Many different software environments are available for the type of scientific computation that underpins data analysis. Some are more applicable and others less applicable to inverse theory problems, but among the applicable ones, none has a decisive advantage. Nevertheless, we have chosen MATLAB® or Python—we use the word "or" because only one or the other is needed—as the book's software environments, for several reasons, some having to do with their designs and others more practical. The most persuasive design reason is that both fully support linear algebra, which is needed by almost every inverse theory method. Furthermore, both support *scripts*, that is, sequences of data analysis commands that are communicated in written form and which serve to document the data analysis process. Practical considerations include the following: they are long-lived and stable platforms, available for several decades; implementations are available for most commonly used types of computers; and they are widely used in university settings.

In both MATLAB's and Python's scripting languages, data are presented as one or more named variables (in the same sense that c and d in the formula $c = \pi d$ are named variables). Data are manipulated by typing formulas that create new variables from old ones and by running scripts, that is, sequences of formulas stored in a file. Much of inverse theory is simply the application of well-known formulas to novel data, so the great advantage of this approach is that the formulas that are typed usually have a strong similarity to those printed in a textbook. Furthermore, scripts provide both a way of documenting the sequence of a formula used to analyze a particular data set and a way to transfer the overall data analysis procedure from one data set to another. However, one disadvantage is that the parallel between the syntax of the scripting language and the syntax of standard mathematical notation is nowhere near perfect. A person needs to learn to translate one into the other.

Part A. MATLAB® as a tool for learning inverse theory

1A.1 Getting started with MATLAB®

Unfortunately, this book must avoid discussion of the installation of MATLAB® and its appearance on your computer screen, for procedures and appearances vary from computer to computer and quickly become outdated, anyway. We will assume that you have successfully installed it and that you can identify the Command Window, the place where MATLAB® formula and commands are typed. Once you have identified the Command Window, try typing:

```
date
```

[gdama01_01]

MATLAB® should respond by displaying today's date.

You should copy the gdama folder to some convenient place on your computer's file system. It contains all files needed to perform this book's MATLAB® examples.

Although commands can be executed by typing them in the Command Window, we will seldom type them there, because the sequence of commands is lost when MATLAB® is closed at the end of session. Instead, we will write *scripts*; that is, sequences of commands written first to a file, and then executed as many times as we wish. The great advantage of a script is that it documents the sequence of steps used in the data analysis process.

All the scripts that are used in this book are freely available. This one is named gdama01_01 and is in Live Script file gdama01.mlx, which contains all the script for Chapter 1. Live Scripts are MATLAB's version of a research notebook and conventionally have filenames that end with ".mlx". Scripts, their output (including graphics), and text are comingled within a single document, which can be viewed, edited, and executed within MATLAB®; exported to a PDF file; etc.

1A.2 Effective use of folders

Files proliferate at an astonishing rate, even in the most trivial data analysis project. Data, notes, Live Scripts, intermediate results, and final results will all be contained in files, and their numbers will grow during the project. These files need to be organized through a system of folders (directories), subfolders (subdirectories), and filenames that are sufficiently systematic that files can be located easily and so that they are not confused with one another. Predictability both in the pattern of filenames and in the arrangement of folders and subfolders is an extremely important part of the design.

The MATLAB®-related files associated with this book are in a folder/subfolder/filename structure modeled on the format of the book itself (Fig. 1A.1). The main folder is named gdama. It contains several subfolders, LiveScripts (containing Live Scripts), Data (containing data), TestFolder (for test purposes), and Html (for hypertext). The Live Scripts are named gdama01.mlx, gdama02.mlx, etc., one file for each of the book's chapters. We have chosen to use leading zeros in the naming scheme (e.g., 01) so that filenames appear in the correct order when they are sorted alphabetically (as when listing the contents of a folder). A given Live Scripts file contains all the scripts for the corresponding chapter, one per section. Name-tags at the top of each section are used to identify each script, with names of the form gdamaNN_MM.mlx, where the chapter number NN and the script number MM are sequential integers.

You should now open gdama01.mlx and scroll through it, noting that it is divided into sections and that these sections start with comments giving a brief description of its function, followed by MATLAB® commands. Comments, which begin with the % (percent) character, serve to document a script, but have no effect on its function; MATLAB®

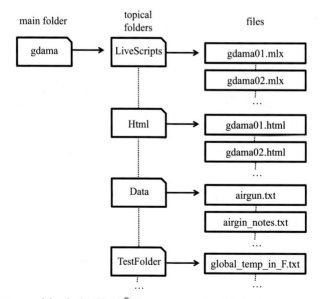

FIG. 1A.1 Folder (directory) structure used for the MATLAB® files accompanying this book.

ignores them when performing calculations. The gdama01_01 script is very short, because it contains just one command, date, together with a couple of comment lines:

```
% gdama01_01
% displays the current date
date
```
[gdama01_0])

The current date will be displayed within the Live Script when the gdama01_01 script is run (say, by using the "run section" button).

MATLAB® supports a number of commands that enable you to navigate from folder to folder, list the contents of folders, etc. For example, typing the command

```
pwd
```
[gdama01_02]

(for "print working directory") in the Command Window displays the name of the current folder. Initially, this is almost invariably the wrong folder, so you will need to cd (for "change directory") to the folder where you want to be—the LiveScript folder, when running the book's scripts. The pathname will depend upon where you copied the gdama folder but will end in gdama/LiveScripts. On the author's computer, typing

```
cd c:/menke/docs/gdama/LiveScripts
```
[gdama01_02]

in the command window does the trick. If you have spaces in your pathname, just surround it with single quotes:

```
cd 'c:/menke/my docs/gdama/LiveScripts'
```
[gdama01_02]

Changing to the directory above the working one is accomplished by the command

```
cd ..
```
[gdama01_02]

and to one below it by giving just the folder name, for example

```
cd LiveScipts
```
[gdama01_02]

Finally, the command dir (for "directory") lists the files and subfolders in the current directory.

```
dir
```
[gdama01_02]

As we will discuss in more detail later in this book, all these commands can be used within Live Scripts to facilitate processing data files located in multiple folders.

1A.3 Simple arithmetic

The MATLAB® commands for simple arithmetic and algebra closely parallel standard mathematical notation. For instance, the command sequence

```
a=4.5;
b=5.1;
c=a+b;
disp(c);
```
[gdama01_03]

evaluates the formula $c = a + b$ for the case $a = 4.5$ and $b = 5.1$ to obtain $c = 9.6$. By default, MATLAB® displays the value of every formula after it is executed. A semicolon at the end of the formula suppresses the display. Furthermore, the semicolon serves to indicate the end of the command. The final command, `disp(c)`, causes MATLAB® to display the final result c.

Note that MATLAB® variables are *static*, meaning that they persist in MATLAB's Workspace until you explicitly delete them or exit the program. Variables created by one script can be used by subsequent scripts. At any time, the value of a variable can be examined, either by displaying it in the Live Script (as we have done before) or by using the spreadsheet-like display tools available through MATLAB's Workspace Window. The persistence of MATLAB's variables can sometimes lead to scripting errors, such as when the definition of a variable in a script is inadvertently omitted, but MATLAB® uses the value defined in a previous script. The command `clear all` deletes all previously defined variables in the Workspace. Placing this command at the beginning of a script causes it to delete any previously defined variables every time that it is run, ensuring that it cannot be affected by them.

A somewhat more complicated formula is

$$c = \sqrt{a^2 + b^2} \ \text{ with } a = 6 \ \text{ and } \ b = 8 \tag{1A.1}$$

The corresponding script is

```
a=6;
b=8;
c = sqrt(a^2 + b^2);
disp(c);
```
                                                                                            [gdama01_04]

Note that MATLAB's syntax for a^2 is `a^2` and that the square root is computed using the `sqrt()` function. This is an example of MATLAB's syntax differing from standard mathematical notation.

A final example is

$$c = \sin\left(\frac{n\pi(x - x_0)}{L}\right) \ \text{ with } \ n = 3, x = 4, x_0 = 1 \ \text{ and } \ L = 6 \tag{1A.2}$$

The corresponding script is

```
n=3; x=4; x0=1; L=6;
c = sin(n*pi*(x-x0)/L);
disp(c);
```
                                                                                            [gdama01_05]

Note that several formulas, separated by semicolons, can be typed on the same line. Variables, such as `x0` and `pi`, given before, can have names consisting of more than one character and can contain numerals as well as letters (though they must start with a letter). MATLAB® has a variety of predefined mathematical constants, including `pi`, which is the usual mathematical constant π.

Many MATLAB® manuals, guides, and tutorials are available, both in the printed form (e.g., Menke, 2022; Part-Enander et al., 1996; Pratap, 2009) and on the Web (e.g., at www.mathworks.com). The reader may find that they complement this book by providing more detailed information about MATLAB® as a scientific computing environment.

1A.4 Vectors and matrices and their representation in MATLAB®

Vectors and matrices are fundamental to inverse theory both because they provide a convenient way to organize data and because many important operations on data can be very succinctly expressed using linear algebra (i.e., the algebra of vectors and matrices). In the simplest interpretation, a vector is just a list of numbers that are treated as a unit and given a symbolic name. The list can be organized horizontally, as a row, in which case it is called a *row vector*. Alternately, the list can be organized vertically, as a column, in which case it is called a *column vector*. We will use lower-case bold letters to represent both kinds of vector. An exemplary 1×3 row vector **r** and a 3×1 column vector **c** are

$$\mathbf{r} = [2, 4, 6] \quad \text{and} \quad \mathbf{c} = \begin{bmatrix} 1 \\ 3 \\ 5 \end{bmatrix} \tag{1A.3}$$

In MATLAB®, a row vector and a column vector can be created with the commands

```
r = [2, 4, 6];
c = [1, 3, 5]';
```

[gdama01_06]

This syntax is most effective when the lengths of the vectors are short. We will discuss a variety of more effective means of creating long vectors later in the text, but mention two very simple, yet extremely useful, methods here. Length-N row and column vectors of zeros are formed by

```
N=3;
r = zeros(1,N);
c = zeros(N,1);
```

[gdama01_06]

Similarly, length-N row and column vectors of ones are formed by

```
r = ones(1,N);
c = ones(N,1);
```

[gdama01_06]

The length of a vector can be determined using the command

```
N = length(c);
```

[gdama01_06]

A row vector can be turned into a column vector, and vice versa, by the transpose operation, denoted by superscript T. Thus

$$\mathbf{r}^T = \begin{bmatrix} 2 \\ 4 \\ 6 \end{bmatrix} \quad \text{and} \quad \mathbf{c}^T = [1, 3, 5] \tag{1A.4}$$

In MATLAB®, the transpose is denoted by the single quote, so the column vector c in the script given before is created by transposing a row vector. Although both column vectors and row vectors are useful, our experience is that defining both in the same script creates serious opportunities for error. A formula that requires a column vector will usually yield incorrect results if a row vector is substituted into it, and vice versa. Consequently, we will adhere to a protocol where all vectors defined in this book are column vectors. Row vectors will be created when needed—and as close as possible to where they are used in the script—by transposing the equivalent column vector.

An individual number within a vector is called an *element* (or, sometimes, *component*) and is denoted with an integer index, written as a subscript, with the index 1 in the leftmost element of the row vector and the topmost element of the column vector. Thus $r_2 = 4$ and $c_3 = 5$ in the example given before. In MATLAB®, the index is written inside parentheses, as in

```
a=r(2);
b=c(3);
```

[gdma01_06]

Sometimes, we will wish to indicate a generic element of the vector, in which case we will give it a variable index, as in r_i and c_j, with the understanding that i and j are integer variables.

In the simplest interpretation, a matrix is just a rectangular table of numbers. We will use bold uppercase names to denote matrices, as in

$$\mathbf{A} = \begin{bmatrix} 1 & 2 & 3 \\ 4 & 5 & 6 \\ 7 & 8 & 9 \end{bmatrix} \tag{1A.5}$$

In this example, the number of rows and number of columns are equal, but this property is not required; matrices can also be rectangular. Thus row vectors and column vectors are just special cases of rectangular matrices. The transposition operation can also be performed on a matrix, in which case its rows and columns are interchanged

$$\mathbf{A}^T = \begin{bmatrix} 1 & 4 & 7 \\ 2 & 5 & 8 \\ 3 & 6 & 9 \end{bmatrix} \tag{1A.6}$$

A square matrix $\mathbf{A}$ is said to be *symmetric* if $\mathbf{A}^T = \mathbf{A}$. In MATLAB®, a matrix is defined by

```
A=[ [1, 4, 7]', [2, 5, 8]', [3, 6, 9]'];
```
[gdama01_06]

or, alternately, by

```
A = [1, 2, 3; 4, 5, 6; 7, 8, 9];
```
[gdama01_06]

The first case, the matrix M, is constructed from a "row vector of column vectors" and in the second case, as a "column vector of row vectors." This syntax is most effective when the dimensions of the vectors are small. We will discuss a variety of more effective means of creating larger matrices later in the text, but mention two very simple, yet extremely useful, methods here. An $N \times M$ matrices $\mathbf{A}$ of zeros and $\mathbf{B}$ of ones are formed, respectively, by

```
A = zeros(N,M);
B = ones(N,M);
```
[gdama01_06]

The size of a matrix can be determined using the command

```
[N,M] = size(A);
```
[gdama01_06]

Note that the size() function returns two parameters, the number of rows N, and number of columns M in the matrix.

The individual elements of a matrix are denoted with two integer indices, the first indicating the row and the second the column, starting in the upper left. Thus, in the earlier example, $A_{31} = 7$. Note that transposition swaps indices; that is, A_{ji} is the transpose of A_{ij}. In MATLAB®, the indices are written inside parentheses, as in

```
s = A(1,3);
```
[gdama01_06]

One of the key properties of vectors and matrices is that they can be manipulated symbolically—as entities—according to specific rules that are similar to normal arithmetic. This allows tremendous simplification of data processing formulas, since all the details of what happens to individual elements within those entities are hidden from view and automatically performed.

In order to be added, two matrices (or vectors, viewing them as a special case of a rectangular matrix) must have the same number of rows and columns. Their sum is then just the matrix that results from summing corresponding elements. Thus if

$$\mathbf{A} = \begin{bmatrix} 1 & 0 & 2 \\ 0 & 1 & 0 \\ 2 & 0 & 1 \end{bmatrix} \text{ and } \mathbf{B} = \begin{bmatrix} 1 & 0 & -1 \\ 0 & 2 & 0 \\ 1 & 0 & 3 \end{bmatrix}$$

(1A.7)

$$\mathbf{S} = \mathbf{A} + \mathbf{B} = \begin{bmatrix} 1+1 & 0+0 & 2-1 \\ 0+0 & 1+2 & 0+0 \\ 2+1 & 0+0 & 1+3 \end{bmatrix} = \begin{bmatrix} 2 & 0 & 1 \\ 0 & 3 & 0 \\ 3 & 0 & 4 \end{bmatrix}$$

Subtraction is performed in an analogous manner. In terms of the components, addition and subtraction are written as

$$S_{ij} = A_{ij} + B_{ij} \text{ and } D_{ij} = A_{ij} - B_{ij}$$

(1A.8)

Note that addition is commutative (i.e., $\mathbf{A} + \mathbf{B} = \mathbf{B} + \mathbf{A}$) and associative (i.e., $(\mathbf{A} + \mathbf{B}) + \mathbf{C} = \mathbf{A} + (\mathbf{B} + \mathbf{C})$). In MATLAB®, addition and subtraction are written as

```
S = A+B;
D = A-B;
```

[gdama01_06]

Multiplication of two matrices is a more complicated operation and requires that the number of columns of the left-hand matrix equals the number of rows of the right-hand matrix. Thus if the matrix $\mathbf{A}$ is $N \times K$ and the matrix $\mathbf{B}$ is $K \times M$, the product $\mathbf{P} = \mathbf{AB}$ is an $N \times M$ matrix defined according to the rule:

$$P_{ij} = \sum_{k=1}^{K} A_{ik} B_{kj}$$

(1A.9)

In MATLAB®, multiplication is written as:

```
P = A*B;
```

[gdama01_06]

The order of the indices is important. Matrix multiplication is in its standard form when all summations involve neighboring indices of adjacent quantities. Thus, for instance, the two instances of the summed variable k are not neighboring in the equation

$$Q_{ij} = \sum_{k=1}^{K} A_{ki} B_{kj}$$

(1A.10)

and so the equation corresponds to $\mathbf{Q} = \mathbf{A}^T \mathbf{B}$ and not $\mathbf{Q} = \mathbf{AB}$. Matrix multiplication is not commutative (i.e., $\mathbf{AB} \neq \mathbf{BA}$) but is associative (i.e., $(\mathbf{AB})\mathbf{C} = \mathbf{A}(\mathbf{BC})$) and distributive (i.e., $\mathbf{A}(\mathbf{B} + \mathbf{C}) = \mathbf{AB} + \mathbf{AC}$). An important rule involving the matrix transpose is $(\mathbf{AB})^T = \mathbf{B}^T \mathbf{A}^T$ (note the reversal of the order).

Several special cases of multiplication involving vectors are noteworthy. Suppose that $\mathbf{a}$ and $\mathbf{b}$ are length-N column vectors. The combination $s = \mathbf{a}^T \mathbf{b}$ is a scalar number s and is called the *dot product* (or sometimes, *inner product*) of the vectors. It obeys the rule $\mathbf{a}^T \mathbf{b} = \mathbf{b}^T \mathbf{a}$. The dot product of a vector with itself is the square of its Euclidian length; that is, $\mathbf{a}^T \mathbf{a}$ is the sum of its squared elements of $\mathbf{a}$. The vector $\mathbf{a}$ is said to be a *unit vector* when $\mathbf{a}^T \mathbf{a} = 1$. The combination $\mathbf{ab}^T$ is an $N \times N$ matrix; it is called the *outer product*. The product of a matrix and a vector is another vector, as in $\mathbf{c} = \mathbf{Ba}$. One interpretation of this relationship is that the matrix $\mathbf{B}$ "turns one vector into another." Note that the vectors $\mathbf{a}$ and $\mathbf{c}$ can be of different length. An $M \times N$ matrix $\mathbf{B}$ turns the length-N vector $\mathbf{a}$ into a length-M vector $\mathbf{c}$. The combination $s = \mathbf{a}^T \mathbf{Ba}$ is a scalar and is called a *quadratic form*, as it contains terms quadratic in the elements of $\mathbf{a}$. Matrix multiplication $\mathbf{P} = \mathbf{AB}$ has a useful interpretation in terms of dot products: P_{ij} is the dot product of the ith row of $\mathbf{A}$ with the jth column of $\mathbf{B}$.

Any matrix is unchanged when multiplied by the *identity matrix*, conventionally denoted $\mathbf{I}$. Thus $\mathbf{a} = \mathbf{Ia}$, $\mathbf{A} = \mathbf{IA} = \mathbf{AI}$, etc. This matrix has ones along its main diagonal, and zeroes elsewhere, as in

$$\mathbf{I} = \begin{bmatrix} 1 & 0 & 0 \\ 0 & 1 & 0 \\ 0 & 0 & 1 \end{bmatrix} \tag{1A.11}$$

The elements of the identity matrix are usually written δ_{ij} and not I_{ij}, and the symbol δ_{ij} is usually called the *Kronecker delta symbol*, not the elements of the identity matrix (though that is exactly what it is). The equation $\mathbf{A} = \mathbf{IA}$ for an $N \times N$ matrix $\mathbf{A}$ is written componentwise as

$$A_{ij} = \sum_{k=1}^{N} \delta_{ik} A_{kj} \tag{1A.12}$$

This equation indicates that any summation containing a Kronecker delta symbol can be performed trivially. To obtain the result, one first identifies the variable that is being summed over (k in this case) and the variable that the summed variable is paired within the Kronecker delta symbol (i in this case). The summation and the Kronecker delta symbol then are deleted from the equation, and all occurrences of the summed variable are replaced with the paired variable (all ks are replaced by is in this case). In MATLAB®, an $N \times N$ identity matrix can be created with the command

```
I = eye(N);
```

[gdama01_07]

MATLAB® performs all multiplicative operations with ease. For example, suppose column vectors $\mathbf{a}$ and $\mathbf{b}$ and matrices $\mathbf{M}$ and $\mathbf{N}$ are defined as

$$\mathbf{a} = \begin{bmatrix} 1 \\ 3 \\ 5 \end{bmatrix} \quad \text{and} \quad \mathbf{b} = \begin{bmatrix} 2 \\ 4 \\ 6 \end{bmatrix} \quad \text{and} \quad \mathbf{A} = \begin{bmatrix} 1 & 0 & 2 \\ 0 & 1 & 0 \\ 2 & 0 & 1 \end{bmatrix} \quad \text{and} \quad \mathbf{B} = \begin{bmatrix} 1 & 0 & -1 \\ 0 & 2 & 0 \\ -1 & 0 & 3 \end{bmatrix} \tag{1A.13}$$

Then

$$s = \mathbf{a}^T \mathbf{b} = \begin{bmatrix} 1 \\ 3 \\ 5 \end{bmatrix}^T \begin{bmatrix} 2 \\ 4 \\ 6 \end{bmatrix} = \begin{bmatrix} 1 & 3 & 5 \end{bmatrix} \begin{bmatrix} 2 \\ 4 \\ 6 \end{bmatrix} = 1 \times 2 + 3 \times 4 + 5 \times 6 = 44$$

$$\mathbf{T} = \mathbf{a}\mathbf{b}^T = \begin{bmatrix} 1 \\ 3 \\ 5 \end{bmatrix} \begin{bmatrix} 2 & 4 & 6 \end{bmatrix} = \begin{bmatrix} 1 \times 2 & 1 \times 4 & 1 \times 6 \\ 3 \times 2 & 3 \times 4 & 3 \times 6 \\ 5 \times 2 & 5 \times 4 & 5 \times 6 \end{bmatrix} = \begin{bmatrix} 2 & 4 & 6 \\ 6 & 12 & 18 \\ 10 & 20 & 30 \end{bmatrix}$$

$$\mathbf{c} = \mathbf{Ab} = \begin{bmatrix} 1 & 0 & 2 \\ 0 & 1 & 0 \\ 2 & 0 & 1 \end{bmatrix} \begin{bmatrix} 2 \\ 4 \\ 6 \end{bmatrix} = \begin{bmatrix} 1 \times 2 + 0 \times 4 + 2 \times 6 \\ 0 \times 2 + 1 \times 4 + 0 \times 6 \\ 2 \times 2 + 0 \times 4 + 1 \times 6 \end{bmatrix} = \begin{bmatrix} 14 \\ 4 \\ 10 \end{bmatrix} \tag{1A.14}$$

$$\mathbf{P} = \mathbf{AB} = \begin{bmatrix} 1 & 0 & 2 \\ 0 & 1 & 0 \\ 2 & 0 & 1 \end{bmatrix} \begin{bmatrix} 1 & 0 & -1 \\ 0 & 2 & 0 \\ -1 & 0 & 3 \end{bmatrix} = \begin{bmatrix} -1 & 0 & 5 \\ 0 & 2 & 0 \\ 1 & 0 & 1 \end{bmatrix}$$

correspond to

```
s = a'*b;
T = a*b';
c = A*b;
P = M*N;
```

[gdama01_07]

In MATLAB®, matrix multiplication is signified using the multiplications sign * (the asterisk). There are cases, however, where one needs to violate the rules and multiply the quantities element-wise (e.g., create a column vector **d** with elements $d_i = a_i b_i$). MATLAB® provides a special element-wise version of the multiplication sign, denoted .* (a period followed by an asterisk)

```
d = a.*b;
```

[gdama01_07]

Individual elements of matrices can be accessed by specifying the relevant row and column indices, in parentheses, e.g., and M(2,3) is the second row, third column element of the matrix M.

Ranges of rows and columns can be specified using the : (colon) operator, e.g., M(:,2) is the second column of matrix M; M(2,:) is the second row of matrix M; and M(2:3,2:3) is the 2×2 submatrix in the lower right-hand corner of the 3×3 matrix M (the expression M(2:end,2:end) would work as well). These operations are further illustrated as follows:

$$\mathbf{a} = \begin{bmatrix} 1 \\ 2 \\ 3 \end{bmatrix} \text{ and } \mathbf{B} = \begin{bmatrix} 1 & 2 & 3 \\ 4 & 5 & 6 \\ 7 & 8 & 9 \end{bmatrix} \tag{1A.15}$$

$$\mathbf{x} = \begin{bmatrix} a_1 \\ a_2 \end{bmatrix} = \begin{bmatrix} 1 \\ 2 \end{bmatrix} \text{ and } \mathbf{y} = \begin{bmatrix} B_{12} \\ B_{22} \\ B_{32} \end{bmatrix} = \begin{bmatrix} 2 \\ 5 \\ 8 \end{bmatrix} \tag{1A.16}$$

$$\text{and } \mathbf{z} = \begin{bmatrix} B_{21} & B_{22} & B_{23} \end{bmatrix} = \begin{bmatrix} 4 & 5 & 6 \end{bmatrix} \text{ and } \mathbf{T} = \begin{bmatrix} B_{22} & B_{23} \\ B_{32} & B_{33} \end{bmatrix} = \begin{bmatrix} 5 & 6 \\ 8 & 9 \end{bmatrix}$$

correspond to

```
x = a(1:2);
y = B(:,2);
z = B(2,:);
T = B(2:3,2:3);
```

[gdama01_08]

The colon notation can be used in other contexts in MATLAB® as well. For instance, [1:4] is the row vector [1,2,3,4]. The syntax 1:4, which omits the square brackets, and the syntax (1:4), which used parentheses, work as well. However, we often will use square brackets, because they draw attention to the presence of a vector. Finally, we note that two colons can be used in sequence to indicate the spacing of elements in the resulting row vector. For example, the expression [1:2:9] is the row vector [1,3,4,7,9] and that the expression [10:-1:1] is a row vector whose elements are in the reverse order from [1:10].

A very useful length-N column vector, say **t**, has elements that increment by a constant amount Δt from zero to $(N-1)\Delta t$, that is, $t_i = (i-1)\Delta t$. It can be formed as

```
N=11;
Dt = 2;
t = Dt*[0:N-1]';
```

[gdama01_09]

Here, the increment Dt has been set to 2. One use of such a vector is for a *time axis*; that is, a vector that represents time increasing by regular increments.

Matrix division is defined in analogy to reciprocals. For a scalar number s, multiplication by the reciprocal s^{-1} is equivalent to division by s. Here, the reciprocal obeys $ss^{-1} = s^{-1}s = 1$. The matrix analog to the reciprocal is called the matrix *inverse* and obeys

$$\mathbf{A}\mathbf{A}^{-1} = \mathbf{A}^{-1}\mathbf{A} = \mathbf{I} \tag{1A.17}$$

It is defined only for square matrices. The calculation of the inverse of a matrix is complicated, and we will not describe it here, except to mention the 2×2 case

$$\mathbf{A} = \begin{bmatrix} a & b \\ c & d \end{bmatrix} \text{ and } \mathbf{A}^{-1} = \frac{1}{ad - bc} \begin{bmatrix} d & -b \\ -c & a \end{bmatrix} \tag{1A.18}$$

Just as the reciprocal s^{-1} is defined only when $s \neq 0$, the matrix inverse $\mathbf{A}^{-1}$ is defined only when a quantity called the *determinant* of $\mathbf{A}$, denoted det($\mathbf{A}$) (or sometimes $|\mathbf{A}|$), is not equal to zero. The determinant of the 2×2 matrix $\mathbf{A}$ is det($\mathbf{A}$) $= ad - bc$. It consists of the sum of products of two elements of the matrix. The determinant of a square $N \times N$ matrix $\mathbf{M}$ consists of the sum of products of N elements of the matrix and is given by

$$\det(\mathbf{M}) = \sum_{n_1=1}^{N} \sum_{n_2=1}^{N} \sum_{n_3=1}^{N} \cdots \sum_{n_N=1}^{N} \varepsilon_{n_1, n_2, n_3 \cdots n_N} M_{1n_1} M_{2n_2} M_{3n_3} \cdots M_{Nn_N} \tag{1A.19}$$

Here the quantity $\varepsilon_{n_1, n_2, n_3 \cdots n_N}$ is $+1$ when $(n_1, n_2, n_3 \cdots n_N)$ is an even permutation of $(1, 2, 3, \cdots N)$, -1 when it is an odd permutation, and zero otherwise.

In MATLAB®, the matrix inverse and determinant of a square matrix A are computed as

```
B = inv(A);
d = det(A);
```
[gdama01_10]

In many of the formulas of inverse theory, the matrix inverse either premultiplies or postmultiplies other quantities, for instance:

$$\mathbf{c} = \mathbf{A}^{-1}\mathbf{b} \text{ and } \mathbf{D} = \mathbf{B}\mathbf{A}^{-1} \tag{1A.20}$$

These cases do not actually require the explicit calculation of $\mathbf{A}^{-1}$, just the combinations $\mathbf{A}^{-1}\mathbf{b}$ and $\mathbf{B}\mathbf{A}^{-1}$, which are computationally simpler. MATLAB® provides generalizations of the division operator that implement these two cases:

```
c = A\b;
D = B/A;
```
[gdama01_10]

A surprising amount of information on the structure of a matrix can be gained by studying how it affects a column vector that it multiplies. Suppose that $\mathbf{A}$ is an $N \times N$ square matrix and that it multiplies an *input* column vector $\mathbf{v}$, producing an *output* column vector $\mathbf{w} = \mathbf{A}\mathbf{v}$. We can examine how the output $\mathbf{w}$ compares to the input $\mathbf{v}$, as $\mathbf{v}$ is varied. One question of particular importance is

When is the output parallel to the input? (1A.21)

This question is called the *algebraic eigenvalue problem*. If $\mathbf{w}$ is parallel to $\mathbf{v}$, then $\mathbf{w} = \lambda\mathbf{v}$, where λ is a scalar proportionality factor. The parallel vectors satisfy the following equation:

$$\mathbf{A}\mathbf{v} = \lambda\mathbf{v} \text{ or } (\mathbf{A} - \lambda\mathbf{I})\mathbf{v} = 0 \tag{1A.22}$$

The trivial solution $\mathbf{v} = (\mathbf{A} - \lambda\mathbf{I})^{-1}0 = 0$ is not very interesting. A nontrivial solution is only possible when the matrix inverse $(\mathbf{A} - \lambda\mathbf{I})^{-1}$ does not exist. This is the case where the parameter λ is specifically chosen to make the determinant det$(\mathbf{A} - \lambda\mathbf{I})$ exactly zero, as a matrix with zero determinant has no inverse. The determinant is calculated by adding together terms, each of which contains the product of N elements of the matrix. Since each element of the matrix contains, at most, one instance of λ, the product will contain powers of λ up to λ^N. Thus the equation det$(\mathbf{A} - \lambda\mathbf{I}) = 0$ is an

Nth order polynomial equation for λ. An Nth order polynomial equation has N solutions, so we conclude that there must be N different proportionality factors, say λ_i, and N corresponding column vectors, say $\mathbf{v}^{(i)}$, that solve $\mathbf{Av} = \lambda\mathbf{v}$. The column vectors $\mathbf{v}^{(i)}$ are called the *characteristic vectors* (or eigenvectors) of the matrix $\mathbf{A}$, and the proportionality factors λ_i are called the *characteristic values* (or eigenvalues). Eigenvectors are determined only up to an arbitrary multiplicative factor s, since if $\mathbf{v}^{(i)}$ is an eigenvector, so is $s\mathbf{v}^{(i)}$. Consequently, they are conventionally chosen to be unit vectors.

In the special case where the matrix $\mathbf{A}$ is symmetric, it can be shown that the eigenvalues λ_i are real and the eigenvectors are mutually perpendicular $\mathbf{v}^{(i)T}\mathbf{v}^{(j)} = 0$ for $i \neq j$. The N eigenvalues can be arranged into a diagonal matrix $\mathbf{\Lambda}$, whose elements are $[\mathbf{\Lambda}]_{ij} = \lambda_i\delta_{ij}$, where δ_{ij} is the Kronecker delta. The corresponding N eigenvectors $\mathbf{v}^{(i)}$ can be arranged as the columns of an $N \times N$ matrix $\mathbf{V}$, which satisfies $\mathbf{VV}^T = \mathbf{V}^T\mathbf{V} = \mathbf{I}$. The eigenvalue equation $\mathbf{Av}^{(i)} = \lambda_i\mathbf{v}^{(i)}$ can then be succinctly written as $\mathbf{AV} = \mathbf{V\Lambda}$. Postmultiplying by $\mathbf{V}^T$, and applying $\mathbf{VV}^T = \mathbf{I}$, yields:

$$\mathbf{A} = \mathbf{V\Lambda V}^T \tag{1A.23}$$

That is, a symmetric matrix $\mathbf{A}$ is completely specified by its eigenvalues and eigenvectors.

We now note an interesting property of a matrix that can be written as $\mathbf{A} = \mathbf{B}^T\mathbf{B}$, where $\mathbf{B}$ is some matrix. By premultiplying the eigenvalue equation by $\mathbf{v}^{(i)T}$, and noting that $\mathbf{v}^{(i)T}\mathbf{v}^{(i)} = 1$, we obtain $\lambda_i = \mathbf{v}^{(i)T}\mathbf{Av}^{(i)}$. Substituting $\mathbf{B}^T\mathbf{B}$ for $\mathbf{A}$, we have $\lambda = \mathbf{v}^{(i)T}\mathbf{B}^T\mathbf{Bv}^{(i)} = [\mathbf{Bv}^{(i)}]^T\mathbf{Bv}^{(i)}$. Consequently, λ_i is the squared length of the vector $\mathbf{Bv}^{(i)}$ and cannot be negative. In this case, the matrix $\mathbf{A}$ is said to be *nonnegative definite*.

In MATLAB®, the diagonal matrix of eigenvalues $\mathbf{\Lambda}$ and matrix of eigenvectors $\mathbf{V}$ of a symmetric matrix $\mathbf{A}$ are computed as

```
[V,LAMBDA] = eig(A);
```

[gdama01_10]

Here the eigenvector matrix is called LAMBDA.

1A.5 Matrix differentiation

Many of the derivations of inverse theory require that a column vector $\mathbf{v}$ be considered a function of an independent variable, say x, and then differentiated with respect to that variable to yield the derivative $d\mathbf{v}/dx$. Such a derivative represents the fact that the vector changes from $\mathbf{v}$ to $\mathbf{v} + \mathbf{\Delta v}$ as the independent variable changes from x to $x + \Delta x$. Note that the resulting change $\mathbf{\Delta v}$ is itself a vector. Derivatives are performed element-wise, that is,

$$\left[\frac{d\mathbf{v}}{dx}\right]_i \equiv \lim_{\Delta x \to 0} \frac{v_i(x + \Delta x) - v_i(x)}{\Delta x} = \frac{dv_i}{dx} \tag{1A.24}$$

A somewhat more complicated situation is where the column vector $\mathbf{v}$ is a function of another column vector, say $\mathbf{y}$, that is, $\mathbf{v(y)}$. The partial derivative

$$\frac{\partial v_i}{\partial y_j} \tag{1A.25}$$

represents the change in the ith component of $\mathbf{v}$ caused by a change in the jth component of $\mathbf{y}$. Frequently, we will need to differentiate the linear function $\mathbf{v} = \mathbf{Ay}$, where $\mathbf{A}$ is a matrix, with respect to $\mathbf{y}$:

$$\frac{\partial v_i}{\partial y_j} = \frac{\partial}{\partial y_j}\sum_{k=1}^{N} A_{ik}y_k = \sum_{k=1}^{N} A_{ik}\frac{\partial y_k}{\partial y_j} = \sum_{k=1}^{N} A_{ik}\delta_{kj} = A_{ij} \tag{1A.26}$$

Since the components of $\mathbf{y}$ are assumed to be independent, the derivative $\partial y_k/\partial y_j$ is zero except when $k = j$, in which case it is unity, which is to say $\partial y_k/\partial y_j = \delta_{kj}$, where δ_{kj} is the Kronecker delta. Thus the derivative of the linear function $\mathbf{v} = \mathbf{Ay}$ is the matrix $\mathbf{A}$. This relationship is the vector analog to the scalar case, where the derivative of the linear function $v = ay$ is the constant a.

1A.6 Character strings and lists

The numerical variable x created with the command x=2.5 has a close relationship to variables that are encountered in elementary mathematics. They are very useful for performing the calculations essential to data analysis. However, they are not very useful during file manipulation, which is usually a vital prerequisite to data analysis. Furthermore, the results of data analysis are most understandable if described in a combination of words and numbers—and not just numbers.

MATLAB® has a type of variable, called the *string variable*, the value of which can be set to a *character string* (sequence of alpha-numerical characters). For instance, a string variable with name myfilename and with value 'mydata.txt' is created by

```
myfilename='mydata.txt';
```
[gdama01_11]

Note that the value of the variable is quoted to prevent MATLAB® from trying to interpret it as a command.

Combining numerical variables with character strings can be very useful in file manipulation, where a sequence of filenames may contain a number that sequentially increments, like 'myfile_1.txt', 'myfile_2.txt', or when one wants to display the value of a variable together with some text that explains it, e.g., display the sentence 'position x is 10.40 meters'.

The sprintf() function (for "string print formatted") creates a character string with a value that includes both text and the value of a variable. Thus, for instance,

```
i=1;
myfilename = sprintf('myfile_%d.txt',i);
```
[gdama01_11]

creates a character string myfilename with value 'myfile_1.txt', where the 1 is taken from the variable i. Similarly,

```
x=10.40;
mysentence = sprintf('position x is %.2f meters',x);
```
[gdama01_11]

creates a character string mysentence with value 'position x is 10.40 meters', where the 10.4 is taken from the variable x.

Unfortunately, the sprintf() function is fairly inscrutable, and we refer readers to the MATLAB® help pages for a detailed description. Briefly, it uses a *format string* (in this case 'position x is %.2f meters'), which contains *placeholders* that start with the character % to indicate where in the character string the value of the variable should be placed. Thus sprintf('myfile_%d.txt',i) creates the character string 'myfile_1.txt' (because the value of i is 1). The %d is the placeholder for an integer. It is replaced with 1 the value of i. When the variable can have fractional, as contrasted to integer, values, the floating-point placeholder %f is used instead. Thus sprintf('position x is %.2f meters',x) creates the character string 'position x is 10.40 meters' (because the value of x is 10.4). The precision to which x is to be written is indicated by the .2 inside the %.2f; in this case, two significant digits.

The command disp(mysentence) can be used to display a character string. However, a commonly used alternative that combines the disp() and sprintf() functions is

```
fprintf('position x is %.2f meters\n',x);
```
[gdama01_11]

The \n (for *newline*) at the end of the format string 'a=%f\n' indicates that subsequent characters should be written to a new line in the command window, rather than being appended onto the end of the current line.

MATLAB® provides a mechanism for creating a *list* of character strings, which can be very useful when automating complicated tasks. For instance,

```
mycolorlist = {'red', 'crimson', 'chartreuse', 'teal'};
```
[gdama01_11]

This type of variable, called a *cellstr* (for cell string) in MATLAB®, functions something like a row vector, except that every element is a character string, as contrasted to a number. Individual elements are accessed via standard indexing; that is, mycolorlist(4) is cell number three containing the character string 'teal'. However, in order to actually use an element as a character string, one must also *cast* it to a character variable, using the char() function. That is,

```
myfavoritecolor = char(mycolorlist(4));
```

[gdama01_11]

1A.7 Loops

MATLAB® provides a looping mechanism, the for command, which can be useful when the need arises to sequentially access the elements of vectors and matrices. Thus, for example,

```
A = [ 1, 2, 3; 4, 5, 6; 7, 8, 9];
b = [0, 0, 0]';
for i = (1:3)
        b(i) = A(i,i);
end
```

[gdama01_12]

executes the b(i)=A(i,i) formula three times, each time with a different value of i (in this case i=1, i=2, and i=3). The net effect is to copy the diagonal elements of the matrix A to the vector a, that is, $b_i = A_{ii}$. Note that the *end statement* indicates the position of the bottom of the loop. Subsequent commands are not part of the loop and are executed only once. Loops can be nested, that is, one loop can be inside another. Such an arrangement is necessary for accessing all the elements of a matrix in sequence. For example,

```
A = [1, 2, 3; 4, 5, 6; 7, 8, 9];
B = zeros(3,3);
for i = (1:3)
        for j = (1:3)
            B(i,4-j) = A(i,j);
        end
end
```

[gdama01_13]

copies the elements of the matrix A to the matrix B, but reverses the order of the elements in each row, that is, $B_{i,\,4-j}=A_{i,\,j}$. Loops are especially useful in conjunction with *conditional commands*. For example,

```
a = [ 1, 2, 1, 4, 3, 2, 6, 4, 9, 2, 1, 4 ]';
N = length(a);
b = zeros(N,1);
for i = (1:N)
        if ( a(i) >= 6 )
            b(i) = 6;
        else
            b(i) = a(i);
        end
end
```

[gdama01_14]

sets $b_i=a_i$ when $a_i<6$ and sets $b_i=6$ otherwise (a process called *clipping* a vector, for it clips off parts of the vector that are larger than 6).

A purist might point out that MATLAB® syntax is so flexible that for loops are almost never really necessary. In fact, all three examples, given before, can be computed with one-line formulas that omit for loops:

```
a = diag(A);
B = fliplr(A);
b=a; b(a>6)=6;
```

[gdama01_15]

The first two formulas are very simple, but rely upon the MATLAB® functions `diag()` (for "diagonal") and `fliplr()` (for "flip-left-right"), whose existence we have not hitherto mentioned. The third formula, which used logical addressing, requires further explanation. The vector a is first copied to b. Then just those elements of b that are greater than 6 are reset to 6, using a technique called *logical addressing*. The expression a>6 returns a vector of true's and false's, depending upon whether the elements of the column vector a satisfy the inequality or not. The command b(a>6)=6 resets the elements of b to 6 only when the value is true.

One of the problems of a script-based environment is that learning the complete syntax of the scripting language can be pretty daunting. Writing a long script, such as one containing a `for` loop, will often be faster than searching through MATLAB® help files for a predefined function that implements the desired functionality in a single line of the script. When deciding between alternative ways of implementing a given functionality, you should always choose the one which you find clearest. Scripts that are terse or even computationally efficient are not necessarily a virtue, especially if they are difficult to debug. You should avoid creating formulas that are so inscrutable that you are not sure whether they will function correctly. Of course, the degree of inscrutability of any given formula will depend upon your level of familiarity with MATLAB®. Your repertoire of techniques will grow as you become more practiced.

1A.8 Loading data from a file

MATLAB® can read files with a variety of formats, but we start here with the simplest and most common, the text file. As an example, we load a global temperature data set compiled by the National Aeronautics and Space Administration. The author's recommendation is that you always keep a file of notes about any data set that you work with, and that these notes include information on where you obtained the data set and any modifications that you subsequently made to it.

The text file `global_temp.txt` contains global temperature change data from NASA's web site http://data.giss. nasa.gov/gistemp. It has two columns of data, time (in calendar years) and temperature anomaly (in degrees C) and is 57 lines long. Information about the data is in the file global_temp_notes.txt. The citation for this data is Hansen et al. (2010).

We reproduce the first few lines of global_temp.txt, here:

```
1965 -0.11
1966 -0.06
1967 -0.02
...  ...
```

The data are read into MATLAB® as follows:

```
D = load('../Data/global_temp.txt');
t = D(:,1);
d = D(:,2);
N=length(d);
```

[gdama01_16]

The `load()` function reads the data into a 46×24 matrix **D**. Note that the filename is given as `'../Data/global_temp.txt'`, as contrasted to just `'global_temp.txt'`, since the script is run from the `LiveScripts` folder while the data are in the `Data` folder (which is "up and over" from `LiveScripts`). The filename is surrounded by single quotes to indicate that it is a character string. The subsequent two lines break out D into two separate column vectors t of time and d of temperature data. This step is not strictly speaking necessary, but fewer mistakes will be made if the different variables in the data set have each their own name.

1A.9 Writing data to a file

MATLAB® can write files with a variety of formats, but we start here with the simplest and most common, the text file. Suppose that we wanted a file containing global temperature anomaly in degrees Fahrenheit, using the fact that $1°C = 1.8°F$. Starting with time t and temperature anomaly d from the previous section, we use the command

```
dF = 1.8*d;
newfilename = '../TestFolder/global_temp_in_F.txt';
DF = [t,dF];
dlmwrite(newfilename, DF, 'delimiter', '\t' );
```

[gdama01_17]

The `dlmwrite()` function writes a single matrix (in this case, `DF`) to a file (in this case, `newfilename`) with a tab (denoted `'\t'`) between the columns of data. We form the matrix `DF` by row-wise concatenating the time column vector `t` and the Fahrenheit temperature data `dF` via `[t,dF]`. Note that the new filename `'global_temp_in_F.txt'` is chosen to indicate that the units have been changed.

1A.10 Plotting data

MATLAB's plotting commands are very powerful, but they are also very complicated. We present here a set of commands for making a simple x–y plot that is intermediate between a very crude, unlabeled plot and an extremely artistic one. The reader may wish to adopt either a simpler or a more complicated version of this set, depending upon need and personal preference. The plot of the global temperature data (Fig. 1A.2) was created with the commands:

```
figure();
clf;
hold on;
set(gca,'LineWidth',3);
set(gca,'FontSize',14);
axis( [1965, 2020, -0.5, 1.0] );
plot(t,d,'r-','LineWidth',3);
plot(t,d,'ko','LineWidth',3);
xlabel('calendar year');
ylabel('temperature anomaly, deg C');
title('global temperature data');
```

[gdama01_16]

The `figure()` command directs MATLAB® to create a new figure. The `clf` (for clear figure) command erases any previous plots in the figure window, ensuring that it is blank. The `hold on` command instructs MATLAB® to overlay

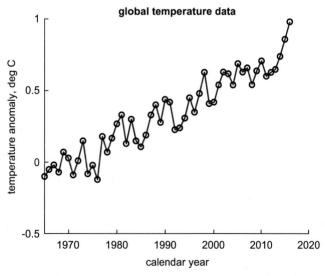

FIG. 1A.2 MATLAB® plot of deviations of global temperature from average for the time period, 1965–2021. Script gdama01_16. *Data from Hansen, J., Ruedy, R., Sato, M., Lo, K., 2010. Global surface temperature change. Rev. Geophys. 48, RG4004. https://doi.org/10.1029/2010RG000.*

multiple plots in the same window, overriding the default that a second plot erases the first. We prefer axes with thicker lines and bigger font than MATLAB's default, and so reset them to 3 and 14, respectively, with two calls to the `set(gca,...)` command. The `axis([1965,2020,-0.5,1.0])` command manually sets the axes to an appropriate range. If this command is omitted, MATLAB® chooses the scaling of the axes based on the range of the data. The two `plot(...)` commands plot the time `t` and temperature `d` data in two different ways: first as a red line (indicated by the `'r-'`) and the second with black circles (the `'ko'`). We also increase the width of the lines with the `'LineWidth', 3` directive. Finally, the horizontal and vertical axes are labeled using the `xlabel()` and `ylabel()` commands, and the plot is titled with the `title()` command. Note that the text is surrounded with single quotes, to indicate that it is a character string.

Part B. Python as a tool for learning inverse theory

1B.1 Getting started with Python

The Python environment that we will be using is assembled from the following freely available software elements:

python, itself;

jupyter notebook (or jupyter lab), a "research notebook" environment for developing and running Python scripts and viewing and storing their results;

numpy, (pronounced *num-pie*), the Python Numbers module, which extends Python by adding vectors, matrices and basic linear algebra;

scipy, (pronounced *sci-pie*), the Scientific module, which extends Python by adding advanced linear algebraic functions;

matplotlib, the plotting module, which expands Python by adding graphics;

gdapy, a folder of Jupyter Notebooks containing exemplary Python scripts referenced by this book.

We will also be using the Anaconda Powershell environment to install the Python packages, and (in some cases) to launch Jupyter Notebook (Or Jupyter Lab). Finally, Jupyter Notebook (and Jupyter Lab) uses a web browser, such as Chrome or Firefox, so you will need to have one of those installed on your computer. Almost everyone already has one of them installed, anyway.

You first should copy the gdapy folder to some convenient place on your computer's file system. It contains all files needed to perform this book's Python examples. Procedures for installing the other packages vary from computer to computer and quickly become outdated. We do not discuss them in this text, but instead provide a document `InstallationHelp.pdf` within the gdapy folder that provides a step-by-step installation procedure.

Launch the Jupyter Notebook (or Jupyter Lab) and open the file `gdapy_01.ipynb`. It contains a collection of Python scripts related to this chapter. As you scroll through the Notebook, you will observe that it is organized in sections, called *cells*, and that sometimes text and graphics appear between the cells. Each cell contains a Python script, and the text and graphics following a cell are the results (or "output") associated with it.

In this book's notebooks, the first cell performs all required initializations. It always must be run *first*, immediately after the Notebook is opened, and before running any of the other cells. Lines starting with the # character, such as # gdapy_01_00, are comment lines; their only function is to provide information about the script; they do not influence its function. Our convention is that the first line of a cell contains a name-tag, in this case gdapy_01_01, which is used in the text to reference the script. The line

```
%reset -f
```

[gdapy01_00]

directs the notebook to completely clear its memory, thus deleting any variables that may have been defined previously. The practice of starting with a fresh slate every time one opens a Notebook is a good one, since it eliminates the possibility that old (and possibly incorrect) results will be comingled with new ones. The next section of the first cell "imports" various modules that expand the rather bare-bones functionality of Python. The command

```
import numpy as np
```

[gdapy01_00]

imports the complete Python Numbers module numpy and abbreviates its name as np. This module adds vectors and matrices and methods for manipulating them. Alternatively, just part of a module can be imported. For instance, the command

```
from matplotlib import pyplot as plt
```
[gdapy01_00]

imports just the Python Plotting module `pyplot` from the larger `matplotlib` module, and abbreviates it as `plt`. Finally, functions defined in a module can be imported individually. For instance, the command

```
from math import exp, pi, sin, sqrt
```
[gdapy01_00]

imports the exponential function `exp()`, the sine function `sin()`, the square root function `sqrt()`, and the number π from Python's `math` module. This is a very frugal style of importing, but has the advantage that you can never accidentally use any function that is not in your list. For instance, a slip of the typing finger that adds an extraneous "h" can never accidentally turn the sine function `sin()` into the hyperbolic-sine function `sinh()`. On the other hand, should a modification of the script introduce the cosine function `cos()`, then that function needs to be added to the import command.

Incidentally, when you read Python documentation, you will often encounter the word *method* to describe a component of a module. Methods are similar to, but not quite the same as, functions, but we will not draw any distinction between them right now.

Most of the functions and methods used in this book are not part of Python itself, but instead are part of modules. They must be imported before they can be used. This point brings out several problems encountered by all of us who write Python scripts: how to decide what modules to import, how to discover whether a module that adds sought-after functionality exists, and how to learn to use an unfamiliar module. This book will introduce you to a wide selection of useful modules—enough for you to perform fairly complicated data analysis. On the other hand, your work might be expedited by specialized modules not discussed here. Fortunately, because Python is so popular, web searches will often turn up documentation and tutorials that offer helpful advice.

Notwithstanding that most of the functions and methods we will be using are not part of Python itself, but rather of some module that extends Python, we will use colloquial and inexact language and call everything "Python"—unless we want to draw attention to some specific module.

The rest of the first cell defines several functions that are not part of any module, but rather were written by the author. For instance, the line

```
def gda_cvec(v):
```
[gdapy01_00]

(and the many lines of code that follow it) defines a function `gda_cvec()` that expands Python's ability to manipulate vectors. Its use will be discussed later in this chapter. You should run the first cell once you have opened `gdapy_01.ipynb` in the Jupyter Notebook. You should check that it successfully executed by checking for error messages, which would be written immediately following the end of the cell. Hopefully, there are none!

The second cell in `gdapy_01.ipynb` contains the short script that prints today's date:

```
# gdapy01_01 print the date
thedate = date.today(),
print(thedate);
```
[gdapy01_01]

The first line, starting with the character #, is a *comment* that includes our name of the script gdapy01_01 (which stands for Geophysical Data Analysis, Python, Chapter 1, Script 1). The second line gets today's date and puts it into a variable thedate. The notation `date.today()` means the method `today()` from the `date` module. (The method takes no arguments, so nothing is enclosed by the parentheses.) Once the second line is executed, the variable `thedate` contains the date. The third line prints it at the end of the cell as a character string. For example:
2022-11-01
The `print()` function is extremely useful when debugging scripts, for it can print more or less anything and easily can be inserted at strategic points in scripts to print the values of variables. However, often its output is not very aesthetic.

1B.2 Effective use of folders

Files proliferate at an astonishing rate, even in the most trivial data analysis project. Data, notes, Live Scripts, intermediate results, and final results will all be contained in files, and their numbers will grow during the project. These files need to be organized through a system of folders (directories), subfolders (subdirectories), and filenames that are sufficiently systematic that files can be located easily and so that they are not confused with one another. Predictability both in the pattern of filenames and in the arrangement of folders and subfolders is an extremely important part of the design.

The Python-related files associated with this book are in a folder/subfolder/filename structure modeled on the format of the book itself (Fig. 1B.1). The main folder is named gdapy. It contains several subfolders, Notebooks (containing Jupyter Notebooks), Data (containing data), TestFolder (for test purposes), and Html (for hypertext). The Jupyter Notebooks are named gdapy01.ipynb, gdapy02.ipynb, etc., one file for each of the book's chapters. We have chosen to use leading zeros in the naming scheme (e.g., 01) so that filenames appear in the correct order when they are sorted alphabetically (as when listing the contents of a folder). A given Notebook file contains all the scripts for the corresponding chapter, one per cell. Comments at the top of each cell are used to identify each script, with name-tags of the form gdapyNN_MM, where the chapter number NN and the script number MM are sequential integers.

Python supports a number of commands that enable you to navigate from folder to folder, list the contents of folders, etc. For example, the commands

```
mydir=os.getcwd();
print(mydir);
```

[gdapy01_02]

get and print the name of the current folder (or current working directory). Initially, this should be the Notebook directory, which on the author's computer is

```
c:\bill\GDA-5thEd\gdapy\Notebooks
```

Changing the working directory that is one level above Notebooks is accomplished by the command:

```
os.chdir("..");
```

[gdapy01_02]

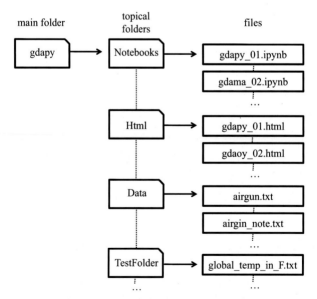

FIG. 1B.1   Folder (directory) structure used for the Python files accompanying this book.

The .. signifies one folder up (which is the gdapy folder). Note that the pathname is quoted to signify that it is a character string (discussed further later). Changing the working directory to one that is a level below gdapy, such as Notebooks, is accomplished by

```
os.chdir("Notebooks");
```

[gdapy01_02]

Finally, a listing of the files and subfolders in the current working directory can be created using the commands:

```
mydirlist=os.listdir();
print(mydirlist);
```

[gdapy01_02]

As we will discuss in more detail later in this book, these commands can be used to facilitate processing data files located in multiple folders.

1B.3 Simple arithmetic

The Python commands for simple arithmetic and algebra closely parallel standard mathematical notation. For instance, the command sequence

```
a=4.5;
b=5.1;
c=a+b;
print(c);
```

[gdapy01_03]

evaluates the formula $c=a+b$ for the case $a=4.5$ and $b=5.1$ to obtain $c=9.6$. A semicolon at the end of the formula serves to indicate the end of the command. The final command print(c) causes Python to display the final result c.

Note that Python variables are *static*, meaning that they persist until you explicitly delete them or exit the program. Variables created in one cell can be used by subsequent cells. At any time, the value of a variable can be examined, by printing or plotting it. The persistence of variables can sometimes lead to scripting errors, such as when the definition of a variable in a cell is inadvertently omitted, but Python uses the value defined in a previously executed cell. Although the command

```
%reset -f
```

[gdapy01_00]

deletes all previously defined variables, it also deletes all imported modules, which makes its use cumbersome, at best.

A somewhat more complicated formula is

$$c = \sqrt{a^2 + b^2} \text{ with } a = 6 \text{ and } b = 8 \tag{1B.1}$$

The corresponding script is

```
a=6.0;
b=8.0;
c = sqrt(a**2 + b**2);
disp(c);
```

[gdapy01_04]

Note that Python's syntax for a^2 is a**2 and that the square root is computed using the sqrt() function. This is an example of Python's syntax differing from standard mathematical notation.

A final example is

$$c = \sin\left(\frac{n\pi(x - x_0)}{L}\right) \quad \text{with } n = 3, x = 4, x_0 = 1 \text{ and } L = 6 \tag{1B.2}$$

The corresponding script is

```
n=3.0; x=4.0; x0=1.0; L=6.0;
c = sin(n*pi*(x-x0)/L);
print(c);
```
[gdapy01_05]

Note that several formulas, separated by semicolons, can be typed on the same line. Variables, such as x0 and pi, given before, can have names consisting of more than one character and can contain numerals as well as letters (though they must start with a letter). Python's math module defines a variety of predefined mathematical constants, including pi (the usual mathematical constant π).

Many Python manuals, guides, and tutorials are available, both in the printed form (e.g., Menke, 2022; Sundnes, 2020; VanderPlas, 2016) and on the Web (e.g., at https://docs.python.org/3/tutorial/, https://numpy.org/). The reader may find that they complement this book by providing more detailed information about Python's functionality and script writing in general.

1B.4 Lists, tuples, vectors, and matrices

In the simplest interpretation, a *vector* is just a list of numbers that is treated as unit and given a symbolic name. Similarly, a *matrix* is just a table of numbers that is treated as unit and given a symbolic name. Both are essential to data analysis, because they allow large amounts of data to be processed as units. Most programming languages have vector and matrix data types—but not Python! That functionality is added to it by the Numpy module.

Python does, however, have two types of list-like variables that are somewhat similar to vectors. One is called the *list* and the other the *tuple*. A list of the numbers 1, 0, 7.0, 2.0, and 6.0 can be created with the command

```
mylist = [1.0, 7.0, 2.0, 6.0];
```
[gdapy01_06]

and an individual element within the list can be accessed by *indexing* it. For example,

```
b=mylist[2];
```
[gdapy01_06]

sets b to the third element of the list, which is 2.0. Similarly, a tuple of the same numbers can be created with the command

```
mytuple = (1.0, 7.0, 2.0, 6.0);
```
[gdapy01_06]

A one-element tuple is specified with the notation (1.0,). An individual element within the tuple can be accessed by indexing it. For example,

```
b=mytuple[2];
```
[gdapy01_06]

sets b to the third element of the tuple, which is 2.0. Note that indexing starts at 0, not 1, in Python. Notwithstanding this example, lists and tuples are of limited use for representing numerical data. However, they are often used to organize character data (discussed further later). Furthermore, they are commonly used to pass information to and from methods. Because they do not have exactly the same functionality, lists and tuples are not interchangeable; many methods that expect tuples will not accept lists and vice versa.

Vectors and matrices are fundamental to inverse theory both because they provide a convenient way to organize data and because many important operations on data can be very succinctly expressed using linear algebra (i.e., the algebra of vectors and matrices). When a vector is organized horizontally, as a row, it is called a *row vector* and when it is organized vertically, as a column, in which case it is called a *column vector*.

We will use lowercase bold letters to represent both kinds of vector. An exemplary 1×3 row vector $\mathbf{r}$ and a 3×1 column vector $\mathbf{c}$ are

$$\mathbf{r} = [2, 4, 6] \text{ and } \mathbf{c} = \begin{bmatrix} 1 \\ 3 \\ 5 \end{bmatrix} \tag{1B.3}$$

In Python, a row vector and a column vector can be created with the commands

```
r = np.array([[2, 4, 6]]);
c = np.array([[1], [3], [5]]);
```
[gdapy01_06]

This syntax is most effective when the lengths of the vectors are short. We will discuss a variety of more effective means of creating long vectors later in the text, but mention two very simple, yet extremely useful, methods here. Length-N row and column vectors of zeros are formed by

```
r = np.zeros((1,N));
c = np.zeros((N,1));
```
[gdapy01_06]

Note that the quantity $(1,N)$ is a tuple. Similarly, length-N row and column vectors of ones are formed by

```
r = np.ones((1,N));
c = np.ones((N,1));
```
[gdapy01_06]

A row vector can be turned into a column vector, and vice versa, by the transpose operation, denoted by superscript T. Thus

$$\mathbf{r}^T = \begin{bmatrix} 2 \\ 4 \\ 6 \end{bmatrix} \text{ and } \mathbf{c}^T = [1, 3, 5] \tag{1B.4}$$

In Python, the transpose is computed by applying the T method to the vector; that is r.T is a column vector and c.T is a row vector. Although both column vectors and row vectors are useful, our experience is that defining both in the same script creates serious opportunities for error. A formula that requires a column vector will usually yield incorrect results if a row vector is substituted into it, and vice versa. Consequently, we will adhere to a protocol where all vectors defined in this book are column vectors. Row vectors will be created when needed—and as close as possible to where they are used in the script—by transposing the equivalent column vector.

A column vector has size $N \times 1$ and row vector has size $1 \times N$. They are specified by the tuples $(N,1)$ and $(1,N)$, respectively. Python also permits vectors of size $N \times 0$, specified by the tuple $(N,)$. Insofar as is possible, we do not use them in this book, for we feel that they create even more opportunity for error. However, a few important methods require them. In these cases, we convert a column vector into a $N \times 0$ vector using the ravel() method

```
c0 = c.ravel();
```
[gdapy01_06]

Finally, we provide the function gda_cvec() that converts ordinary numbers, $N \times 0$ vectors, row vectors, lists, and tuples into a column vector. For example,

```
c = gda_cvec(1.0,3.0,5.0);
```
[gdapy01_06]

An individual number within a vector is called an *element* (or, sometimes, *component*) and is denoted with an integer index, written as a subscript, with the index 0 in the leftmost element of the row vector and the topmost element of the column vector. Thus $r_1 = 4$ and $c_2 = 5$ in the example given before. In Python, the index is written inside square brackets, as in

```
a=r[0,1];
b=c[2,0];
```

[gdapy01_06]

Note that we always use two indices, with the 0 indicating row zero of the row vector and column zero of the column vector.

Sometimes, we will wish to indicate a generic element of the vector, in which case we will give it a variable index, as in r_i and c_j, with the understanding that i and j are integers (whole numbers). We will use bold uppercase names to denote matrices, as in

$$\mathbf{A} = \begin{bmatrix} 1 & 2 & 3 \\ 4 & 5 & 6 \\ 7 & 8 & 9 \end{bmatrix} \tag{1B.5}$$

In this example, the number of rows and number of columns are equal, but this property is not required; matrices can also be rectangular. Thus row vectors and column vectors are just special cases of rectangular matrices. The transposition operation can also be performed on a matrix, in which case its rows and columns are interchanged

$$\mathbf{A}^T = \begin{bmatrix} 1 & 4 & 7 \\ 2 & 5 & 8 \\ 3 & 6 & 9 \end{bmatrix} \tag{1B.6}$$

A square matrix $\mathbf{A}$ is said to be *symmetric* if $\mathbf{A}^T = \mathbf{A}$. In Python, a matrix is defined by

```
A = np.array([[1.,4.,7.],[2.,5.,8.],[3.,6.,9.]]);
```

This syntax is most effective when the dimensions of the array are small. We will discuss a variety of more effective means of creating larger matrices later in the text, but mention two very simple, yet extremely useful, methods here. An $N \times M$ matrices $\mathbf{A}$ of zeros and $\mathbf{B}$ of ones are formed, respectively, by

```
B = np.zeros((N,M));
C = np.ones((N,M));
```

[gdapy01_06]

The individual elements of a matrix are denoted with two integer indices, the first indicating the row and the second the column, starting with zeros in the upper left. Thus, in the earlier example $A_{20} = 7$. Note that transposition swaps indices; that is, M_{ji} is the transpose of M_{ij}. In Python, the indices are written inside square brackets, as in

```
s = M[0,2];
```

[gdapy01_06]

One of the key properties of vectors and matrices is that they can be manipulated symbolically—as entities—according to specific rules that are similar to normal arithmetic. This allows tremendous simplification of data processing formulas, since all the details of what happens to individual elements within those entities are hidden from view and automatically performed.

In order to be added, two matrices (or vectors, viewing them as a special case of a rectangular matrix) must have the same number of rows and columns. Their sum is then just the matrix that results from summing corresponding elements. Thus if

$$\mathbf{A} = \begin{bmatrix} 1 & 0 & 2 \\ 0 & 1 & 0 \\ 2 & 0 & 1 \end{bmatrix} \text{ and } \mathbf{B} = \begin{bmatrix} 1 & 0 & -1 \\ 0 & 2 & 0 \\ 1 & 0 & 3 \end{bmatrix}$$

$$\tag{1B.7}$$

$$\mathbf{S} = \mathbf{A} + \mathbf{B} = \begin{bmatrix} 1+1 & 0+0 & 2-1 \\ 0+0 & 1+2 & 0+0 \\ 2+1 & 0+0 & 1+3 \end{bmatrix} = \begin{bmatrix} 2 & 0 & 1 \\ 0 & 3 & 0 \\ 3 & 0 & 4 \end{bmatrix}$$

Subtraction is performed in an analogous manner. In terms of the components, addition and subtraction are written as

$$S_{ij} = M_{ij} + N_{ij} \text{ and } D_{ij} = M_{ij} - N_{ij} \tag{1B.8}$$

Note that addition is commutative (i.e., $\mathbf{N}+\mathbf{M}=\mathbf{M}+\mathbf{N}$) and associative (i.e., $(\mathbf{A}+\mathbf{B})+\mathbf{C}=\mathbf{A}+(\mathbf{B}+\mathbf{C})$. In Python, addition and subtraction are written as

```
S = M+N;
D = M-N;
```

[gdapy01_06]

Multiplication of two matrices is a more complicated operation and requires that the number of columns of the left-hand matrix equals the number of rows of the right-hand matrix. Thus if the matrix $\mathbf{A}$ is $N \times K$ and the matrix $\mathbf{B}$ is $K \times M$, the product $\mathbf{P}=\mathbf{AB}$ is an $N \times M$ matrix defined according to the rule:

$$P_{ij} = \sum_{k=0}^{K-1} A_{ik}B_{kj} \tag{1B.9}$$

The order of the indices is important. Matrix multiplication is in its standard form when all summations involve neighboring indices of adjacent quantities. Thus, for instance, the two instances of the summed variable k are not neighboring in the equation

$$Q_{ij} = \sum_{k=0}^{K-1} A_{ki}B_{kj} \tag{1B.10}$$

and so the equation corresponds to $\mathbf{Q}=\mathbf{A}^T\mathbf{B}$ and not $\mathbf{Q}=\mathbf{AB}$. Matrix multiplication is not commutative (i.e., $\mathbf{AB} \neq \mathbf{BA}$) but is associative (i.e., $(\mathbf{AB})\mathbf{C}=\mathbf{A}(\mathbf{BC})$ and distributive (i.e., $\mathbf{A}(\mathbf{B}+\mathbf{C})=\mathbf{AB}+\mathbf{AC}$). An important rule involving the matrix transpose is $(\mathbf{AB})^T=\mathbf{B}^T\mathbf{A}^T$ (note the reversal of the order).

Several special cases of multiplication involving vectors are noteworthy. Suppose that $\mathbf{a}$ and $\mathbf{b}$ are length-N column vectors. The combination $s=\mathbf{a}^T\mathbf{b}$ is a scalar number s and is called the *dot product* (or sometimes, *inner product*) of the vectors. It obeys the rule $\mathbf{a}^T\mathbf{b}=\mathbf{b}^T\mathbf{a}$. The dot product of a vector with itself is the square of its Euclidian length; that is, $\mathbf{a}^T\mathbf{a}$ is the sum of its squared elements of $\mathbf{a}$. The vector $\mathbf{a}$ is said to be a *unit vector* when $\mathbf{a}^T\mathbf{a}=1$. The combination $\mathbf{ab}^T$ is an $N \times N$ matrix; it is called the *outer product*. The product of a matrix and a vector is another vector, as in $\mathbf{c}=\mathbf{Ba}$. One interpretation of this relationship is that the matrix $\mathbf{B}$ "turns one vector into another." Note that the vectors $\mathbf{a}$ and $\mathbf{c}$ can be of different length. An $M \times N$ matrix $\mathbf{B}$ turns the length-N vector a into a length-M vector $\mathbf{c}$. The combination $s=\mathbf{a}^T\mathbf{Ba}$ is a scalar and is called a *quadratic form*, as it contains terms quadratic in the elements of $\mathbf{a}$. Matrix multiplication $\mathbf{P}=\mathbf{AB}$ has a useful interpretation in terms of dot products: P_{ij} is the dot product of the ith row of $\mathbf{A}$ with the jth column of $\mathbf{B}$.

Any matrix is unchanged when multiplied by the *identity matrix*, conventionally denoted $\mathbf{I}$. Thus $\mathbf{a}=\mathbf{Ia}$, $\mathbf{A}=\mathbf{IA}=\mathbf{AI}$, etc. This matrix has ones along its main diagonal, and zeroes elsewhere, as in

$$\mathbf{I} = \begin{bmatrix} 1 & 0 & 0 \\ 0 & 1 & 0 \\ 0 & 0 & 1 \end{bmatrix} \tag{1B.11}$$

The elements of the identity matrix are usually written δ_{ij} and not I_{ij}, and the symbol δ_{ij} is usually called the *Kronecker delta symbol*, not the elements of the identity matrix (though that is exactly what it is). The equation $\mathbf{A}=\mathbf{IA}$ for an $N \times N$ matrix $\mathbf{A}$ is written componentwise as

$$A_{ij} = \sum_{k=0}^{N-1} \delta_{ik}A_{kj} \tag{1B.12}$$

This equation indicates that any summation containing a Kronecker delta symbol can be performed trivially. To obtain the result, one first identifies the variable that is being summed over (k in this case) and the variable that the summed variable is paired within the Kronecker delta symbol (i in this case). The summation and the Kronecker delta symbol then are deleted from the equation, and all occurrences of the summed variable are replaced with the paired variable (all ks are replaced by is in this case). In Python, an $N \times N$ identity matrix can be created with the command

```
I = np.identity(N);
```

[gdapy01_07]

Several different syntaxes can be used to multiply vectors and matrices in Python—but in the author's opinion, all of them are flawed because they have attributes that can lead to coding errors. We have chosen to use a "function call" syntax, because we believe it is the alternative least prone to error. We illustrate matrix multiplication by considering the column vectors **a** and **b** and matrices **A** and **B**

$$\mathbf{a} = \begin{bmatrix} 1 \\ 3 \\ 5 \end{bmatrix} \text{ and } \mathbf{b} = \begin{bmatrix} 2 \\ 4 \\ 6 \end{bmatrix} \text{ and } \mathbf{A} = \begin{bmatrix} 1 & 0 & 2 \\ 0 & 1 & 0 \\ 2 & 0 & 1 \end{bmatrix} \text{ and } \mathbf{B} = \begin{bmatrix} 1 & 0 & -1 \\ 0 & 2 & 0 \\ -1 & 0 & 3 \end{bmatrix} \tag{1B.13}$$

Then, the products:

$$s = \mathbf{a}^T \mathbf{b} = \begin{bmatrix} 1 \\ 3 \\ 5 \end{bmatrix}^T \begin{bmatrix} 2 \\ 4 \\ 6 \end{bmatrix} = \begin{bmatrix} 1 & 3 & 5 \end{bmatrix} \begin{bmatrix} 2 \\ 4 \\ 6 \end{bmatrix} = 1 \times 2 + 3 \times 4 + 5 \times 6 = 44$$

$$\mathbf{T} = \mathbf{a} \mathbf{b}^T = \begin{bmatrix} 1 \\ 3 \\ 5 \end{bmatrix} \begin{bmatrix} 2 & 4 & 6 \end{bmatrix} = \begin{bmatrix} 1 \times 2 & 1 \times 4 & 1 \times 6 \\ 3 \times 2 & 3 \times 4 & 3 \times 6 \\ 5 \times 2 & 5 \times 4 & 5 \times 6 \end{bmatrix} = \begin{bmatrix} 2 & 4 & 6 \\ 6 & 12 & 18 \\ 10 & 20 & 30 \end{bmatrix}$$

$$\mathbf{c} = \mathbf{A} \mathbf{b} = \begin{bmatrix} 1 & 0 & 2 \\ 0 & 1 & 0 \\ 2 & 0 & 1 \end{bmatrix} \begin{bmatrix} 2 \\ 4 \\ 6 \end{bmatrix} = \begin{bmatrix} 1 \times 2 + 0 \times 4 + 2 \times 6 \\ 0 \times 2 + 1 \times 4 + 0 \times 6 \\ 2 \times 2 + 0 \times 4 + 1 \times 6 \end{bmatrix} = \begin{bmatrix} 14 \\ 4 \\ 10 \end{bmatrix}$$

$$\mathbf{P} = \mathbf{A} \mathbf{B} = \begin{bmatrix} 1 & 0 & 2 \\ 0 & 1 & 0 \\ 2 & 0 & 1 \end{bmatrix} \begin{bmatrix} 1 & 0 & -1 \\ 0 & 2 & 0 \\ -1 & 0 & 3 \end{bmatrix} = \begin{bmatrix} -1 & 0 & 5 \\ 0 & 2 & 0 \\ 1 & 0 & 1 \end{bmatrix}$$

(1B.14)

correspond to

```
s = np.matmul(a.T,b);
T = np.matmul(a,b.T);
c = np.matmul(A,b);
P = np.matmul(A,B);
```

[gdapy01_07]

In our preferred syntax, the `np.matmul(A,B)` method always is provided with the two quantities A and B that are to be multiplied. However, although we have chosen this syntax advisedly, readers should feel free to use one of the other supported syntaxes if they feel it works better for them.

Cases occur where the rules of matrix multiplication need to be overridden and the matrices multiplied *element-wise* (e.g., create a column vector **d** with elements $d_i = a_i b_i$):

```
d = np.multiply(a,b);
```

[gdapy01_07]

Individual elements of matrices can be accessed by specifying the relevant row and column indices, in square brackets, e.g., and `A[1,2]` is the second row, third column element of the matrix `A`.

Ranges of rows and columns can be specified by *slicing* an array using the `:` (colon) operator—but its syntax is tricky! The range `i:j` means the range from `i` to `j-1` (e.g., `0:3` is from 0 to 2). For the 3×3 matrix **B**, the second column is `B[0:3,1:2]`, the second row is `M[1:2,0:3]`, and the 2×2 submatrix in the lower right-hand corner is `B[1,3,1:3]`. These operations are further illustrated for

$$\mathbf{a} = \begin{bmatrix} 1 \\ 2 \\ 3 \end{bmatrix} \text{ and } \mathbf{B} = \begin{bmatrix} 1 & 2 & 3 \\ 4 & 5 & 6 \\ 7 & 8 & 9 \end{bmatrix} \tag{1B.15}$$

$$\mathbf{x} = \begin{bmatrix} a_0 \\ a_1 \end{bmatrix} = \begin{bmatrix} 1 \\ 2 \end{bmatrix} \text{ and } \mathbf{y} = \begin{bmatrix} B_{01} \\ B_{11} \\ B_{21} \end{bmatrix} = \begin{bmatrix} 2 \\ 5 \\ 8 \end{bmatrix} \tag{1B.16}$$

$$\text{and } \mathbf{z} = \begin{bmatrix} B_{10} & B_{11} & B_{12} \end{bmatrix} = \begin{bmatrix} 4 & 5 & 6 \end{bmatrix} \text{ and } \mathbf{T} = \begin{bmatrix} B_{11} & B_{12} \\ B_{21} & B_{22} \end{bmatrix} = \begin{bmatrix} 5 & 6 \\ 8 & 9 \end{bmatrix}$$

The corresponding commands are

```
x = a[0:2,0:1];
y = B[0:3,1:2];
c = B[1:2,0:3];
T = B[1:3,1:3];
```
[gdapy01_08]

In Python, the variables x, y, c, and T are not independent, but rather *views* into the larger variables a and B. Consequently, if a and M are modified, x, y, c, and T become modified, too. Views must be copied in order to become independent of the original variable

```
x = np.copy( a[0:2,0:1] );
y = np.copy( B[0:3,1:2] );
c = np.copy( B[1:2,0:3] );
T = np.copy( B[1:3,1:3] );
```
[gdapy01_08]

Two colons can be used in sequence to indicate the spacing of elements in the resulting row vector. For example, the expression `1:10:2` is the sequence `1,3,5,7,9` and that the expression `10:5:-1` is the sequence `9,8,7,6`.

Although Python's slicing syntax is very rich, we use only a very restricted subset in this book. In particular, we always specify both the start and end of the slice of every index, and we only equate views that have exactly the same shape—because our sense is that these restrictions lead to fewer errors, especially among novice scriptwriters.

A very useful length-N column vector, say **t**, has elements that increment by a constant amount Δt from zero to $(N-1)\Delta t$, that is, $t_i = i\Delta t$. It can be formed as

```
Dt = 2.0;
t = gda_cvec(np.linspace(0,Dt*(N-1),N));
```
[gdapy01_09]

Here, the increment Dt has been set to 2. The method `np.linspace()` returns an $N \times 0$ vector, so its result is converted into a column vector using `gda_cvec()`. One use of such a vector is for a *time axis*; that is, a vector that represents time increasing by regular increments.

Matrix division is defined in analogy to reciprocals. For a scalar number s, multiplication by the reciprocal s^{-1} is equivalent to division by s. Here, the reciprocal obeys $ss^{-1} = s^{-1}s = 1$. The matrix analog to the reciprocal is called the matrix *inverse* and obeys

$$\mathbf{A}\mathbf{A}^{-1} = \mathbf{A}^{-1}\mathbf{A} = \mathbf{I} \tag{1B.17}$$

It is defined only for square matrices. The calculation of the inverse of a matrix is complicated, and we will not describe it here, except to mention the 2×2 case

$$\mathbf{A} = \begin{bmatrix} a & b \\ c & d \end{bmatrix} \text{ and } \mathbf{A}^{-1} = \frac{1}{ad-bc} \begin{bmatrix} d & -b \\ -c & a \end{bmatrix} \tag{1B.18}$$

Just as the reciprocal s^{-1} is defined only when $s \neq 0$, the matrix inverse $\mathbf{A}^{-1}$ is defined only when a quantity called the *determinant* of $\mathbf{A}$, denoted $\det(\mathbf{A})$ (or sometimes $|\mathbf{A}|$), is not equal to zero. The determinant of the 2×2 matrix $\mathbf{A}$ is $\det(\mathbf{A}) = ad - bc$. It consists of the sum of products of two elements of the matrix. The determinant of a square $N \times N$ matrix $\mathbf{A}$ consists of the sum of products of N elements of the matrix and is given by

$$\det(\mathbf{A}) = \sum_{n_0=0}^{N-1} \sum_{n_1=0}^{N-1} \sum_{n_2=1}^{N-1} \cdots \sum_{n_{N-1}=1}^{N-1} \varepsilon_{n_0,n_1,n_2\cdots n_{N-1}} A_{0n_0} A_{1n_1} A_{2n_2} \cdots A_{N-1,n_{N-1}} \tag{1B.19}$$

Here the quantity $\varepsilon_{n_0,n_1,n_2\cdots n_{N-1}}$ is $+1$ when $(n_0, n_1, n_2, \cdots, n_{N-1})$ is an even permutation of $(0, 1, 2, \cdots, N-1)$, -1 when it is an odd permutation, and zero otherwise.

In Python, the matrix inverse and determinant of a square matrix A are computed as

```
B = la.inv(A);
d = la.det(A);
```
[gdapy01_10]

Here `la` is an abbreviation for the `scipy.linalg` module. In many of the formulas of inverse theory, the matrix inverse either premultiplies or postmultiplies other quantities, for instance:

$$\mathbf{c} = \mathbf{A}^{-1}\mathbf{b} \text{ and } \mathbf{C} = \mathbf{A}^{-1}\mathbf{B} \text{ and } \mathbf{D} = \mathbf{B}\mathbf{A}^{-1} \tag{1B.20}$$

These cases do not actually require the explicit calculation of $\mathbf{A}^{-1}$, just the combinations $\mathbf{A}^{-1}\mathbf{b}$, $\mathbf{A}^{-1}\mathbf{B}$, and $\mathbf{B}\mathbf{A}^{-1}$, which are computationally simpler. They can be computed as

```
c = la.solve(A,b);
C = la.solve(A,B);
D = la.solve(A.T,B.T).T;
```
[gdapy01_10]

The third equation follows from transposing $\mathbf{D} = \mathbf{B}\mathbf{A}^{-1}$ to $\mathbf{D}^T = [\mathbf{A}^T]^{-1}\mathbf{B}^T$.

A surprising amount of information on the structure of a matrix can be gained by studying how it affects a column vector that it multiplies. Suppose that $\mathbf{A}$ is an $N \times N$ square matrix and that it multiplies an *input* column vector $\mathbf{v}$, producing an *output* column vector $\mathbf{w} = \mathbf{A}\mathbf{v}$. We can examine how the output $\mathbf{w}$ compares to the input $\mathbf{v}$, as $\mathbf{v}$ is varied. One question of particular importance is

When is the output parallel to the input? (1B.21)

This question is called the *algebraic eigenvalue problem*. If $\mathbf{w}$ is parallel to $\mathbf{v}$, then $\mathbf{w} = \lambda\mathbf{v}$, where λ is a scalar proportionality factor. The parallel vectors satisfy the following equation:

$$\mathbf{A}\mathbf{v} = \lambda\mathbf{v} \text{ or } (\mathbf{A} - \lambda\mathbf{I})\mathbf{v} = 0 \tag{1B.22}$$

The trivial solution $\mathbf{v} = (\mathbf{A} - \lambda\mathbf{I})^{-1}0 = 0$ is not very interesting. A nontrivial solution is only possible when the matrix inverse $(\mathbf{A} - \lambda\mathbf{I})^{-1}$ does not exist. This is the case where the parameter λ is specifically chosen to make the determinant $\det(\mathbf{A} - \lambda\mathbf{I})$ exactly zero, because a matrix with zero determinant has no inverse. The determinant is calculated by adding together terms, each of which contains the product of N elements of the matrix. Since each element of the matrix contains, at most, one instance of λ, the product will contain powers of λ up to λ^N. Thus the equation $\det(\mathbf{A} - \lambda\mathbf{I}) = 0$ is an Nth order polynomial equation for λ. An Nth order polynomial equation has N solutions, so we conclude that there must be N different proportionality factors, say λ_i, and N corresponding column vectors, say $\mathbf{v}^{(i)}$, that solve $\mathbf{A}\mathbf{v} = \lambda\mathbf{v}$. The column vectors $\mathbf{v}^{(i)}$ are called the *characteristic vectors* (or eigenvectors) of the matrix $\mathbf{A}$, and the proportionality factors λ_i are called the *characteristic values* (or eigenvalues). Eigenvectors are determined only up to an arbitrary multiplicative factor s, since if $\mathbf{v}^{(i)}$ is an eigenvector, so is $s\mathbf{v}^{(i)}$. Consequently, they are conventionally chosen to be unit vectors.

In the special case where the matrix $\mathbf{A}$ is symmetric, it can be shown that the eigenvalues λ_i are real and the eigenvectors are mutually perpendicular, $\mathbf{v}^{(i)T}\mathbf{v}^{(j)}=0$ for $i \neq j$. The N eigenvalues can be arranged into a diagonal matrix $\mathbf{\Lambda}$, whose elements are $[\mathbf{\Lambda}]_{ij}=\lambda_i\delta_{ij}$, where δ_{ij} is the Kronecker delta. The corresponding N eigenvectors $\mathbf{v}^{(i)}$ can be arranged as the columns of an $N \times N$ matrix $\mathbf{V}$, which satisfies $\mathbf{VV}^T = \mathbf{V}^T\mathbf{V} = \mathbf{I}$. The eigenvalue equation $\mathbf{Av}^{(i)} = \lambda_i\mathbf{v}^{(i)}$ can then be succinctly written as $\mathbf{AV} = \mathbf{V\Lambda}$. Postmultiplying by $\mathbf{V}^T$, and applying $\mathbf{VV}^T = \mathbf{I}$, yields:

$$\mathbf{A} = \mathbf{V\Lambda V}^T \tag{1B.23}$$

That is, a symmetric matrix $\mathbf{A}$ is completely specified by its eigenvalue and eigenvectors.

We now note an interesting property of a matrix that can be written as $\mathbf{A} = \mathbf{B}^T\mathbf{B}$, where $\mathbf{B}$ is some matrix. By premultiplying the eigenvalue equation by $\mathbf{v}^{(i)T}$, and noting that $\mathbf{v}^{(i)T}\mathbf{v}^{(i)}=1$, we obtain $\lambda_i=\mathbf{v}^{(i)T}\mathbf{Av}^{(i)}$. Substituting $\mathbf{B}^T\mathbf{B}$ for $\mathbf{A}$, we have $\lambda=\mathbf{v}^{(i)T}\mathbf{B}^T\mathbf{Bv}^{(i)} = [\mathbf{Bv}^{(i)}]^T\mathbf{Bv}^{(i)}$. Consequently, λ_i is the squared length of the vector $\mathbf{Bv}^{(i)}$ and cannot be negative. In this case, the matrix $\mathbf{A}$ is said to be *nonnegative definite*.

In Python, the diagonal matrix of eigenvalues $\mathbf{\Lambda}$ and matrix of eigenvectors $\mathbf{V}$ of a symmetric matrix $\mathbf{A}$ are computed as

```
LAMBDA, V = la.eigh(A);
```
[gdapy01_10]

Here LAMBDA is an $N \times 0$ vector of eigenvalues.

1B.5 Matrix differentiation

Many of the derivations of inverse theory require that a column vector $\mathbf{v}$ be considered a function of an independent variable, say x, and then differentiated with respect to that variable to yield the derivative $d\mathbf{v}/dx$. Such a derivative represents the fact that the vector changes from $\mathbf{v}$ to $\mathbf{v} + \mathbf{\Delta v}$ as the independent variable changes from x to $x + \Delta x$. Note that the resulting change $\mathbf{\Delta v}$ is itself a vector. Derivatives are performed element-wise, that is,

$$\left[\frac{d\mathbf{v}}{dx}\right]_i = \lim_{\Delta x \to 0} \frac{v_i(x + \Delta x) - v_i(x)}{\Delta x} = \frac{dv_i}{dx} \tag{1B.24}$$

A somewhat more complicated situation is where the column vector $\mathbf{v}$ is a function of another column vector, say $\mathbf{y}$, that is, $\mathbf{v}(\mathbf{y})$. The partial derivative

$$\frac{\partial v_i}{\partial y_j} \tag{1B.25}$$

represents the change in the ith component of $\mathbf{v}$ caused by a change in the jth component of $\mathbf{y}$. Frequently, we will need to differentiate the linear function $\mathbf{v} = \mathbf{Ay}$, where $\mathbf{A}$ is a matrix with respect to $\mathbf{y}$:

$$\frac{\partial v_i}{\partial y_j} = \frac{\partial}{\partial y_j}\sum_{k=0}^{N-1}A_{ik}y_k = \sum_{k=0}^{N-1}A_{ik}\frac{\partial y_k}{\partial y_j} = \sum_{k=0}^{N-1}A_{ik}\delta_{kj} = A_{ij} \tag{1B.26}$$

Since the components of $\mathbf{y}$ are assumed to be independent, the derivative $\partial y_k/\partial y_j$ is zero except when $k=j$, in which case it is unity, which is to say $\partial y_k/\partial y_j=\delta_{kj}$, where δ_{kj} is the Kronecker delta. Thus the derivative of the linear function $\mathbf{v} = \mathbf{Ay}$ is the matrix $\mathbf{A}$. This relationship is the vector analog to the scalar case, where the derivative of the linear function $v=ay$ is the constant a.

1B.6 Character strings and lists

The numerical variable x created with the command x=2.5 has a close relationship to variables that are encountered in elementary mathematics. They are very useful to performing the calculations essential to data analysis. However, they are not very useful during file manipulation, which is usually a vital prerequisite to data analysis. Furthermore, the results of data analysis are most understandable if described in a combination of words and numbers—and not just numbers.

Python has a type of variable, called the *string variable*, the value of which can be set to a *character string* (sequence of alpha-numerical characters). For instance, a string variable with name `myfile` and with value `"mydata.txt"` is created by

```
myfile="mydata.txt";
```
[gdapy01_11]

Note that the value of the variable is quoted to prevent Python from trying to interpret it as a command.

Combining numerical variables with character strings can be very useful in file manipulation, where a sequence of filenames may contain a number that sequentially increments, like `"myfile_1.txt"`, `"myfile_2.txt"`, or when one wants to display the value of a variable together with some text that explains it; e.g., display the sentence `"position x is 10.40 meters"`.

The `%` string formatting operator is used to create a character string with a value that includes both text and the value of a variable. Thus, for instance,

```
i=1;
myfilename = "myfile_%d.txt" % (i);
```
[gdapy01_11]

creates a character string `myfilename` with value `"myfile_1.txt"`, where the 1 is taken from the variable i. Similarly,

```
x=10.40;
mysentence = "position x is %.2f meters" % (x);
```
[gdapy01_11)]

creates a character string `mysentence` with value `"position x is 10.40 meters"`, where the 10.4 is taken from the variable x.

Unfortunately, the `%` operator is fairly inscrutable, and we refer readers to the Python help pages for a detailed description. Briefly, it uses a *format string* (in this case `"position x is %.2f meters"`), which contains *placeholders* that start with the character `%` to indicate where in the character string the value of the variable should be placed. Thus `"myfile_%d.txt" % (i)` creates the character string `'myfile_1.txt'` (because the value of i is 1). The %d is the placeholder for an integer. It is replaced with 1 the value of i. When the variable can have fractional, as contrasted to integer, values, the floating-point placeholder %f is used instead. Thus `"position x is %.2f meters" % (x)` creates the character string `"position x is 10.40 meters"` (because the value of x is 10.4). The precision to which x is to be written is indicated by the .2 inside the `%.2f`; in this case, two significant digits.

The command `print(mysentence)` can be used to display the character string.

Lists of character strings are very useful when automating complicated tasks. For instance,

```
mycolorlist = ["red", "crimson", "chartreuse", "teal"];
```
[gdapy01_11]

Individual elements are accessed via standard indexing; that is, `mycolorlist[3]` is element number three containing the character string `"teal"`.

```
myfavoritecolor = mycolorlist[3];
```
[gdapy01_11]

1B.7 Loops

Python provides a looping mechanism, the `for` command, which can be useful when the need arises to sequentially access the elements of vectors and matrices. Thus, for example,

```
M = np.array( [[1, 2, 3],[4, 5, 6], [7, 8, 9]] );
for i in range(3):
    a[i,0] = M[i,i];
```

[gdapy01_12]

executes the a(i)=M(i,i) formula three times, each time with a different value of i, according to the specifications of the range() function. Here, range(3) means i=0, i=1, and i=2. The net effect is to copy the diagonal elements of the matrix M to the vector a, that is, $a_i = M_{ii}$. Indentation is a crucial part of the for-loop syntax, for it indicates which statements are inside the loop. Unindented commands are not part of the loop and are executed only once. Loops can be nested, that is, one loop can be inside another. Such an arrangement is necessary for accessing all the elements of a matrix in sequence. For example,

```
M = [1, 2, 3; 4, 5, 6; 7, 8, 9];
for i in range(3):
    for j in range(3):
        N[i,3-j] = M[i,j];
```

[gdapy01_13]

copies the elements of the matrix M to the matrix N, but reverses the order of the elements in each row, that is, $N_{i, 3-j} = M_{i, j}$. Loops are especially useful in conjunction with *conditional commands*. For example,

```
a=[1.0, 2.0, 1.0, 4.0, 3.0, 2.0, 6.0, 4.0, 9.0, 2.0, 1.0, 4.0];
for i in range(12):
    if ( a[i,0] >= 6.0 ):
        b[i,0] = 6.0;
    else:
        b[i,0] = a[i,0];
```

[gdapy01_14]

sets $b_i = a_i$ when $a_i < 6$ and sets $b_i = 6$ otherwise (a process called *clipping* a vector, for it clips off parts of the vector that are larger than 6).

A purist might point out that Python syntax is so flexible that for loops are almost never really necessary. In fact, all three examples, given before, can be computed with one-line formulas that omit for loops:

```
a = np.diag(M);
N = np.fliplr(M);
b=np.copy(a); b[a>6]=6;
```

[gdapy01_15]

The first two formulas are very simple, but rely upon the Python methods np.diag() (for "diagonal") and np.fliplr() (for "flip left-right"), whose existence we have not hitherto mentioned. The third formula, which used logical addressing, requires further explanation. The vector a is first copied to b. Then just those elements of b that are greater than 6 are reset to 6, using a technique called *logical addressing*. The expression a>6 returns a vector of true's and false's, depending upon whether the elements of the column vector a satisfy the inequality or not. The command b[a>6]=6 resets the elements of b to 6 only when the value is true.

One of the problems of a script-based environment is that learning the complete syntax of the scripting language can be pretty daunting. Writing a long script, such as one containing a for loop, will often be faster than searching through Python help files for a predefined function that implements the desired functionality in a single line of the script. When deciding between alternative ways of implementing a given functionality, you should always choose the one which you find clearest. Scripts that are terse or even computationally efficient are not necessarily a virtue, especially if they are difficult to debug. You should avoid creating formulas that are so inscrutable that you are not sure whether they will function correctly. Of course, the degree of inscrutability of any given formula will depend upon your level of familiarity with Python. Your repertoire of techniques will grow as you become more practiced.

1B.8 Loading data from a file

Python can read files with a variety of formats, but we start here with the simplest and most common, the text file. As an example, we load a global temperature data set compiled by the National Aeronautics and Space Administration. The author's recommendation is that you always keep a file of notes about any data set that you work with, and that these notes include information on where you obtained the data set and any modifications that you subsequently made to it:

The text file global_temp.txt contains global temperature change data from NASA's web site http://data.giss.nasa.gov/gistemp. It has two columns of data, time (in calendar years) and temperature anomaly (in degrees C) and is 57 lines long. Information about the data is in the file global_temp_notes.txt. The citation for this data is Hansen et al. (2010).

We reproduce the first few lines of global_temp.txt, here:

```
1965 -0.11
1966 -0.06
1967 -0.02
...  ...
```

The data are read into Python as follows:

```
D = np.genfromtxt("../Data/global_temp.txt",delimiter="\t");
N,M = np.shape(D);
t = np.copy( D[0:N,0:1] );  # time in column 1
d = np.copy( D[0:N,1:2] );  # temperature in column 2
```
                                                                    [gdapy01_16]

The `np.genfromt` function reads the data into a matrix **D**. Note that the filename is given as `'../Data/global_temp.txt'`, as contrasted to just `'global_temp.txt'`, since the script is run from the `LiveScripts` folder while the data are in the `Data` folder (which is "up and over" from `LiveScripts`). The filename is surrounded by quotes to indicate that it is a character string. The subsequent two lines break out D into two separate column vectors t of time and d of temperature data. This step is not strictly speaking necessary, but fewer mistakes will be made if the different variables in the data set have each their own name.

1B.9 Writing data to a file

Python can write files with a variety of formats, but we start here with the simplest and most common, the text file. Suppose that we wanted a file containing global temperature anomaly in degrees Fahrenheit, using the fact that $1°C = 1.8°F$. Starting with time t and temperature anomaly d from the previous section, we use the command

```
cf = 1.8;
dF = cf*d;
DF = np.concatenate((t,dF),axis=1);
newfilename="../TestFolder/global_temp_in_F.txt";
np.savetxt(newfilename,DF,delimiter="\t");
```
                                                                    [gdapy01_16]

The data is converted to degrees Fahrenheit by multiplying by the conversion factor cf. A single matrix DF is formed containing both time and temperature, by size-by-size concatenating the time column vector t and the Fahrenheit temperature data dF. The `np.savetxt()` method writes the matrix DF to a file (in this case `newfilename`) with a tab (denoted `"\t"`) between the columns of data. Note that the new filename `'global_temp_in_F.txt'` is chosen to indicate that the units have been changed.

1B.10 Plotting data

Python's plotting commands are very powerful, but they are also very complicated. We present here a set of commands for making a simple x–y plot that is intermediate between a very crude, unlabeled plot and an extremely artistic one. The reader may wish to adopt either a simpler or a more complicated version of this set, depending upon need and

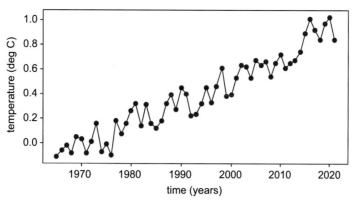

FIG. 1B.2 Python plot of deviations of global temperature from average for the time period, 1965–2021. Script gdapy01_16. *Data from Hansen, J., Ruedy, R., Sato, M., Lo, K., 2010. Global surface temperature change. Rev. Geophys. 48, RG4004. https://doi.org/10.1029/2010RG000.*

personal preference. The plot of the global temperature data (Fig. 1B.2) was created with the Pyplot (`plt`) commands:

```
fig1 = plt.figure(1,figsize=(8,4));    # smallish figure
ax1 = plt.subplot(1, 1, 1);            # only one subplot
plt.plot(t,d,"k-");                    # plot d(t) with black line
plt.plot(t,d,"ro");                    # plot d(t) with black line
plt.xlabel("time (years)");            # label time axis
plt.ylabel("temperature (deg C)");     # label data axis
plt.show();                            # display plot
                                              [gdapy01_16]
```

The `plt.figure()` method creates a new figure window, labels it as Figure 1, and sets its size. Although Pyplot allows several subplots within the figure, we specify that there be only in the call to the `plt.subplot()` method. The two `plt.plot(...)` commands plot the time axis `t` and temperature data `d` in two different ways: first as a black line (indicated by the `"k-"`) and the second with red circles (the `"ko"`). Finally, the horizontal and vertical axes are labeled using the `plt.xlabel()` and `plt.ylabel()` commands. The final command `plt.show()` indicates that the plot is finished and can be displayed.

References

Hansen, J., Ruedy, R., Sato, M., Lo, K., 2010. Global surface temperature change. Rev. Geophys. 48. RG4004.

Menke, W., 2022. Environmental Data Analysis with MATLAB® or Python, third ed. Elsevier, Oxford, ISBN: 978-0-323-95576-8. 444 pp.

Part-Enander, E., Sjoberg, A., Melin, B., Isaksson, P., 1996. The MATLAB® Handbook. Addison-Wesley, New York. 436 pp., ISBN: 0201877570.

Pratap, R., 2009. Getting Started With MATLAB®: A Quick Introduction for Scientists and Engineers. Oxford University Press, Oxford. 256 pp., ISBN 978-0190602062.

Sundnes, J., 2020. Introduction to Scientific Programming with Python. Springer International (Switzerland), ISBN. 978-3-030-50355-0. 147 pp.

VanderPlas, J., 2016. Python Data Science Handbook: Essential Tools for Working With Data. O'Reilly Media, Sebastopol, CA, ISBN: 978-1-491-91205-8. 517 pp.

Describing inverse problems

Inverse theory is an organized set of mathematical techniques for reducing data to obtain knowledge about the physical world on the basis of inferences drawn from observations. Inverse theory, as we shall consider it in this book, is limited to observations and questions that can be represented numerically. The observations of the world will consist of a tabulation of measurements or data. The questions we want to answer will be stated in terms of the numerical values (and statistics) of specific (but not necessarily directly measurable) properties of the world. These properties will be called *model parameters* for reasons that will become apparent. Usually, some specific mathematical *theory* (or *model*) has been identified that relates the model parameters to the data.

2.1 Forward and inverse theories

The question, "what causes the motion of the planets?," for example, is not one to which inverse theory can be applied. Even though it is perfectly scientific and historically important, its answer is not numerical in nature. On the other hand, inverse theory can be applied to the question, assuming that Newtonian mechanics applies, determine the number and orbits of the planets on the basis of the observed orbit of Halley's comet. The number of planets and their orbital ephemerides are numerical in nature. Another important difference between these two problems is that the first asks us to determine the reason for the orbital motions, and the second presupposes the reason and asks us only to determine certain details. Inverse theory rarely supplies the kind of insight demanded by the first question; it always requires that the physical model or theory be specified beforehand.

The term *inverse theory* is used in contrast to *forward theory*, which is defined as the process of predicting the results of measurements (predicting data) on the basis of some general principle or model and a set of specific conditions relevant to the problem at hand. Inverse theory, roughly speaking, addresses the reverse problem: starting with data and a general principle, theory, or quantitative model, it determines estimates of the model parameters. In the earlier example, predicting the orbit of Halley's comet from the presumably well-known orbital ephemerides of the planets is a problem for forward theory.

Another comparison of forward and inverse problems is provided by the phenomenon of temperature variation as a function of depth beneath the Earth's surface. Let us assume that the temperature increases linearly with depth in the Earth, that is, temperature T is related to depth z by the rule $T(z) = a + bz$, where a and b are numerical constants that we will refer to as *model parameters*. If one knows that $a = 25\,°C$ and $b = 0.1\,°C/km$, then one can solve the forward problem simply by evaluating the formula for any desired depth. The inverse problem would be to determine a and b on the basis of a suite of temperature measurements made at different depths in, say, a bore hole. One may recognize that this is the problem of fitting a straight line to data, which is a substantially harder problem than the forward problem of evaluating a first-degree polynomial. This brings out a property of most inverse problems: that they are substantially harder to solve than their corresponding forward problems.

$$\begin{array}{c} \text{Forward problem:} \\ \text{estimates of model parameters} \to \text{quantitative model} \to \text{predictions of data} \\ \text{Inverse problem:} \\ \text{observations of data} \to \text{quantitative model} \to \text{estimates of model parameters} \end{array} \qquad (2.1)$$

Note that the role of inverse theory is to provide information about unknown numerical parameters that go into the model, not to provide the model itself. Nevertheless, inverse theory can often provide a means for assessing the correctness of a given model or of discriminating between several possible models.

The model parameters one encounters in inverse theory vary from discrete numerical quantities to continuous functions of one or more variables. The intercept and slope of the straight line mentioned earlier are examples of discrete parameters. Temperature, which varies continuously with position, is an example of a continuous function. This book deals mainly with discrete inverse theory, in which the model parameters are represented as a set of a finite number of numerical values. This limitation does not, in practice, exclude the study of continuous functions, since they usually can be adequately approximated by a finite number of discrete parameters. Temperature, for example, might be represented by its value at a finite number of closely spaced points or by a set of splines with a finite number of coefficients. This approach does, however, limit the rigor with which continuous functions can be studied. Parameterizations of continuous functions are always both approximate and, to some degree, arbitrary properties, which cast a certain amount of imprecision into the theory. Nevertheless, discrete inverse theory is a good starting place for the study of inverse theory, in general, since it relies mainly on the theory of vectors and matrices rather than on the somewhat more complicated theory of continuous functions and operators. Furthermore, careful application of discrete inverse theory can often yield considerable insight, even when applied to problems involving continuous parameters.

Although the main purpose of inverse theory is to provide estimates of model parameters, the theory has a considerably larger scope. Even in cases in which the model parameters are the only desired results, there is a plethora of related information that can be extracted to help determine the "goodness" of the solution to the inverse problem. The actual values of the model parameters are indeed irrelevant in cases when we are mainly interested in using inverse theory as a tool in experimental design or in summarizing the data. Some of the questions inverse theory can help answer are the following:

1. What are the underlying similarities among inverse problems?
2. How are estimates of model parameters made?
3. How much of the error in the measurements shows up as error in the estimates of the model parameters?
4. Given a particular experimental design, can a certain set of model parameters really be determined?

These questions emphasize that there are many different kinds of answers to inverse problems and many different criteria by which the goodness of those answers can be judged. Much of the subject of inverse theory is concerned with recognizing when certain criteria are more applicable than others, as well as detecting and avoiding (if possible) the various pitfalls that can arise.

Inverse problems arise in many branches of the physical sciences. An incomplete list might include such entries as:

1. medical and seismic tomography,
2. image enhancement,
3. curve fitting,
4. earthquake location,
5. oceanographic and meteorological data assimilation,
6. factor analysis,
7. determination of Earth structure from geophysical data,
8. satellite navigation and Global Positioning System (GPS) geodesy,
9. mapping of celestial radio sources with interferometry, and
10. analysis of molecular structure by X-ray diffraction.

Inverse theory was developed by scientists and mathematicians having various backgrounds and goals. Thus, although the resulting versions of the theory possess strong and fundamental similarities, they have tended to look, superficially, very different. One of the goals of this book is to present the various aspects of inverse theory in such a way that both the individual viewpoints and the "big picture" can be clearly understood.

There are perhaps three major viewpoints from which inverse theory can be approached. The first and oldest sprang from probability theory—a natural starting place for such "noisy" quantities as observations of the real world. In this version of inverse theory, the data and model parameters are treated as random variables, and a great deal of emphasis is placed on determining the probability density functions that they follow. This viewpoint leads very naturally to the analysis of error and to tests of the significance of answers.

The second viewpoint developed from that part of the physical sciences that retains a deterministic stance and avoids the explicit use of probability theory. This approach has tended to deal only with estimates of model parameters (and perhaps with their error bars) rather than with probability density functions per se. Yet what one means by an estimate is often nothing more than the expected value of a probability density function; the difference is only one of emphasis.

The third viewpoint arose from a consideration of model parameters that are inherently continuous functions. Whereas the other two viewpoints handled this problem by approximating continuous functions with a finite number of discrete parameters, the third developed methods for handling continuous functions explicitly. Although continuous inverse theory is not the primary focus of this book, many of the concepts originally developed for it have application to discrete inverse theory, especially when it is used with discretized continuous functions.

2.2 Formulating inverse problems

The starting place in most inverse problems is a description of the data. Since in most inverse problems the data are simply a list of numerical values, a vector provides a convenient means of their representation. If N measurements are performed in a particular experiment, for instance, one might consider these numbers as the elements of a vector $\mathbf{d}$ of length N.

$$\text{data} = \text{list of } N \text{ numerical values}$$
$$\mathbf{d} = [d_1 \ d_2 \ d_3 \ \cdots \ d_N]^T \tag{2.2}$$

Here, T signifies transpose.

The purpose of the data analysis is to gain knowledge through systematic examination of data. While knowledge can take many forms, we assume here that it is primarily numerical in nature. We analyze data so to infer, as best we can, the values of numerical quantities—model parameters. Model parameters are chosen to be meaningful; that is, they are chosen to capture the essential character of the processes that are being studied.

The notion that data are limited in number, so that they can be described by a length-N vector, clearly corresponds to the practice of making observations. Each observation has a numerical value—perhaps one handwritten in a laboratory notebook or automatically generated by a laboratory instrument and written to a computer file. Whether model parameters are also limited in number, so that they can be described by a vector, is less clear. Some geophysically interesting model parameters, such as the distance of each of the planets from the Sun, do adhere to this notion of being *discrete*. Others, like spatially (and possibly temporally) varying density within Earth, would seem to be continuously varying and better described by a *function*, not a vector.

$$\text{discrete case:}$$
$$\text{model parameters} = \text{list of } M \text{ numerical values}$$
$$\mathbf{m} = [m_1 \ m_2 \ m_3 \ \cdots \ m_M]^T$$
$$\text{continuous case:} \tag{2.3}$$
$$\text{model parameters} = \text{spatially and temporally varying function}$$
$$m = m(x, y, z, t)$$

Here, (x, y, z) are the spatial coordinates and t is the time. In the first part of this book, we will focus on discrete model parameters, that is, those that can be represented as a length-M vector $\mathbf{m}$. This is not as big a limitation as it might appear, because a continuous function—when sufficiently smoothly varying—can be represented by their values on a grid of point, or by other methods (such as Fourier series) that involve a finite number of discrete parameters. Furthermore, although a function, say $m(x)$, allows variation at every *distance scale* of x, the applicability of this behavior to the real world is dubious. For instance, although density varies smoothly within the Earth at kilometer scales, it varies very disjointedly at millimeter scales, where crystal domains are important. And at even smaller scales, where quantum effects dominate, the notion of density is not well defined.

The fundamental statement of an inverse problem is that the model parameters and the data are in some way related. This relationship is called the quantitative model (or model, or theory, for short). Usually, the model takes the form of one or more formulas that the data and model parameters are expected to follow.

If, for instance, one were attempting to determine the density of an object, such as a rock, by measuring its mass and volume, there would be $N=2$ data—mass and volume (say, d_1 and d_2, respectively)—and $M=1$ unknown model parameter, density (say, d_1). The model would be the statement that density times volume equals mass, which can be written compactly by the vector equation $d_2 m_1 = d_1$. Note that the model parameter, density, is more meaningful than either mass or volume, in that it represents an intrinsic property of a substance that is related to its chemistry. The data—mass and volume—are easy to measure, but they are less fundamental because they depend on the size of the object, which is usually incidental.

In more realistic situations, the data and model parameters are related in more complicated ways. Most generally, the data and model parameters might be related by one or more implicit equations such as

nonlinear functions of data and model parameters are zero

$$\left.\begin{array}{l} f_1(\mathbf{d}, \mathbf{m}) = 0 \\ f_2(\mathbf{d}, \mathbf{m}) = 0 \\ \qquad \cdots \\ f_L(\mathbf{d}, \mathbf{m}) = 0 \end{array}\right\} \text{ same as } \mathbf{f}(\mathbf{d}, \mathbf{m}) = 0 \qquad (2.4)$$

where L is the number of equations. In this example concerning the measuring of density, $L=1$ and $d_2 m_1 - d_1 = 0$ would constitute the one equation of the form $f_1(\mathbf{d}, \mathbf{m}) = 0$. These implicit equations, which can be compactly written as the vector equation $\mathbf{f}(\mathbf{d}, \mathbf{m}) = 0$, summarize what is known about how the measured data and the unknown model parameters are related. The purpose of inverse theory is to solve, or "invert," these equations for the model parameters, or whatever kinds of answers might be possible or desirable in any given situation.

No claims are made either that the equations $\mathbf{f}(\mathbf{d}, \mathbf{m}) = 0$ contain enough information to specify the model parameters uniquely or that they are even consistent. One of the purposes of inverse theory is to answer these kinds of questions and provide means of dealing with the problems that they imply. In general, $\mathbf{f}(\mathbf{d}, \mathbf{m}) = 0$ can consist of arbitrarily complicated (nonlinear) functions of the data and model parameters. In many problems, however, the equation takes on one of several simple forms. It is convenient to give names to some of these special cases, since they commonly arise in practical problems; we shall give them special consideration in later chapters.

2.3 Special forms

When the function $\mathbf{f}$ is linear in both data and model parameters the equation $\mathbf{f}(\mathbf{d}, \mathbf{m}) = 0$ simplifies to the *implicit linear form*

linear functions of data and model parameters are zero

$$\mathbf{X}\mathbf{x} = 0 \text{ with } \mathbf{x} = \begin{bmatrix} \mathbf{d} \\ \mathbf{m} \end{bmatrix} \qquad (2.5)$$

Here, $\mathbf{X}$ is an $L \times (N+M)$ constant matrix and $\mathbf{x}$ is a vector that concatenates the data $\mathbf{d}$ and the model parameters $\mathbf{m}$, that is, $\mathbf{x} = [d_1, d_2, d_3, \cdots d_N, m_1, m_2, m_3, \cdots m_M]^T$.

In many instances, it is possible to separate the data from the model parameters and to form $L=N$ equations that are linear in the data (but still nonlinear in the model parameters). This *explicit nonlinear form* is then

data minus nonlinear function of model parameters is zero

$$\mathbf{d} - \mathbf{g}(\mathbf{m}) = 0 \qquad (2.6)$$

In this case, $L=N$ and the nonlinear function $\mathbf{g}(\mathbf{m})$ describes how the model parameters *predict* the data. Finally, the function $\mathbf{g}(\mathbf{m})$ may be linear. This is the *explicit linear form*

data minus linear function of model parameters is zero

$$\mathbf{d} - \mathbf{G}\mathbf{m} = 0 \qquad (2.7)$$

The $N \times M$ matrix $\mathbf{G}$ is called the *data kernel*. The quantity $\mathbf{G}\mathbf{m}$ describes how the model parameters *predict* the data. The explicit linear form is a special case of the other forms, with

$$\mathbf{X} = [\mathbf{I} \quad -\mathbf{G}] \text{ and } \mathbf{g}(\mathbf{m}) = \mathbf{G}\mathbf{m} \qquad (2.8)$$

2.4 The linear inverse problem

The simplest and best-understood inverse problems are those that can be represented with the explicit linear equation $\mathbf{d} = \mathbf{G}\mathbf{m}$. This equation, therefore, forms the foundation of the study of inverse theory. As will be shown later, many important inverse problems that arise in the physical sciences involve precisely this equation. Others, while involving more complicated equations, can often be solved through linear approximations.

The matrix $\mathbf{G}$ is called the data kernel, in analogy to the theory of integral equations, in which the analogs of the data and model parameters are two continuous functions $d(x)$ and $m(x)$, where x is some independent variable. Between these two extremes, are problems with discrete data but a continuous model function.

Discrete inverse theory:

$$d_i = \sum_{j=1}^{M} G_{ij} m_j \tag{2.9a}$$

Continuous inverse theory:

$$d_i = \int G_i(x)\, m(x)\, dx \tag{2.9b}$$

Integral equation theory:

$$d(y) = \int G(y, x)\, m(x)\, dx \tag{2.9c}$$

The main difference among discrete inverse theory, continuous inverse theory, and integral equation theory is whether the data and model are treated as continuous functions or discrete parameters. The data d_i in inverse theory are necessarily discrete, because inverse theory is concerned with deducing knowledge from observational data, which always has a discrete character. Both continuous inverse problems and integral equations can be converted to discrete inverse problems by approximating the model $m(x)$ as a vector of its values at a set of M closely spaced points

$$\mathbf{m} = [m(x_1), m(x_2), m(x_3), \cdots, m(x_M)]^T \tag{2.10}$$

and the integral as a Riemann summation (or by some other quadrature formula).

$$d_i = \int G_i(x)\, m(x)\, dx \approx \sum_{j=1}^{M} G_i(x_j) m(x_j) \Delta x = \sum_{j=1}^{M} G_{ij} m_j \ \text{ with } \ G_{ij} \equiv \Delta x\, G_i(x_j) \tag{2.11}$$

Here Δx is a small increment in x. The symbol $\equiv$ is read as "is defined as."

2.5 Example: Fitting a straight line

Suppose that N temperature measurements T_i are made at times t_i in the atmosphere (Fig. 2.1) (Hansen et al., 2010). The data are then a vector $\mathbf{d}$ of N measurements of temperature, where $\mathbf{d} = [T_1, T_2, T_3, \cdots, T_N]^T$. The times t_i are not, strictly speaking, data. Instead, they provide some *auxiliary information* that describes the geometry of the experiment. This distinction will be further clarified later.

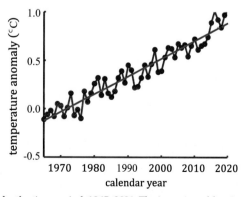

FIG. 2.1 (Red) Average global temperature for the time period, 1965–2021. The inverse problem is to determine the rate of increase of temperature and its confidence interval. (Blue) Straight-line fit to data. The slope of the line is 0.0183 ± 0.0015 (95%) $^\circ$C/year. Scripts gdama02_01 and gdapy02_01. *Data from Hansen, J., Ruedy, R., Sato, M., Lo, K., 2010. Global surface temperature change. Rev. Geophys. 48, RG4004. https://doi.org/10.1029/2010RG000345.*

Suppose that we assume a model in which temperature is a linear function of time, that is, $T = a + bt$. The intercept a and slope b then form the two model parameters of the problem $\mathbf{m} = [a, b]^T$. According to the model, each temperature observation must satisfy $T_i = a + bt_i$

$$
\begin{aligned}
T_1 &= a + bt_1 \\
T_2 &= a + bt_2 \\
&\vdots \\
T_N &= a + bt_N
\end{aligned}
\tag{2.12}
$$

These equations can be arranged as a matrix equation of the form $\mathbf{d} = \mathbf{Gm}$

$$
\begin{bmatrix} T_1 \\ T_2 \\ \vdots \\ T_N \end{bmatrix} = \begin{bmatrix} 1 & t_1 \\ 1 & t_2 \\ \vdots & \vdots \\ 1 & t_N \end{bmatrix} \begin{bmatrix} a \\ b \end{bmatrix}
\tag{2.13}
$$

The matrix $\mathbf{G}$ is computed as

MATLAB®
| ```
G=[ones(N,1), t];
```                                          [gdama02_01] |

Python
| ```
G=np.zeros((N,2));
G[0:N,0:1] = np.ones((N,1));
G[0:N,1:2] = t;
```                                          [gdapy02_01] |

Here t is a column vector $\mathbf{t}$ of times. Alternatively, the `np.concatenate()` method can be used to construct the data kernel $\mathbf{G}$:

Python
| ```
G = np.concatenate( (np.ones((N,1)),t), axis=1);
```                                          [gdapy02_01] |

The concatenation is side by side with `axis=1` (as is needed here), and one atop another, with `axis=0`.

2.6 Example: Fitting a parabola

If the model in Example 1 is changed to assume a quadratic variation of temperature with time of the form $T = a + bt + ct^2$, then a new model parameter c is added to the problem and $\mathbf{m} = [a, b, c]^T$. The number of model parameters is now $M = 3$. The data and model parameters are supposed to satisfy

$$
T_1 = a + bt_1 + ct_1^2
$$

$$
T_2 = a + bt_2 + ct_2^2
$$

$$
\vdots
\tag{2.14}
$$

$$
T_N = a + bt_N + ct_N^2
$$

These equations can be arranged into the matrix equation

$$
\begin{bmatrix} T_1 \\ T_2 \\ \vdots \\ T_N \end{bmatrix} = \begin{bmatrix} 1 & t_1 & t_1^2 \\ 1 & t_2 & t_2^2 \\ \vdots & \vdots & \vdots \\ 1 & t_N & t_N^2 \end{bmatrix} \begin{bmatrix} a \\ b \\ c \end{bmatrix}
\tag{2.15}
$$
```

This matrix equation has the form $\mathbf{d} = \mathbf{Gm}$. Note that, although the equation is linear in the data and model parameters, it is not linear in the auxiliary variable t.

The equation has a very similar form to the equation of the previous example, which brings out one of the underlying reasons for employing matrix notation: it can often emphasize similarities between superficially different problems. The matrix $\mathbf{G}$ is computed as

MATLAB®
`G=[ones(N,1), t, t.^2];`
[gdama02_02]

Python

`G=np.zeros((N,3));`
`G[0:N,0:1] = np.ones((N,1));`
`G[0:N,1:2] = t;`
`G[0:N,2:3] = np.power(t,2);`

[gdapy02_02]

Note the use of the element-wise power to compute t_i^2.

2.7 Example: Acoustic tomography

Suppose that a wall is assembled from a rectangular array of bricks (Fig. 2.2) and that each brick is composed of a different type of clay. If the acoustic velocities of the different clays differ, one might attempt to distinguish the different kinds of bricks by measuring the travel time of sound across the various rows and columns of bricks in the wall. The data in this problem are $N = 8$ measurements of travel times $\mathbf{d} = [T_1, T_2, T_3, \cdots, T_8]^T$. The model assumes that each brick is composed of a uniform material and that the travel time of sound across each brick is proportional to the width and height of the brick. The proportionality factor is the brick's *slowness* s_i, thus giving $M = 16$ model parameters, $\mathbf{m} = [s_1, s_2, s_3, \cdots, s_{16}]^T$, where the ordering is according to the numbering scheme of the figure. The data and model parameters are related by

$$
\begin{aligned}
\text{row 1}: \ & T_1 = hs_1 + hs_2 + hs_3 + hs_4 \\
\text{row 2}: \ & T_2 = hs_5 + hs_6 + hs_7 + hs_8 \\
\text{row 3}: \ & T_3 = hs_9 + hs_{10} + hs_{11} + hs_{12} \\
\text{row 4}: \ & T_4 = hs_{13} + hs_{14} + hs_{15} + hs_{16} \\
\text{column 1}: \ & T_5 = hs_1 + hs_5 + hs_9 + hs_{13} \\
\text{column 2}: \ & T_6 = hs_2 + hs_6 + hs_{10} + hs_{14} \\
\text{column 3}: \ & T_7 = hs_3 + hs_7 + hs_{11} + hs_{15} \\
\text{column 4}: \ & T_8 = hs_4 + hs_8 + hs_{12} + hs_{16}
\end{aligned}
\tag{2.16}
$$

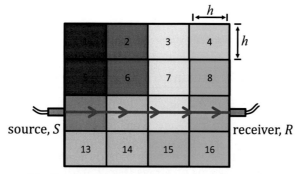

FIG. 2.2 The travel time of acoustic waves (blue line) through the rows and columns of a square array of bricks is measured with acoustic source S and receiver R placed on the edges of the square. The inverse problem is to infer the acoustic properties of the bricks, here depicted by the colors. Although the overall pattern is spatially variable, individual bricks are assumed to be homogeneous.

and the matrix equation is:

$$
\begin{bmatrix} T_1 \\ T_2 \\ \vdots \\ T_4 \\ T_5 \\ \vdots \\ T_8 \end{bmatrix} = h \begin{bmatrix} 1 & 1 & 1 & 1 & 0 & 0 & 0 & 0 & 0 & 0 & 0 & 0 & 0 & 0 & 0 & 0 \\ 0 & 0 & 0 & 0 & 1 & 1 & 1 & 1 & 0 & 0 & 0 & 0 & 0 & 0 & 0 & 0 \\ \vdots & \vdots & \vdots & \vdots & \vdots & \vdots & \vdots & \vdots & \vdots & \vdots & \vdots & \vdots & \vdots & \vdots & \vdots & \vdots \\ 0 & 0 & 0 & 0 & 0 & 0 & 0 & 0 & 0 & 0 & 0 & 0 & 1 & 1 & 1 & 1 \\ 1 & 0 & 0 & 0 & 1 & 0 & 0 & 0 & 1 & 0 & 0 & 0 & 1 & 0 & 0 & 0 \\ \vdots & \vdots & \vdots & \vdots & \vdots & \vdots & \vdots & \vdots & \vdots & \vdots & \vdots & \vdots & \vdots & \vdots & \vdots & \vdots \\ 0 & 0 & 0 & 1 & 0 & 0 & 0 & 1 & 0 & 0 & 0 & 1 & 0 & 0 & 0 & 1 \end{bmatrix} \begin{bmatrix} s_1 \\ \vdots \\ s_{16} \end{bmatrix}
\tag{2.17}
$$

Here, the bricks are assumed to be of width *and* height *h*. The matrix **G** is computed as

MATLAB®
| ```
N=8;  M=16;
G=zeros(N,M);
for i = (1:4) % loop over group of four data
    for j = (1:4)
        % measurements over rows
        k = (i-1)*4 + j;
        G(i,k)=h;
        % measurements over columns
        k = (j-1)*4 + i;
        G(i+4,k)=h;
    end
end
                                        [gdama02_03]
``` |

Python
| ```
N=8;  M=16;
G=zeros((N,M));
for i in range(4): # loop over group of four data
    for j in range(4):
        # measurements over rows
        k = i*4 + j;
        G[i,k]=h;
        % measurements over columns
        k = j*4 + i;
        G[i+4,k]=h;
                                        [gdapy02_03]
``` |

2.8 Example: X-ray imaging

Tomography is the process of forming images of the interior of an object from measurements made along rays passed through that object ("tomo" comes from the Greek word for "slice"). The computerized tomography scanner is an X-ray imaging device that has revolutionized the diagnosis of brain tumors and many other medical conditions. The scanner solves an inverse problem for the X-ray opacity of body tissues using measurements of the amount of radiation absorbed from many crisscrossing beams of X-rays (Fig. 2.3) (de Lastours et al., 2008).

The physical model underlying this device is the idea that the intensity of X-rays diminishes with the distance traveled, at a rate proportional to the intensity of the X-ray beam, and an absorption coefficient that depends on the type of tissue

$$
\frac{\mathrm{d}I}{\mathrm{d}s} = -c(x,y)I
\tag{2.18}
$$

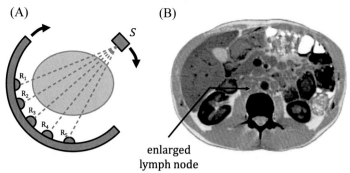

FIG. 2.3 (A) An idealized computed tomography (CT) medical scanner measures the X-ray absorption along lines (blue) passing through the body of the patient (orange). After a set of measurements are made, the source S and receiver R are rotated, and the measurements are repeated so that data along many crisscrossing lines are collected. The inverse problem is to determine the X-ray opacity as a function of position in the body. (B) Actual CT image of a patient infected with *Mycobacterium genavense*. From de Lastours, V., Guillemain, R., Mainardi, J.-L., Aubert, A., Chevalier, P., Lefort, A., Podglajen, I., 2008. Early diagnosis of disseminated Mycobacterium genavense infection. Emerg. Infect. Dis. 14(2), 346.

Here, I is the intensity of the beam, s the distance along the beam, and $c(x,y)$ the absorption coefficient, which varies with position (x,y). If the X-ray source has intensity I_0, then the intensity at the ith detector is

$$I_i = I_0 \exp\left\{-\int_{\text{beam } i} c(x,y)\, ds\right\} = I_0\left\{1 - \int_{\text{beam } i} c(x,y)\, ds\right\} \tag{2.19a}$$

$$\frac{I_0 - I_i}{I_0} = \int_{\text{beam } i} c(x,y)\, ds \tag{2.19b}$$

Eq. (2.19a) is a nonlinear function of the unknown absorption coefficient $c(x,y)$, which varies continuously along the beam. Consequently, this is a nonlinear problem in continuous inverse theory. However, it can be linearized, for small net absorption, by approximating the exponential with the first two terms in its Taylor series expansion, that is, $\exp(-x) \approx 1 - x$.

We now convert this problem to a discrete inverse problem of the form $\mathbf{d} = \mathbf{Gm}$. We assume that the continuously varying absorption coefficient can be adequately represented by a grid of many small square boxes (or pixels), each of which has a constant absorption coefficient. With these pixels numbered 1 through M, the model parameters are then the vector $\mathbf{m} = [c_1, c_2, c_3, \cdots, c_M]^T$. The integral can then be written as the sum

$$\Delta I_i = \sum_{j=1}^{M} \Delta s_{ij} c_j \quad \text{with } \Delta I_i = \frac{I_0 - I_i}{I_0} \tag{2.20}$$

Here, the data $d_i = \Delta I_i$ represents the fractional change in X-ray intensity between the source and at the detector, and the data kernel $G_{ij} = \Delta s_{ij}$ is the distance the ith beam travels in the jth pixel. The inverse problem is summarized by the matrix equation

$$\begin{bmatrix} \Delta I_1 \\ \Delta I_2 \\ \vdots \\ \Delta I_N \end{bmatrix} = \begin{bmatrix} \Delta s_{11} & \Delta s_{12} & \cdots & \Delta s_{1M} \\ \Delta s_{21} & \Delta s_{22} & \cdots & \Delta s_{2M} \\ \vdots & \vdots & \ddots & \vdots \\ \Delta s_{N1} & \Delta s_{N2} & \cdots & \Delta s_{NM} \end{bmatrix} \begin{bmatrix} c_1 \\ c_2 \\ \vdots \\ c_M \end{bmatrix} \tag{2.21}$$

As each beam passes through only a few of the many boxes, many of the Δs_{ij} are zero. Such a matrix is said to be sparse.

Computations with sparse matrices can be made extremely efficient by storing only the nonzero elements and by never explicitly multiplying or adding the zero elements (since the result is a foregone conclusion). However, special software support is necessary to gain this efficiency, as the computer must keep track of the zero elements. Once so defined, many normal matrix operations, including addition and multiplication, are efficiently computed without further user intervention. We will discuss this technique further in subsequent chapters, for its use makes practical the solving of very large inverse problems (say, with millions of model parameters). Further examples are given in Menke (2022).

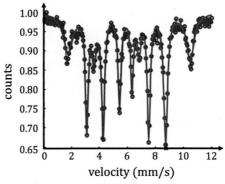

FIG. 2.4 Example of a Mossbauer spectroscopy experiment performed by the Spirit rover on Martian soil. (Red) Absorption peaks reflect the concentration of different iron-bearing minerals in the soil. The inverse problem is to determine the position and area of each peak, which can be used to determine the concentration of the minerals. (Blue) The sum of $K=10$ Lorentzian curves fit to the data. Script `gdama02_04` and `gdapy02_04`. *Data courtesy of NASA and the University of Mainz.*

2.9 Example: Spectral curve fitting

Not every inverse problem can be adequately represented by the discrete linear equation $\mathbf{d} = \mathbf{Gm}$. Consider, for example, a spectrogram containing a set of emission or absorption peaks that vary with some auxiliary variable z (Fig. 2.4). The positions f, area A, and width c of the peaks are of interest because they reflect the chemical composition of the sample. Denoting the shape of the jth peak as $p(z, f_j, A_j, c_j)$, the model is that the spectrum consists of a sum of, say, K such peaks.

$$d_i = \sum_{j=1}^{K} p\left(z_i, f_j, A_j, c_j\right) \ \text{ with } \ p\left(z_i, f_j, A_j, c_j\right) \equiv \frac{A_j c_j^2}{\left(z_i - f_j\right)^2 + c_j^2} \tag{2.22}$$

Here, the shape $p(z_i, f_j, A_j, c_j)$ of the peak is taken to be a *Lorentzian*. The data and model are related by a nonlinear explicit equation of the form $\mathbf{d} = \mathbf{g}(\mathbf{m})$, where $\mathbf{m}$ is a vector of length $M = 3K$ of the positions, areas, and widths of all the peaks. This equation is inherently nonlinear.

2.10 Example: Factor analysis

Another example of a nonlinear inverse problem is that of determining the composition of chemical end-members on the basis of the chemistry of a suite of mixtures of the end-members. Consider a simplified "ocean" (Fig. 2.5) in which sediments are composed of mixtures of several chemically distinct rocks eroded from the continents. One expects the amount S_{ij} of the chemical j in the ith sediment sample to be related to the amount C_{ik} of kth end-member of the ith sample and the amount F_{kj} of the chemical in that rock

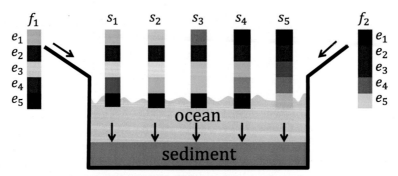

FIG. 2.5 Sediment on the floor of this idealized ocean is a mixture of rocks eroded from several sources f_i. The sources are characterized by chemical elements e_1 through e_5 depicted here with color bars. The chemical composition s_i of the sediments is a simple mixture of the composition of the sources. The inverse problem is to determine the number and composition of sources from observations of the composition of the sediments. Scripts `gdama02_05` and `gdapy02_05`.

$$\begin{bmatrix} \text{sample} \\ \text{composition} \end{bmatrix} = \sum_{\text{end-members}} \begin{bmatrix} \text{amount of} \\ \text{end-member} \end{bmatrix} \begin{bmatrix} \text{end-member} \\ \text{composition} \end{bmatrix}$$

$$S_{ij} = \sum_{k=1}^{P} C_{ik}F_{kj} \text{ or } \mathbf{S} = \mathbf{CF}$$

(2.23)

In a typical experiment, the number of end-members P, the end-member composition $\mathbf{F}$, and the amount of end-members in the samples $\mathbf{C}$ are all unknown model parameters. This problem has an explicit form because the data $\mathbf{S}$ are on one side of the equations and the unknowns are on the other, and is nonlinear, because it involves products of the unknowns. Note that the basic problem is to *factor* the matrix $\mathbf{S}$ into two other matrices $\mathbf{C}$ and $\mathbf{F}$. This factoring problem is a well-studied part of the theory of matrices, and methods are available to solve it. As will be discussed in Chapter 13, this problem (which is often called *factor analysis*) is very closely related to the algebraic eigenvalue problem.

2.11 Example: Correcting for an instrument response

Because of physical limitations, many types of geophysical sensors output a time-varying signal $d(t)$ (where t is time), which differs from the true input value of the parameter $m(t)$ being measured. Consequently, the inverse problem of estimating $m(t)$ from observations $d(t)$ is a common one; it must be solved before the output of the sensor can be fully understood and interpreted.

A seismometer is a good example of a sensor with an output that does not exactly match its input. Most seismometer models are much more sensitive to short period ground vibrations than to long period ones. Consequently, a simple spike in ground motion is output from the sensor as a pulse of more complicated shape (Fig. 2.6). When the input spike has unit area, the output pulse is called the instrument's *response* $g(t)$. Seismometers (and many other geophysical sensors) behave linearly. The amplitude of the output pulse is proportional to the amplitude of the input spike and, when several spikes occur in rapid succession, the sensor outputs several overlapping response functions.

When represented as a time series $\mathbf{m}$, smoothly varying ground motion can be thought of as superposition of a sequence of spikes of amplitude m_i. The seismometer output is a sequence of overlapping response functions of corresponding amplitude (Fig. 2.7):

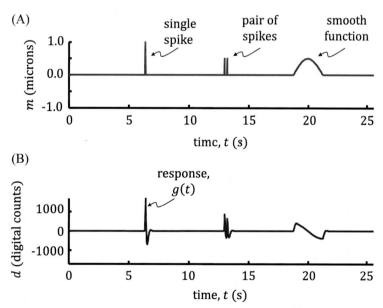

FIG. 2.6 (A) Hypothetical ground displacement $m(t)$ (where t is time) consisting of a single spike, followed by a pair of closely spaced spikes, followed by a smooth function. (B) Output $d(t)$ of a typical seismometer to the ground displacement. Although the output has some correspondence to the input, the shapes of features are different. The output associated with a single spike is called the seismometer's response $g(t)$. Scripts gdama02_06 and gdapy02_06.

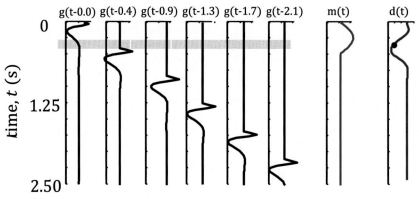

FIG. 2.7 Graphical depiction of the convolution relationship $\mathbf{d} = \mathbf{g} * \mathbf{m}$. A horizontal band (gray bar) through time-shifted versions of the response $g(t)$ (black curves) is "dotted into" the input $m(t)$ (blue curve) to yield one point (black circle) of the output $d(t)$ (red curve). Scripts gdama02_07 and gdapy02_07.

$$d_1 = g_1 m_1$$
$$d_2 = g_1 m_2 + g_2 m_1$$
$$d_3 = g_1 m_3 + g_2 m_2 + g_3 m_1 \qquad\qquad (2.24)$$
$$d_4 = g_1 m_4 + g_2 m_3 + g_3 m_2 + g_4 m_1$$
$$\vdots$$

In this problem, the number N of data equals the number M of unknowns. The current datum d_i always includes the term $g_1 m_i$, that is, the first element of the response g_1 scaled by the current input m_i. The other terms represent the contribution of the past inputs $m_{i-1}, m_{i-2}, m_{i-3}, \cdots$. Eq. (2.24) can be written succinctly as:

$$d_i = \sum_{j=1}^{M} g_{i-j+1} m_j \qquad\qquad (2.25)$$

and is called the *convolution* of $\mathbf{g}$ and $\mathbf{m}$. It is written as $\mathbf{d} = \mathbf{g} * \mathbf{m}$, where the asterisk $*$ denotes the *convolution operation* (not multiplication). The convolution can be put into the standard linear form $\mathbf{d} = \mathbf{G}\mathbf{m}$ by defining a data kernel with elements $G_{ij} \equiv g_{i-j+1}$

$$
\begin{bmatrix} d_1 \\ d_2 \\ d_3 \\ \vdots \\ d_N \end{bmatrix}
=
\begin{bmatrix}
g_1 & 0 & 0 & 0 & 0 \\
g_2 & g_1 & 0 & 0 & 0 \\
g_3 & g_2 & g_1 & 0 & 0 \\
\vdots & \vdots & \vdots & \ddots & \vdots \\
g_N & g_{N-1} & g_{N-3} & \cdots & g_1
\end{bmatrix}
\begin{bmatrix} m_1 \\ m_2 \\ m_3 \\ \vdots \\ m_N \end{bmatrix}
\qquad\qquad (2.26)
$$

Matrices with constant diagonals, like $\mathbf{G}$ in Eq. (2.26), are called *Toeplitz* matrices (Fig. 2.8). The process of estimating $\mathbf{m}$ from observations of $\mathbf{d}$ is called *deconvolution* and is an important type of inverse problem.

2.12 Solutions to inverse problems

We shall use the term *solution* (or *answer*) to indicate broadly whatever information we are able to determine about the problem under consideration. As we shall see, there are many different points of view regarding what constitutes a solution to an inverse problem. Of course, one generally wants to know the numerical values of the model parameters (we call this kind of answer an *estimate* of the model parameters). Unfortunately, this issue is made complicated by the ubiquitous presence of measurement error and also by the possibility that some model parameters are not constrained by any observation. The solution of inverse problems rarely leads to exact information about the values of the model parameters. More typically, practitioners of inverse theory are forced to make various compromises between the kind of information they actually want and the kind of information that can in fact be obtained from any given data set. These compromises lead to new kinds of answers that are more abstract than simple estimates of the model parameters. Part of the practice of inverse theory is identifying what features of a solution are most valuable and making the compromises that emphasize these features.

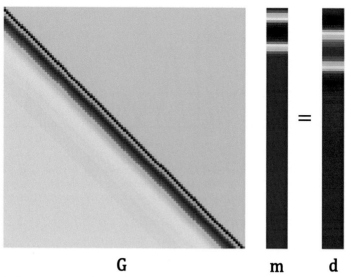

FIG. 2.8 Graphical depiction of the linear equation $\mathbf{d} = \mathbf{Gm}$ for a convolution. Note that the data kernel $\mathbf{G}$ is Toeplitz (meaning that it has constant diagonals). Scripts `gdama02_07` and `gdapy02_07`.

2.13 Estimates as solutions

The simplest kind of solution to an inverse problem is an estimate $\mathbf{m}^{est}$ of the model parameters. An estimate is simply a set of numerical values for the model parameters, for example, $\mathbf{m}^{est} = [1.4, 2.9, \cdots, 1.0]^T$. Estimates are generally the most useful kind of solution to an inverse problem. Nevertheless, in many situations, they can be very misleading. For instance, estimates in themselves give no insight into the *quality* of the solution. Depending on the structure of the particular problem, measurement errors might be averaged out (in which case the estimates might be meaningful) or amplified (in which case the estimates might be nonsense). In other problems, many solutions might exist. To single out arbitrarily only one of these solutions and call it $\mathbf{m}^{est}$ gives the false impression that a unique solution has been obtained.

2.14 Bounding values as solutions

One remedy to the problem of defining the quality of an estimate is to state additionally some bounds that define its *certainty*. These bounds can be either absolute or probabilistic. Absolute bounds imply that the true value of the model parameter lies between two stated values, for example, $1.3 \leq m_1^{true} \leq 1.5$. Probabilistic bounds imply that the true value is *likely* (but not guaranteed) to be between two bounds, with some given degree of certainty (Fig. 2.9A). For instance, $m_1^{true} = 1.4 \pm 0.1$ (95%) means that there is a 95% probability that the true value of the model parameter lies between 1.3 and 1.5 (and a 5% probability of lying outside that range).

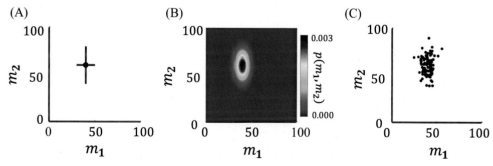

FIG. 2.9 Three types of solutions to an inverse problem, for the case of two model parameters m_1 and m_2. (A) Estimated model parameters (circle), together with confidence limits (red bars). (B) Probability density function $p(m_1, m_2)$. (C) An ensemble of 100 probable solutions, randomly drawn from $p(m_1, m_2)$. Scripts `gdama02_08` and `gdapy02_08`.

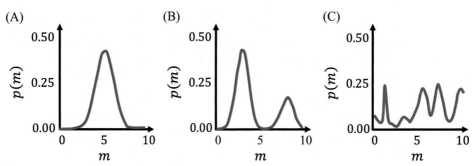

FIG. 2.10   Three hypothetical probability density functions $p(m)$ for a model parameter m. (A) The first is so simple that its properties can be summarized by its central position, at $m=5$, and the width of its peak. (B) The second implies that the model parameter has two probable ranges of values, one near $m=3$ and the other near $m=8$. (C) The third is so complicated that it provides no easily interpretable information about the model parameter. Scripts gdama02_09 and gdama02_09.

When they exist, bounding values can often provide the supplementary information needed to interpret properly the solution to an inverse problem. There are, however, many instances in which bounding values do not exist.

2.15 Probability density functions as solutions

A generalization of the stating of bounding values is the stating of the complete *probability density function (or pdf)* $p(\mathbf{m})$ for a model parameter, either as an analytic function or as values on an M-dimensional grid (Fig. 2.9B). The usefulness of this technique depends, in part, on the complexity of $p(\mathbf{m})$. If the probability density function $p(m_i)$ for an individual model parameter m_i has only one peak (Fig. 2.10A), then it provides little more information than an estimate based on the position of the peak's center with error bounds based on the peak's shape. Such an estimate would be written as $m_1^{true} = m_1^{est} \pm \frac{1}{2}w$ (95%), where m_1^{est} is the position of the peak and w is the interval of the m_1-axis centered on the peak that encloses 95% of the area.

On the other hand, if the probability density function is very complicated (Fig. 2.10C), it is basically uninterpretable (except in the sense that it implies that the model parameter cannot be well estimated). Only in those exceptional instances in which it has some intermediate complexity (Fig. 2.10B) does it really provide information toward the solution of an inverse problem.

2.16 Ensembles of realizations as solutions

Except in the well-understood Normal (Gaussian) case, which we will discuss later in the book, most probability density functions are exceedingly difficult to compute. A large set (or *ensemble*) of realizations $\mathbf{m}^{(i)}$, $i=1, 2, 3, \cdots, K$ of model parameter vectors drawn from $p(\mathbf{m})$ are somewhat easier to compute and can serve as an alternative. The set, itself, might be considered the solution to the inverse problem (Fig. 2.9C), as many of the properties of the probability density function can be inferred from it. For instance, the question, "How likely is it that $m_2 > m_6$," can be answered by examining each member of the ensemble and counting up the number of times the inequality holds. The number, expressed as a percentage of K, answers the question. However, an ensemble needs to be extremely large to capture the properties of $p(\mathbf{m})$, when the number M of model parameters is large.

Deriving useful knowledge from, say, a billion examples of *probable* **m**s is a challenging task.

2.17 Weighted averages of model parameters as solutions

In many instances, it is possible to identify combinations or averages of the model parameters that are in some sense better determined than the model parameters themselves. For instance, given model parameters $\mathbf{m}=[m_1, m_2, m_3]^T$, it may turn out that the average $\langle m \rangle = 0.1m_1 + 0.8m_2 + 0.1m_3$, where $\langle \, \rangle$ mean "weighted average" is better determined than any of m_1, m_2, and m_3 considered individually.

Averages can be of considerable interest when the model parameters represent a discretized version of some continuous function. For example, m_1, m_2, and m_3 might be the values of the function $m(x)$ at points x_1, x_2, and x_3, respectively. If the weights are large only for a few physically adjacent parameters, then the average is said to be localized. For example, the average $\langle m \rangle = 0.1m_1 + 0.8m_2 + 0.1m_3$ is localized around position x_2. The meaning of the average in such a case is that, although the data cannot resolve the model parameters at a particular point, they can resolve the average of the model parameters in the neighborhood of that point.

In other cases, one might not have the slightest interest in a weighted average of the model parameters, be it well determined or not, because it may not have physical significance. For instance, the weighted average of the three coefficients of a quadratic fit $d_i = m_1 + m_2 x_i + m_3 x_i^2$ does not contribute much useful knowledge about the behavior of the quadratic.

In the following chapters, we shall derive methods for determining each of these different kinds of solutions to inverse problems. We note here, however, that there is a great deal of underlying similarity between these types of "answers." In fact, it will turn out that the same numerical "answer" will be interpretable as any of several classes of solutions.

2.18 Problems

2.1 Suppose that you determine the masses of 100 objects by weighing the first, then weighing the first and second together, and then weighing the rest in triplets: the first, second, and third; the second, third, and fourth; and so forth. (A) Identify the data and model parameters in this problem. How many of each are there? (B) Write down the matrix $\mathbf{G}$ in the equation $\mathbf{d} = \mathbf{Gm}$ that relates the data to the model parameters. (C) How sparse is $\mathbf{G}$? What percent of it is zero?

2.2 Suppose that you determine the height of 50 objects by measuring the first, and then stacking the second on top of the first and measuring their combined height, stacking the third on top of the first two and measuring their combined height, and so forth. (A) Identify the data and model parameters in this problem. How many of each are there? (B) Write down the matrix $\mathbf{G}$ in the equation $\mathbf{d} = \mathbf{Gm}$ that relates the data to the model parameters. (C) How sparse is $\mathbf{G}$? What percent of it is zero?

2.3 Write a MATLAB® or Python script to compute $\mathbf{G}$ in the case of the cubic equation $T = T = a + bz + cz^2 + dz^3$. Assume that 11 zs are equally spaced from 0 to 10.

2.4 Let the data $\mathbf{d}$ be the running average of the model parameters $\mathbf{m}$ computed by averaging groups of three neighboring points, that is, $d_i = \frac{1}{3}m_{i-1} + \frac{1}{3}m_i + \frac{1}{3}m_{i+1}$. (A) What is the matrix $\mathbf{G}$ in the equation $\mathbf{d} = \mathbf{Gm}$ in this case? (B) What problems arise at the top and bottom rows of the matrix and how can you deal with them? (C) How sparse is $\mathbf{G}$? What percent of it is zero?

2.5 Simplify Eq. (2.23) by assuming that there is only one sample S_{1j}, whose composition is measured. Consider the case where the composition of the P factors is known, but their proportions in the sample are unknown, and rewrite the equation in the form $\mathbf{d} = \mathbf{Gm}$. Hint: You might start by taking the transpose of Eq. (2.23).

References

de Lastours, V., Guillemain, R., Mainardi, J.-L., Aubert, A., Chevalier, P., Lefort, A., Podglajen, I., 2008. Early diagnosis of disseminated Mycobacterium genavense infection. Emerg. Infect. Dis. 14 (2), 346.

Hansen, J., Ruedy, R., Sato, M., Lo, K., 2010. Global surface temperature change. Rev. Geophys. 48. https://doi.org/10.1029/2010RG000345. RG4004.

Menke, W., 2022. Environmental Data Analysis with MATLAB® or Python, third ed. Elsevier, Oxford, ISBN: 978-0-323-95576-8. 444 pp.

3

Using probability to describe random variation

In the preceding chapter, we represented the results of an experiment as a vector **d**, whose elements were individual measurements. Usually, however, a single number is insufficient to represent a single observation. Measurements contain noise, and if an observation were to be performed several times, each measurement would be different (Fig. 3.1). Information about the range and shape of this scatter is needed to characterize the data completely.

The concept of a random variable is used to describe this property. Each random variable has definite and precise properties, governing the range and shape of the scatter of values one observes. These properties cannot be measured directly; however, one can only make individual measurements, or *realizations*, of the random variable and try to estimate its true properties from these data.

3.1 Noise and random variables

The true properties of the random variable d are specified by a *probability density function* (abbreviated pdf), $p(d)$. This function gives the probability that a particular realization of the random variable will have a value in the neighborhood of d. The probability that the measurement is between d and $d + dd$ is $p(d)dd$ (Fig. 3.2). (Our choice of the variable name "d" for "data" makes the differential "dd" look a bit funny, but we will just have to live with it.)

Since each measurement must have some value, the probability that d lies somewhere between $-\infty$ and $+\infty$ is complete certainty (usually given the value of 100% or unity), which is written as

$$\int_{-\infty}^{+\infty} p(d) \, dd = 1 \tag{3.1}$$

The probability P that d lies in some specific range, say between d_1 and d_2, is the integral of $p(d)$ over that range

$$P(d_1, d_2) \equiv \int_{d_1}^{d_2} p(d) \, dd \tag{3.2}$$

The special case, $d_1 = -\infty$ and $d_2 = d$, represents the probability that the value of the random variable is less than or equal to a given value d and is called the *cumulative distribution function $P(d)$* (abbreviated cdf). Note that the numerical value of P represents an actual probability, while the numerical value of p does not. The pdf is the slope of P, that is, $p(d) = dP/dd$.

In a script, we use a column vector d of evenly spaced values with sampling Dd to represent possible values of the random variable and we use a vector p to represent the probability density function at corresponding values of d. The total probability Ptotal (which should be unity) and the cumulative probability distribution P are calculated as

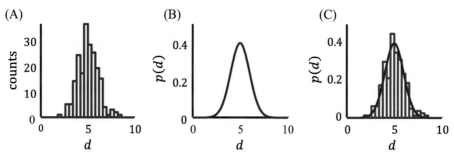

FIG. 3.1   (A) Histogram showing data from 200 repetitions of an experiment in which datum d is measured. Noise causes observations to scatter about their mean value $\langle d \rangle = 5$, (B) Probability density function (pdf) $p(d)$ of the data. (C) Histogram (blue) and pdf (red) superimposed. Note that the histogram has a shape similar to the pdf. Scripts gdama03_01 and gdapy03_01.

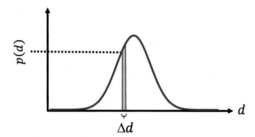

FIG. 3.2   The shaded area $p(d)\Delta d$ of the probability density function $p(d)$ gives the probability P that the observation will fall between d and $d + \Delta d$. Scripts gdama03_02 and gdapy03_02.

MATLAB®
| ```
Ptotal = Dd * sum(p);
P = Dd * cumsum(p);
```                                                                      [gdama03_03] |

Python
| ```
Ptotal = Dd * np.sum(p);
P = gda_cvec(Dd*np.cumsum(p));
```                                                               [gdapy03_03] |

Here, we are employing the Riemann approximation for an integral $\int p(d)\mathrm{d}d \approx \Delta d \sum_i p(d_i)$. Note the distinction between a sum of all the elements of p, which is a scalar, and the *cumulative sum* (running sum) of elements of p, which is a vector.

The probability density function $p(d)$ completely describes the random variable d. Unfortunately, it is a continuous function that may be very complicated. A few numbers that summarize its major properties can be very helpful. One such kind of number indicates the *typical* numerical value of a measurement. The most likely measurement is the one with the highest probability, that is, the value of d at which $p(d)$ is peaked (Fig. 3.3). However, if the pdf is skewed, this *maximum likelihood*

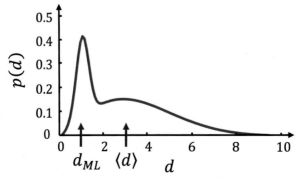

FIG. 3.3   The maximum likelihood point d_{ML} of the probability density function $p(d)$ gives the most probable value of the datum d. In general, this value can be different than the mean datum $\langle d \rangle$, which is at the "balancing point" of the pdf. Scripts gdama02_04 and gdapy02_04.

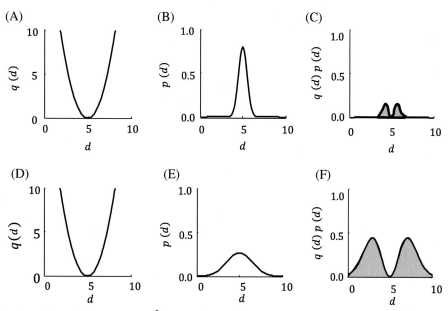

FIG. 3.4 (A and D) Parabola of the form $q(d) = (d - \langle d \rangle)^2$ is used to measure the width of two probability density functions $p(d)$ (B and E), which have the same mean $\langle d \rangle$ but different widths. The product qp is everywhere small for the narrow function (C) but had two large peaks for the wider function (F). The area (shaded orange) under qp is a measure of the width of the pdf and is called the variance. The variances of (A) and (F) are $(0.5)^2$ and $(1.5)^2$, respectively. Scripts gdama03_05 and gdapy03_05.

point may not be a good indication of the typical measurement, as a wide range of other values also have high probability. In such instances, the *mean*, or *expected value* $\langle d \rangle$ (sometimes written as $E(d)$), is a better characterization of a typical measurement. This number is the "balancing point" of the pdf and is given by

$$\langle d \rangle \equiv \int_{-\infty}^{+\infty} d\, p(d)\, \mathrm{d}d \tag{3.3}$$

Here, the $\equiv$ symbol should be read as "is defined as."

Another property of a pdf is its overall width. Wide pdfs imply very noisy data, and narrow ones imply relatively noise-free data. One way of measuring the width of a pdf is to multiply it by a function, say $q(d)$, that is zero near the center (peak) and that grows on either side of the peak (Fig. 3.4). If the pdf is narrow, then the product $q(d)p(d)$ will be everywhere small; if it is wide, then the result will be large.

A quantitative measure of the width of the peak is the area under the product. If one chooses the parabola $q(d) = (d - \langle d \rangle)^2$ as the function, where $\langle d \rangle$ is the expected value of the random variable, then this measure is called the *variance* σ_d^2 of the pdf and is written as

$$\sigma_d^2 \equiv \int_{-\infty}^{+\infty} (d - \langle d \rangle)^2\, p(d)\, \mathrm{d}d \tag{3.4}$$

The square root of the variance σ_d is a measure of the width of the pdf. In scripts, the expected value and variance are computed as

MATLAB®
| ```
Ed = Dd * sum(d.*p);
q = (d-Ed).^2;
sd2est = Dd * sum(q.*p);
``` |
| [gdama03_05] |

Python
| ```
Ed = Dd * np.sum(np.multiply(d,p)); # expectation
q = np.power((d-Ed),2); # quadratic
qp = np.multiply(q,p); # product
sd2est = Dd * np.sum(qp); # variance
``` |
| [gdapy03_05] |

Here, d is a vector of equally spaced values of the random variable d with spacing Dd, p is the corresponding value of the probability density function and $q = (d - \langle d \rangle)^2$.

As we will discuss further in Chapter 5, the mean and variance can be estimated from a set of N realizations of data d_i as

$$\langle d \rangle^{est} = \frac{1}{N} \sum_{i=1}^{N} d_i \quad \text{and} \quad (\sigma_d^2)^{est} = \frac{1}{N-1} \sum_{i=1}^{N} \left(d_i - \langle d \rangle^{est} \right)^2 \tag{3.5}$$

The quantity $\langle d \rangle^{est}$ is called the *sample mean*, the quantity $(\sigma_d^2)^{est}$ is called the *sample variance*, and the quantity σ_d^{est} is called the *sample standard deviation*. We refer to them as *estimates* because they are based on noisy data and will not be equal to the true mean and true variance of $p(d)$ (though we hope that they will be close). In scripts, these estimates can be computed as

MATLAB®
| ```
dbarest = mean(dr);
sigmadest = std(dr);
``` |
| [gdama03_05] |

Python
| ```
dbarest = np.mean(dr);
sigmaest = np.std(dr);
``` |
| [gdapy03_10] |

Here, dr is a vector of N realizations of the random variable d.

3.2 Correlated data

Experiments usually involve the collection of more than one datum. We therefore need to quantify the probability that a *set* of random variables will take on a given value. The *joint* probability density function $p(\mathbf{d})$ is the probability that the first datum will be in the neighborhood of d_1, that the second will be in the neighborhood of d_2, etc. If the data are independent—that is, if there are no patterns in the occurrence of the values between pairs of random variables— then this joint pdf is just the product of the individual pdfs (Fig. 3.5)

$$p(\mathbf{d}) = p(d_1)p(d_2)p(d_3)\cdots p(d_N) \tag{3.6}$$

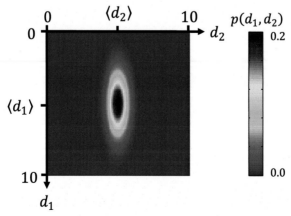

FIG. 3.5 The multivariate probability density function $p(d_1, d_2)$ is displayed as an image, with values given by the accompanying color bar. These data are uncorrelated, as especially large values of d_2 are no more or less likely if d_1 is large or small. In this example, the variances of d_1 and d_2 are $(1.5)^2$ and $(0.5)^2$, respectively. Scripts gdama03_06 and gdapy03_06.

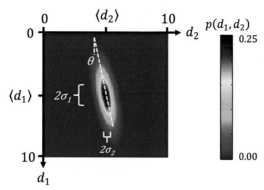

FIG. 3.6 The multivariate probability density function $p(d_1, d_2)$ is displayed as an image, with values given by the accompanying color bar. These data are positively correlated, as large values of d_2 are especially probable if d_1 is large. The function has means $\langle d_1 \rangle = \langle d_1 \rangle = 5$ and widths in the coordinate directions $\sigma_1 = 1.5$ and $\sigma_2 = 0.5$. The angle θ is a measure of the degree of correlation and is related to the covariance $\text{cov}(d_1, d_2) = 0.4$. Scripts gdama03_07 and gdapy03_07.

The probability density function for a single random variable, say d_i, irrespective of all the others, is computed by integrating $p(\mathbf{d})$ over all the other variables:

$$p(d_i) = \underbrace{\int_{-\infty}^{+\infty} \cdots \int_{-\infty}^{+\infty}}_{(N-1 \text{ integrals})} p(\mathbf{d}) \, \mathrm{d}d_j \mathrm{d}d_k \cdots \mathrm{d}d_l \tag{3.7}$$

In some experiments, measurements are correlated. High values of one datum tend to occur consistently with either high or low values of another datum (Fig. 3.6). The joint pdf for such data must be constructed to take this correlation into account. Given a joint pdf $p(d_1, d_2)$ for two random variables d_1 and d_2, one can test for correlation by selecting a function that divides the (d_1, d_2) plane into four quadrants of alternating sign, centered on the mean of the pdf (Fig. 3.7). If one multiplies the pdf by this function, and then sums up the area, the result will be zero for uncorrelated pdfs, since they tend to lie equally in all four quadrants. Correlated pdfs will have either positive or negative area, since they tend to be concentrated in two opposite quadrants (Fig. 3.8). When $(d_1 - \langle d_1 \rangle)(d_2 - \langle d_2 \rangle)$ used as the function, the resulting measure of correlation is called the *covariance*

$$\text{cov}(d_1, d_2) \equiv \int_{-\infty}^{+\infty} \int_{-\infty}^{+\infty} (d_1 - \langle d_1 \rangle)(d_2 - \langle d_2 \rangle) \, p(d_1, d_2) \, \mathrm{d}d_1 \mathrm{d}d_2 \tag{3.8}$$

Note that the covariance of a datum with itself is just the variance. The covariance, therefore, characterizes the basic shape of the joint pdf. When there are many data given by the vector $\mathbf{d}$, it is convenient to define a vector of expected values and a matrix of covariances as

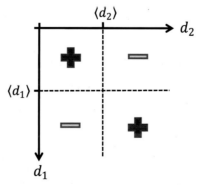

FIG. 3.7 The function $q(d_1, d_2) \equiv (d_1 - \langle d_1 \rangle)(d_2 - \langle d_2 \rangle)$ divides the (d_1, d_2) plane into four quadrants of alternating sign.

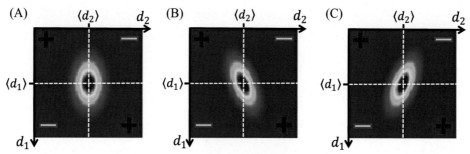

FIG. 3.8 The probability density function $p(d_1, d_2)$ displayed as an image, when the data are (A) uncorrelated, (B) positively correlated, and (C) negatively correlated. The dashed lines indicate the four quadrants of alternating sign used to determine the correlation (see Fig. 3.7). Scripts gdama02_08 and gdapy02_08.

$$\langle d_i \rangle \equiv \int_{-\infty}^{+\infty} \cdots \int_{-\infty}^{+\infty} d_i \, p(\mathbf{d}) \, \mathrm{d}d_1 \cdots \mathrm{d}d_N$$

$$[\mathrm{cov}\,\mathbf{d}]_{ij} \equiv \int_{-\infty}^{+\infty} \cdots \int_{-\infty}^{+\infty} (d_i - \langle d_i \rangle)(d_j - \langle d_j \rangle) \, p(\mathbf{d}) \, \mathrm{d}d_1 \cdots \mathrm{d}d_N \tag{3.9}$$

Henceforth, we will abbreviate these multidimensional integrals as $\int \mathrm{d}^N d$. The diagonal elements of the *covariance matrix* are variances. They are measures of the scatter in the data. The off-diagonal elements are covariances. They indicate the degree to which pairs of data are correlated. Note that the integral for the mean can be written in terms of the univariate probability density function $p(d_i)$ and the integral for the variance can be written in terms of the bivariate probability density function $p(d_i, d_j)$, as the other dimensions of $p(\mathbf{d})$ are just "integrated away" to unity:

$$\langle d_i \rangle \equiv \int_{-\infty}^{+\infty} d_i \, p(d_i) \, \mathrm{d}d_i$$

$$[\mathrm{cov}\,\mathbf{d}]_{ij} \equiv \int_{-\infty}^{+\infty} \int_{-\infty}^{+\infty} (d_i - \langle d_i \rangle)(d_j - \langle d_j \rangle) \, p(d_i, d_j) \, \mathrm{d}d_i \mathrm{d}d_j \tag{3.10}$$

The covariance matrix can be estimated from a set of realizations of data. Suppose that there are N different types of data and that K realizations of them have been observed. The data can be organized into a *sample matrix* $\mathbf{S}$, with the N columns referring to the different data types and the K rows to the different realizations. The *sample covariance* is then

$$[\mathrm{cov}\,\mathbf{d}]_{ij}^{est} = \frac{1}{K} \sum_{k=1}^{K} \left(S_{ki} - \langle d_i \rangle^{est}\right)\left(S_{kj} - \langle d_j \rangle^{est}\right) \quad \text{with} \quad \langle d_i \rangle^{est} = \frac{1}{K} \sum_{k=1}^{K} S_{ki} \tag{3.11}$$

Here, $\langle d_i \rangle^{est}$ is the sample mean of the ith data type. In scripts, the sample covariance is calculated as

MATLAB®
`Cest = cov(S);`
[gdama03_10]

Python
`Cest = np.cov(S.T);`
[gdapy03_10]

3.3 Functions of random variables

The basic premise of inverse theory is that the data and model parameters are related. Any method that *solves* the inverse problem—that estimates a model parameter on the basis of data—will map errors from the data to the estimated model parameters. Thus the estimates of the model parameters are themselves random variables and are described by a pdf $p(\mathbf{m}^{est})$. Whether or not the true model parameters are random variables depends on the problem.

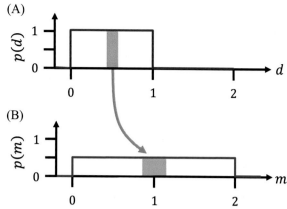

FIG. 3.9 (A) The uniform probability density function $p(d)=1$ on the interval, $0 \le d \le 1$. (B) The transformed probability density function $p(m)$, given the relationship $m=2d$. Note that a patch (shaded rectangle) of probability in m is wider and lower than the equivalent patch in d.

It is appropriate to consider them deterministic quantities in some problems and random variables in others. Estimates of the model parameters, however, are always random variables.

We need the tools to transform probability density functions from $p(\mathbf{d})$ to $p(\mathbf{m})$ when the relationship $\mathbf{m}(\mathbf{d})$ is known. We start simply and consider just one datum and one model parameter, related by the simple function $m(d)=2d$. Now suppose that $p(d)$ is *uniform* on the interval $(0,1)$; that is, d has equal probability of being anywhere in this range. The probability density function is constant and must have amplitude $p(d)=1$, as the total probability must be unity (width $\times$ height $=1 \times 1=1$). The probability density function $p(m)$ is also uniform, but on the interval $(0,2)$, since m is twice d. Thus $p(d)=\frac{1}{2}$, as its total probability must also be unity (width $\times$ height $=2 \times \frac{1}{2}=1$) (Fig. 3.9). This result shows that $p(m)$ is not merely $p[d(m)]$, but rather must include a factor that accounts for the stretching (or shrinking) of the m-axis with respect to the d-axis.

This stretching factor can be derived by transforming the integral for total probability:

$$1 = \int_{d_{min}}^{d_{max}} p(d)\, \mathrm{d}d = \int_{m(d_{min})}^{m(d_{max})} p[d(m)]\, \frac{\mathrm{d}d}{\mathrm{d}m}\mathrm{d}m = \int_{m_{min}}^{m_{max}} p(m)\, \mathrm{d}m \tag{3.12}$$

By inspection, $p(m)=p[d(m)]\, \mathrm{d}d/\mathrm{d}m$, so the stretching factor is $\mathrm{d}d/\mathrm{d}m$. The limits (d_{min}, d_{max}) transform to (m_{min}, m_{max}). However, depending upon the function $m(d)$, we may find that $m_{min} > m_{max}$, that is, the direction of integration might be reversed ($m(d)=1/d$ would be one such case). We handle this problem by adding an absolute value sign

$$p(m) \equiv p[d(m)] \left| \frac{\mathrm{d}d}{\mathrm{d}m} \right| \tag{3.13}$$

together with the understanding that the integration is always performed in the direction of positive m. Note that in the case before, with $p(d)=1$ and $m(d)=2d$, we find $\mathrm{d}d/\mathrm{d}m=\frac{1}{2}$ and (as expected) $p(m)=1 \times \frac{1}{2}=\frac{1}{2}$.

In general, probability density functions change shape when transformed from d to m. Consider, for example, the uniform probability density function $p(d)=1$ on the interval $(0,1)$ together with the function $m(d)=d^2$ (Fig. 3.10). We find $d=m^{1/2}$, $\mathrm{d}d/\mathrm{d}m=\frac{1}{2}m^{-1/2}$, and $p(m)=\frac{1}{2}m^{-1/2}$, with m defined on the interval $(0,1)$. Thus, while $p(d)$ is uniform, $p(m)$ has a peak (actually an integrable singularity) at $m=0$ (Fig. 3.10B).

The general case of transforming $p(\mathbf{d})$ to $p(\mathbf{m})$, given the functional relationship $\mathbf{m}(\mathbf{d})$, is more complicated, but is derived using the rule for transforming multidimensional integrals that is analogous to Eq. (3.13). This rule states that the volume element transforms as $\mathrm{d}^N d=J(\mathbf{m})\mathrm{d}^N m$ where $J(\mathbf{m}) \equiv |\det(\partial \mathbf{d}/\partial \mathbf{m})|$ is the Jacobian determinant, that is, the absolute value of the determinant of the matrix whose elements are $[\partial \mathbf{d}/\partial \mathbf{m}]_{ij} \equiv \partial d_i/\partial m_j$

$$1 = \int p(\mathbf{d})\mathrm{d}^N d = \int p[\mathbf{d}(\mathbf{m})]\, J(\mathbf{m})\, \mathrm{d}^N m = \int p[\mathbf{d}(\mathbf{m})] \left| \det\left(\frac{\partial \mathbf{d}}{\partial \mathbf{m}}\right) \right| \mathrm{d}^N m = \int p(\mathbf{m})\mathrm{d}^N m \tag{3.14}$$

Hence, by inspection, we find that the probability density functions transform as

$$p(\mathbf{m}) = p[\mathbf{d}(\mathbf{m})]\, J(\mathbf{m}) \text{ with } J(\mathbf{m}) \equiv \left| \det\left(\frac{\partial \mathbf{d}}{\partial \mathbf{m}}\right) \right|$$

$$p(\mathbf{d}) = p[\mathbf{m}(\mathbf{d})]\, J(\mathbf{d}) \text{ with } J(\mathbf{d}) \equiv \left| \det\left(\frac{\partial \mathbf{m}}{\partial \mathbf{d}}\right) \right| \tag{3.15}$$

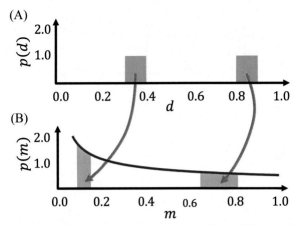

FIG. 3.10　(A) The uniform probability density function $p(d)=1$ on the interval, $0 \leq d \leq 1$. (B) The transformed probability density function $p(m)$, given the relationship $m = d^2$. Note regions of equal probability (shaded), which are of equal height and width in the variable d, are transformed into regions of equal area but unequal height and width in the variable m. Scripts gdama03_09 and gdapy03_09.

The Jacobian is constant for the linear transformation $\mathbf{m} = \mathbf{Md}$, with the value $J = |\det \mathbf{M}^{-1}| = 1/|\det \mathbf{M}|$. As an example, consider a two-dimensional probability density function that is uniform on the intervals $(0,1)$ for d_1 and $(0,1)$ for d_2, together with the transformation $m_1 = d_1 + d_2$, $m_2 = d_1 - d_2$. As is shown in Fig. 3.11, $p(\mathbf{d})$ corresponds to a square of unit area in the (d_1, d_2) plane and $p(\mathbf{m})$ corresponds to a square of area 2 in the (m_1, m_2) plane. In order that the total area be unity in both cases, we must have $p(\mathbf{d}) = 1$ and $p(\mathbf{m}) = \frac{1}{2}$. The transformation matrix is

$$\mathbf{M} = \begin{bmatrix} 1 & 1 \\ 1 & -1 \end{bmatrix} \text{ so } |\det \mathbf{M}| = 2 \text{ and } J = \frac{1}{2} \tag{3.16}$$

Thus, by Eq. (3.15), we find that $p(m) = p(d)J = 1 \times \frac{1}{2} = \frac{1}{2}$, which agrees with our expectations.

Note that we can convert $p(\mathbf{d})$ to a univariate pdf $p(d_1)$ by integrating over d_2. As the sides of the square are parallel to the coordinate axes, the integration yields the uniform probability density function $p(d_1) = 1$ (Fig. 3.11C). Similarly, we can convert $p(\mathbf{m})$ to a univariate pdf $p(m_1)$ by integrating over m_2. However, because the sides of the square are oblique to the coordinate axes, $p(m_1)$ is a *triangular*—not a uniform—probability density function (Fig. 3.11D).

Transforming a probability density function $p(\mathbf{d})$ is straightforward, but tedious. Fortunately, in the case of the linear function $\mathbf{m} = \mathbf{Md} + \mathbf{v}$, where $\mathbf{M}$ and $\mathbf{v}$ are an arbitrary matrix and vector, respectively, it is possible to make some statements about the properties of the results without explicitly calculating the transformed probability density function $p(\mathbf{m})$. In particular, the mean and covariance are

$$\langle \mathbf{m} \rangle = \mathbf{M} \langle \mathbf{d} \rangle + \mathbf{v} \tag{3.17a}$$

$$[\text{cov } \mathbf{m}] = \mathbf{M} [\text{cov } \mathbf{d}] \mathbf{M}^T \tag{3.17b}$$

These rules are derived by transforming the definition of the mean and variance:

$$\langle m \rangle_i = \int m_i p(\mathbf{m}) d^N m = \int \left[\sum_j M_{ij} d_j + v_i \right] p[\mathbf{m}(\mathbf{d})] \det \left[\frac{\partial \mathbf{m}}{\partial \mathbf{d}} \right] d^N d =$$
$$= \sum_j M_{ij} \int d_j p(\mathbf{d}) d^N d + v_i \int p(\mathbf{d}) d^N d = \sum_j M_{ij} \langle d_j \rangle + v_i \tag{3.18}$$

$$[\text{cov } \mathbf{m}]_{ij} = \int (m_i - \langle m_i \rangle)(m_j - \langle m_j \rangle) p(\mathbf{m}) d^N m$$
$$= \int \left[\sum_p M_{ip} d_p - M_{jp} \langle d_p \rangle \right] \left[\sum_q M_{iq} d_q - M_{jq} \langle d_q \rangle \right] p[\mathbf{m}(\mathbf{d})] \det \left[\frac{\partial \mathbf{m}}{\partial \mathbf{d}} \right] d^N d$$
$$= \sum_p M_{ip} \sum_q M_{jq} \int (d_p - \langle d_p \rangle)(d_q - \langle d_q \rangle) p(\mathbf{d}) d^N d$$
$$= \sum_p M_{ip} \sum_q M_{jq} [\text{cov } \mathbf{d}]_{pq} \tag{3.19}$$

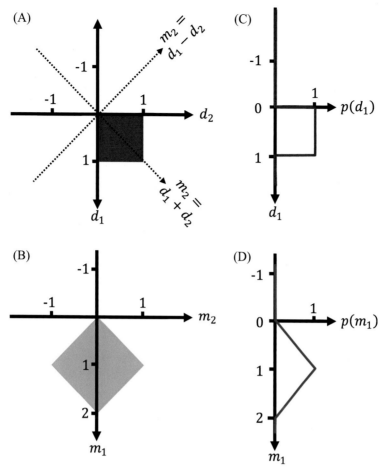

FIG. 3.11 (A) The uniform probability density function $p(d_1, d_2) = 1$ on the interval, $0 \leq d_1 \leq 1$ and $0 \leq d_2 \leq 1$. Also shown are the (m_1, m_2) axes, where $m_1 = d_1 + d_2$ and $m_2 = d_1 - d_2$. (B) The transformed probability density function $p(m_1, m_2) = 0.5$. (C) The univariate pdf $p(d_1)$ is formed by integrating $p(d_1, d_2)$ over d_2. It is a uniform pdf of amplitude 1. (D) The univariate pdf $p(m_1)$ is formed by integrating $p(m_1, m_2)$ over m_2. It is a triangular pdf of peak amplitude 1.

Eq. (3.17b) is *very* important, because it constitutes a *rule for error propagation*. The covariance of the data [cov **d**] is a measure of the amount of measurement error. The covariance of the model parameters [cov **m**] is a measure of the confidence by which they are determined. The rule connects the two; that is, given [cov **d**] representing measurement error, it provides a way to compute representing the corresponding uncertainty in the model parameters. While the rule requires that the data and the model parameters be linearly related, it is independent of the functional form of the probability density function $p(\mathbf{d})$. Furthermore, it can be shown to be correct even when the matrix **M** is not square.

As an example, consider a model parameter m_1, which is linearly related to the data by

$$m_1 = \frac{1}{N} \sum_{i=1}^{N} d_i = \frac{1}{N} [1, 1, 1, \cdots, 1] \mathbf{d} \tag{3.20}$$

Note that this formula is the sample mean, as defined in Eq. (3.5). This formula implies that matrix $\mathbf{M} = N^{-1}[1, 1, 1, \cdots, 1]$ and vector, $\mathbf{v} = 0$. Suppose that the data are uncorrelated and all have the same mean $\langle d \rangle$ and variance σ_d^2. Then, we find that $\langle m_1 \rangle = \mathbf{M}\langle \mathbf{d} \rangle + \mathbf{v} = \langle d \rangle$ and [cov **m**] $= \mathbf{M}$[cov **d**]$\mathbf{M}^T = \sigma_d^2/N$; that is, the variance of m_1 is σ_d^2/N, which is less than the variance of d. The square root of the variance, which is a measure of the width of $p(m_1)$, is proportional to $N^{-1/2}$. Thus accuracy of determining the mean of a group of data increases as the number of observations increases, albeit slowly (because of the square root).

In the case of uncorrelated data with uniform variance (i.e., [cov **d**] $= \sigma_d^2 \mathbf{I}$), the covariance of the model parameters is [cov **m**] $= \mathbf{M}$[cov **d**]$\mathbf{M}^T = \sigma_d^2 \mathbf{M}\mathbf{M}^T$. In general, $\mathbf{M}\mathbf{M}^T$, while symmetric, is not diagonal. Not only do the model parameters have unequal variance, but they are correlated. Strongly correlated model parameters are usually undesirable, but (as we will discuss later) good experimental design can sometimes eliminate them.

3.4 Normal (Gaussian) probability density functions

The probability density function for a particular random variable can be arbitrarily complicated, but in many instances, data possess the simple Normal (or Gaussian) probability density function

$$p(d) = \frac{1}{(2\pi)^{1/2}\sigma_d} \exp\left\{ -\frac{(d - \langle d \rangle)^2}{2\sigma_d^2} \right\} \tag{3.21}$$

This probability density function has mean $\langle d \rangle$ and variance σ_d^2 (Fig. 3.12). The Normal pdf is so common because it is the limiting probability density function for the sum of random variables. The *central limit theorem* shows (with certain limitations) that regardless of the probability density function of a set of independent random variables, the probability density function of their sum tends to be a Normal pdf as the number of summed variables increases. As long as the noise in the data comes from several sources of similar size, it will tend to follow a Normal probability density function. This behavior is exemplified by the sum of the two uniform probability density functions in Section 2.3. The probability density function of their sum is more nearly Normal than the individual probability density functions (it being triangular instead of rectangular).

The joint probability density function for two independent Normally distributed variables is just the product of two univariate probability density functions. When the data are correlated (say, with mean $\langle \mathbf{d} \rangle$ and covariance [cov $\mathbf{d}$]), the joint probability density function is more complicated, since it must express the degree of correlation. The appropriate generalization can be shown to be

$$p(\mathbf{d}) = \frac{1}{(2\pi)^{N/2}(\det [\text{cov } \mathbf{d}])^{1/2}} \exp\left\{ -\frac{1}{2}[\mathbf{d} - \langle \mathbf{d} \rangle]^T[\text{cov } \mathbf{d}]^{-1}[\mathbf{d} - \langle \mathbf{d} \rangle] \right\} \tag{3.22}$$

This probability density function reduces to Eq. (3.21) in the special case of $N=1$ (where [cov $\mathbf{d}$] becomes σ_d^2). It is not apparent that the $p(\mathbf{d})$ in Eq. (3.22) has an area of unity, a mean of $\langle \mathbf{d} \rangle$, and a covariance of [cov $\mathbf{d}$]. However, these properties easily can be derived by inserting Eq. (3.22) into the relevant integral and by transforming to the new variable $\mathbf{y} = [\text{cov } \mathbf{d}]^{-1/2}[\mathbf{d} - \langle \mathbf{d} \rangle]$ (whence the integral becomes substantially simplified).

When $p(\mathbf{d})$ (Eq. 3.22) is transformed using the linear rule $\mathbf{m} = \mathbf{Md}$, the resulting $p(\mathbf{m})$ is also Normal with mean $\langle \mathbf{m} \rangle = \mathbf{M} \langle \mathbf{d} \rangle$ and covariance matrix [cov $\mathbf{m}$] = $\mathbf{M}$[cov $\mathbf{d}$]$\mathbf{M}^T$. Thus all linear functions of Normal random variables are themselves Normal.

In Chapter 5, we will show that the information contained in each of two probability density functions can be combined by multiplying the two pdfs. Interestingly, the product of two Normal probability density functions is itself Normal (Fig. 3.13). Given Normal $p_A(\mathbf{d})$ with mean $\langle \mathbf{d}_A \rangle$ and covariance [cov $\mathbf{d}]_A$ and Normal $p_B(\mathbf{d})$ with mean $\langle \mathbf{d}_B \rangle$ and covariance [cov $\mathbf{d}]_B$, the product $p_C(\mathbf{d}) = p_A(\mathbf{d})p_B(\mathbf{d})$ is Normal with mean and covariance

$$\langle \mathbf{d}_C \rangle = \left([\text{cov } \mathbf{d}]_A^{-1} + [\text{cov } \mathbf{d}]_B^{-1} \right)^{-1} \left([\text{cov } \mathbf{d}]_A^{-1}\langle \mathbf{d}_A \rangle + [\text{cov } \mathbf{d}]_B^{-1}\langle \mathbf{d}_B \rangle \right)$$

$$[\text{cov } \mathbf{d}]_C = \left([\text{cov } \mathbf{d}]_A^{-1} + [\text{cov } \mathbf{d}]_B^{-1} \right)^{-1} \tag{3.23}$$

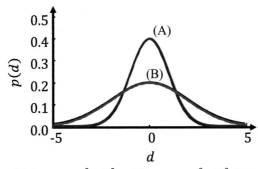

FIG. 3.12   Normal pdfs with zero mean and (A) variance $\sigma^2 = (1)^2$ and (B) variance $\sigma^2 = (2)^2$. Scripts `gdama03_10` and `gdapy03_10`.

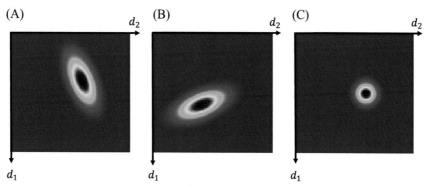

FIG. 3.13 (A) A Normal probability density function $p_A(d_1, d_2)$. (B) Another Normal probability density function $p_B(d_1, d_2)$, (C) The product $p_C(d_1, d_2) = p_A(d_1, d_2)p_B(d_1, d_2)$ is also Normal. Scripts `gdama03_11` and `gdapy03_11`.

The idea that the model and data are related by an explicit relationship $\mathbf{d} = \mathbf{g}(\mathbf{m})$ can be reinterpreted in light of this probabilistic description of the data. We no longer can assert that this relationship can hold for the data themselves, as they are random variables. Instead, we assert that this relationship holds for the mean of the data, that is, $\langle \mathbf{d} \rangle = \mathbf{g}(\mathbf{m})$. The pdf for the data can then be written as

$$p(\mathbf{d}) = \frac{1}{(2\pi)^{N/2}(\det [\text{cov } \mathbf{d}])^{\frac{1}{2}}} \exp\left\{ -\frac{1}{2}[\mathbf{d} - \mathbf{g}(\mathbf{m})]^T [\text{cov } \mathbf{d}]^{-1}[\mathbf{d} - \mathbf{g}(\mathbf{m})] \right\} \tag{3.24}$$

The model parameters now have the interpretation of a set of unknown quantities that define the shape of the pdf for the data. One approach to inverse theory (which will be pursued in Chapter 6) is to use the data to determine the pdf and thus the values of the model parameters.

For the Normal pdf (Eq. 3.24) to be sensible, $\mathbf{g}(\mathbf{m})$ must not be a function of any random variables. This is why we differentiated between data and auxiliary variables in Section 2.5; the latter must be known exactly. If the auxiliary variables are themselves uncertain, then they must be treated as data and the inverse problem becomes an implicit one with a much more complicated pdf than the problem given before exhibits.

As an example of constructing the pdf for a set of data, consider an experiment in which the temperature d_i in some small volume of space is measured N times. If the temperature is assumed not to be a function of time and space, the experiment can be viewed as the measurement of N realizations of the same random variable or as the measurement of one realization of N distinct random variables that all have the same pdf. We adopt the second viewpoint.

If the data are independent Normally distributed random variables with mean $\langle \mathbf{d} \rangle$ and variance σ_d^2, so that $[\text{cov } \mathbf{d}] = \sigma_d^2 \mathbf{I}$, then we can represent the assumption that all the data have the same mean by an equation of the form $\mathbf{Gm} = \mathbf{d}$:

$$\begin{bmatrix} 1 \\ 1 \\ \vdots \\ 1 \end{bmatrix} [m_1] = \begin{bmatrix} d_1 \\ d_2 \\ \vdots \\ d_N \end{bmatrix} \tag{3.25}$$

where m_1 is a single model parameter. We can then compute explicit formulas for the expressions in $p(\mathbf{d})$ as

$$\begin{aligned} (\det [\text{cov } \mathbf{d}])^{-\frac{1}{2}} &= \left(\sigma_d^{2N}\right)^{-\frac{1}{2}} = \sigma_d^{-N} \\ [\mathbf{d} - \mathbf{Gm}]^T [\text{cov } \mathbf{d}]^{-1}[\mathbf{d} - \mathbf{Gm}] &= \sigma_d^{-2} \sum_{i=1}^{N} (d_i - m_1)^2 \end{aligned} \tag{3.26}$$

The probability density function for the data is

$$p(\mathbf{d}) = \frac{\sigma_d^{-N}}{(2\pi)^{N/2}} \exp\left\{ -\frac{1}{2}\sigma_d^{-2} \sum_{i=1}^{N} (d_i - m_1)^2 \right\} \tag{3.27}$$

3.5 Testing the assumption of normal statistics

In the following chapters, we shall derive methods of solving inverse problems that are applicable whenever the data are Normally distributed. In many instances, the assumption that the data follow a Normal pdf is a reasonable one; nevertheless, having some means to test it is important.

First, consider a set of K Normal random variables x_i each with zero mean and unit variance. Suppose we construct a new random variable

$$\chi_K^2 = \sum_{i=1}^{K} x_i^2 \tag{3.28}$$

by summing squares of x_i. The function relating the x_i to χ_K^2 is nonlinear, so χ_K^2 does not have a Normal probability density function, but rather a different one (which we will not derive here) with the functional form

$$p(\chi_K^2) = \frac{1}{2^{K/2}(\frac{K}{2}-1)!} (\chi_K^2)^{(K/2)-1} \exp(-\tfrac{1}{2}\chi_K^2) \tag{3.29}$$

Here, ! denotes the *factorial*, defined as $n! \equiv n(n-1)(n-2)\cdots 1$. Eq. (3.39) is called the *chi-squared* probability density function, and the quantity K is called its *degrees of freedom*. It can be shown to be unimodal with mean K and variance $2K$. We shall make use of it in the discussion to follow.

We begin by supposing that we have some method of solving the inverse problem for the estimated model parameters. Assuming further that the model is explicit, we can compute the variation of the data about its estimated mean—a quantity we refer to as the error $\mathbf{e} = \mathbf{d} - \mathbf{g}(\mathbf{m}^{est})$. We now test whether the error follows an uncorrelated Normal pdf with uniform variance.

We begin by making a histogram of the N errors e_i, in which the histogram intervals have been chosen so that there are about the same number of errors in each interval. This histogram is then normalized to unit area, and the area A_i^{est} of each of the, say, p intervals, is noted. We then compare these areas (which are all in the range from zero to unity) with the areas A_i predicted by a Normal pdf with the same mean and variance as the e_i. The overall difference between these areas is quantified by using

$$(\chi_K^2)^{est} = N \sum_{i=1}^{p} \frac{(A_i^{est} - A_i)^2}{A_i} \tag{3.30}$$

If the data followed a Normal pdf exactly, then $(\chi_K^2)^{est}$ should be close to zero (it will not be zero since there are always random fluctuations). We need to inquire whether the $(\chi_K^2)^{est}$ measured for any particular data set is sufficiently far from zero that it is improbable that the data follow a Normal pdf. Only when values greater than or equal to $(\chi_K^2)^{est}$ occur less than 5% of the time—we are imagining that many realizations of the entire experiment were performed—does the pdf pass the test.

The quantity $(\chi_K^2)^{est}$ can be shown to follow approximately a chi-squared pdf with $K = p - 3$ degrees of freedom, irrespective of the type of pdf that is being tested. The reason that the degrees of freedom are $p - 3$ rather than p is that three constraints have been introduced into the problem: that the area of the histogram is unity and that the mean and variance of the Normal pdf match those of the data. This test is known as *Pearson's chi-squared test* (Fig. 3.14). The probability P that χ_K^2 is greater than or equal to $(\chi_K^2)^{est}$ is computed as

MATLAB®
`P = 1-chi2cdf(x2est, K);` <div align="right">[gdama03_12]</div>

Python
`P = 1-st.chi2.cdf( x2est, K );` <div align="right">[gdapy03_12]</div>

These scripts use the *cumulative chi-squared distribution*; that is, the probability that χ_K^2 is less than $(\chi_K^2)^{est}$. In the Python version, `st` is an abbreviation for `numpy.stats`.

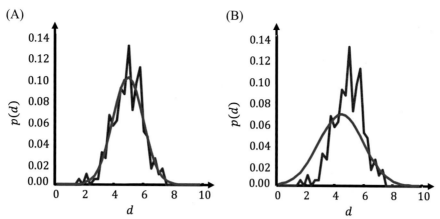

(A)   (B)

FIG. 3.14   Example of the Pearson's chi-squared test. The red curve is a probability density function (pdf) estimated by binning 200 realizations of a random variable d drawn from a Normally distributed population with a mean of 5 and a variance of $(1)^2$. (A) Normal pdf with the same mean and variance as the population. (B) Normal pdf with a mean of 4.5 and a variance of $(1.5)^2$. According to the test, χ^2 values exceeding the observed value occur extremely frequently (75% of the time) for (A) but extremely infrequently (0.003%) for (B). Scripts gdama03_12 and gdapy03_12.

3.6 Conditional probability density functions

Consider a scenario in which we are measuring the diameter d_1 and weight d_2 of sand grains drawn randomly from a pile of sand. We can consider d_1 and d_2 random variables described by a joint probability density function $p(d_1, d_2)$. The variables d_1 and d_2 will be correlated, as large grains will tend to be heavy. Now, suppose that $p(d_1, d_2)$ is known. Once we draw a sand grain from the pile and weigh it, we already know something about its diameter, as diameter is correlated with weight. The quantity that embodies this information is called the *conditional probability density function* of d_1, given d_2, and is written $p(d_1 | d_2)$.

The conditional probability density function of $p(d_1 | d_2)$ is not the same as $p(d_1, d_2)$, although it is related to it. The key difference is that $p(d_1 | d_2)$ is only a probability density function in the variable d_1, with the variable d_2 just providing auxiliary information. Thus the integral of $p(d_1 | d_2)$ with respect to d_1 needs to be unity, regardless of the value of d_2. Thus we must normalize $p(d_1, d_2)$ by dividing it by the total probability that d_1 occurs, given a specific value for d_2

$$p(d_1 | d_2) = \frac{p(d_1, d_2)}{\int p(d_1, d_2) \mathrm{d}d_1} = \frac{p(d_1, d_2)}{p(d_2)} \tag{3.31}$$

Here, we have used the fact that $p(d_2) = \int p(d_1, d_2) \mathrm{d}d_1$. The same logic allows us to calculate the conditional probability density function for d_2 given d_1

$$p(d_2 | d_1) = \frac{p(d_1, d_2)}{\int p(d_1, d_2) \mathrm{d}d_2} = \frac{p(d_1, d_2)}{p(d_1)} \tag{3.32}$$

See Fig. 3.15 for an example. Combining these two equations yields

$$p(d_1, d_2) = p(d_1 | d_2) p(d_2) = p(d_2 | d_1) p(d_1) \tag{3.33}$$

This result shows that the two conditional probability density functions are related but that they are not equal, that is, $p(d_2 | d_1) \neq p(d_1 | d_2)$. The two equations can be further rearranged into a result called *Bayes theorem* (Bayes and Price, 1763; Jeffreys, 1973)

$$p(d_1 | d_2) = \frac{p(d_2 | d_1) p(d_1)}{p(d_2)} = \frac{p(d_2 | d_1) p(d_1)}{\int p(d_1, d_2) \mathrm{d}d_1} = \frac{p(d_2 | d_1) p(d_1)}{\int p(d_2 | d_1) p(d_1) \mathrm{d}d_1}$$

$$p(d_2 | d_1) = \frac{p(d_1 | d_2) p(d_2)}{p(d_1)} = \frac{p(d_1 | d_2) p(d_2)}{\int p(d_1, d_2) \mathrm{d}d_2} = \frac{p(d_1 | d_2) p(d_2)}{\int p(d_1 | d_2) p(d_2) \mathrm{d}d_2} \tag{3.34}$$

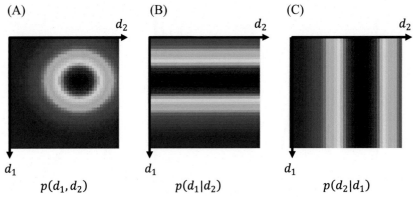

FIG. 3.15 Example of conditional probability density functions. (A) A Normal joint probability density function $p(d_1, d_2)$. (B) The corresponding conditional probability density function $p(d_1|d_2)$. (C) The corresponding conditional probability density function $p(d_2|d_1)$. Scripts `gdama03_13` and `gdapy03_13`.

Note that only the denominators of the three fractions in each equation are different. They correspond to three different but equivalent ways of writing $p(d_1)$ and $p(d_2)$.

As an example, consider the case where diameter can take on only two discrete values, S for small and B for big, and when weight can take on only two values, L for light and H for heavy. A hypothetical joint probability function is

$$P(d_1, d_2) = \begin{bmatrix} d_1 \backslash d_2 & L & H \\ S & 0.8000 & 0.0010 \\ B & 0.1000 & 0.0990 \end{bmatrix} \tag{3.35}$$

In this scenario, about 90% of the small sand grains are light, about 99% of the large grains are heavy, and small/light grains are much more common than big/heavy ones. Univariate distributions are computed by summing over rows or columns, and in Eq. (3.7):

$$P(d_1) = \begin{bmatrix} d_1 \\ S & 0.8010 \\ B & 0.1990 \end{bmatrix} \text{ and } P(d_2) = \begin{bmatrix} d_2 & L & H \\ & 0.9000 & 0.0100 \end{bmatrix} \tag{3.36}$$

According to Eq. (3.34), the conditional pdfs are

$$P(d_1|d_2) = \begin{bmatrix} d_1 \backslash d_2 & L & H \\ S & 0.8888 & 0.0100 \\ B & 0.1111 & 0.9900 \end{bmatrix} \text{ and } P(d_2|d_1) = \begin{bmatrix} d_1 \backslash d_2 & L & H \\ S & 0.9986 & 0.0012 \\ B & 0.5025 & 0.4974 \end{bmatrix} \tag{3.37}$$

Now suppose that we pick one sand grain from the pile, measure its diameter, and determine that it is big. What is the probability that it is heavy? We may be tempted to think that the probability is very high, since weight is highly correlated to size. But this reasoning is incorrect because heavy grains are about equally divided between the big and small size categories. The correct probability is given by $P(H|B)$, which is 49.74%.

Bayes theorem offers some insight into what is happening. Eq. (3.34), adapted for discrete values by interpreting the integral as a sum, becomes

$$P(H|B) = \frac{P(B|H)P(H)}{P(B|L)P(L) + P(B|H)P(H)} = \frac{0.9900 \times 0.1000}{0.1111 \times 0.9000 + 0.9900 \times 0.1000}$$
$$= \frac{0.0990}{0.1000 + 0.0990} = \frac{0.0990}{0.1990} = 0.4974 \tag{3.38}$$

The numerator of Eq. (3.38) represents the big, heavy grains and the denominator represents all the ways that one can get big grains, that is, the sum of big, heavy grains and big, light grains. In the scenario, light grains are extremely common, and although only a small fraction of them are heavy, their number affects the probability very significantly.

This analysis, called *Bayesian Inference*, allows us to assess the importance of any given measurement. Before having measured the size of the sand grain, our best estimate of whether it is heavy is 10%, because heavy grains make up 10% of the total population (i.e., $P(H) = 0.10$). After the measurement, the probability rises to 49.74%, which is about a factor of five more certain. As we will see in Section 6.5, Bayesian Inference plays an important role in the solution of inverse problems.

3.7 Confidence intervals

The confidence of a particular observation is the probability that one realization of the random variable falls within a specified interval centered on the true mean. Confidence is therefore related to the distribution of area in $p(d)$. If most of the area is concentrated near the mean, then the interval for, say, 95% confidence will be very small; otherwise, the confidence interval will be large. The width of the confidence interval is related to the variance. Pdfs with large variances will also tend to have large confidence intervals. Nevertheless, the relationship is not direct, as variance is a measure of width, not area. The relationship is easy to quantify for the simplest univariate pdfs. For instance, Normal probability density functions have 68% confidence intervals 1σ wide and 95% confidence intervals 2σ wide. Other types of simple pdfs have analogous relationships. If one knows that a particular Normal random variable has $\sigma = 1$, then if a realization of that variable has the value 50, one can state that there is a 95% chance that the mean of the random variable lies between 48 and 52. One might symbolize this by $\langle d \rangle = 50 \pm 2$ (95%).

The concept of confidence intervals is more difficult to work with when one is dealing with joint probability density functions of several correlated random variables. One must define some volume in the space of data and compute the probability that the true means of the data are within the volume. One must also specify the shape of that volume. The more complicated the pdf, the more difficult it is to choose an appropriate shape and calculate the probability within it.

Even in the case of the multivariate Normal probability density functions, statements about confidence levels need to be made carefully, as is illustrated by the following scenario. Suppose that the Normal probability density function $p(d_1, d_2)$ represents two measurements, say the length and diameter of a cylinder, and suppose that these measurements are uncorrelated with equal variance σ_d^2. As we might expect, the univariate probability density function $p(d_1) = \int p(d_1, d_2) dd_2$ has variance σ_d^2 and so the probability P_1 that d_1 falls between $\langle d_1 \rangle - \sigma_d$ and $\langle d_1 \rangle + \sigma_d$ is 0.68 or 68%. Similarly, the probability P_2 that d_2 falls between $\langle d_2 \rangle - \sigma_d$ and $\langle d_2 \rangle + \sigma_d$ is 0.68 or 68%. But P_1 represents the probability of d_1, irrespective of the value of d_2, and P_2 represents the probability of d_2, irrespective of the value of d_1. The probability that both d_1 and d_2 *simultaneously* fall within their respective one-sigma confidence intervals is $P = P_1 P_2 = (0.68)^2 = 0.46$ or 46%, which is significantly smaller than 68%.

One occasionally encounters a journal article containing a table of many (say 100) estimated parameters, each one with a stated 2σ confidence bound. The probability that all one hundred measurements fall within their respective bounds is $(0.95)^{100}$ or 0.6%—which is pretty close to zero!

3.8 Computing realizations of random variables

The ability to create a vector of realizations of a random variable is very important. For instance, it can be used to simulate noise when testing a data analysis method on synthetic data (i.e., artificially prepared data with well-controlled properties). And it can be used to generate a suite of possible models, to test against data. Both MATLAB® and Python provide methods that can generate realizations drawn from many different probability density functions. For instance,

MATLAB®
```m = random('Normal',mbar,sigmam,N,1);``` [gdama03_14]

Python
```m = np.random.normal(loc=mbar,scale=sigma,size=(N,1));``` [gdapy03_14]

creates a column vector m of N Normally distributed elements with mean mbar and standard deviation sigmam.

In cases where no predefined method is available, it is possible to transform an available pdf, say $p(d)$, into the desired pdf, say $q(m)$, using the transformation rule

$$p[d(m)] \frac{dd}{dm} = q(m) \tag{3.39}$$

Most software environments provide a predefined method for generating realizations of a uniform pdf on the interval $(0,1)$, including MATLAB® and Python. Then, as $p[d(m)] = 1$, Eq. (3.39) is a differential equation for the required transformation $m(d)$—or rather its inverse $d(m)$

$$\frac{\mathrm{d}d}{\mathrm{d}m} = q(m) \quad \text{or} \quad d(m) = \int_{-\infty}^{m} q(m')\,\mathrm{d}m' = Q(m) \tag{3.40}$$

Here, $Q(m)$ is the cumulative probability distribution corresponding to $q(m)$. The transformation is then $m(d) = Q^{-1}(d)$; that is, one must invert the cumulative probability distribution to find the value of m for which the probability d occurs. Thus the transformation requires that the inverse cumulative probability distribution be known. Presuming a method of calculating $Q^{-1}(d)$ is available, one obtains realizations of $p(m)$ by first computing realizations of the uniform pdf $p(d)$ and then applying $Q^{-1}(d)$ to them.

MATLAB®

```d = random('unif',0,1,N,1);```
```m = norminv(d,mbar,sigmam);```
<div align="right">[gdama03_14]</div>

Python

```d = np.random.uniform(low=0.0,high=1.0,size=(N,1));```
```m = st.norm.ppf(duniform,loc=mbar,scale=sigma);```
<div align="right">[gdapy03_14]</div>

This approach offers no advantage in the Normal case, because direct methods are available to compute realizations of the Normal pdf (Fig. 3.16). It is of practical use in cases where no direct method is available, as long as an appropriate Q^{-1} function can be developed.

Another method of producing a vector of realizations of a random variable is the *Metropolis- Hastings algorithm* (Metropolis et al., 1953; Hastings, 1970). It is a useful alternative to the transformation method described earlier, especially since it requires evaluating only the probability density function $p(d)$ and not its cumulative inverse. It is an iterative algorithm that builds the vector $\mathbf{d}$ of realizations element by element. The first element d_1 is set to an arbitrary number, such as zero. Subsequent elements are generated in sequence, with an element d_i generating a successor d_{i+1} according to this algorithm:

Step 1: Randomly draw a *proposed successor* d' from a conditional probability density function $q(d'|d_i)$. The exact form of $q(d'|d_i)$ is arbitrary; however, it must be chosen so that d' is in the neighborhood of d_i. One possible choice is the Normal function

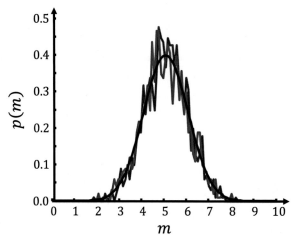

FIG. 3.16 Normal probability density function $p(m)$, with a mean of 5 and variance of $(1)^2$. (black curve). Pdf estimated by binning 1000 realizations of a Normally distributed random variable, computed directly via a standard random number generator (red). Pdf estimated by binning 1000 realizations of a random variable generated by transforming a uniform pdf (blue). Scripts gdama03_14 and gdapy03_14.

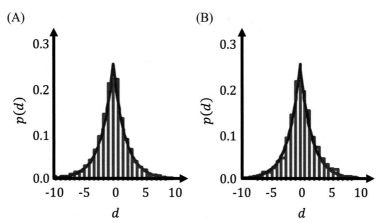

(A)

(B)

FIG. 3.17 Histograms (blue curves) of 5000 realizations of a random variable d for the probability density function $p(d) = \frac{1}{2}c\,\exp\{-|d|/c\}$ (red curves) with $c = 2$. (A) Realizations computed by transforming data drawn from a uniform pdf and (B) realizations computed using the Metropolis-Hastings algorithm. Scripts gdama03_15 and gdapy03_15.

$$q(d'|d_i) = \frac{1}{(2\pi)^{\frac{1}{2}}\sigma_q} \exp\left\{ -\frac{(d'-d_i)^2}{2\sigma_q^2} \right\} \tag{3.41}$$

Here, σ_q represents the size of the neighborhood, that is, the typical deviation of d' away from d_i.

Step 2: Generate a random number r drawn from a uniform pdf on the interval $(0,1)$.

Step 3: Accept the proposed successor and set $d_{i+1} = d'$ if

$$r < \alpha \quad \text{with} \quad \alpha \equiv \frac{p(d')\,q(d_i|d')}{p(d_i)\,q(d'|d_i)} \tag{3.42}$$

Otherwise, set $d_{i+1} = d_i$. Note that the test implies that the proposed successor is always accepted when $\alpha > 1$.

When repeated many times, this algorithm leads to a vector $\mathbf{d}$, whose elements approximately sample the probability density function $p(d)$ (Fig. 3.17). However, it is advisable to delete the first part of vector, which can be unduly influenced by the starting value.

The conditional probability density functions cancel from Eq. (3.42) when $q(d_i|d') = q(d'|d_i)$, as is the case for the Normal pdf in Eq. (3.41). The formula for α in Eq. (3.42) can be numerically unstable when its denominator is near-zero. A good strategy is to compute $\gamma \equiv \log(\alpha)$ instead of α. The proposed successor is always accepted when $\gamma > 0$, obviating the need to compute $\alpha = \exp(\gamma)$ in this case (thus circumventing the instability), and α only is calculated when $\gamma \leq 0$.

As we will describe in detail in Section 12.4, the Metropolis-Hastings algorithm is widely used to create ensemble solutions to inverse problems.

3.9 Problems

3.1 What is the mean and variance of the uniform pdf $p(d) = 1$ on the interval $(0,1)$?

3.2 Suppose d is a Normally distributed random variable with zero mean and unit variance. What is the probability density function of $E = d^2$? Hint: Since the sign of d gets lost when it is squared, you can assume that $p(d)$ is one sided, that is, defined for only $d \geq 0$ and with twice the amplitude of the usual Normal pdf.

3.3 Write a script that uses MATLAB's random('Normal',...) function or Python's np.random.normal(...) method to create a column vector d of N=1000 realizations of a Normally distributed random variable with mean dbar=4 and standard deviation sigmad=2. Count up the number of instances where $d > \langle d \rangle + 2\sigma_d$. Is this about the number you expected?

3.4 Suppose that the data are uncorrelated with uniform variance [cov $\mathbf{d}] = \sigma_d^2 \mathbf{I}$ and that the model parameters are linear functions of the data $\mathbf{m} = \mathbf{M}\mathbf{d}$, where $\mathbf{M}$ is a matrix. (A) What property must $\mathbf{M}$ have for the model parameters to be uncorrelated with uniform variance σ_m^2? (B) Express this property in terms of the rows of $\mathbf{M}$.

3.5 Use the transformation method to compute realizations of the probability density function $p(m) = 3m^2$ on the interval $(0,1)$, starting from realizations of the uniform pdf $p(d) = 1$. Check your results by plotting a histogram and comparing it to the pdf.

References

Bayes, T., Price, R., 1763. An essay towards solving a problem in the doctrine of chance. By the late Rev. Mr. Bayes, communicated by Mr. Price, in a letter to John Canton, A. M. F. R. S. Philos. Trans. R. Soc. Lond. 53, 370–418. https://doi.org/10.1098/rstl.1763.0053.

Hastings, W.K., 1970. Monte Carlo sampling methods using Markov chains and their applications. Biometrika 57, 97–109. https://doi.org/10.1093/biomet/57.1.97.

Jeffreys, H., 1973. Scientific Inference, third ed. Cambridge University Press, Cambridge, United Kingdom. 273 pp. ISBN-13-978-0521084468.

Metropolis, N., Rosenbluth, A.W., Rosenbluth, M.N., Teller, A.H., Teller, E., 1953. Equation of state calculations by fast computing machines. J. Chem. Phys. 21, 1087–1092. https://doi.org/10.1063/1.1699114.

4

Solution of the linear, Normal inverse problem, viewpoint 1: The length method

4.1 The lengths of estimates

The simplest of methods for solving the linear inverse problem $\mathbf{d} = \mathbf{Gm}$ is based on measures of the size, or *length*, of the estimated model parameters $\mathbf{m}^{est}$ and of the predicted data $\mathbf{d}^{pre} = \mathbf{Gm}^{est}$. To see that measures of length can be relevant to the solution of inverse problems, consider the simple problem of fitting a straight line to data (Fig. 4.1). This problem is often solved by the so-called *method of least squares*. In this method, one tries to pick the model parameters (intercept and slope) so that the predicted data are as close as possible to the observed data.

For each observation, one defines a *prediction error*, or misfit, $e_i = d_i^{obs} - d_i^{pre}$. The best-fit line is then the one with model parameters that lead to the smallest overall error E, defined as

$$E \equiv \sum_{i=1}^{N} e_i^2 = \mathbf{e}^T \mathbf{e} \tag{4.1}$$

The total error E is the sum of the squares of the individual errors and is exactly the squared Euclidean length of the misfit vector $\mathbf{e}$, that is, $E = \mathbf{e}^T \mathbf{e}$.

The method of least squares estimates the solution of an inverse problem by finding the model parameters that minimize a particular measure of the length of the prediction error, $\mathbf{e} = \mathbf{d}^{obs} - \mathbf{d}^{pre}$, namely, its Euclidean length E. As will be detailed later, it is the simplest of the methods that use measures of length as the guiding principle in solving an inverse problem.

4.2 Measures of length

Although the Euclidean length is one way of quantifying the size or length of a vector, it is by no means the only possible measure. For instance, one could equally well quantify length by summing the absolute values of the elements of the vector.

The term *norm* is used to refer to some measure of length or size and is indicated by a set of double vertical bars, that is, $\|\mathbf{e}\|$ is the norm of the vector $\mathbf{e}$. The most commonly employed norms are those based on the sum of some power of the elements of a vector and are given the name L_n, where n is the power:

$$L_1 \text{ norm} : \|\mathbf{e}\|_1 = \left[\sum_{i=1}^{N} |e_i| \right]$$

$$L_2 \text{ norm} : \|\mathbf{e}\|_2 = \left[\sum_{i=1}^{N} |e_i|^2 \right]^{\frac{1}{2}} \tag{4.2}$$

$$L_n \text{ norm} : \|\mathbf{e}\|_n = \left[\sum_{i=1}^{N} |e_i|^n \right]^{1/n}$$

FIG. 4.1 (A) Least-squares fitting of a straight line to (z,d) data. (B) The error e_i for each observation is the difference between the observed and predicted datum: $e_i = d_i^{obs} - d_i^{pre}$. Scripts gdama04_01 and gdapy04_01.

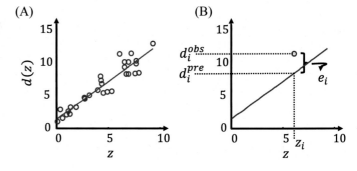

Successively higher norms give the largest element of **e** successively larger weight. The limiting case $n \to \infty$ gives non-zero weight to only the largest element (Fig. 4.2); therefore, it is equivalent to the selection of the vector element with largest absolute value as the measure of length and is written as

$$L_\infty \text{ norm} : \|\mathbf{e}\|_\infty = \max_i |e_i| \tag{4.3}$$

The number of nonzero elements in a vector is a measure of its simplicity and is referred to as its *sparseness*. It is sometimes denoted $\|\mathbf{e}\|_0^0$, as:

$$\|\mathbf{e}\|_0^0 = \lim_{n \to 0} \sum_{i=1}^{N} |e_i|^n = \text{number of nonzero elements in } \mathbf{e} \tag{4.4}$$

(at least if one defines 0^0 as zero). However, this usage is understood as an informal shorthand, for $\|\mathbf{e}\|_0^0$ does not obey all the rules of a norm (such as Eq. (4.5b)).

The method of least squares uses the L_2 norm to quantify length. It is appropriate to inquire why this, and not some other choice of norm, is used. The answer involves the way in which one chooses to weight data outliers that fall far from the average trend (Fig. 4.3). If the data are very accurate, then the fact that one prediction falls far from its observed value is important. A high-order norm is used, since it weights the larger errors preferentially. On the other hand, if the data are expected to scatter widely about the trend, then no significance can be placed upon a few large prediction errors. A low-order norm is used, since it gives more equal weight to errors of different sizes.

As will be discussed in more detail later, the L_2 norm implies that the data obey Normal statistics. Gaussians are rather short-tailed functions, so it is appropriate to place considerable weight on any data that have a large prediction error. The likelihood of an observed datum falling far from the trend depends on the shape of the distribution for that

FIG. 4.2 Hypothetical prediction error $e_i(z_i)$ and its absolute value, raised to the powers of 1, 2, and 10. While most elements of e_i are numerically significant, only a few elements of $|e_i|^{10}$ are. Scripts gdama04_02 and gdapy04_02.

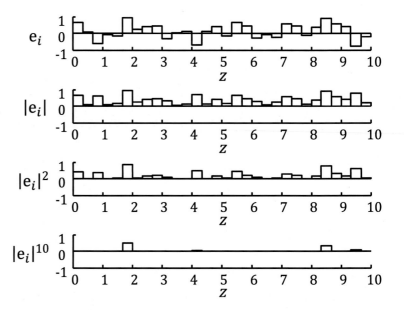

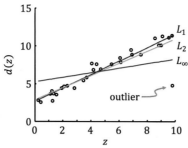

FIG. 4.3 Straight line fits to (z, d) pairs where the error is measured under the L_1, L_2, and L_∞ norms. The L_1 norm gives the least weight to the single outlier. Script `gdama04_03` and `gdapy04_03`.

(A) (B)

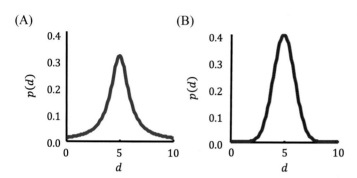

FIG. 4.4 (A) Long-tailed probability density function. (B) Short-tailed probability density function. Scripts `gdama04_04` and `gdapy04_04`.

datum. Long-tailed distributions imply many scattered (improbable) points. Short-tailed distributions imply very few scattered points (Fig. 4.4). The choice of a norm, therefore, implies an assertion that the data obey a particular type of statistics.

Even though many measurements have approximately Normal statistics, most data sets generally have a few outliers that are wildly improbable. The occurrence of these points demonstrates that the assumption of Normal statistics is in error, especially in the tails of the distribution. If one applies least squares to this kind of problem, the estimates of the model parameters can be completely erroneous. Least squares weights large errors so heavily that even one "bad" data point can completely throw off the result. In these situations, methods based on the L_1 norm give more reliable estimates. Methods that can tolerate a few bad data are said to be *robust*.

Matrix norms can be defined in a manner similar to vector norms. Vector and matrix norms obey the following relationships:

Vector norms:

$$\|\mathbf{x}\| > 0 \text{ as long as } \mathbf{x} \neq \mathbf{0} \tag{4.5a}$$

$$\|a\mathbf{x}\| = |a| \, \|\mathbf{x}\| \tag{4.5b}$$

$$\|\mathbf{x} + \mathbf{y}\| < \|\mathbf{x}\| + \|\mathbf{y}\| \tag{4.5c}$$

Matrix norms:

$$\|\mathbf{A}\|_2 = \left[\sum_{i=1}^{N} \sum_{j=1}^{N} A_{ij}^2 \right]^{1/2} \tag{4.5d}$$

$$\|c\mathbf{A}\| = |c| \, \|\mathbf{A}\| \tag{4.5e}$$

$$\|\mathbf{A}\mathbf{x}\| \leq \|\mathbf{A}\| \|\mathbf{x}\| \tag{4.5f}$$

$$\|\mathbf{A}\mathbf{B}\| \leq \|\mathbf{A}\| \|\mathbf{B}\| \tag{4.5g}$$

$$\|\mathbf{A} + \mathbf{B}\| \leq \|\mathbf{A}\| + \|\mathbf{B}\| \tag{4.5h}$$

Eqs. (4.5c) and (4.5h) are called triangle inequalities because of their similarity to Pythagoras's law for right triangles.

4.3 Least squares for a straight line

The elementary problem of fitting a straight line to data illustrates the basic procedures applied in this technique. The model is the assertion that the data can be described by the linear equation $d_i = m_1 + m_2 z_i$. Note that there are two model parameters, so $M = 2$, and that typically there are many more than two data, $N > M$. As a line is defined by precisely two points, it is clearly impossible to choose a straight line that passes through every one of the data, except in the instance that they all lie precisely on the same straight line. Collinearity rarely occurs when measurements are influenced by noise.

As we shall discuss in more detail later, the fact that the equation $d_i = m_1 + m_2 z_i$ cannot be satisfied for every i means that the inverse problem is *overdetermined*, that is, it has no solution for which $\mathbf{e} = 0$. One therefore seeks values of the model parameters that solve $d_i = m_1 + m_2 z_i$ approximately, where the goodness of the approximation is defined by the L_2 error

$$E = \mathbf{e}^T \mathbf{e} = \sum_{i=1}^{N} \left(d_i^{obs} - m_1 - m_2 z_i \right)^2 \tag{4.6}$$

This is then the elementary calculus problem of locating the minimum of the function $E(m_1, m_2)$ and is solved by setting the derivatives of E to zero and solving the resulting equations.

$$\frac{\partial E}{\partial m_1} = \frac{\partial}{\partial m_1} \sum_{i=1}^{N} \left(d_i^{obs} - m_1 - m_2 z_i \right)^2 = 2Nm_1 + 2m_2 \sum_{i=1}^{N} z_i - 2 \sum_{i=1}^{N} d_i^{obs} = 0$$

$$\frac{\partial E}{\partial m_2} = \frac{\partial}{\partial m_2} \sum_{i=1}^{N} \left(d_i^{obs} - m_1 - m_2 z_i \right)^2 = 2m_1 \sum_{i=1}^{N} z_i + 2m_2 \sum_{i=1}^{N} z_i^2 - 2 \sum_{i=1}^{N} z_i d_i^{obs} = 0 \tag{4.7}$$

These two equations are then solved simultaneously for m_1 and m_2, yielding the classic formulas for the least squares fitting of a line.

4.4 The least-squares solution of the linear inverse problem

Least squares can be extended to the general linear inverse problem in a very straightforward manner. Again, one computes the derivative of the error E with respect to one of the model parameters, say, m_q, and sets the result to zero. The error E is

$$E = \mathbf{e}^T \mathbf{e} = \sum_{i=1}^{N} \left[d_i^{obs} - \sum_{j=1}^{M} G_{ij} m_j \right] \left[d_i^{obs} - \sum_{k=1}^{M} G_{ik} m_k \right] \tag{4.8}$$

Note that the indices on the sums within the parentheses are different *dummy* variables, to prevent confusion. Multiplying out the terms and reversing the order of the summations lead to

$$E = \sum_{j=1}^{M} \sum_{k=1}^{M} m_j m_k \sum_{i=1}^{N} G_{ij} G_{ik} - 2 \sum_{j=1}^{M} m_j \sum_{i=1}^{N} G_{ij} d_i^{obs} + \sum_{i=1}^{N} d_i^{obs} d_i^{obs} \tag{4.9}$$

The derivatives $\partial E / \partial m_q$ are now computed. Performing this differentiation term by term gives

$$\frac{\partial}{\partial m_q} \sum_{j=1}^{M} \sum_{k=1}^{M} m_j m_k \sum_{i=1}^{N} G_{ij} G_{ik} = \sum_{j=1}^{M} \sum_{k=1}^{M} \left[\delta_{jq} m_k + m_j \delta_{kq} \right] \sum_{i=1}^{N} G_{ij} G_{ik}$$

$$= 2 \sum_{k=1}^{M} m_k \sum_{i=1}^{N} G_{iq} G_{ik} \tag{4.10}$$

for the first term. Since the model parameters are independent variables, derivatives of the form $\partial m_i / \partial m_j$ are either unity, when $i = j$, or zero, when $i \neq j$. Thus $\partial m_i / \partial m_j$ is just the Kronecker delta δ_{ij} (Eqs. 1A.12 and 1B.12) and the formula containing it can be simplified trivially.

The second term gives

$$-2\frac{\partial}{\partial m_q}\sum_{j=1}^{M}m_j\sum_{i=1}^{N}G_{ij}d_i^{obs}=-2\sum_{j=1}^{M}\delta_{jq}\sum_{i=1}^{N}G_{ij}d_i^{obs}=-2\sum_{i=1}^{N}G_{iq}d_i^{obs} \tag{4.11}$$

The third term is zero, as it does not contain any ms

$$\frac{\partial}{\partial m_q}\sum_{i=1}^{N}d_i^{obs}d_i^{obs}=0 \tag{4.12}$$

Combining the three terms gives

$$\frac{\partial}{\partial m_q}=0=2\sum_{k=1}^{M}m_k\sum_{i=1}^{N}G_{iq}G_{ik}-2\sum_{i=1}^{N}G_{iq}d_i^{obs} \tag{4.13}$$

Writing this equation in matrix notation yields

$$\mathbf{G}^T\mathbf{G}\mathbf{m}-\mathbf{G}^T\mathbf{d}^{obs}=0 \tag{4.14}$$

Note that the quantity $\mathbf{G}^T\mathbf{G}$ is a square symmetric $M\times M$ matrix and that it multiplies a vector $\mathbf{m}$ of length M. It is often referred to as the *Gram* matrix. The quantity $\mathbf{G}^T\mathbf{d}^{obs}$ is a vector of length M. It is often referred to as the *back-projected data*. Eq. (4.14) is a square matrix equation for the unknown model parameters $\mathbf{m}$. Presuming that $[\mathbf{G}^T\mathbf{G}]^{-1}$ exists (an important issue that we shall return to later), we have the following estimate for the model parameters:

$$\mathbf{m}^{est}=[\mathbf{G}^T\mathbf{G}]^{-1}\mathbf{G}^T\mathbf{d}^{obs} \tag{4.15}$$

In Eq. (3.17b), we showed that measurement error propagates to model parameter uncertainty by the rule $[\text{cov }\mathbf{m}]=\sigma_d^2\mathbf{M}\mathbf{M}^T$ when the data were uncorrelated and with uniform variance σ_d^2 and when the model parameters were related to the data by the linear equation $\mathbf{m}=\mathbf{M}\mathbf{d}$. In a least-squares problem $\mathbf{M}=[\mathbf{G}^T\mathbf{G}]^{-1}\mathbf{G}^T$, so the covariance is

$$[\text{cov }\mathbf{m}]=\sigma_d^2[\mathbf{G}^T\mathbf{G}]^{-1}\mathbf{G}^T\left[[\mathbf{G}^T\mathbf{G}]^{-1}\mathbf{G}^T\right]^T=\sigma_d^2[\mathbf{G}^T\mathbf{G}]^{-1}\mathbf{G}^T\mathbf{G}[\mathbf{G}^T\mathbf{G}]^{-1}=\sigma_d^2[\mathbf{G}^T\mathbf{G}]^{-1} \tag{4.16}$$

Here, we have used the rule $(\mathbf{A}\mathbf{B})^T=\mathbf{B}^T\mathbf{A}^T$ to simplify the expression, along with the rule that a symmetric matrix has a symmetric inverse. Consequently, the least-squares solution and its 95% confidence limits are

$$m_i^{est}\pm 2\sqrt{[\text{cov }\mathbf{m}]_{ii}}\ (95\%) \tag{4.17}$$

When the dimensions of $\mathbf{G}$ are small (say, N and M less than a few hundred), the least-squares solution and its covariance are computed as

MATLAB®
| ```
GTG = G'*G;
mest = GTG\(G'*dobs);
covm = sigmad2 * inv(GTG)
```                                    [gdama04_01] |

Python
| ```
GTG = np.matmul(G.T,G);
mest = la.solve(GTG,np.matmul(G.T,dobs));
covm = sd2 * la.inv(GTG);
```                                    [gdapy04_01] |

Here, dobs is column vector of the observed data and mest is a column vector of the estimated model parameters.

4.5 Example: Fitting a straight line

In the straight-line problem, the model is $d_i^{obs} = m_1 + m_2 z_i$, so the equation $\mathbf{Gm} = \mathbf{d}^{obs}$ has the form

$$\begin{bmatrix} 1 & z_1 \\ 1 & z_2 \\ \vdots & \vdots \\ 1 & z_N \end{bmatrix} \begin{bmatrix} m_1 \\ m_2 \end{bmatrix} = \begin{bmatrix} d_1^{obs} \\ d_2^{obs} \\ \vdots \\ d_N^{obs} \end{bmatrix} \tag{4.18}$$

The matrix $\mathbf{G}$ is created as

MATLAB®
| ```
M=2;
G=[ones(N,1), z];
``` |
| [gdama04_01] |

Python
| ```
M=2;
G = np.zeros((N,M));
G[0:N,0:1] = np.ones((N,1));
G[0:N,1:2] = z;
``` |
| [gdapy04_01] |

Or alternatively,

Python
| ```
M=2;
G = np.concatenate( (np.ones((N,1)), z), axis=1);
``` |
| [gdapy04_01] |

Here, `axis=1` specifies that the arrays are to be concatenated size by side (as contrasted to `axis=0`, which means one atop the other). Although `np.concatenate()` is the more elegant method, it becomes increasingly unwieldy as the number of items being concatenated grows. For this reason, we prefer the first method.

Usually, a numerical solution is sufficient, but in this simple case, the analytic solution easily can be constructed and is very instructive. The matrix products required by the least-squares solution are

$$\mathbf{G}^T\mathbf{G} = \begin{bmatrix} 1 & 1 & \cdots & 1 \\ z_1 & z_2 & \cdots & z_N \end{bmatrix} \begin{bmatrix} 1 & z_1 \\ 1 & z_2 \\ \vdots & \vdots \\ 1 & z_N \end{bmatrix} = \begin{bmatrix} N & \sum_{i=1}^{N} z_i \\ \sum_{i=1}^{N} z_i & \sum_{i=1}^{N} z_i^2 \end{bmatrix}$$

$$[\mathbf{G}^T\mathbf{G}]^{-1} = \frac{1}{N\sum_{i=1}^{N} z_i^2 - \left(\sum_{i=1}^{N} z_i\right)^2} \begin{bmatrix} \sum_{i=1}^{N} z_i^2 & -\sum_{i=1}^{N} z_i \\ -\sum_{i=1}^{N} z_i & N \end{bmatrix} \tag{4.19}$$

$$\mathbf{G}^T\mathbf{d}^{obs} = \begin{bmatrix} 1 & 1 & \cdots & 1 \\ z_1 & z_2 & \cdots & z_N \end{bmatrix} \begin{bmatrix} d_1^{obs} \\ d_2^{obs} \\ \vdots \\ d_N^{obs} \end{bmatrix} = \begin{bmatrix} \sum_{i=1}^{N} d_i^{obs} \\ \sum_{i=1}^{N} z_i d_i^{obs} \end{bmatrix}$$
```

Here, we have used the formula for the inverse of a 2×2 matrix (Eqs. 1A.18 and 1B.18). The least-squares solution and its covariance are

$$
\begin{bmatrix} m_1^{est} \\ m_2^{est} \end{bmatrix} = \frac{1}{N \sum_{i=1}^{N} z_i^2 - \left(\sum_{i=1}^{N} z_i \right)^2} \begin{bmatrix} \sum_{i=1}^{N} z_i^2 & -\sum_{i=1}^{N} z_i \\ -\sum_{i=1}^{N} z_i & N \end{bmatrix} \begin{bmatrix} \sum_{i=1}^{N} d_i^{obs} \\ \sum_{i=1}^{N} z_i d_i^{obs} \end{bmatrix}
$$

$$
= \frac{1}{N \sum_{i=1}^{N} z_i^2 - \left(\sum_{i=1}^{N} z_I \right)^2} \begin{bmatrix} \sum_{i=1}^{N} z_i^2 \sum_{i=1}^{N} d_i^{obs} - \sum_{i=1}^{N} z_i \sum_{i=1}^{N} z_i d_i^{obs} \\ -\sum_{i=1}^{N} z_I \sum_{i=1}^{N} d_i^{obs} + N \sum_{i=1}^{N} z_i d_i^{obs} \end{bmatrix}
$$

(4.20)

$$
[\text{cov } \mathbf{m}] = \sigma_d^2 [\mathbf{G}^T \mathbf{G}]^{-1} = \frac{\sigma_d^2}{N \sum_{i=1}^{N} z_i^2 - \left(\sum_{i=1}^{N} z_i \right)^2} \begin{bmatrix} \sum_{i=1}^{N} z_i^2 & -\sum_{i=1}^{N} z_i \\ -\sum_{i=1}^{N} z_i & N \end{bmatrix}
$$

(4.21)

These equations sometimes are derived in advanced calculus courses, but without recourse to the matrix formalism that so expedites their derivation. Note that model parameters are correlated, except when the sum of the auxiliary variable $\mathbf{z}$ is zero, that is, when the data scatter evenly about the origin.

4.6 Example: Fitting a parabola

The problem of fitting a parabola is a trivial generalization of fitting a straight line (Fig. 4.4). Now the model is $d_i^{obs} = m_1 + m_2 z_i + m_3 z_i^2$, so the equation $\mathbf{Gm} = \mathbf{d}^{obs}$ has the form

$$
\begin{bmatrix} 1 & z_1 & z_1^2 \\ 1 & z_2 & z_2^2 \\ \vdots & \vdots & \vdots \\ 1 & z_N & z_N^2 \end{bmatrix} \begin{bmatrix} m_1 \\ m_2 \\ m_3 \end{bmatrix} = \begin{bmatrix} d_1^{obs} \\ d_2^{obs} \\ \vdots \\ d_N^{obs} \end{bmatrix}
$$

(4.22)

The matrix $\mathbf{G}$ is created as

MATLAB®
| ```
M=3;
G=[ones(N,1), z, z.^2];
```                                                    [gdama04_05] |

Python
| ```
M=3;
G = np.zeros((N,M));
G[0:N,0:1] = np.ones((N,1));
G[0:N,1:2] = z;
G[0:N,2:3] = np.power(z,2);
```                                                    [gdapy04_05] |

An example of using a quadratic fit to examine Kepler's third law is shown in Fig. 4.5.

FIG. 4.5 Test of Kepler's third law, which states that the cube of the orbital radius of a planet equals the square of its orbital period. (A) Data (red circles) for our solar system are least-squares fit with a quadratic formula $d_i = m_1 + m_2 z_i + m_3 z_i^2$, where d_i is radius cubed and z_i is period. (B) Error of fit. A separate graph is used so that the error can be plotted at a meaningful scale. Scripts gdama04_05 and gdapy04_05. *Data courtesy of Wikipedia.*

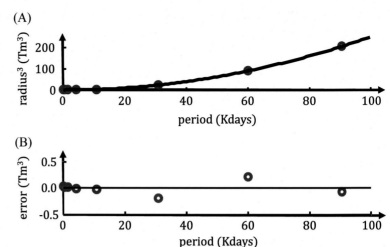

4.7 Example: Fitting of a planar surface

To fit a plane surface, two auxiliary variables, say, x and y, are needed. The model is $d_i^{obs} = m_1 + m_2 x_i + m_3 y_i$, so the equation $\mathbf{Gm} = \mathbf{d}^{obs}$ has the form

$$\begin{bmatrix} 1 & x_1 & y_1 \\ 1 & x_2 & y_2 \\ \vdots & \vdots & \vdots \\ 1 & x_N & y_N \end{bmatrix} \begin{bmatrix} m_1 \\ m_2 \\ m_3 \end{bmatrix} = \begin{bmatrix} d_1^{obs} \\ d_2^{obs} \\ \vdots \\ d_N^{obs} \end{bmatrix} \tag{4.23}$$

The matrix $\mathbf{G}$ is created as

MATLAB®
| ```
M=3;
G = [ones(N,1), x, y];
```  [gdama04_06] |

Python
| ```
M=3;
G = np.zeros((N,M));
G[0:N,0:1] = np.ones((N,1));
G[0:N,1:2] = x;
G[0:N,2:3] = y;
```  [gdapy04_06] |

An example of using a planar fit to examine earthquakes on a geologic fault is shown in Fig. 4.6.

4.8 Example: Inverting reflection coefficient for interface properties

The seismological properties of an elastic material, such as the Earth, can be described by three material parameters, density ρ, compressional velocity α, and shear velocity β. When two homogenous layers are in contact, their contrast can be characterized by the change $\Delta\rho$, $\Delta\alpha$, and $\Delta\beta$ in these material properties, with respect to their average values of ρ, α, and β. The interface can be studied by directing a compressional wave ("P" wave) or a shear wave ("S" wave) at it and observing the P and S waves that are *reflected* from it. When a unit amplitude P wave is incident on the interface, the coefficients R_{PP} and R_{PS} give the amplitudes of the reflected P and S waves, respectively (Fig. 4.7A).

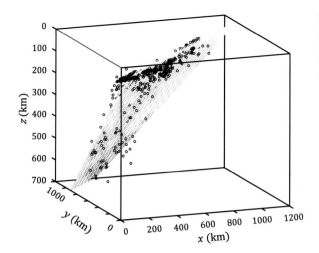

FIG. 4.6 (Circles) Earthquakes in the Kurile subduction zone, northwest Pacific Ocean. The x-axis points north and the y-axis east. The earthquakes scatter about a dipping planar surface (colored grid), determined using least squares. Scripts gdama04_06 and gdapy04_06. *Data courtesy of the U.S. Geological Survey.*

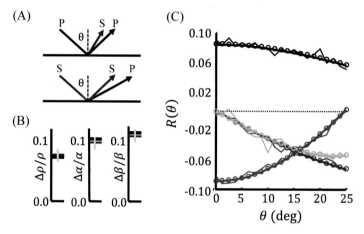

FIG. 4.7 (A) Schematic diagrams of P and S wave reflected from a layer interface. In this scenario, measurements of reflection coefficients are used to infer interface properties. (B) True jumps in density ρ, compressional velocity α, and shear velocity β across the interface (black bars) together with corresponding estimates (red bars) and 95% confidence intervals (orange bars). (C) True reflection coefficients, $R_{PP}(\theta)$ (bold black curve), $R_{PS}(\theta)$ (bold red curve), $R_{SP}(\theta)$ (bold green curve), and $R_{SS}(\theta)$ (bold blue curve) as a function of angle of incidence θ together with corresponding noisy versions (thin curves) and the predictions of the model (circles). Scripts gdama04_07 and gdapy04_07.

When a unit-amplitude S wave is incident on the interface, the coefficients R_{SP} and R_{SS} give the amplitudes of the reflected P and S waves, respectively. These coefficients depend upon the angle of incidence of the incident waves, say θ_P and θ_S, as well as the changes in the material parameters across the interface.

A common problem in seismology is estimating the properties of the interface using measurements of the reflection coefficients made at a suite of angles of incidence. This inverse problem has $M = 3$ unknowns that represent the fractional change in material properties across the interface:

$$\mathbf{m} = \begin{bmatrix} m_1 \\ m_2 \\ m_3 \end{bmatrix} \equiv \begin{bmatrix} \Delta\rho/\rho \\ \Delta\alpha/\alpha \\ \Delta\beta/\beta \end{bmatrix} \tag{4.24}$$

For a weak interface, the reflection coefficients can be shown to be linear functions of these model parameters (Aki and Richards, 2009, their Eq. 5.46):

$$R_{PP} = R_{PP}^{(1)}m_1 + R_{PP}^{(2)}m_2 + R_{PP}^{(3)}m_3$$
$$R_{PS} = R_{PS}^{(1)}m_1 + R_{PS}^{(2)}m_2 + R_{PS}^{(3)}m_3$$
$$R_{SP} = R_{SP}^{(1)}m_1 + R_{SP}^{(2)}m_2 + R_{SP}^{(3)}m_3 \tag{4.25}$$
$$R_{SS} = R_{SS}^{(1)}m_1 + R_{SS}^{(2)}m_2 + R_{SS}^{(3)}m_3$$

Each of the twelve coefficients $R_{PP}^{(1)}, R_{PP}^{(2)}, \cdots R_{SS}^{(3)}$ quantifies the contribution of one model parameter to one reflection coefficient. They are complicated functions of angle of incidence and the layer properties:

$$R_{PP}^{(1)} = \tfrac{1}{2}\left(1 - 4\beta^2 p^2\right) \text{ and } R_{PP}^{(2)} = \frac{1}{2\cos^2\theta_P} \text{ and } R_{PP}^{(3)} = -4\beta^2 p^2 \text{ and}$$

$$R_{PS}^{(1)} = C\left(1 - 2\beta^2 p^2 - 2\beta^2 \frac{\cos\theta_P}{\alpha}\frac{\cos\theta_S}{\beta}\right) \text{ and } R_{PS}^{(2)} = 0 \text{ and } R_{PS}^{(3)} = 4C\beta^2\left(p^2 - \frac{\cos\theta_P}{\alpha}\frac{\cos\theta_S}{\beta}\right)$$

$$R_{SP}^{(1)} = DR_{PS}^{(1)} \text{ and } R_{SP}^{(2)} = DR_{SP}^{(2)} \text{ and } R_{SP}^{(3)} = DR_{PS}^{(3)}$$

$$R_{SS}^{(1)} = -\tfrac{1}{2}\left(1 - 4\beta^2 p^2\right) \text{ and } R_{SS}^{(2)} = 0 \text{ and } R_{SS}^{(3)} = -\left(\frac{1}{2\cos^2\theta_S} - 4\beta^2 p^2\right)$$

$$\text{and } p = \frac{\sin\theta_P}{\alpha} = \frac{\sin\theta_S}{\beta} \text{ and } \cos\theta_P = \sqrt{1 - \alpha^2 p^2} \text{ and } \cos\theta_S = \sqrt{1 - \beta^2 p^2}$$

$$\text{and } C = \frac{-p\alpha}{2\cos\theta_S} \text{ and } D = \frac{\cos\theta_S}{\alpha}\frac{\beta}{\cos\theta_P}$$

(4.26)

An experiment in which each reflection coefficient is measured at a suite of K angles (Fig. 4.7C) has $N = 4K$ data and $M = 3$ unknowns. The data vector is formed by concatenating the measurements of each of the reflection coefficients, and the data kernel is built from Equation (4.25), ordering its rows in a corresponding fashion:

$$\mathbf{d}^{obs} = \begin{bmatrix} R_{PP}(\theta_1) \\ \vdots \\ R_{PP}(\theta_K) \\ R_{PS}(\theta_1) \\ \vdots \\ R_{PS}(\theta_K) \\ R_{SP}(\theta_1) \\ \vdots \\ R_{SP}(\theta_K) \\ R_{SS}(\theta_1) \\ \vdots \\ R_{SS}(\theta_K) \end{bmatrix} \text{ and } \mathbf{G} = \begin{bmatrix} R_{PP}^{(1)}(\theta_1) & R_{PP}^{(2)}(\theta_1) & R_{PP}^{(3)}(\theta_1) \\ \vdots & \vdots & \vdots \\ R_{PP}^{(1)}(\theta_K) & R_{PP}^{(2)}(\theta_K) & R_{PP}^{(3)}(\theta_K) \\ R_{PS}^{(1)}(\theta_1) & R_{PS}^{(2)}(\theta_1) & R_{PS}^{(3)}(\theta_1) \\ \vdots & \vdots & \vdots \\ R_{PS}^{(1)}(\theta_K) & R_{PS}^{(2)}(\theta_K) & R_{PS}^{(3)}(\theta_K) \\ R_{SP}^{(1)}(\theta_1) & R_{SP}^{(2)}(\theta_1) & R_{SP}^{(3)}(\theta_1) \\ \vdots & \vdots & \vdots \\ R_{SP}^{(1)}(\theta_K) & R_{SP}^{(2)}(\theta_K) & R_{SP}^{(3)}(\theta_K) \\ R_{SS}^{(1)}(\theta_1) & R_{SS}^{(2)}(\theta_1) & R_{SS}^{(3)}(\theta_1) \\ \vdots & \vdots & \vdots \\ R_{SS}^{(1)}(\theta_K) & R_{SS}^{(2)}(\theta_K) & R_{SS}^{(3)}(\theta_K) \end{bmatrix}$$

(4.27)

The least-squares estimate of the model parameters is then $\mathbf{m}^{est} = [\mathbf{G}^T\mathbf{G}]^{-1}\mathbf{G}^T\mathbf{d}^{obs}$ (Fig. 4.7B).

4.9 The existence of the least-squares solution

The least-squares solution arose from consideration of an inverse problem that had no exact solution. As there was no exact solution to find, we chose to do the next best thing: to estimate the solution as those values of the model parameters that gave the best approximate solution (where "best" meant minimizing the L_2 prediction error). By writing a single formula $\mathbf{m}^{est} = [\mathbf{G}^T\mathbf{G}]^{-1}\mathbf{G}^T\mathbf{d}^{obs}$, we implicitly assumed that there was only one such "best" solution.

As we shall prove later, least squares fails when the number of solutions that give the same minimum prediction error is greater than one.

To see that least squares fails for problems with nonunique solutions, consider the straight-line problem with only one data point (Fig. 4.8). It is clear that this problem is nonunique; many possible lines can pass through the point, and each has zero prediction error. The solution then contains the following expression:

$$[\mathbf{G}^T\mathbf{G}]^{-1} = \begin{bmatrix} N & \sum_{i=1}^{N} z_i \\ \sum_{i=1}^{N} z_i & \sum_{i=1}^{N} z_i^2 \end{bmatrix}^{-1} \rightarrow \begin{bmatrix} 1 & z_1 \\ z_1 & z_1^2 \end{bmatrix}^{-1}$$

(4.28)

The inverse of a matrix is proportional to the reciprocal of the determinant of the matrix so that

$$[\mathbf{G}^T\mathbf{G}]^{-1} \propto \left[\det\begin{bmatrix} 1 & z_1 \\ z_1 & z_1^2 \end{bmatrix}\right]^{-1} = \frac{1}{z_1^2 - z_1^2}$$

(4.29)

This expression clearly is singular. The formula for the least-squares solution fails.

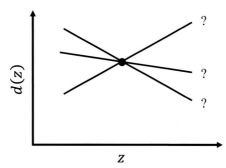

FIG. 4.8 An infinity of different lines can pass through a single point. The prediction error for each is $E=0$.

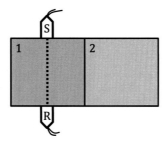

FIG. 4.9 An acoustic travel time experiment with source S and receiver R in a medium consisting of two discrete bricks. Because of poor experimental geometry, the acoustic waves (dashed line) pass only through brick 1. The slowness of brick 2 is completely underdetermined.

The question of whether the equation $\mathbf{Gm}=\mathbf{d}$ provides enough information to specify *uniquely* the model parameters serves as a basis for classifying inverse problems.

Underdetermined problems occur when the equation $\mathbf{Gm}=\mathbf{d}^{obs}$ does not provide enough information to determine uniquely all the model parameters. As we saw in the earlier example, this can happen if there are several solutions that have zero prediction error. From elementary linear algebra, we know that underdetermined problems occur when there are more unknowns than data, that is, when $M>N$. We must note, however, that there is no special reason why the prediction error must be zero for an underdetermined problem. Frequently, the data uniquely determine some of the model parameters but not others. For example, consider the acoustic experiment in Fig. 4.9. Since no measurements are made of the acoustic slowness s_2 in the second brick, it is clear that this model parameter is completely unconstrained by the data. In contrast, the acoustic slowness of the first brick is overdetermined, since in the presence of measurement noise, no choice of s_1 can satisfy the data exactly. The equation describing this experiment is

$$
h\begin{bmatrix} 1 & 0 \\ 1 & 0 \\ \vdots & \vdots \\ 1 & 0 \end{bmatrix}\begin{bmatrix} s_1 \\ s_2 \end{bmatrix} = \begin{bmatrix} d_1^{obs} \\ d_2^{obs} \\ \vdots \\ d_N^{obs} \end{bmatrix} \tag{4.30}
$$

where s_i is the slowness in the ith brick, h the brick width, and the d_i^{obs} the measurements of travel time. If one were to attempt to solve this problem with least squares, one would find that

$$
\det \mathbf{G}^T\mathbf{G} = h^2 \det \begin{bmatrix} N & 0 \\ 0 & 0 \end{bmatrix} = 0 \tag{4.31}
$$

so that $[\mathbf{G}^T\mathbf{G}]^{-1}$ is singular. Even though $M<N$, the problem is still underdetermined, because the data kernel has a very poor structure. Although this is a rather trivial case in which only some of the model parameters are underdetermined, the problem arises in subtler forms in realistic experiments.

We shall refer to underdetermined problems that have nonzero prediction error as *mixed-determined* problems, to distinguish them from purely underdetermined problems that have zero prediction error.

Even-determined problems are those in which exactly enough information to determine the model parameters. There is only one solution, and it has zero prediction error.

Overdetermined problems are those in which too much information contained in the equation $\mathbf{Gm}=\mathbf{d}^{obs}$ for it to possess an exact solution. This is the case in which we can employ least squares to select a "best" approximate solution. Overdetermined problems typically have more data than unknowns, that is, $N>M$, although for the reasons discussed earlier it is possible to have problems that are to some degree overdetermined even when $N<M$ and to have problems that are to some degree underdetermined even when $N>M$.

To deal successfully with the full range of inverse problems, we shall need to be able to characterize whether an inverse problem is under- or overdetermined (or some combination of the two). We shall develop quantitative methods for making this characterization in Chapter 9. For the moment, we assume that it is possible to characterize the problem intuitively on the basis of the kind of experiment the problem represents.

4.10 The purely underdetermined problem

Suppose that an inverse problem of the form $\mathbf{Gm} = \mathbf{d}^{obs}$ has been identified as one that is purely underdetermined. For simplicity, assume that there are fewer equations than unknown model parameters, that is, $N < M$, and that there are no inconsistencies in these equations. It is therefore possible to find more than one solution for which the prediction error E is zero. (In fact, we shall show that underdetermined linear inverse problems have an infinite number of such solutions.) Although the data provide some information about the model parameters, they do not provide enough to determine them uniquely.

We must have some means of singling out precisely one of the infinite number of solutions with zero prediction error E to obtain a unique solution $\mathbf{m}^{est}$ to the inverse problem. To do this, we must add to the problem some information not contained in the equation $\mathbf{Gm} = \mathbf{d}^{obs}$. This extra information is called *prior* (or *a priori*) information (Jackson, 1979). Prior information can take many forms, but in each case, it quantifies expectations about the character of the solution that are not based on the actual data.

For instance, in the case of fitting a straight line through a single data point, one might have the expectation that the line also passes through the origin. This prior information now provides enough information to solve the inverse problem uniquely, since two points (one datum, one prior) determine a line.

Another example of prior information concerns expectations that the model parameters possess a given sign or lie in a given range. For instance, suppose the model parameters represent density at different points in the Earth. Even without making any measurements, one can state with certainty that the density is everywhere positive, as density is an inherently positive quantity. Furthermore, as the interior of the Earth can reasonably be assumed to be rock, its density must have values in some range known to characterize rock, say, between 1000 and 10,000 kg/m^3. If one can use this prior information when solving the inverse problem, it may greatly reduce the range of possible solutions—or even cause the solution to be unique.

There is something unsatisfying about having to add prior information to an inverse problem to single out a solution. Where does this information come from, and how certain is it? There are no firm answers to these questions. In certain instances, one might be able to identify reasonable prior assumptions; in other instances, one might not. Clearly, the importance of the prior information depends greatly on the use one plans for the estimated model parameters. If one simply wants one example of a solution to the problem, the choice of prior information is unimportant. However, if one wants to develop arguments that depend on the uniqueness of the estimates, the validity of the prior assumptions is of paramount importance.

These issues are the price one must pay for estimating the model parameters of a nonunique inverse problem. As will be shown in Chapter 8, there are other kinds of "answers" to inverse problems that do not depend on prior information (localized averages, for example). However, these "answers" invariably are more abstract and not as easily interpretable as estimates of model parameters.

The first kind of prior assumption we shall consider is the expectation that the solution to the inverse problem is *simple*, where the notion of *simplicity* is quantified by some measure of the length of the solution. One such measure is simply the Euclidean length of the solution $L = \mathbf{m}^T\mathbf{m} = \sum_i m_i^2$. A solution is defined to be simple if it is small when measured under the L_2 norm. Admittedly, this measure is perhaps not a particularly realistic measure of simplicity. It can be useful occasionally, and we shall describe shortly how it can be generalized to more realistic measures.

One instance in which solution length may be realistic is when the model parameters describe the velocity of various points in a moving fluid. The length L is then a measure of the kinetic energy of the fluid. In certain instances, it may be appropriate to find that velocity field in the fluid that has the smallest possible kinetic energy of those solutions satisfying the data.

We pose the following problem: Find the solution $\mathbf{m}^{est}$ that minimizes $L = \mathbf{m}^T\mathbf{m}$ subject to the constraint that the prediction error $\mathbf{e} = \mathbf{d}^{obs} - \mathbf{Gm} = \mathbf{0}$. This problem can easily be solved by the *method of Lagrange multipliers* (see Section 17.1). We minimize the function,

$$\Phi(\mathbf{m}) = L + \sum_{i=1}^{N} \lambda_i e_i = \sum_{i=1}^{M} m_i^2 + \sum_{i=1}^{N} \lambda_i \left[d_i^{obs} - \sum_{j=1}^{M} G_{ij} m_j \right] \tag{4.32}$$

with respect to model parameter m_q, where the λs are the Lagrange multipliers. Taking the derivatives and setting the result to zero yields

$$\frac{\partial}{\partial m_q} \Phi(\mathbf{m}) = 0 = \sum_{i=1}^{N} 2 \frac{\partial m_i}{\partial m_q} m_i + \sum_{i=1}^{N} \lambda_i \left[-\sum_{j=1}^{M} G_{ij} \frac{\partial m_j}{\partial m_q} \right] = 2m_q - \sum_{i=1}^{N} \lambda_i G_{iq} \qquad (4.33)$$

Rewriting this equation in matrix notation yields $0 = 2\mathbf{m} - \mathbf{G}^T \boldsymbol{\lambda}$, which must be solved along with the constraint equation $\mathbf{d}^{obs} - \mathbf{Gm} = \mathbf{0}$. Plugging the first equation into the second gives $\mathbf{d}^{obs} = \mathbf{Gm} = \frac{1}{2}\mathbf{GG}^T\boldsymbol{\lambda}$. We note that the matrix $\mathbf{GG}^T$ is a square $N \times N$ matrix. If its inverse exists, we can then solve this equation for the Lagrange multipliers $\boldsymbol{\lambda} = 2[\mathbf{GG}^T]^{-1}\mathbf{d}^{obs}$. Then inserting this expression into the first equation yields the *minimum-length solution*

$$\mathbf{m}^{est} = \mathbf{G}^T[\mathbf{GG}^T]^{-1}\mathbf{d}^{obs} \qquad (4.34)$$

We shall discuss the conditions under which this solution exists later. As we shall see, one condition is that the equation $\mathbf{d}^{obs} = \mathbf{Gm}$ be purely underdetermined—that it contains no inconsistencies. For uncorrelated data with uniform variance σ_d^2, the covariance of the estimated solution is

$$[\text{cov } \mathbf{m}] = \sigma_d^2 \mathbf{G}^T[\mathbf{GG}^T]^{-2}\mathbf{G} \qquad (4.35)$$

Here, we have used the rule $(\mathbf{AB})^T = \mathbf{B}^T\mathbf{A}^T$ to simplify the expression, along with the rule that a symmetric matrix has a symmetric inverse.

4.11 Mixed-determined problems

Most inverse problems that arise in practice are neither completely overdetermined nor completely underdetermined. For instance, in the X-ray tomography problem, there may be one *voxel* (box-shaped volume element) through which several rays pass (Fig. 4.10A). The X-ray opacity of this voxel is clearly overdetermined. On the other hand, there may be voxels that have been missed entirely (Fig. 4.10B). These voxels are completely undetermined. There may also be voxels that cannot be individually resolved because every ray that passes through one also passes through an equal distance of the other (Fig. 4.10C). These boxes are also underdetermined, as only their mean opacity is determined.

Ideally, we would like to sort the unknown model parameters into two groups: those that are overdetermined and those that are underdetermined. Actually, to do this, we need to form a new set of model parameters that are linear combinations of the old. For example, in the two-box problem before, the average opacity $m_1' = \langle \mathbf{m} \rangle = \frac{1}{2}(m_1 + m_2)$ is completely overdetermined, whereas the difference in opacity $m_2' = \Delta m = m_1 - m_2$ is completely underdetermined. We want to perform this partitioning from an arbitrary equation $\mathbf{Gm} = \mathbf{d}^{obs}$ to a transformed equation $\mathbf{G'm'} = \mathbf{d'}$, where $\mathbf{m'}$ is partitioned into an upper part $\mathbf{m}'^o$ that is overdetermined and a lower part $\mathbf{m}'^u$ that is underdetermined:

$$\begin{bmatrix} \mathbf{G}'^o & 0 \\ 0 & \mathbf{G}'^u \end{bmatrix} \begin{bmatrix} \mathbf{m}'^o \\ \mathbf{m}'^u \end{bmatrix} = \begin{bmatrix} \mathbf{d}'^o \\ \mathbf{d}'^u \end{bmatrix} \qquad (4.36)$$

If this can be achieved, we could determine the overdetermined model parameters by solving the upper equations in the least-squares sense and determine the underdetermined model parameters by finding those that have minimum L_2 solution length. In addition, we would have found a solution that added as little prior information to the inverse problem as possible.

This partitioning process can be accomplished through *singular-value decomposition* of the data kernel, a process that we shall discuss in Chapter 9. As it is a relatively time consuming process, we first examine an approximate process that works if the inverse problem is not too underdetermined.

Instead of partitioning $\mathbf{m}$, suppose that we determine a solution that minimizes some combination Φ of the prediction error and the solution length for the unpartitioned model parameters

$$\Phi = E + \varepsilon^2 L = \mathbf{e}^T\mathbf{e} + \varepsilon^2\mathbf{m}^T\mathbf{m} \qquad (4.37)$$

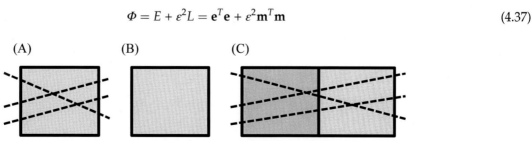

(A)   (B)   (C)

FIG. 4.10   (A) The X-ray opacity of the brick is overdetermined, as measurements of X-ray intensity are made along three different paths (dashed lines). (B) The opacity is underdetermined, since no measurements have been made. (C) The average opacity of these two bricks is overdetermined, but as each path has an equal length in either brick, the individual opacities are underdetermined.

where the weighting factor ε^2 determines the relative importance given to the prediction error and solution length. If ε^2 is made large enough, this procedure will clearly minimize the underdetermined part of the solution. Unfortunately, it also tends to minimize the overdetermined part of the solution. As a result, the solution will not minimize the prediction error E and will not be a very good estimate of the true model parameters. If ε is set to zero, the prediction error will be minimized, but no prior information will be provided to control the underdetermined model parameters. It may be possible, however, to find some compromise value for ε^2 that will approximately minimize E while approximately minimizing the length of the underdetermined part of the solution. There is no simple method of determining what this compromise ε^2 should be (without solving the partitioned problem); it must be determined by trial and error. By minimizing $\Phi(\mathbf{m})$ with respect to the model parameters in a manner exactly analogous to the least-squares derivation, we obtain

$$\left[\mathbf{G}^T\mathbf{G} + \varepsilon^2\mathbf{I}\right]\mathbf{m}^{est} = \mathbf{G}^T\mathbf{d}^{obs} \quad\text{or}\quad \mathbf{m}^{est} = \left[\mathbf{G}^T\mathbf{G} + \varepsilon^2\mathbf{I}\right]^{-1}\mathbf{G}^T\mathbf{d}^{obs} \tag{4.38}$$

By the standard error propagation rule, its covariance is

$$[\text{cov}\,\mathbf{m}] = \sigma_d^2\left[\mathbf{G}^T\mathbf{G} + \varepsilon^2\mathbf{I}\right]^{-1}\mathbf{G}^T\mathbf{G}\left[\mathbf{G}^T\mathbf{G} + \varepsilon^2\mathbf{I}\right]^{-1} \tag{4.39}$$

This estimate of the model parameters is called the *damped least-squares* solution. In the limit $\varepsilon^2 \to 0$, the solution and its covariance reduce to the standard least-squares versions (compare with Eqs. (4.15) and (4.16)). The concept of error has been generalized to include not only prediction error but also error in fitting the prior information (that the solution length is zero, in this case). The underdeterminacy of the inverse problem is said to have been *damped* and the inverse problem is said to have been *regularized*.

4.12 Weighted measures of length as a type of prior information

There are many instances in which $L = \mathbf{m}^T\mathbf{m}$ is not a very good measure of solution simplicity. For instance, suppose that one were solving an inverse problem for density fluctuations in the ocean. One may not want to find a solution that is smallest in the sense of closest to zero but one that is smallest in the sense that it is closest to some other value, such as the typical density of sea water. The obvious generalization of L is then

$$L = (\langle\mathbf{m}\rangle - \mathbf{m})^T(\langle\mathbf{m}\rangle - \mathbf{m}) \tag{4.40}$$

where $\langle\mathbf{m}\rangle$ is the prior value of the model parameters (a known typical value of sea water, in this case).

Sometimes the whole idea of length as a measure of simplicity is inappropriate. For instance, one may feel that a solution is simple if it is flat or if it is smooth. These measures may be particularly appropriate when the model parameters represent a discretized continuous function such as density or X-ray opacity. One may have the expectation that these parameters vary only slowly with position. Fortunately, properties such as flatness can be easily quantified by measures that are generalizations of length. For example, the flatness of a continuous function of space can be quantified by the norm of its first derivative, which is a measure of steepness (the opposite of flatness). For discrete model parameters, one can use the difference between physically adjacent model parameters as approximations of a derivative. The steepness $\mathbf{l}$ of a vector $\mathbf{m}$ is then

$$\mathbf{l} = \frac{1}{\Delta x}\begin{bmatrix} -1 & 1 & & & \\ & -1 & 1 & & \\ & & \ddots & \ddots & \\ & & & -1 & 1 \end{bmatrix}\begin{bmatrix} m_1 \\ m_2 \\ \vdots \\ m_M \end{bmatrix} \equiv \mathbf{D}^{(1)}\mathbf{m} \tag{4.41}$$

Here, the *first-difference* matrix $\mathbf{D}^{(1)}$ is used to quantify the *steepness* of the solution. Other methods of simplicity can also be represented by a matrix multiplying the model parameters. For instance, solution smoothness can be implemented by quantifying the *roughness* (the opposite of smoothness) using the second-difference matrix $\mathbf{D}^{(2)}$ (which is an approximation for the second derivative)

$$\mathbf{D}^{(2)} \equiv \frac{1}{(\Delta x)^2}\begin{bmatrix} 1 & -2 & 1 & & \\ & 1 & -2 & 1 & \\ & & \ddots & \ddots & \ddots \\ & & & 1 & -2 & 1 \end{bmatrix} \tag{4.42}$$

The overall steepness or roughness of the solution is then just the length

$$L = \mathbf{l}^T\mathbf{l} = (\mathbf{Dm})^T(\mathbf{Dm}) = \mathbf{m}^T(\mathbf{D}^T\mathbf{D})\mathbf{m} = \mathbf{m}^T\mathbf{W}_m\mathbf{m} \quad\text{with}\quad \mathbf{W}_m \equiv \mathbf{D}^T\mathbf{D} \tag{4.43}$$

The matrix $\mathbf{W}_m$ can be interpreted as a *weighting factor* that enters into the calculation of the length of the vector $\mathbf{m}$. Note, however, that $\|\mathbf{m}\|^2_{weighted} \ \mathbf{m}^T\mathbf{W}_m\mathbf{m}$ is not a proper norm, because it violates the positivity condition given in Eq. (4.3a), that is, $\|\mathbf{m}\|^2_{weighted} = 0$ for some nonzero vectors, such as the constant vector. This behavior usually poses no insurmountable problems, but it can cause solutions based on minimizing this norm to be nonunique.

The measure of solution simplicity can therefore be generalized to include both weighting and a typical value:

$$L = (\langle\mathbf{m}\rangle - \mathbf{m})^T\mathbf{W}_m(\langle\mathbf{m}\rangle - \mathbf{m}) \tag{4.44}$$

By suitably choosing the prior model vector $\langle\mathbf{m}\rangle$ and the weighting matrix $\mathbf{W}_m$, we can quantify a wide variety of measures of simplicity.

Weighted measures of the prediction error can also be useful. Frequently, some observations are made with more accuracy than others. In this case, one would like the prediction error e_i of the more accurate observations to have a greater weight in the quantification of the overall error E than the inaccurate observations. To accomplish this weighting, we define a generalized prediction error

$$E = \mathbf{e}^T\mathbf{W}_e\mathbf{e} \tag{4.45}$$

where the matrix $\mathbf{W}_e$ defines the relative contribution of each individual error to the total prediction error. Normally we would choose this matrix to be diagonal. For example, if $N = 5$ and the third observation is known to be twice as accurately determined as the others, one might use

$$\mathbf{W}_e = \begin{bmatrix} 1 & & & & \\ & 1 & & & \\ & & 4 & & \\ & & & 1 & \\ & & & & 1 \end{bmatrix} \tag{4.46}$$

Here, the weight $4 = (2)^2$ is used because the square $(e_3)^2$ of the error appears in the formula for E.

The solutions stated earlier can be modified to take into account these new measures of prediction error and solution simplicity. The derivations are substantially the same as for the unweighted cases but the algebra is lengthy. We present only the results in the next several sections.

4.13 Weighted least squares

If the equation $\mathbf{Gm} = \mathbf{d}^{obs}$ is completely overdetermined, then one can estimate the model parameters by minimizing the generalized prediction error $E = \mathbf{e}^T\mathbf{W}_e\mathbf{e}$. This procedure leads to the solution and covariance

$$\mathbf{m}^{est} = [\mathbf{G}^T\mathbf{W}_e\mathbf{G}]^{-1}\mathbf{G}^T\mathbf{W}_e\mathbf{d}^{obs} \text{ and } [\text{cov } \mathbf{m}] = \sigma_d^2[\mathbf{G}^T\mathbf{W}_e\mathbf{G}]^{-1} \tag{4.47}$$

4.14 Weighted minimum length

If the equation $\mathbf{Gm} - \mathbf{d}^{obs}$ is completely underdetermined, then one can estimate the model parameters by choosing the solution that is simplest, where simplicity is defined by the generalized length $L = (\mathbf{m} - \langle\mathbf{m}\rangle)^T\mathbf{W}_m(\mathbf{m} - \langle\mathbf{m}\rangle)$. This procedure leads to

$$\mathbf{m}^{est} = \langle\mathbf{m}\rangle + \mathbf{W}_m^{-1}\mathbf{G}^T[\mathbf{G}\mathbf{W}_m^{-1}\mathbf{G}^T]^{-1}(\mathbf{d}^{obs} - \mathbf{G}\langle\mathbf{m}\rangle) \tag{4.48}$$

We put off discussing the covariance of this and subsequent solutions until the next chapter.

4.15 Weighted damped least squares

If the equation $\mathbf{Gm} = \mathbf{d}^{obs}$ is mix determined, it can be solved by minimizing a combination of prediction error and solution length $\Phi = E + \varepsilon^2 L = \mathbf{e}^T\mathbf{W}_e\mathbf{e} + \varepsilon^2(\mathbf{m} - \langle\mathbf{m}\rangle)^T\mathbf{W}_m(\mathbf{m} - \langle\mathbf{m}\rangle)$ (Franklin, 1970; Jackson, 1972; Jordan and Franklin, 1971). The parameter ε^2 is chosen by trial and error to yield a solution that has a reasonably small prediction error. The solution obtained by minimizing Φ with respect to $\mathbf{m}$ is

$$\left[\mathbf{G}^T\mathbf{W}_e\mathbf{G} + \varepsilon^2\mathbf{W}_m\right]\mathbf{m}^{\text{est}} = \mathbf{G}^T\mathbf{W}_e\mathbf{d}^{obs} + \varepsilon^2\mathbf{W}_m\langle\mathbf{m}\rangle$$
$$\mathbf{m}^{\text{est}} = \left[\mathbf{G}^T\mathbf{W}_e\mathbf{G} + \varepsilon^2\mathbf{W}_m\right]^{-1}\left(\mathbf{G}^T\mathbf{W}_e\mathbf{d}^{obs} + \varepsilon^2\mathbf{W}_m\langle\mathbf{m}\rangle\right) \tag{4.49}$$

This equation appears to be rather complicated. However, it can be vastly simplified by noting that it is equivalent to solving the $L \times M$ equation

$$\mathbf{Fm}^{est} = \mathbf{f} \quad \text{with} \quad \mathbf{F} \equiv \begin{bmatrix} \mathbf{W}_e^{\frac{1}{2}}\mathbf{G} \\ \varepsilon\mathbf{D} \end{bmatrix} \quad \text{and} \quad \mathbf{f} \equiv \begin{bmatrix} \mathbf{W}_e^{\frac{1}{2}}\mathbf{d}^{obs} \\ \varepsilon\mathbf{D}\langle\mathbf{m}\rangle \end{bmatrix} \quad \text{and} \quad \mathbf{W}_m \equiv \mathbf{D}^T\mathbf{D} \tag{4.50}$$

by simple least squares, that is, the equation $\mathbf{F}^T\mathbf{Fm}^{est} = \mathbf{F}^T\mathbf{f}$, when multiplied out, is identical to Eq. (4.50). As explained previously, the weight matrix $\mathbf{W}_e$ typically is diagonal. In that case, its square root $\mathbf{W}_e^{\frac{1}{2}}$ is also diagonal with elements that are the square roots of the corresponding elements of $\mathbf{W}_e$.

Eq. (4.50) has a very simple interpretation: its top row is the data equation $\mathbf{Gm}^{est} = \mathbf{d}^{obs}$ with both sides multiplied by the weight matrix $\mathbf{W}_e^{\frac{1}{2}}$ and its bottom row is the prior equation $\mathbf{m} = \langle\mathbf{m}\rangle$ with both sides multiplied by the weight matrix $\varepsilon\mathbf{D}$. Note that the data and prior information play completely symmetric roles in this equation.

Eq. (4.49) can be manipulated into another useful form by subtracting $[\mathbf{G}^T\mathbf{W}_e\mathbf{G} + \varepsilon^2\mathbf{W}_m]\langle\mathbf{m}\rangle$ from both sides of the equation and rearranging to obtain:

$$\Delta\mathbf{m} = \mathbf{M}\Delta\mathbf{d} \quad \text{with} \quad \mathbf{M} \equiv \left[\mathbf{G}^T\mathbf{W}_e\mathbf{G} + \varepsilon^2\mathbf{W}_m\right]^{-1}\mathbf{G}^T\mathbf{W}_e$$
$$\text{and} \quad \Delta\mathbf{m} \equiv \mathbf{m}^{\text{est}} - \langle\mathbf{m}\rangle \quad \text{and} \quad \Delta\mathbf{d} \equiv \mathbf{d}^{obs} - \mathbf{G}\langle\mathbf{m}\rangle \tag{4.51}$$

This form emphasizes that the deviation $\Delta\mathbf{m}$ of the estimated solution from the prior value is a linear function of the deviation $\Delta\mathbf{d}$ of the data from the value predicted by the prior model.

Finally, we note that an alternative form of $\mathbf{M}$ in Eq. (4.49), reminiscent of the minimum-length solution, is

$$\mathbf{M} \equiv \mathbf{W}_m^{-1}\mathbf{G}^T\left[\mathbf{GW}_m^{-1}\mathbf{G}^T + \varepsilon^2\mathbf{W}_e^{-1}\right]^{-1} \tag{4.52}$$

The equivalence of these two forms can be established by equating them and manipulating terms

$$\mathbf{W}_m^{-1}\mathbf{G}^T\left[\mathbf{GW}_m^{-1}\mathbf{G}^T + \varepsilon^2\mathbf{W}_e^{-1}\right]^{-1} \overset{?}{=} \left[\mathbf{G}^T\mathbf{W}_e\mathbf{G} + \varepsilon^2\mathbf{W}_m\right]^{-1}\mathbf{G}^T\mathbf{W}_e$$
$$\mathbf{G}^T\mathbf{W}_e\mathbf{GW}_m^{-1}\mathbf{G}^T + \varepsilon^2\mathbf{W}_m\mathbf{W}_m^{-1}\mathbf{G}^T \overset{?}{=} \mathbf{G}^T\mathbf{W}_e\mathbf{GW}_m^{-1}\mathbf{G}^T + \varepsilon^2\mathbf{G}^T\mathbf{W}_e\mathbf{W}_e^{-1} \tag{4.53}$$
$$\mathbf{G}^T\mathbf{W}_e\mathbf{GW}_m^{-1}\mathbf{G}^T + \varepsilon^2\mathbf{G}^T = \mathbf{G}^T\mathbf{W}_e\mathbf{GW}_m^{-1}\mathbf{G}^T + \varepsilon^2\mathbf{G}^T$$

In both instances, one must take care to ascertain whether the inverses actually exist. Depending on the choice of the weighting matrices, sufficient prior information may or may not have been added to the problem to damp the underdeterminacy.

4.16 Generalized least squares

The weighted damped least-squares solution easily can be adapted to the completely general prior information equation $\mathbf{Hm} = \mathbf{h}$, where the solution simplicity L is now interpreted as the L_2 error in prior information

$$L = \mathbf{l}^T\mathbf{l} \quad \text{with} \quad \mathbf{l} = \mathbf{h} - \mathbf{Hm} \tag{4.54}$$

The solution corresponding to the minimization of $\Phi = E + \varepsilon^2 L$ is called *generalized least squares* and corresponds to

$$\mathbf{Fm}^{est} = \mathbf{f} \quad \text{with} \quad \mathbf{F} \equiv \begin{bmatrix} \mathbf{W}_e^{\frac{1}{2}}\mathbf{G} \\ \varepsilon\mathbf{H} \end{bmatrix} \quad \text{and} \quad \mathbf{f} \equiv \begin{bmatrix} \mathbf{W}_e^{\frac{1}{2}}\mathbf{d}^{obs} \\ \varepsilon\mathbf{h} \end{bmatrix}$$

$$\text{with GLS equation} \quad \left[\mathbf{F}^T\mathbf{F}\right]\mathbf{m}^{est} = \mathbf{F}^T\mathbf{f} \tag{4.55}$$

$$\text{and solution} \quad \mathbf{m}^{est} = \left[\mathbf{F}^T\mathbf{F}\right]^{-1}\mathbf{F}^T\mathbf{f}$$

As an example, consider an inverse problem in which the model parameters represent the percentages of chemical elements in a rock sample. The prior information that the sum of the model parameters is 100%. This case can be implemented by choosing $\mathbf{H} = [1, 1, 1, \cdots, 1]$ and $\mathbf{h} = 100$. In typical cases, this prior information is very certain, suggesting that ε^2 should be chosen to be a large number (say 10^6).

Eq. (4.55) is extremely well suited to computations, especially if $\mathbf{F}$ is a sparse matrix.

4.17 Use of sparse matrices in MATLAB® and Python

With larger problems, the computational cost of computing the $\mathbf{F}^T\mathbf{F}$ that appears in the generalized least-squares solution can be prohibitive. Furthermore, $\mathbf{F}^T\mathbf{F}$ is rarely as sparse as $\mathbf{F}$ itself. In such cases, an iterative matrix solver, such as the *biconjugate gradient algorithm*, is preferred. This algorithm only requires products of the form $\mathbf{F}^T\mathbf{F}\mathbf{v}$, where $\mathbf{v}$ is a vector constructed by the algorithm, and this product can be performed as $\mathbf{F}^T(\mathbf{F}\mathbf{v})$ so that $\mathbf{F}^T\mathbf{F}$ is not explicitly calculated.

For example, the $N=5$, $M=5$ *antiidentity matrix*

$$\mathbf{F} = \begin{bmatrix} 0 & 0 & 0 & 0 & 1.0 \\ 0 & 0 & 0 & 1.0 & 0 \\ 0 & 0 & 1.0 & 0 & 0 \\ 0 & 1.0 & 0 & 0 & 0 \\ 1.0 & 0 & 0 & 0 & 0 \end{bmatrix} \tag{4.56}$$

is sparse, because the number $P=5$ of nonzero elements is small compared to the total number of elements $N \times M = 25$. Most scientific-computing software environments, including MATLAB® and Python, provide software support for such matrices, allowing them to defined and manipulated with the minimum of memory and computation time.

Both MATLAB® and Python provide many ways to define sparse matrices, but we present only one very simple method here, because (as we discuss further later) we have found it to work well in a wide variety of inverse theory settings. The process begins by preparing three column vectors, say $\mathbf{F}_r$, $\mathbf{F}_c$, and $\mathbf{F}_v$, each of length P, which enumerate the row index, column index, and value of each of the nonzero elements of the sparse matrix $\mathbf{F}$. The column vectors corresponding to Eq. (4.56) are

$$\mathbf{F}_r = \begin{bmatrix} 1 \\ 2 \\ 3 \\ 4 \\ 5 \end{bmatrix} \text{ and } \mathbf{F}_c = \begin{bmatrix} 5 \\ 4 \\ 3 \\ 2 \\ 1 \end{bmatrix} \text{ and } \mathbf{F}_v = \begin{bmatrix} 1.0 \\ 1.0 \\ 1.0 \\ 1.0 \\ 1.0 \end{bmatrix} \tag{4.57}$$

These arrays are then passed to an MATLAB® function or Python method, which creates the sparse version of the matrix $\mathbf{F}$. The column vectors $\mathbf{F}_r$, $\mathbf{F}_c$, and $\mathbf{F}_v$ are first initialized to zero

MATLAB®
| ```
P=0;
Pmax = 100;
Fr = zeros(Pmax,1);
Fc = zeros(Pmax,1);
Fv = zeros(Pmax,1);
```                                             [gdama04_08] |

Python
| ```
P=0;          # element counter
Pmax = 100; # maximum number of elements
Fr = np.zeros((Pmax,1),dtype=int);
Fc = np.zeros((Pmax,1),dtype=int);
Fv = np.zeros((Pmax,1));
```                                             [gdapy04_08] |

In most cases, the exact number of nonzero elements P is not initially known, but an upper bound P_{max} can be established for it. The column vectors of length P_{max} can be created and the value of P is updated as their values are set. The column vectors are then truncated to length P at the end of the process. Note that the index arrays Fr and Fc are defined as holding integer values (because array indices are integers).

The column vectors $\mathbf{F}_r$, $\mathbf{F}_c$, and $\mathbf{F}_v$ for the matrix in Eq. (4.56) can be set up by traversing the matrix, row-wise:

MATLAB®

```
% build array row-wide
for i = (1:L)      % row i
    j=N-i+1;       % column j
    P=P+1;         % increment element number
    Fr(P,1)=i;     % row index
    Fc(P,1)=j;     % column index
    Fv(P,1)=1.0;   % value
end
```
[gdama04_08]

Python

```
for i in range(L):  # row i
    j=M-i-1;         # column j
    Fr[P,0]=i;       # row index
    Fc[P,0]=j;       # column index
    Fv[P,0]=1.0;     # value
    P=P+1;           # increment element number
```
[gdapy04_08]

The sparse matrix is then built

MATLAB®

```
L=N;
Fr=Fr(1:P); Fc=Fc(1:P); Fv=Fv(1:P); % truncate
gdaFsparse = sparse(Fr,Fc,Fv,L,M); % build sparse matrix
clear Fr Fc Fv;  % delete no longer needed arrays
```
[gdama04_08]

Python

```
L=5;
M=5;
gdaFsparse=sparse.coo_matrix((Fv[0:P,0],
            (Fr[0:P,0],Fc[0:P,0])),shape=(L,M));
del Fr;   # delete no longer needed arrays
del Fc;
del Fv;
```
[gdapy04_08]

For reasons that will be explained later, the sparse matrix version of $\mathbf{F}$ is given the standard name, gdaFsparse (and not F). Note that the column vectors $\mathbf{F}_r$, $\mathbf{F}_c$, and $\mathbf{F}_v$ are deleted after the sparse matrix is created, for they are no longer needed.

Although both MATLAB® and Python provide specialty functions that create sparse matrices with simple structures, such as a one with a few nonzero diagonals, their application is limited to one-dimensional problems. The sparse matrices that underpin the multidimensional problems so common in geophysical data analysis have complicated structures for which specialty functions typically are not available. In MATLAB® (but not Python) an alternative method is to create an empty sparse matrix, say M, with the spalloc() function and then to set its values with ordinary assignments of the form M(i,j)=something. Unfortunately, the method is too slow to be useful when the number of nonzero elements is large (say, millions). For these reasons, we advocate the *row-column-value method* described before.

In some cases, the number of nonzero elements in a row of **F** is difficult to determine in advance. In this case, **F** is processed row-wise and a nonsparse version of each row is first built. After it is complete, its nonzero elements are extracted. The process is then repeated on the next row.

MATLAB®

```
for i = (1:L) % row i
    Frow = zeros(M,1); % i-th row of F
    % now set elements of Frow here
    ...
    for j = (1:M)
        if( Frow(j) ~= 0.0 )
            P=P+1; % increment element number
            Fr(P,1)=i; % row index
            Fc(P,1)=j; % column index
            Fv(P,1)=Frow(j,1); % value
        end
    end
end
```

[gdama04_09]

Python

```
for i in range(L): # row i
    Frow = np.zeros((M,1));  # one row of matrix
    % now set elements of Frow here
    ...
    for j in range(M):
        if( Frow[j,0] != 0.0 ):
            Fr[P,0]=i;            # row index
            Fc[P,0]=j;            # column index
            Fv[P,0]=Frow[j,0]; # value
            P=P+1; # increment element number
```

[gdapy04_09]

Once the sparse version of **G** is built, the equation $\mathbf{Gm}^{est} = \mathbf{d}^{obs}$ can be solved for $\mathbf{m}^{est}$ using the biconjugate gradient solver:

MATLAB®

```
# at the beginning of your code, you must put
# these two commands:
clear gdaFsparse;
global gdaFsparse;

# later in the code, you can then call the solver
tol = 1e-12; % tolerance
maxit = 3*(M+N); % maximum number of iterations allowed
FTf = gda_FTFrhs(dobs);
mest=bicg(@gda_FTFmul,FTf,tol,maxit);
```

[gdama04_09]

```
                                    Python
# define linear operator needed for biconjugate gradient solver
LO=las.LinearOperator(shape=(M,M),
                        matvec=gdaFTFmul,rmatvec=gdaFTFmul);
# solve using conjugate gradient algorithm
tol = 1e-12;      # tolerance
maxit = 3*(L+M); # maximum number of iterations allowed
FTh = gdaFsparse.T*hobs;
q=las.bicg(LO,FTh,tol=tol,maxiter= maxit);
mest = gda_cvec(q[0]);
                                                      [gdapy04_09]
```

The MATLAB® function gda_FTFrhs() and analogous Python method gda_FTFrhs() compute the product $\mathbf{F}^T\mathbf{v}$.

The biconjugate gradient algorithm involves iteratively improving a solution, with the iterations terminating when the error is less than a tolerance tol or when maxit iterations have been performed (whichever comes first). In Python, the * operator correctly multiplies spare matrices, so expressions like FTh = gdaFsparse.T*hobs work fine.

The biconjugate gradient solver calls a short function gda_FTFmul(), which performs the multiplication $\mathbf{G}^T(\mathbf{Gv})$. The MABLAB® version is in a separate file gda_FTFmul.m, which must be in the same directory as the LiveScript (as must be gda_FTFrhs.m). The Python version is defined in cell zero. In the MATLAB® script, the syntax @gda_FTFmul means "a *handle* to the function gda_FTFmul." The matrix $\mathbf{G}$ is passed to this function via a *global variable*, which is why its name is standardized to gdaFsparse.

```
                                   MATLAB®
% gda_FTFmul: used in conjunction with biconj()
% to solve FTF m = FTf
function y = gda_FTFmul(v,transp_flag)
global gdaFsparse;
temp = gdaFsparse*v;
y = gdaFsparse'*temp;
                                           return [gda_FTFmul.m]
```

```
                                    Python
# function to perform the multiplication FT (F v)
def gdaFTFmul(v):
    # this function is used by bicongugate gradient to
    # solve the generalized least squares problem Fm=F.
    # Note that "F" must be called "gdaFsparse".
    global gdaFsparse;
    s = np.shape(v);
    if(len(s)==1):
        vv = np.zeros((s[0],1));
        vv[:,0] = v;
    else:
        vv=v;
    # weirdly, "*" multiplies sparse matrices just fine
    temp = gdaFsparse*vv;
    return gdaFsparse.transpose()*temp;
                                                      [gdapy04_00]
```

4.18 Example: Using generalized least squares to fill in data gaps

As an example, suppose that $\mathbf{m}$ represents the values of a function $m(x)$ at evenly spaced xs (say with spacing Δx but that only a few of these ms have been observed). The data equation is then just $m_i = d_j$, where the indices i and j "match up" the observation with the corresponding model parameter. The ith row of the data kernel $\mathbf{G}$ is all zero, except for a single one in the jth column.

As the observations are insufficient to determine all the model parameters, we add prior information of smoothness using a second-difference matrix $\mathbf{D}^{(2)}$ to quantify roughness. Each row of $\mathbf{D}$ is mostly zeros, except for the sequence $1, -2, 1$, with the -2 centered on the model parameter whose second derivative is being computed. We can only form $M - 2$ of these rows, because computing the second derivative of m_1 or m_M would require model parameters off the ends of the vector $\mathbf{m}$. We choose to add prior information of flatness at these two points, with two rows drawn from the top and bottom of the first-difference matrix $\mathbf{D}^{(1)}$. Both $\mathbf{D}^{(1)}\langle\mathbf{m}\rangle$ and $\mathbf{D}^{(2)}\langle\mathbf{m}\rangle$ are taken to be zero, as the solution is assumed to be smooth and flat. This leads to an equation $\mathbf{Fm} = \mathbf{f}$ with

$$\mathbf{F} = \begin{bmatrix} & 1 & & & & & & & & & \\ \cdots & \cdots & \cdots & \cdots & \cdots & \cdots & \cdots & \cdots & \cdots & \cdots & \cdots \\ & & & & & & & & 1 & & \\ - & - & - & - & - & - & - & - & - & - & - \\ a & -2a & a & & & & & & & & \\ & a & -2a & a & & & & & & & \\ & & \ddots & \ddots & \ddots & & & & & & \\ & & & & & & a & -2a & a & & \\ - & - & - & - & - & - & - & - & - & - & - \\ -b & b & & & & & & & & & \\ & & & & & & & & -b & b \end{bmatrix} \quad \text{and} \quad \mathbf{f} = \begin{bmatrix} d_1 \\ \vdots \\ d_N \\ - \\ 0 \\ 0 \\ \vdots \\ 0 \\ - \\ 0 \\ 0 \end{bmatrix} \quad (4.58)$$

Here, $a \equiv \varepsilon(\Delta z)^{-2}$ and $b \equiv \varepsilon(\Delta x)^{-1}$. As with many inverse problems, the bookkeeping needed to build the matrices is the hardest part of the problem, often requiring several pages of complex code. The solution requires just a few lines of code. An exemplary problem and its solution are shown in Fig. 4.11.

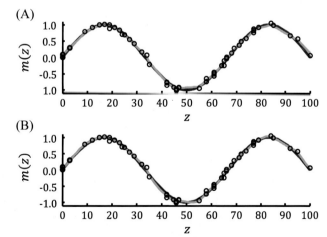

FIG. 4.11 Examples of weighted damped least squares applied to the problem of filling in data gaps, for two different types of prior information. (A) (Red curve) The true model is a sinusoid sampled with distance spacing $\Delta z = 1$. (Black circles) The data are noisy observations of the model at just a few points. (Green curve) The estimated model is reconstructed from the data using the prior information of flatness (small first derivative). (B) Same as (A) except using the prior information of smoothness (small second derivative) in the interior of the (0,100) interval and flatness at its ends. Scripts gdama04_11 and gdapy04_11.

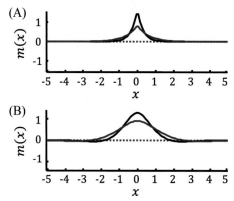

FIG. 4.12   Smoothing of a unit spike at position $x=0$. (A) First derivative smoothing for damping $\varepsilon=0.3$ (black) and 0.6 (red). (B) Second derivative smoothing for $\varepsilon=0.3$ (black) and 0.6 (red). Scripts gdama04_12 and gdapy04_12.

4.19 Choosing between prior information of flatness and smoothness

Prior information of flatness (small first derivative) and smoothness (small second derivative) are not the same, but both result in a solution that is qualitatively "smooth." The difference between them can be understood by studying the data smoothing problem, in which a smooth estimate $m(x)$ of a function is reconstructed from noisy observations $d(x)$ of the same function. This inverse problem is a special case of the one in the previous example, with no missing data, so that the data kernel is simply $\mathbf{G}=\mathbf{I}$. The prior information has the form $\mathbf{Dm}=0$, where $\mathbf{D}$ is either the first- or second-difference matrix. The generalized least-squares solution is then:

$$\mathbf{m}^{est} = \left[\mathbf{I} + \varepsilon^2\mathbf{D}^T\mathbf{D}\right]^{-1}\mathbf{d}^{obs} \tag{4.59}$$

where ε^2 quantifies the strength of the prior information, and where, for simplicity, we have assumed that $\mathbf{W}_e=\mathbf{I}$. Menke and Eilon (2015) derive an approximate solution to this equation when the data vector consists of a single spike at the origin, that is, $d(x)=\delta(x)$. The solution $m(x)$ then represents how the spike is smoothed by the prior information. In the case of first derivative smoothing, the solution is

$$m(x) = \frac{\varepsilon^{-1}}{2}\exp\left\{-\varepsilon^{-1}|x|\right\} \tag{4.60}$$

and in the case of second derivative smoothing, it is

$$m(x) = \left(\frac{c^3}{8\varepsilon^2}\right)\exp\left\{-\frac{|x|}{c}\right\}\left\{\cos\left(\frac{|x|}{c}\right) + \sin\left(\frac{|x|}{c}\right)\right\} \quad \text{with} \quad c = (2\varepsilon)^{1/2} \tag{4.61}$$

Both *smoothing operators* have the desirable properties of being symmetric about $x=0$ and having unit area and so conserving the area under $d(x)$. The first derivative version (Fig. 4.12A) is everywhere positive (a good feature, because negative values cause overshoots in the smoothing) but has a cusp at the origin (a bad feature, since it adversely affects the power spectrum of the smoothed data). The second derivative version (Fig. 4.12B) has negative side lobes (bad) but lacks cusps (good). While both versions qualitatively smooth the data, their properties are subtly different and the choice of one over the other should take into consideration these differences.

4.20 Other types of prior information

One commonly encountered type of prior information is the knowledge that some function of the model parameters equals a constant. Linear equality constraints of the form $\mathbf{Hm}=\mathbf{h}$ are particularly easy to implement. For example, one such linear constraint requires that the mean of the model parameters must equal some value h_1

$$\mathbf{Hm} = \frac{1}{M}[1\ 1\ \cdots\ 1]\begin{bmatrix} m_1 \\ m_2 \\ \vdots \\ m_M \end{bmatrix} = [h_1] = \mathbf{h} \tag{4.62}$$

Another such constraint requires that a particular model parameter m_k equals a given value

$$\mathbf{Hm} = [0 \; \cdots \; 0 \; 1 \; 0 \; \cdots \; 0] \begin{bmatrix} m_1 \\ \vdots \\ m_{k-1} \\ m_k \\ m_{k+1} \\ \vdots \\ m_M \end{bmatrix} = [h_k] = \mathbf{h} \tag{4.63}$$

One problem that frequently arises is to solve an inverse problem $\mathbf{d}^{obs} = \mathbf{Gm}$ in the least-squares sense with the prior constraint that linear relationships between the model parameters of the form $\mathbf{h} = \mathbf{Hm}$ are satisfied exactly.

One way to implement this constraint is to use generalized least squares (Eq. 4.49), and with the weighting factor ε^2 chosen to be very large, so that the prior equations are given much more weight than the data equations (Lanczos, 1961). This method is well suited for computation, but it does require the value of ε^2 to be chosen with some care—too big and the solution will suffer from numerical noise; too small and the constraint will be only very approximately satisfied.

Another method of implementing the constraints is through the use of Lagrange multipliers. One minimizes the error $E = \mathbf{e}^T\mathbf{e}$ with the constraint that $\mathbf{h} - \mathbf{Hm} = 0$ by forming the function

$$\Phi(\mathbf{m}) = \sum_{i=1}^{N} \left[d_i^{obs} - \sum_{j=1}^{M} G_{ij}m_j \right]^2 - 2\sum_{i=1}^{K} \lambda_i \left[h_i - \sum_{j=1}^{M} H_{ij}m_j \right] \tag{4.64}$$

where there are K constraints and the $-2\lambda_i$ are the Lagrange multipliers. Differentiating Φ with respect to m_q and setting the results to zero yields

$$\frac{\partial \Phi}{\partial m_q} = 0 = -\sum_{i=1}^{N} 2 \sum_{j=1}^{M} G_{ij} \frac{\partial m_j}{\partial m_q} \left[d_i^{obs} - \sum_{j=1}^{M} G_{ij}m_j \right] + 2\sum_{i=1}^{K} \lambda_i \sum_{j=1}^{M} H_{ij} \frac{\partial m_j}{\partial m_q} \tag{4.65}$$

The relationship $\partial m_j / \partial m_q = \delta_{iq}$ is now used to simplifty this expression:

$$0 = -\sum_{i=1}^{N} 2G_{iq} \left[d_i^{obs} - \sum_{k=1}^{M} G_{ik}m_k \right] + 2\sum_{i=1}^{K} \lambda_i H_{iq}$$

$$0 = -2\sum_{i=1}^{N} G_{iq}d_i^{obs} + 2\sum_{i=1}^{N} G_{iq} \sum_{k=1}^{M} G_{ik}m_k - 2\sum_{i=1}^{K} \lambda_i H_{iq} \tag{4.66}$$

$$\mathbf{G}^T\mathbf{Gm} + \mathbf{H}^T\boldsymbol{\lambda} = \mathbf{G}^T\mathbf{d}^{obs}$$

This equation must be solved simultaneously with the constraint equation $\mathbf{h} - \mathbf{Hm} = 0$ to yield the estimated solution. These equations, in matrix form, are

$$\begin{bmatrix} \mathbf{G}^T\mathbf{G} & \mathbf{H}^T \\ \mathbf{H} & \mathbf{0} \end{bmatrix} \begin{bmatrix} \mathbf{m}^{est} \\ \boldsymbol{\lambda} \end{bmatrix} = \begin{bmatrix} \mathbf{G}^T\mathbf{d}^{obs} \\ \mathbf{h} \end{bmatrix} \tag{4.67}$$

This equation can be manipulated to yield the explicit formula $\mathbf{m}^{est} = \mathbf{m}^{LS} + [\mathbf{G}^T\mathbf{G}]^{-1}\mathbf{H}^T[\mathbf{H}\mathbf{G}^T\mathbf{G}\mathbf{H}^T]^{-1}(\mathbf{h} - \mathbf{Hm}^{LS})$ with least squared solution $\mathbf{m}^{LS} \equiv [\mathbf{G}^T\mathbf{G}]^{-1}\mathbf{G}^T\mathbf{d}^{obs}$. Here, the second term on the r.h.s. is a correction to the least-squares solution whose amplitude depends on the misfit between the prior information $\mathbf{h}$ and the value $\mathbf{Hm}^{LS}$ predicted by the least-squares solution. However, direct solution of the $(M+P) \times (M+P)$ system of equations in Eq. (4.67) often is to be preferred, especially if $\mathbf{G}$ and/or $\mathbf{H}$ are sparse and sparse matrix techniques are applied.

4.21 Example: Constrained fitting of a straight line

Consider the problem of fitting the straight line $d_i^{obs} = m_1 + m_2z_i$, where one has prior information that the line must pass through the point (z', d') (Fig. 4.13). There are two model parameters: intercept m_1 and slope m_2. The $P=1$ constraint is that $d' = m_1 + m_2z'$, or

$$\mathbf{Hm} = [1 \; z'] \begin{bmatrix} m_1 \\ m_2 \end{bmatrix} = [d'] = \mathbf{h} \tag{4.68}$$

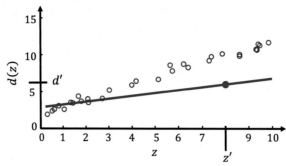

FIG. 4.13   Least-squares fitting of a straight line to (z, d) data, where the line is constrained to pass through the point $(z', d') = (8, 6)$. Scripts gdama04_13 and gdapy04_13.

Using the $\mathbf{G}^T\mathbf{G}$ and $\mathbf{G}^T\mathbf{d}$ computed in Section 4.5, the solution is

$$
\begin{bmatrix} m_1^{est} \\ m_2^{est} \\ \lambda_1 \end{bmatrix} =
\begin{bmatrix} N & \sum\limits_{i=1}^{N} z_i & 1 \\ \sum\limits_{i=1}^{N} z_i & \sum\limits_{i=1}^{N} z_i^2 & z' \\ 1 & z' & 0 \end{bmatrix}^{-1}
\begin{bmatrix} \sum\limits_{i=1}^{N} d_i^{obs} \\ \sum\limits_{i=1}^{N} z_i d_i^{obs} \\ d' \end{bmatrix}
\tag{4.69}
$$

Another kind of prior constraint is the linear inequality constraint, which we can write as $\mathbf{Hm} \geq \mathbf{h}$ (the inequality being interpreted component by component). This form can also include $\leq$ inequalities by multiplying the inequality relation by -1. This kind of prior constraint has application to problems in which the model parameters are inherently positive quantities, $m_i \geq 0$, and to other cases when the solution is known to possess some kind of bounds. One could therefore propose a new kind of constrained least-squares solution of overdetermined problems, one that minimizes the error subject to the given inequality constraints. Prior inequality constraints also have application to underdetermined problems. One can find the smallest solution that solves both $\mathbf{d}^{obs} = \mathbf{Gm}$ and $\mathbf{Hm} \geq \mathbf{h}$. These problems can be solved in a straightforward fashion, which will be discussed in Chapter 9.

4.22 Prior and posterior estimates of the variance of the data

The data invariably contain noise that causes errors in the estimates of the model parameters. We can calculate how this measurement error maps into errors in $\mathbf{m}^{est}$ by noting that all of the formulas derived before for estimates of the model parameters are linear functions of the data, of the form $\mathbf{m}^{est} = \mathbf{Md}^{obs} + \mathbf{v}$, where $\mathbf{M}$ is some matrix and $\mathbf{v}$ some vector. Therefore if we assume that the data have a distribution characterized by some covariance matrix [cov $\mathbf{d}$], the estimates of the model parameters have a distribution characterized by a covariance matrix [cov $\mathbf{m}$] = $\mathbf{M}$[cov $\mathbf{d}$]$\mathbf{M}^T$. The covariance of the solution can therefore be calculated in a straightforward fashion. If the data are uncorrelated and of equal variance σ_d^2, so that [cov $\mathbf{d}$] = $\sigma_d^2 \mathbf{I}$, then as has been shown before, very simple formulas are obtained for the covariance of some of the more simple inverse problem solutions (Eqs. 4.16 and 4.35).

An important issue is how to arrive at an estimate of the variance of the data σ_d^2 that can be used in these equations. One possibility is to base it upon knowledge about the inherent accuracy of the measurement process, in which case it is termed a *prior variance*. For instance, if lengths are being measured with a ruler with 1-mm divisions, the estimate $\sigma_d = \frac{1}{2}$ mm would be reasonable. Another possibility is to base the estimate upon the size distribution of the prediction error $\mathbf{e}$ determined by fitting a model to the data, in which case it is termed a *posterior* (or *a posteriori*) variance. A reasonable estimate is

$$
\sigma_d^2 = \frac{1}{N - M} \sum_{i=1}^{N} e_i^2
\tag{4.70}
$$

Its theoretical justification will be discussed later. This estimate essentially is the mean-squared error $N^{-1}\sum_{i=1}^{N} e_i^2$, except that N has been replaced by $N - M$ to account for the ability of a model with M model parameters to exactly fit M data. Posterior estimates are usually overestimates because inaccuracies in the model contribute to the size of the prediction error.

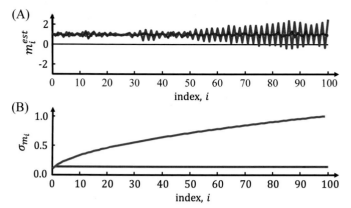

FIG. 4.14 Two hypothetical experiments to measure the weight m_i of each of 100 bricks. In experiment 1 (red), the bricks are accumulated on a scale so that an observation d_i is the sum of the weight of the first i bricks. In experiment 2 (blue), the first brick is weighed, and then subsequently, pairs or bricks (the first and the second, the second and the third, and so forth). (A) Least-squares solution for weights m_i. (B) Corresponding error σ_{m_i}. Note that the first experiment has the lower error. Scripts gdama04_14 and gdapy04_14.

The simple least-squares rule for error propagation $[\text{cov } \mathbf{m}] = \sigma_d^2[\mathbf{G}^T\mathbf{G}]^{-1}$ indicates that the model parameters can be correlated and can be of unequal variance even when the data are uncorrelated and are of equal variance. Whether observational error is attenuated or amplified by the inversion process is critically dependent upon the structure of the data kernel $\mathbf{G}$. In the problem for the mean of N data, discussed earlier, observational error is attenuated, but this desirable behavior is not common to all inverse problems (Fig. 4.14).

The reflection coefficient problem discussed in Section 4.8 has true model parameters $\mathbf{m}^{\text{true}} = [0.07, 0.10, 0.11]^T$ and a priori data variance $\sigma_d^2 = 0.0050$. After solving the inverse problem, the estimated model parameters, a posteriori data variance, and model covariance matrix are found to be:

$$\mathbf{m}^{est} = \begin{bmatrix} 0.08 \\ 0.09 \\ 0.10 \end{bmatrix} \text{ and } \sigma_d^2 = 0.0048 \text{ and } [\text{cov } \mathbf{m}] = (10^{-5}) \begin{bmatrix} 7 & -7 & -8 \\ -7 & 8 & 8 \\ -8 & 8 & 9 \end{bmatrix} \tag{4.71}$$

As will be discussed further in Chapter 5, the 95% confidence intervals of estimated model parameter m_i is $\pm 2\sigma_{m_i}$, where the σ_{m_i}'s are the square roots of the diagonal elements of $[\text{cov } \mathbf{m}]$. The solution, with its confidence intervals, is written as:

$$\begin{aligned} m_1^{est} &= 0.08 \pm 0.02 \ (95\%) \\ m_2^{est} &= 0.09 \pm 0.02 \ (95\%) \\ m_3^{est} &= 0.10 \pm 0.02 \ (95\%) \end{aligned} \tag{4.72}$$

Alternatively, these confidence limits can be illustrated as error bars, as in Fig. 4.7B.

4.23 Variance and prediction error of the least-squares solution

If the prediction error $E(\mathbf{m}) = \mathbf{e}^T\mathbf{e}$ of an overdetermined problem has a very sharp minimum in the vicinity of the estimated solution $\mathbf{m}^{est}$, we expect that the solution is well determined in the sense that it has small variance. Small errors in determining the shape of $E(\mathbf{m})$ due to random fluctuations in the data lead to only small errors in $\mathbf{m}^{est}$. Conversely, if $E(\mathbf{m})$ has a broad minimum, we expect that $\mathbf{m}^{est}$ has a large variance. As the curvature of a function is a measure of the sharpness of its minimum, we expect that the variance of the solution is related to the curvature of $E(\mathbf{m})$ at its minimum, which in turn depends on the structure of the data kernel $\mathbf{G}$ (Fig. 4.15).

The curvature of the prediction error can be measured by its second derivative, as we can see by computing how small changes in the model parameters change the prediction error. Expanding the prediction error in a Taylor series about its minimum and keeping up to second-order terms give

$$\Delta E \equiv E(\mathbf{m}) - E(\mathbf{m}^{est}) = \frac{1}{2} \sum_{i=1}^{M} \sum_{j=1}^{M} \frac{\partial^2 E}{\partial m_i \partial m_j}\bigg|_{\mathbf{m}^{est}} \Delta m_i \Delta m_j = (\Delta \mathbf{m})^T \left[\frac{1}{2}\frac{\partial^2 E}{\partial \mathbf{m}\partial \mathbf{m}}\right]_{\mathbf{m}^{est}} \Delta \mathbf{m}$$

$$\text{with } \Delta m_i \equiv m_i - m_i^{est} \text{ and } \left[\frac{\partial^2 E}{\partial \mathbf{m}\partial \mathbf{m}}\right]_{ij} \equiv \frac{\partial^2 E}{\partial m_i \partial m_j} \tag{4.73}$$

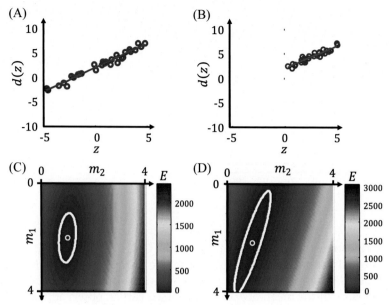

FIG. 4.15 (A) Least-squares fitting of a straight line (blue) to (z, d) data (red). (C) The best estimate of the model parameters (m_1^{est}, m_2^{est}) (white circle) occurs at the minimum of the error surface $E(m_1, m_2)$, which is a function of model parameters, intercept m_1 and slope m_2. The minimum is surrounded by a region of low error (white ellipse) that corresponds to lines that fit "almost as well" as the best estimate. The variance of the estimate is related to the size of the ellipse. In this example, the ellipse is narrowest in the m_2 direction, indicating that the slope m_2 is determined more accurately than intercept m_1. The geometry of the experiment, and not the overall level of observational error, determines the shape of the ellipse, as can be seen from the example in (B) and (D). The tilt of the ellipse indicates that the intercept and slope are negatively correlated. Scripts gdama04_15 and gdapy04_15.

Note that the first-order term is zero, since the expansion is made at a minimum. The second derivative can also be computed by starting with the expression for the error

$$E(\mathbf{m}) = \left\| \mathbf{d}^{obs} - \mathbf{Gm} \right\|_2^2 = \sum_{i=1}^{N} \left(d_i^{obs} \right)^2 - 2 \sum_{i=1}^{N} d_i^{obs} \sum_{j=1}^{M} G_{ij} m_j + \sum_{i=1}^{N} \sum_{j=1}^{M} G_{ij} m_j \sum_{k=1}^{M} G_{ik} m_k \quad (4.74)$$

and twice differentiating it

$$\frac{\partial^2 E}{\partial m_p \partial m_q} = 2 \sum_{i=1}^{N} G_{ip} G_{iq} \text{ or } \tfrac{1}{2} \frac{\partial^2 E}{\partial \mathbf{m} \partial \mathbf{m}} = \mathbf{G}^T \mathbf{G} \quad (4.75)$$

Substituting this expression into $[\text{cov } \mathbf{m}] = \sigma_d^2 [\mathbf{G}^T \mathbf{G}]^{-1}$ yields

$$[\text{cov } \mathbf{m}] = \sigma_d^2 \left[\tfrac{1}{2} \frac{\partial^2 E}{\partial \mathbf{m} \partial \mathbf{m}} \right]_{\mathbf{m}^{est}}^{-1} \quad (4.76)$$

which demonstrates that the covariance of the model parameters is related to the curvature (second derivative) of error.

The prediction error $E = \mathbf{e}^T \mathbf{e}$ is the sum of squares of Normally distributed data. Furthermore, for a well-chosen model, the predicted data will scatter about the observed data, so that the prediction error has zero mean. Consequently, E is a chi-squared distributed random variable with $N - M$ degrees of freedom, implying that it has mean $(N - M)\sigma_d^2$ and variance $2(N - M)\sigma_d^4$. (The degrees of freedom are reduced by M because the model can force M linear combinations of the e_i to zero.) We can use the standard deviation of E

$$\sigma_E = [2(N - M)]^{\frac{1}{2}} \sigma_d^2 \quad (4.77)$$

in the expression for variance as

$$[\text{cov } \mathbf{m}] = \sigma_d^2 [\mathbf{G}^T \mathbf{G}]^{-1} = \frac{\sigma_E}{[2(N - M)]^{\frac{1}{2}}} \left[\frac{1}{2} \frac{\partial^2 E}{\partial \mathbf{m} \partial \mathbf{m}} \right]_{\mathbf{m}^{est}}^{-1} \quad (4.78)$$

The covariance [cov **m**] can be interpreted as being controlled either by the variance of the data times a measure of how error in the data is mapped into error in the model parameters, or by the standard deviation of the total prediction error times a measure of the curvature of the prediction error at its minimum.

4.24 Concluding remarks

The methods of solving inverse problems that have been discussed in this chapter emphasize the data and model parameters themselves. The method of least squares estimates the model parameters with smallest prediction length. The method of minimum length estimates the simplest model parameters. The ideas of data and model parameters are very concrete and straightforward, and the methods based on them are simple and easily understood. Nevertheless, this viewpoint tends to obscure an important aspect of inverse problems: that the nature of the problems depends more on the relationship between the data and model parameters than on the data or model parameters themselves. For instance, it should be possible to tell a well-designed experiment from a poor one without knowing what the numerical values of the data or model parameters are, or even the range in which they fall. In the next chapter, we will begin to explore this kind of problem.

4.25 Problems

4.1 Show that the equations worked out for the straight-line problem in Eq. (4.7) have the solution given in Eq. (4.20).

4.2 This problem builds on Problem 2.1. Suppose that you determine the masses of 100 objects by weighing the first, then weighing the first and second together, and then weighing the rest in triplets: the first, second, and third; the second, third, and fourth; and so forth. Write a script that (A) randomly assigns masses m_i^{true} to the objects in the range of $0-1$ kg; (B) builds the appropriate data kernel **G**; (C) creates synthetic observed data $\mathbf{d}^{obs} = \mathbf{G}\mathbf{m}^{true} + \mathbf{n}$, where **n** is a vector of Normally distributed random numbers with zero mean and $\sigma_d = 0.01$ kg; (D) solves the inverse problem by simple least squares; (E) estimates the posterior variance $\sigma_{m_i}^2$ of each of the estimated model parameters; and (F) counts up the number of estimated model parameters that are within $2\sigma_{m_i}$ of their true value. (G) Make a plot of σ_{m_i} as a function of the index i of the model parameter. Does it decline, remain constant, or grow?

4.3 This problem builds on Problem 2.2. Suppose that you determine the height of 50 objects by measuring the first, and then stacking the second on top of the first and measuring their combined height, stacking the third on top of the first two and measuring their combined height, and so forth. Write a script that (A) randomly assigns heights m_i^{true} to the objects in the range of $0-1$ m; (B) builds the appropriate data kernel **G**; (C) creates synthetic observed data $\mathbf{d}^{obs} = \mathbf{G}\mathbf{m}^{true} + \mathbf{n}$, where **n** is a vector of Normally distributed random numbers with zero mean and $\sigma_d = 0.01$ m; (D) solves the inverse problem by simple least squares; (E) estimates the variance $\sigma_{m_i}^2$ of each of the estimated model parameters; and (F) counts up the number of estimated model parameters that are within $2\sigma_{m_i}$ of their true value. (G) Make a plot of σ_{m_i} as a function of the index i of the model parameter. Does it decline, remain constant, or grow?

4.4 This problem builds on Problem 2.3, which considers the cubic equation $d_i^{obs} = m_1 + m_2 z_i + m_3 z_i^2 + m_4 z_i^3$. Write a script that (A) computes a vector **z** with $N = 11$ elements equally spaced between 0 and 1; (B) randomly assigns the elements of $\mathbf{m}^{true}$ in the range of -1 to 1; (C) builds the appropriate data kernel **G**; (D) creates synthetic observed data $\mathbf{d}^{obs} = \mathbf{G}\mathbf{m}^{true} + \mathbf{n}$, where **n** is a vector of Normally distributed random numbers with zero mean and $\sigma_d = 0.05$; (E) solves the inverse problem by simple least squares; (F) calculates the predicted data $\mathbf{d}^{pre} = \mathbf{G}\mathbf{m}^{est}$; and (G) plots d_i^{obs} and d_i^{pre}. Comment upon the results.

4.5 Modify your solution of Problem 4.4 by adding the constraint that the predicted data pass through the point $(z', d') = (5, 0)$. Comment upon the results.

4.6 Modify the reflection coefficient example of Section 4.8 to include the prior information that the ratio of compressional to shear wave velocities is exactly $r = 1.76$. (A) Convert the equation $(\alpha + \Delta\alpha)/(\beta + \Delta\beta)$ to a linear relationship involving model parameters $\Delta\alpha/\alpha$ and $\Delta\beta/\beta$ using small number approximations. (B) Apply this prior information to the estimation of the model parameters by combining the data equation (Eq. 4.31) with a prior information equation of the form $\mathbf{Hm} = \mathbf{h}$. Omit calculation of confidence intervals. (B) Compare the results to the case where no prior information is used.

References

Aki, K., Richards, P.G., 2009. Quantitative Seismology, second ed. University Science Books, Sausalito, CA. 700 pp.

Franklin, J.N., 1970. Well-posed stochastic extensions of ill-posed linear problems. J. Math. Anal. Appl. 31, 682–716.

Jackson, D.D., 1972. Interpretation of inaccurate, insufficient and inconsistent data. Geophys. J. R. Astron. Soc. 28, 97–110.

Jackson, D.D., 1979. The use of a priori data to resolve non-uniqueness in linear inversion. Geophys. J. R. Astron. Soc. 57, 137–157.

Jordan, T.H., Franklin, J.N., 1971. Optimal solutions to a linear inverse problem in geophysics. Proc. Natl. Acad. Sci. U. S. A. 68, 291–293.

Lanczos, C., 1961. Linear Differential Operators. Van Nostrand-Reinhold, Princeton, NJ.

Menke, W., Eilon, Z., 2015. Relationship between data smoothing and the regularization of inverse problems. Pure Appl. Geophys. 172, 2711–2726.

Solution of the linear, Normal inverse problem, viewpoint 2: Generalized inverses

5.1 Solutions versus operators

In the previous chapter, we derived methods of solving the linear inverse problem $\mathbf{Gm} = \mathbf{d}^{obs}$ that were based on examining two properties of its solution: prediction error and solution simplicity (or length). Most of these solutions had a form that was linear in the data $\mathbf{m}^{est} = \mathbf{Md}^{obs} + \mathbf{v}$, where $\mathbf{M}$ is some matrix and $\mathbf{v}$ is some vector, both of which are independent of the data $\mathbf{d}^{obs}$. This equation indicates that the estimate of the model parameters is controlled by some matrix $\mathbf{M}$ *operating* on the data (i.e., multiplying the data). We therefore shift our emphasis from the estimates $\mathbf{m}^{est}$ to the operator matrix $\mathbf{M}$ with the expectation that by studying it we can learn more about the properties of inverse problems. As the matrix $\mathbf{M}$ solves, or "inverts," the inverse problem $\mathbf{Gm} = \mathbf{d}$, it is often called the *generalized inverse* and given the symbol $\mathbf{G}^{-g}$. The exact form of the generalized inverse depends on the problem at hand. The generalized inverse of the overdetermined least-squares problem is $\mathbf{G}^{-g} = [\mathbf{G}^T\mathbf{G}]^{-1}\mathbf{G}^T$, and for the minimum-length underdetermined solution it is $\mathbf{G}^{-g} = \mathbf{G}^T[\mathbf{GG}^T]^{-1}$.

In some ways the generalized inverse is analogous to the ordinary matrix inverse. The solution to the square (even-determined) matrix equation, $\mathbf{Ax} = \mathbf{y}$, is $\mathbf{x} = \mathbf{A}^{-1}\mathbf{y}$, and the solution to the inverse problem, $\mathbf{Gm} = \mathbf{d}^{obs}$, is $\mathbf{m}^{est} = \mathbf{G}^{-g}\mathbf{d}^{obs}$ (plus some vector, possibly). The analogy is very limited, however. The generalized inverse is not a matrix inverse in the usual sense. It is not square, and neither $\mathbf{G}^{-g}\mathbf{G}$ nor $\mathbf{GG}^{-g}$ need equal an identity matrix.

5.2 The data resolution matrix

Suppose we have found a generalized inverse that in some sense solves the inverse problem $\mathbf{Gm} = \mathbf{d}^{obs}$ yielding an estimate of the model parameters $\mathbf{m}^{est} = \mathbf{G}^{-g}\mathbf{d}^{obs}$ (for the sake of simplicity we assume that there is no additive vector). We can then retrospectively ask how well this estimate of the model parameters fits the data. By plugging our estimate into the equation $\mathbf{d}^{pre} = \mathbf{Gm}^{est}$, we conclude

$$\mathbf{d}^{pre} = \mathbf{Gm}^{est} = \mathbf{GG}^{-g}\mathbf{d}^{obs} = \mathbf{Nd}^{obs} \text{ with } \mathbf{N} \equiv \mathbf{GG}^{-g} \tag{5.1}$$

Here, the superscripts *obs* and *pre* mean observed and predicted, respectively. The $N \times N$ square matrix $\mathbf{N} \equiv \mathbf{GG}^{-g}$ is called the *data resolution matrix*. This matrix describes how well the predictions match the data. If $\mathbf{N} = \mathbf{I}$, then $\mathbf{d}^{pre} = \mathbf{d}^{obs}$ and the prediction error is zero. On the other hand, if the data resolution matrix is not an identity matrix, the prediction error is nonzero.

The data resolution matrix has a simple interpretation when the elements of the data vector $\mathbf{d}$ possess a natural ordering. Consider, for example, the problem of fitting a curve to (z_i, d_i^{obs}) points, where the data have been ordered according to the value of the auxiliary variable z_i. If $\mathbf{N}$ is not an identity matrix but is close to an identity matrix (in the sense that its largest elements are near its main diagonal), then the configuration of the matrix signifies that averages of neighboring data can be predicted, whereas individual data cannot. Consider the ith row of $\mathbf{N}$. If this row contained all zeros except for a one in the ith column, then d_i^{obs} would be predicted exactly. On the other hand, suppose that the row contained the elements

$$[0 \ \cdots \ 0 \ 0.1 \ 0.8 \ 0.1 \ 0 \ \cdots \ 0] \tag{5.2}$$

(A)

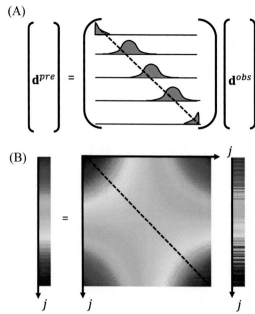

(B)

FIG. 5.1   (A) Plots of selected rows of the data resolution matrix **N** indicate how well the data can be predicted. Narrow peaks occurring near the main diagonal of the matrix (dashed line) indicate that the resolution is good. (B) Actual **N** for the case of fitting a straight line to 100 data, equally spaced along the z-axis. Large values (red colors) occur only near the ends of the main diagonal (dashed line), indicating that the resolution is poor at intermediate values of z. Scripts gdama05_01 and gdapy05_01.

where 0.8 is in the ith column. Then the ith datum is given by

$$d_i^{pre} = \sum_{j=1}^N N_{ij} d_j^{obs} = 0.1\, d_{i-1}^{obs} + 0.8\, d_i^{obs} + 0.1\, d_{i+1}^{obs} \tag{5.3}$$

The predicted value is a weighted average of three neighboring observed data. If the true data vary slowly with the auxiliary variable, then such an average might produce an estimate reasonably close to the observed value. The rows of the data resolution matrix **N** describe how well neighboring data can be independently predicted or *resolved*. If the data have a natural ordering, then a graph of the elements of the rows of **N** against column indices illuminates the sharpness of the resolution (Fig. 5.1A). If the graphs have a single sharp maximum centered about the main diagonal, then the data are well resolved. If the graphs are very broad, then the data are poorly resolved. Even in cases where there is no natural ordering of the data, the resolution matrix still shows how much weight each observation has in influencing the predicted value. There is then no special significance to whether large off-diagonal elements fall near to or far from the main diagonal.

A straight line has only two parameters and so cannot accurately predict many independent data. Consequently, the data resolution matrix for the problem of fitting a straight line to data is not diagonal (Fig. 5.1B). Its largest amplitudes are at its top-right and bottom-left corners, indicating that the points near the ends of the line are controlling the fit.

Because the diagonal elements of the data resolution matrix indicate how much weight a datum has in its own prediction, these diagonal elements are often singled out and called the importance **n** of the data (Minster et al., 1974)

$$\mathbf{n} = \text{diag}(\mathbf{N}) \tag{5.4}$$

The data resolution matrix is not a function of the data but only of the data kernel **G**, which embodies the model and experimental geometry, and any prior information applied to the problem. It can therefore be computed and studied without actually performing the experiment and can be a useful tool in experimental design.

5.3 The model resolution matrix

The data resolution matrix provides an answer to the question of whether the data can be independently predicted; that is, it characterizes whether the data can be independently predicted, or resolved. The same question can be asked about the model parameters. To explore this question, we imagine that there is a *true* but unknown set of model

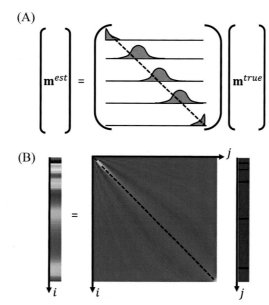

FIG. 5.2 (A) Plots of selected rows of the model resolution matrix $\mathbf{R}$ indicate how well the model parameters can be resolved. Narrow peaks occurring near the main diagonal of the matrix (dashed line) indicate that the resolution is good. (B) Actual $\mathbf{R}$ for the case where the model parameters $m_j(z_j)$ are related to the data through the kernel $G_{ij} = \exp(-c_i z_j)$, where the zs are positions and the cs are constants. Large values (red colors) occur only near the top (small z) of the main diagonal (dashed line), indicating that the resolution is poor at larger values of z. Scripts gdama05_02 and gdapy05_02.

parameters $\mathbf{m}^{true}$ that solve $\mathbf{Gm}^{true} = \mathbf{d}^{obs}$. We then inquire how closely a particular estimate of the model parameters $\mathbf{m}^{est}$ is to this true solution. Plugging the expression for the observed data $\mathbf{d}^{obs} = \mathbf{Gm}^{true}$ into the expression for the estimated model $\mathbf{m}^{est} = \mathbf{G}^{-g}\mathbf{d}^{obs}$ gives

$$\mathbf{m}^{est} = \mathbf{G}^{-g}\mathbf{d}^{obs} = \mathbf{G}^{-g}\mathbf{Gm}^{true} = \mathbf{Rm}^{true} \quad \text{with} \quad \mathbf{R} \equiv \mathbf{G}^{-g}\mathbf{G} \qquad (5.5)$$

(Wiggins, 1972). Here, $\mathbf{R}$ is the $M \times M$ *model resolution matrix*. If $\mathbf{R} = \mathbf{I}$, then each model parameter is uniquely determined. If $\mathbf{R}$ is not an identity matrix, then the estimates of the model parameters are really weighted averages of the true model parameters. If the model parameters have a natural ordering (as they would if they represented a discretized version of a continuous function), then plots of the rows of the resolution matrix can be useful in determining to what scale features in the model can actually be resolved (Fig. 5.2A). Like the data resolution matrix, the model resolution is a function of only the data kernel and the prior information added to the problem. It is therefore independent of the actual values of the data and can therefore be another important tool in experimental design.

As an example, we examine the resolution of the discrete version of the Laplace transform

$$d(c) = \int_0^\infty \exp(-cz)m(z) \, dz \quad \text{so} \quad d(c_i) \approx \sum_{j=1}^M \Delta z \, \exp(-c_i z_j)m(z_j)$$

$$\text{or } d_i = \sum_{j=1}^M G_{ij} m_j \quad \text{with} \quad d_i \equiv d(c_i) \text{ and } m_j \equiv m(z_j) \qquad (5.6)$$

$$\text{and } G_{ij} \equiv \Delta z \, \exp(-c_i z_j)$$

Here, the datum d_i is a weighted average of the model parameters m_j, with weights that decline exponentially with depth z. The decay rate of the exponential is controlled by the constant c_i, so that the smaller cs correspond to averages over a wider range of depths and the larger cs over a shallower range of depths. Not surprisingly, the shallow model parameters are better resolved (Fig. 5.2B).

5.4 The unit covariance matrix

The covariance of the model parameters depends on the covariance of the data and the way in which error is mapped from data to model parameters. This mapping is a function of only the data kernel and the generalized

FIG. 5.3 (A) The method of least squares is used to fit a straight line (red) to data (black circles) with uncorrelated error with uniform variance (vertical bars, 1σ confidence limits). As the data are not well separated in z, the variance of the slope is large, and consequently the variance of the predicted data is large as well (blue curves, 1σ confidence limits). (B) Same as (A) but with the data well separated in z. Although the variance of the data is the same as in (A), the variance of the slope, and consequently the predicted data, is much smaller. Scripts gdama05_03 and gdapy05_03.

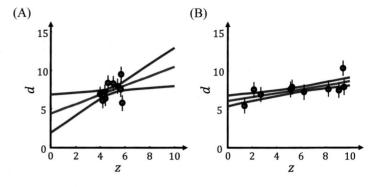

inverse, not of the data itself. A unit covariance matrix can be defined to characterize the degree of error amplification that occurs in the mapping. If the data are assumed to be uncorrelated and to have uniform variance σ_d^2, the unit covariance matrix is given by

$$[\mathrm{cov}_u\,\mathbf{m}] \equiv \sigma_d^{-2}\mathbf{G}^{-g}\left[\sigma_d^2\mathbf{I}\right]\mathbf{G}^{-gT} = \mathbf{G}^{-g}\mathbf{G}^{-gT} \tag{5.7}$$

Even if the data are correlated, one can often find some normalization of the data covariance matrix, so that one can define a unit data covariance matrix $[\mathrm{cov}_u\,\mathbf{d}]$, related to the model covariance matrix by

$$[\mathrm{cov}_u\,\mathbf{m}] = \mathbf{G}^{-g}[\mathrm{cov}_u\,\mathbf{d}]\,\mathbf{G}^{-gT} \tag{5.8}$$

The unit covariance matrix is a useful tool in experimental design because it is independent of the actual values and variances of the data themselves.

As an example, reconsider the problem of fitting a straight line to (z_i, d_i^{obs}) data. The unit covariance matrix for intercept m_1 and slope m_2 is given by

$$[\mathrm{cov}_u\,\mathbf{m}] = \frac{1}{N\sum_i z_i^2 - \left[\sum_i z_i\right]^2}\begin{bmatrix} \sum_i z_i^2 & -\sum_i z_i \\ -\sum_i z_i & N \end{bmatrix} \tag{5.9}$$

(see Eq. 4.19). Note that the estimates of intercept and slope are uncorrelated only when the data are centered about $z = 0$. The overall size of the variance is controlled by the denominator of the fraction. If all the z values are nearly equal, then the denominator of the fraction is small and the variances of the intercept and slope are large (Fig. 5.3A). On the other hand, if the z values have a large spread, the denominator is large, and the variance is small (Fig. 5.3B).

5.5 Resolution and covariance of some generalized inverses

The data and model resolution and unit covariance matrices describe many interesting properties of the solutions to inverse problems. We therefore calculate these quantities for some of the simpler generalized inverses (with $[\mathrm{cov}_u\,\mathbf{d}] = \mathbf{I}$).

Least squares

$$\mathbf{G}^{-g} = \left[\mathbf{G}^T\mathbf{G}\right]^{-1}\mathbf{G}^T$$

$$\mathbf{N} = \mathbf{G}\mathbf{G}^{-g} = \mathbf{G}\left[\mathbf{G}^T\mathbf{G}\right]^{-1}\mathbf{G}^T$$

$$\mathbf{R} = \mathbf{G}^{-g}\mathbf{G} = \left[\mathbf{G}^T\mathbf{G}\right]^{-1}\mathbf{G}^T\mathbf{G} = \mathbf{I} \tag{5.10}$$

$$[\mathrm{cov}_u\,\mathbf{m}] = \mathbf{G}^{-g}\mathbf{G}^{-gT} = \left[\mathbf{G}^T\mathbf{G}\right]^{-1}\mathbf{G}^T\mathbf{G}\left[\mathbf{G}^T\mathbf{G}\right]^{-1} = \left[\mathbf{G}^T\mathbf{G}\right]^{-1}$$

Minimum length

$$\mathbf{G}^{-g} = \mathbf{G}^T\left[\mathbf{G}\mathbf{G}^T\right]^{-1}$$

$$\mathbf{N} = \mathbf{G}\mathbf{G}^{-g} = \mathbf{G}\mathbf{G}^T\left[\mathbf{G}\mathbf{G}^T\right]^{-1} = \mathbf{I}$$

$$\mathbf{R} = \mathbf{G}^{-g}\mathbf{G} = \mathbf{G}^T\left[\mathbf{G}\mathbf{G}^T\right]^{-1}\mathbf{G} \tag{5.11}$$

$$[\mathrm{cov}_u\,\mathbf{m}] = \mathbf{G}^{-g}\mathbf{G}^{-gT} = \mathbf{G}^T\left[\mathbf{G}\mathbf{G}^T\right]^{-1}\left[\mathbf{G}\mathbf{G}^T\right]^{-1}\mathbf{G} = \mathbf{G}^T\left[\mathbf{G}\mathbf{G}^T\right]^{-2}\mathbf{G}$$

There is a great deal of symmetry between the least-squares and minimum-length solutions. Least squares solves the completely overdetermined problem and has perfect model resolution; minimum length solves the completely underdetermined problem and has perfect data resolution. As we shall see later, generalized inverses that solve the intermediate mixed-determined problems will have data and model resolution matrices that are intermediate between these two extremes.

5.6 Measures of goodness of resolution and covariance

Just as we were able to quantify the goodness of the model parameters by measuring their overall prediction error and simplicity, we shall develop techniques that quantify the goodness of data and model resolution matrices and unit covariance matrices. Because the resolution is best when the resolution matrices are identity matrices, one possible measure of resolution is based on the size, or spread, of the off-diagonal elements.

$$\text{spread}(\mathbf{N}) = \|\mathbf{N} - \mathbf{I}\|_2^2 = \sum_{i=1}^{N} \sum_{j=1}^{N} \left(N_{ij} - \delta_{ij}\right)^2$$

$$\text{spread}(\mathbf{R}) = \|\mathbf{R} - \mathbf{I}\|_2^2 = \sum_{i=1}^{M} \sum_{j=1}^{M} \left(R_{ij} - \delta_{ij}\right)^2$$

(5.12)

Here, δ_{ij} are the elements of the identity matrix $\mathbf{I}$. These measures of the goodness of the resolution spread are based on the L_2 norm of the difference between the resolution matrix and an identity matrix. They are sometimes called the *Dirichlet spread functions*. When $\mathbf{R} = \mathbf{I}$, spread($\mathbf{R}$) = 0 (and similarly for $\mathbf{N}$).

The unit standard deviation of the model parameters is a measure of the amount of error amplification mapped from data to model parameters; this quantity can be used to estimate the size of the unit covariance matrix as

$$\text{size}([\text{cov}_u \, \mathbf{m}]) = \left\| \text{diag}([\text{cov}_u \, \mathbf{m}])^{\frac{1}{2}} \right\|_2^2 = \sum_{i=1}^{M} [\text{cov}_u \, \mathbf{m}]_{ii}$$

(5.13)

where the square root is interpreted element by element. Note that this measure of covariance size does not take into account the size of the off-diagonal elements in the unit covariance matrix.

5.7 Generalized inverses with good resolution and covariance

Having found a way to measure quantitatively the goodness of the resolution and covariance of a generalized inverse, we now consider whether it is possible to use these measures as guiding principles for deriving generalized inverses. This procedure is analogous to that of Chapter 4, which involves first defining measures of solution prediction error and simplicity and then using those measures to derive the least-squares and minimum-length estimates of the model parameters.

We first consider a purely overdetermined problem of the form $\mathbf{Gm} = \mathbf{d}^{obs}$. We postulate that this problem has a solution of the form $\mathbf{m}^{est} = \mathbf{G}^{-g}\mathbf{d}^{obs}$ and try to determine $\mathbf{G}^{-g}$ by minimizing some combination of the earlier mentioned measures of goodness. As we previously noted that the overdetermined least-squares solution had perfect model resolution, we shall try to determine $\mathbf{G}^{-g}$ by minimizing only the spread of the data resolution. We begin by writing the spread as the sum of the spread of the row of $\mathbf{N}$, say J_k:

$$\text{spread}(\mathbf{N}) = \sum_{k=1}^{N} J_k \text{ with } J_k = \sum_{i=1}^{N} (N_{ki} - \delta_{ki})^2 \text{ or}$$

$$J_k = \sum_{i=1}^{N} N_{ki}^2 - 2 \sum_{i=1}^{N} N_{ki}\delta_{ki} + \sum_{i=1}^{N} \delta_{ki}^2$$

(5.14)

Because all the Js are positive and independent of one another, the minimization of spread($\mathbf{N}$) can be accomplished by individually minimizing each J_k. We insert the definition of the data resolution matrix $\mathbf{N} = \mathbf{G}\mathbf{G}^{-g}$ into the formula for J_k and minimize it with respect to the elements of the generalized inverse matrix:

$$\frac{\partial J_k}{\partial G_{qr}^{-g}} = 0 \tag{5.15}$$

We shall perform the differentiation separately for each of the three terms of J_k. The first term is given by

$$\frac{\partial}{\partial G_{qr}^{-g}} \sum_{i=1}^{N} \left(\sum_{j=1}^{M} G_{kj}G_{ji}^{-g} \right) \left(\sum_{p=1}^{M} G_{kp}G_{pi}^{-g} \right) = \frac{\partial}{\partial G_{qr}^{-g}} \sum_{i=1}^{N} \sum_{j=1}^{M} \sum_{p=1}^{M} G_{ji}^{-g} G_{pi}^{-g} G_{kj}G_{kp}$$

$$= 2\sum_{i=1}^{N} \sum_{j=1}^{M} \sum_{p=1}^{M} \delta_{jq}\delta_{ir} G_{pi}^{-g} G_{kj}G_{kp} = 2\sum_{p=1}^{M} G_{pr}^{-g} G_{kq}G_{kp} \tag{5.16}$$

The second term is given by

$$-2\frac{\partial}{\partial G_{qr}^{-g}} \sum_{i=1}^{N} \left(\sum_{j=1}^{M} G_{kj}G_{ji}^{-g} \right) \delta_{ki} = -2\sum_{i=1}^{N} \sum_{j=1}^{M} \delta_{jq}\delta_{ir} G_{kj}\delta_{ki} = -2G_{kq}\delta_{kr} \tag{5.17}$$

The third term is zero, since it is not a function of the generalized inverse. The complete equation is $\sum_{p=1}^{M} G_{pr}^{-g} G_{kq}G_{kp} = G_{kq}\delta_{kr}$. After summing over k and converting to matrix notation, we obtain

$$\sum_{k=1}^{M} \sum_{p=1}^{M} G_{kq}G_{kp}\, G_{pr}^{-g} = G_{rq} \quad \text{or} \quad \left[\mathbf{G}^T\mathbf{G}\right]\mathbf{G}^{-g} = \mathbf{G}^T \tag{5.18}$$

As $[\mathbf{G}^T\mathbf{G}]$ is a square matrix, we can premultiply by its inverse (presuming it to exist) to solve for the generalized inverse $\mathbf{G}^{-g} = [\mathbf{G}^T\mathbf{G}]^{-1}\mathbf{G}^T$, which is precisely the same as the formula for the least-squares generalized inverse. The least-squares generalized inverse can be interpreted either as the inverse that minimizes the L_2 norm of the prediction error or as the inverse that minimizes the Dirichlet spread of the data resolution.

The data can be satisfied exactly in a purely underdetermined problem. The data resolution matrix is, therefore, precisely an identity matrix and its spread is zero. We might therefore try to derive a generalized inverse for this problem by minimizing the spread of the model resolution matrix with respect to the elements of the generalized inverse. It is perhaps not particularly surprising that the generalized inverse obtained by this method is exactly the minimum-length generalized inverse $\mathbf{G}^{-g} = \mathbf{G}^T[\mathbf{G}\mathbf{G}^T]^{-1}$. The minimum-length solution can be interpreted either as the inverse that minimizes the L_2 norm of the solution length or as the inverse that minimizes the Dirichlet spread of the model resolution. This is another aspect of the symmetrical relationship between the least-squares and minimum-length solutions.

In the most general case, we could seek the $\mathbf{G}^{-g}$ that minimizes the weighted sum of Dirichlet measures of resolution spread and covariance size.

find the $\mathbf{G}^{-g}$ that minimizes

$$\Phi(\mathbf{G}^{-g}) \equiv \alpha_1 \,\text{spread}(\mathbf{N}) + \alpha_2 \,\text{spread}(\mathbf{R}) + \alpha_3 \,\text{size}([\text{cov}_u\, \mathbf{m}]) \tag{5.19}$$

where the αs are arbitrary weighting factors. This problem is done in exactly the same fashion as the one in Eqs. (5.14)–(5.18), except that there are now three times as much algebra. The result is an equation for the generalized inverse:

$$\alpha_1 \left[\mathbf{G}^T\mathbf{G}\right]\mathbf{G}^{-g} + \mathbf{G}^{-g}\left\{\alpha_2\left[\mathbf{G}\mathbf{G}^T\right] + \alpha_3[\text{cov}_u\, \mathbf{d}]\right\} = (\alpha_1 + \alpha_2)\mathbf{G}^T \tag{5.20}$$

An equation of this form is called a *Sylvester equation*. It is just a set of linear equations in the elements of the generalized inverse $\mathbf{G}^{-g}$ and so could be solved by writing the elements of $\mathbf{G}^{-g}$ as a vector in a huge $NM \times NM$ matrix equation, but it has no explicit solution in terms of algebraic functions of the component matrices. Explicit solutions can be written, however, for a variety of special choices of the weighting factors. The least-squares solution is recovered when $\alpha_1 = 1$ and $\alpha_2 = \alpha_3 = 0$. The minimum-length solution is recovered when $\alpha_2 = 1$ and $\alpha_1 = \alpha_3 = 0$. Of more interest is the case in which $\alpha_1 = 1$, $\alpha_2 = 0$, $\alpha_3 = \varepsilon^2$ (where ε^2 is a weighting constant) and $[\text{cov}_u\, \mathbf{d}] = \mathbf{I}$. The generalized inverse is then given by

$$\mathbf{G}^{-g} = \left[\mathbf{G}^T\mathbf{G} + \varepsilon^2\mathbf{I}\right]\mathbf{G}^T$$

This formula is precisely the damped least-squares inverse, which we derived in the previous chapter (see Eq. 4.38) by minimizing a combination of prediction error and solution length. The damped least-squares solution can also be interpreted as the inverse that minimizes a weighted combination of data resolution spread and covariance size.

Another interesting solution is obtained when a weighted combination of model resolution spread and covariance size is minimized. Setting $\alpha_1 = 0$, $\alpha_2 = 1$, $\alpha_3 = \varepsilon^2$, and $[cov_u\ \mathbf{d}] = \mathbf{I}$, we find

$$\mathbf{G}^{-g} = \mathbf{G}^T\left[\mathbf{G}\mathbf{G}^T + \varepsilon^2\mathbf{I}\right] \tag{5.21}$$

This solution might be termed the *damped minimum-length* generalized inverse. It will be important in the discussion later in this chapter, because it is the Dirichlet analog to the Backus-Gilbert generalized inverse that will be introduced there.

These generalized inverses often possess resolution matrices containing negative off-diagonal elements. Physically, an average makes most sense when it contained only positive weighting factors, so negative elements interfere with the interpretation of the rows of $\mathbf{R}$ as localized averages. In principle, it is possible to include nonnegativity as a constraint when choosing the generalized inverse by minimizing the spread functions. However, in practice, this constraint is never implemented because it makes the calculation of the generalized inverse very difficult. Furthermore, the more constraints that one places on $\mathbf{R}$, the less localized it tends to become.

5.8 Sidelobes and the Backus-Gilbert spread function

The Dirichlet spread function is not a particularly appropriate measure of the goodness of resolution when the data or model parameters have a natural ordering because the off-diagonal elements of the resolution matrix are all weighted equally, regardless of whether they are close or far from the main diagonal. We would much prefer that any large elements be close to the main diagonal when there is a natural ordering (Fig. 5.4) because the rows of the resolution matrix then represent *localized* averaging functions.

If one uses the Dirichlet spread function to compute a generalized inverse, it will often have *sidelobes*, that is, large amplitude regions in the resolution matrices far from the main diagonal. We would prefer to find a generalized inverse without sidelobes, even at the expense of widening the band of nonzero elements near the main diagonal, since a solution with such a resolution matrix is then interpretable as a localized average of physically adjacent model parameters.

We therefore add a weighting factor $w(k,l)$ to the measure of spread that weights the (k,l) element of $\mathbf{R}$ according to its physical distance from the diagonal element. This weighting preferentially selects resolution matrices that are "spiky" or "delta-like." If the natural ordering were a simple linear one, then the choice $w(k,l) = (k-l)^2$ would be reasonable. If the ordering is multidimensional, a more complicated weighting factor is needed. It is usually convenient to choose the spread function so that the diagonal elements have no weight, that is, $w(k,l) = 0$, and so that $w(k,l)$ is always nonnegative and symmetric in k and l. The new spread function, often called the Backus-Gilbert spread function (Backus and Gilbert, 1967, 1968), is then given by

$$\text{spread}(\mathbf{R}) = \sum_{k=1}^{M}\sum_{l=1}^{M} w(k,l)(R_{kl} - \delta_{kl})^2 = \sum_{k=1}^{M}\sum_{l=1}^{M} w(k,l)R_{kl}^2 \tag{5.22}$$

A similar expression holds for the spread of the data resolution. One can now use this measure of spread to derive new generalized inverses. Their sidelobes will be smaller than those based on the Dirichlet spread functions. On the other hand, they are sometimes worse when judged by other criteria. As we shall see, the Backus-Gilbert generalized inverse for the completely underdetermined problem does not exactly satisfy the data, even though the analogous minimum-length generalized inverse does. These facts demonstrate that there are unavoidable trade-offs inherent in finding solutions to inverse problems.

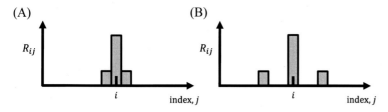

(A) (B)

R_{ij} i index, j R_{ij} i index, j

FIG. 5.4 (A, B) Resolution matrices have the same spread when measured by the Dirichlet spread function. Nevertheless, if the model parameters possess a natural ordering, then (A) is better resolved. The Backus-Gilbert spread function is designed to measure (A) as having a smaller spread than (B).

5.9 The Backus-Gilbert generalized inverse for the underdetermined problem

This problem is analogous to deriving the minimum-length solution by minimizing the Dirichlet spread of model resolution. Since it is very easy to satisfy the data when the problem is underdetermined (so that the data resolution has small spread), we shall find a generalized inverse that minimizes the spread of the model resolution alone.

We seek the generalized inverse $\mathbf{G}^{-g}$ that minimizes the Backus-Gilbert spread of model resolution. Since the diagonal elements of the model resolution matrix are given no weight, we also require that the resulting model resolution matrix satisfies the equation

$$\sum_{i=1}^{M} R_{ki} = [1]_k \tag{5.23}$$

This constraint ensures that the diagonal of the resolution matrix is finite and that the rows are unit averaging functions acting on the true model parameters. Writing the spread of one row of the resolution matrix as J_i and inserting the expression for the resolution matrix, we have

$$\Phi(\mathbf{G}^{-g}) \equiv \text{spread}(\mathbf{R}) = \sum_{k=1}^{M} J_k \text{ with}$$

$$J_k \equiv \sum_{l=1}^{M} w(k,l) R_{kl} R_{kl} = \sum_{l=1}^{M} w(k,l) \sum_{i=1}^{N} G_{ki}^{-g} G_{il} \sum_{j=1}^{N} G_{kj}^{-g} G_{jl} \tag{5.24}$$

$$= \sum_{i=1}^{N} \sum_{j=1}^{N} G_{ki}^{-g} G_{kj}^{-g} \sum_{l=1}^{M} w(k,l) G_{il} G_{jl} = \sum_{i=1}^{N} \sum_{j=1}^{N} G_{ki}^{-g} G_{kj}^{-g} S_{ij}^{(k)}$$

where the quantity $S_{ij}^{(k)}$ is defined as

$$S_{ij}^{(k)} \equiv \sum_{l=1}^{M} w(k,l) G_{il} G_{jl} \quad \text{or} \quad \mathbf{S}^{(k)} \equiv \mathbf{G} \, \text{diag}\left(\mathbf{w}^{(k)}\right) \mathbf{G}^{T} \tag{5.25}$$

where $\text{diag}(\mathbf{w}^{(k)})$ is a diagonal matrix with diagonal elements $w_l^{(k)} = w(k,l)$. The matrix $\mathbf{S}^{(k)}$ is symmetric.

The left-hand side of the constraint equation (Eq. 5.23) can also be written in terms of the generalized inverse

$$\sum_{i=1}^{M} R_{ki} = \sum_{i=1}^{M} \sum_{j=1}^{N} G_{kj}^{-g} G_{ji} = \sum_{j=1}^{N} G_{kj}^{-g} \sum_{i=1}^{M} G_{ji} = \sum_{j=1}^{N} G_{kj}^{-g} u_j \tag{5.26}$$

Here the quantity u_j is defined as

$$u_j \equiv \sum_{i=1}^{M} G_{ji} \quad \text{or} \quad \mathbf{u} \equiv \mathbf{G} \, [1, 1, \cdots, 1]^{T} \tag{5.27}$$

As $\Phi(\mathbf{G}^{-g})$ is the sum on nonnegative and independent quantities J_k, it can be minimized by individually minimizing each J_k. The quantity J_k is minimized with respect to the elements of the generalized inverse (under the given constraints) through the use of Lagrange multipliers. We first define an objective function Ψ

$$\Psi \equiv \sum_{i=1}^{N} \sum_{j=1}^{M} G_{ki}^{-g} G_{kj}^{-g} S_{ij}^{(k)} + 2\lambda \sum_{j=1}^{N} G_{kj}^{-g} u_j \tag{5.28}$$

where 2λ is the Lagrange multiplier. We then differentiate Ψ with respect to the elements of the generalized inverse and set the result equal to zero as

$$\frac{\partial}{\partial G_{kp}^{-g}} \Psi = 2 \sum_{i=1}^{N} S_{pi}^{(k)} G_{ki}^{-g} + 2\lambda u_p = 0 \tag{5.29}$$

(Note that one can solve for each row of the generalized inverse separately, so that it is only necessary to take derivatives with respect to the elements in the kth row.) This equation must be solved along with the original constraint equation. Treating the kth row of $\mathbf{G}^{-g}$ as the transform of a column vector $\mathbf{g}^{(k)}$ and the quantity $S_{ij}^{(k)}$ as the elements of matrices $\mathbf{S}^{(k)}$, we can write these equations as the matrix equation

$$\begin{bmatrix} \mathbf{S}^{(k)} & \mathbf{u} \\ \mathbf{u}^T & 0 \end{bmatrix} \begin{bmatrix} \mathbf{g}^{(k)} \\ \lambda \end{bmatrix} = \begin{bmatrix} \mathbf{0} \\ 1 \end{bmatrix} \tag{5.30}$$

This is a square $(N \times 1) \times (N \times 1)$ system of linear equations that must be solved for the N elements of the kth row of the generalized inverse and for the one Lagrange multiplier λ.

The matrix equation can be solved explicitly using a variant of the *bordering method* of linear algebra, which is used to construct the inverse of a matrix by partitioning it into submatrices with simple properties. Suppose that the inverse of the symmetric matrix in Eq. (5.30) exists and that we partition it into an $N \times N$ symmetric square matrix $\mathbf{A}$, vector $\mathbf{b}$, and scalar c. By assumption, premultiplication by the inverse yields the identity matrix

$$\begin{bmatrix} \mathbf{A} & \mathbf{b} \\ \mathbf{b}^T & c \end{bmatrix} \begin{bmatrix} \mathbf{S}^{(k)} & \mathbf{u} \\ \mathbf{u}^T & 0 \end{bmatrix} = \begin{bmatrix} \mathbf{I} & \mathbf{0} \\ \mathbf{0} & 1 \end{bmatrix} = \begin{bmatrix} \mathbf{A}\mathbf{S}^{(k)} + \mathbf{b}\mathbf{u}^T & \mathbf{A}\mathbf{u} \\ \mathbf{b}^T\mathbf{S}^{(k)} + c\mathbf{u}^T & \mathbf{b}^T\mathbf{u} \end{bmatrix} \tag{5.31}$$

The unknown submatrices $\mathbf{A}$, $\mathbf{b}$, c can now be determined by equating the submatrices

$$\mathbf{A}\mathbf{S}^{(k)} + \mathbf{b}\mathbf{u}^T = \mathbf{I} \text{ so that } \mathbf{A} = \left[\mathbf{S}^{(k)}\right]^{-1} - \mathbf{b}\mathbf{u}^T\left[\mathbf{S}^{(k)}\right]^{-1}$$

$$\mathbf{A}\mathbf{u} = \mathbf{0} \text{ so that } \mathbf{0} = \left[\mathbf{S}^{(k)}\right]^{-1}\mathbf{u} - \mathbf{b}\mathbf{u}^T\left[\mathbf{S}^{(k)}\right]^{-1}\mathbf{u} \text{ and } \mathbf{b} = \frac{\left[\mathbf{S}^{(k)}\right]^{-1}\mathbf{u}}{\mathbf{u}^T\left[\mathbf{S}^{(k)}\right]^{-1}\mathbf{u}} \tag{5.32}$$

$$\mathbf{b}^T\mathbf{S}^{(k)} + c\mathbf{u}^T = 0 \text{ so that } \frac{\mathbf{u}^T}{\mathbf{u}^T\left[\mathbf{S}^{(k)}\right]^{-1}\mathbf{u}} + c\mathbf{u}^T = 0 \text{ so that } c = \frac{-1}{\mathbf{u}^T\left[\mathbf{S}^{(k)}\right]^{-1}\mathbf{u}}$$

See scripts gdama05_04 and gdapy05_04 for an example. Multiplying Eq. (5.31) by the inverse matrix yields

$$\begin{bmatrix} \mathbf{g}^{(k)} \\ \lambda \end{bmatrix} = \begin{bmatrix} \mathbf{A} & \mathbf{b} \\ \mathbf{b}^T & c \end{bmatrix} \begin{bmatrix} \mathbf{0} \\ 1 \end{bmatrix} = \begin{bmatrix} \mathbf{b} \\ c \end{bmatrix} \tag{5.33}$$

The generalized inverse, written with summations, is

$$G_{lk}^{-g} = \frac{\sum_{i=1}^{N} u_i \left[\mathbf{S}^{(k)}\right]_{il}^{-1}}{\sum_{i=1}^{N}\sum_{j=1}^{N} u_i \left[\mathbf{S}^{(k)}\right]_{ij}^{-1} u_j} \tag{5.34}$$

This generalized inverse is the Backus-Gilbert analog to the minimum-length solution.

As an example, we compare the Dirichlet and Backus-Gilbert solutions for the Laplace transform problem discussed in Section 5.3 (Fig. 5.5). The Backus-Gilbert solution is the smoother of the two and has a corresponding model resolution matrix that consists of a single band along the main diagonal. The Dirichlet solution has more details but also more artifacts (such as negative values at $z=3$). They are associated with the large-amplitude sidelobes in the corresponding model resolution matrix.

5.10 Including the covariance size

The measure of goodness that was used to determine the Backus-Gilbert inverse can be modified to include a measure of the covariance size of the model parameters (Backus and Gilbert, 1970). We shall use the same measure as we did when considering the Dirichlet spread functions, so that goodness is measured by

FIG. 5.5 Comparison of the Backus-Gilbert and Dirichlet solutions of the inverse problem described in Fig. 5.2. (A) The true model (red) contains a series of sharp spikes. The estimated model (blue) using the Backus-Gilbert spread function is much smoother, with the width of the smoothing increasing with position z. (B) Corresponding model resolution matrix $\mathbf{R}$. (C, D) Same, but for a Dirichlet spread function. Note that the Backus-Gilbert resolution matrix has the lower intensity sidelobes but a wider central band. Scripts `gdama05_02` and `05` and `gdapy05_02` and `05`.

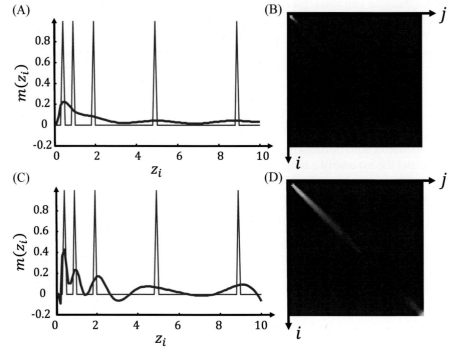

$$\Phi = \alpha \, \text{spread}(\mathbf{R}) + (1 - \alpha)\text{size}([\text{cov}_u \, \mathbf{m}])$$

$$= \alpha \sum_{k=1}^{M} \sum_{l=1}^{M} w(k,l) R_{kl}^2 + (1 - \alpha) \sum_{k=1}^{M} [\text{cov}_u \, \mathbf{m}]_{kk} \tag{5.35}$$

where $0 \leq \alpha \leq 1$ is a weighting factor that determines the relative contribution of model resolution and covariance to the measure of the goodness of the generalized inverse. As with the previous derivation, Φ is the sum of positive and independent contribution J_k' from each row, which can be separately minimized. The J_k' are defined as

$$\Phi \equiv \sum_{k=1}^{M} J_k' \text{ with}$$

$$J_k' \equiv \alpha \sum_{l=1}^{M} w(k,l) R_{kl}^2 + (1 - \alpha)[\text{cov}_u \, \mathbf{m}]_{kk} =$$

$$\alpha \sum_{i=1}^{N} \sum_{j=1}^{N} G_{ki}^{-g} G_{kj}^{-g} S_{ij}^{(k)} + (1 - \alpha) \sum_{i=1}^{N} \sum_{j=1}^{N} G_{ki}^{-g} G_{kj}^{-g} [\text{cov}_u \, \mathbf{d}]_{ij} \tag{5.36}$$

$$= \sum_{i=1}^{N} \sum_{j=1}^{N} G_{ki}^{-g} G_{kj}^{-g} S'^{(k)}_{ij}$$

where the quantity $S'^{(k)}_{ij}$ is defined by

$$S'^{(k)}_{ij} \equiv \alpha S_{ij}^{(k)} + (1 - \alpha)[\text{cov}_u \, \mathbf{d}]_{ij} \tag{5.37}$$

As the function J_k' has exactly the same form as J_k had in the previous section, the generalized inverse is just the previous result with $S_{ij}^{(k)}$ replaced by $S'^{(k)}_{ij}$

$$G_{lk}^{-g} = \frac{\sum_{i=1}^{N} u_i \left[\mathbf{S}'^{(k)}\right]_{il}^{-1}}{\sum_{i=1}^{N} \sum_{j=1}^{N} u_i \left[\mathbf{S}'^{(k)}\right]_{ij}^{-1} u_j} \tag{5.38}$$

This generalized inverse is the Backus-Gilbert analog to the damped minimum-length solution. The one-dimensional Backus-Gilbert generalized inverse $\mathbf{G}^{-g}$ (i.e., for weights, $w(i,j) = (i-j)^2$) is coded as

MATLAB®
| ```
GMG = zeros(M,N);
u = G*ones(M,1);
for k = (1:M)
    S = G * diag(([1:M]-k).^2) * G';
    Sp = alpha*S + (1-alpha)*eye(N,N);
    uSpinv = u'/Sp;
    GMG(k,:) = uSpinv / (uSpinv*u); % generalized inverse
end
                                              [gdama05_06]
``` |

Python
| ```
GMG = np.zeros((M,N)); # generalized inverse
u = np.matmul(G,np.ones((M,1)));
for k in range(M):
    S = np.matmul(np.matmul(G,
        np.diag(np.power(np.arange(M)-k,2))),G.T);
    Sp = alpha*S + (1.0-alpha)*np.identity(N);
    uSpinv = la.solve(Sp,u).T;
    GMG[k:k+1,0:N] = uSpinv / np.matmul(uSpinv,u);
                                              [gdapy05_06]
``` |

Note that a small amount $\varepsilon\mathbf{I}$ has been added to the matrix $\mathbf{S}$ to improve numerical stability.

In two-dimensional problems, a weight function that grows quadratically with physical distance from the target point, say (x_k, y_k), is a reasonable choice, that is, $w_i^{(k)} \equiv w(i,k) = (x_i - x_k)^2 + (y_i - y_k)^2$, where (x_i, y_i) is the position of model parameter m_i. The following code computes the weight vector $\mathbf{w}^{(k)}$ and the matrix $\mathbf{S}^{(k)}$ for a given k:

MATLAB®
| ```
wk=(xm-xm(k)).^2+(ym-ym(k)).^2; % weights
S = G*diag(wk)*G';
                                              [gdama05_10]
``` |

Python
| ```
wk=np.power(xm-xm[k,0],2)+np.power(ym-ym[k,0],2); # weights
S = np.matmul(G,np.matmul(np.diag(wk.ravel()),G.T));
                                              [gdapy05_10]
``` |

Here the xm and ym are length-M column vectors of the (x,y) coordinates of the model parameters.

5.11 The trade-off of resolution and variance

Suppose that one is attempting to determine a set of model parameters that represents a discretized version of a continuous function, such as X-ray opacity in the medical tomography problem (Fig. 5.6). If the discretization is made very fine, then the X-rays will not sample every box; the problem will be underdetermined. If we try to determine the opacity of each box individually, then estimates of opacity will tend to have rather large variance. Few boxes will have several X-rays passing through them, so that little averaging out of the errors will take place. On the other hand, the boxes are very small—and very small features can be detected (the resolution is very good). The large variance can be reduced by increasing the box size (or alternatively, averaging several neighboring boxes). Each of these larger regions will then be crossed by several X-rays, and noise will tend to be averaged out. But because the regions are now larger, small features can no longer be detected and the resolution of the X-ray opacity has become poorer.

This scenario illustrates an important trade-off between model resolution spread and variance size. One can be decreased only at the expense of increasing the other. We can study this trade-off by choosing a generalized inverse that minimizes a weighted sum of resolution spread and covariance size

$$\Phi \equiv \alpha \, \text{spread}(\mathbf{R}) + (1 - \alpha)\text{size}([\text{cov}_u \, \mathbf{m}]) \tag{5.39}$$

If the weighting parameter α is set near 1, then the model resolution matrix of the generalized inverse will have small spread, but the model parameters will have large variance. If α is set close to 0, then the model parameters will have a relatively small variance, but the resolution will have a large spread. A trade-off curve can be defined by varying α on the interval $(0, 1)$ (Fig. 5.7). Such curves can be helpful in choosing a generalized inverse that has an optimum trade-off in model resolution and variance (judged by criteria appropriate to the problem at hand).

FIG. 5.6 Hypothetical tomography experiment with (A) large voxels and (B) small voxels. The small voxel case not only has better spatial resolution but also higher variance, as fewer rays pass through each voxel, leaving less opportunity for measurement error to average out. Scripts gdama05_07 and gdapy05_07.

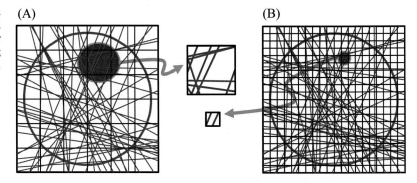

FIG. 5.7 Trade-off curves of resolution and variance for the inverse problem shown in Fig. 5.2. (A) Backus-Gilbert solution, (B) damped minimum-length solution. The larger the parameter α, the more weight resolution is given (relative to variance) when forming the generalized inverse. The details of the trade-off curve depend upon the parameterization. The resolution can be no better than the smallest element in the parameterization and no worse than the sum of all the elements. Scripts gdama05_06 and gdapy05_06.

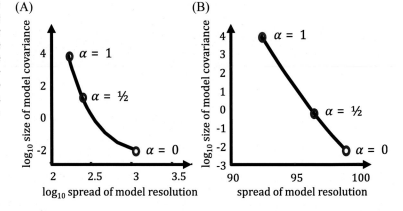

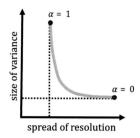

FIG. 5.8 Trade-off curve of resolution and variance has two asymptotes in the case when the model parameter is a continuous function.

Trade-off curves play an important role in continuous inverse theory, where the discretization is (so to speak) infinitely fine, and all problems are underdetermined. It is known that in this continuous limit, the curves are monotonic and possess asymptotes in resolution and variance (Fig. 5.8). The process of approximating a continuous function by a finite set of discrete parameters somewhat complicates this picture. The resolution and variance, and indeed the solution itself, are dependent on the parameterization, so it is difficult to make any definitive statement regarding the properties of the trade-off curves. Nevertheless, if the discretization is sufficiently fine, the discrete trade-off curves are usually close to ones obtained with the use of continuous inverse theory. Therefore discretizations should always be made as fine as computational considerations permit.

5.12 Reorganizing images and 3D models into vectors

Here we take a brief digression from inverse theory and turn our attention to a practical matter that arises when the model parameters represent pixels in a two-dimensional grid or voxels in a three-dimensional grid. The grid of model parameters must be reorganized into a column vector in order to apply inverse theory formulas. Then, after the solution has been found, the column vector will then be reorganized back into an image.

Suppose, for instance, that the model parameter represents slowness pixels in a two-dimensional $L_x \times L_y$ tomography problem. Most intuitively, these slownesses are represented as a matrix. For instance, S_{ij}^{true} represents the true slowness, where (i,j) index two-dimensional position (x,y). However, in order to apply the formulas, S_{ij}^{true} must be reorganized (or *unfolded*) into a length $M = L_x L_y$ column vector m_k^{true} with a single index k. Later, quantities such as the estimated model parameters m_k^{est} must be *refolded* into a matrix S_{ij}^{est}. In the $L_x = L_y = 2$, $M = 4$ case

$$\mathbf{S} = \begin{bmatrix} S_{11} & S_{12} \\ S_{21} & S_{22} \end{bmatrix} \text{ and } \mathbf{m} = \begin{bmatrix} S_{11} \\ S_{12} \\ S_{21} \\ S_{22} \end{bmatrix} \tag{5.40}$$

Many unfolding/refolding strategies are available. One could use formulas, such as

$$i = \text{floor}\left(\frac{k-1}{L_y}\right) + 1 \text{ and } j = k - (i-1)L_y \text{ and } k = (i-1)L_y + j \tag{5.41}$$

Here, floor(x) is the integer part of x. However, in this book we mostly use a very simple alternative approach based on *index tables*. The idea is to create two $M \times 1$ tables, `iofk` and `jofk`, that translate an index k into two indices (i,j) and a $L_x \times L_y$ table, `kofij`, that translates (i,j) into k. These tables are created using for-loops to traverse all the elements of an $L_x \times L_y$ matrix:

MATLAB®
| ```
iofk=zeros(M,1); % backward table, i(k)
iofk=zeros(M,1); % backward table, j(k)
kofij=zeros(Lx,Ly); % forward table k(i,j)
k=0; % counter
for i=(1:Lx)
for j=(1:Ly)
    k=k+1;
    iofk(k)=i;
    jofk(k)=j;
    kofij(i,j)=k;
end
end
``` |
| [gdama05_08] |

Python
| ```
iofk=np.zeros((M,1),dtype=int); # backward table, i(k)
jofk=np.zeros((M,1),dtype=int); # backward table, j(k)
kofij=np.zeros((Lx,Ly),dtype=int); # forward table, k(i,j)

k=0; # counter
for i in range(Lx):
    for j in range(Ly):
        iofk[k,0]=i;
        jofk[k,0]=j;
        kofij[i,j]=k;
        k=k+1;
``` |
| [gdapy05_08] |

In this tomography problem, the $M \times M$ resolution matrix $\mathbf{R}$ actually is a sequence of $L_x \times L_y$ images. In order to display resolution in a meaningful way, one must select just one of its rows (or columns), refold it into an $L_x \times L_y$ image, and display that image. The following code illustrated the creation of the image, here called RicolB from column icolB of the resolution matrix Rres

MATLAB®
| ```
RicolB=zeros(Lx,Ly);
for k=(1:M)
    RicolB(iofk(k),jofk(k))=Rres(k,icolB);
end
``` |
| [gdama05_08] |

Python
| ```
RicolB=np.zeros((Lx,Ly));
for k in range(M):
    RicolB[iofk[k,0],jofk[k,0]]=Rres[k,icolB];
``` |
| [gdapy05_08] |

5.13 Checkerboard tests

In general, the model resolution matrix $\mathbf{R}$ is not symmetric, so its rows are numerically different than its columns; furthermore, their interpretation is different, too. The kth row of the resolution matrix (let's call it the row vector $\mathbf{r}^{(k)}$) relates the kth estimated model parameter to the true model parameters, as $m_k^{est} = \sum_{j=1}^{M} R_{kj} m_j^{true} = \mathbf{r}^{(k)} \mathbf{m}$.

The kth estimated model parameter is a weighted average of the true model parameters, with $\mathbf{r}^{(k)}$ giving the weights. (However, the elements of $\mathbf{r}^{(k)}$ do not necessarily sum to unity, as the weights of a true weighted average should.) The kth column of the resolution matrix (let's call it the column vector $\mathbf{c}^{(k)}$) specifies how each of the estimated model parameters is influenced by the kth true model parameter. This can be seen by setting $\mathbf{m}^{true} = \mathbf{s}^{(k)}$ with $s_i^{(k)} = \delta_{ik}$; that is, all the elements of $\mathbf{s}^{(k)}$ are zero except the kth, which is unity. Denoting the set of estimated model parameters associated with $\mathbf{s}^{(k)}$ as $\mathbf{m}^{(k)} \equiv \mathbf{Rs}^{(k)}$, we have:

$$\mathbf{m}^{(k)} = \mathbf{Rs}^{(k)} = \mathbf{c}^{(k)} \ \text{ or } \ m_i^{(k)} = \sum_{j=1}^{M} R_{ij}\delta_{jk} = R_{ik} = c_i^{(k)} \tag{5.42}$$

Thus the kth column of the resolution matrix quantifies how a single true model parameter spreads out into many estimated model parameters. It is sometimes called a *point-spread function*. Substituting the equation $\mathbf{R} = \mathbf{G}^{-g}\mathbf{G}$ into Eq. (5.42) yields

$$\mathbf{c}^{(k)} = \mathbf{Rs}^{(k)} = \mathbf{G}^{-g}\mathbf{Gs}^{(k)} = \mathbf{G}^{-g}\mathbf{d}^{(k)} \ \text{ with } \ \mathbf{d}^{(k)} \equiv \mathbf{Gs}^{(k)} \tag{5.43}$$

Thus the kth column of the model resolution matrix solves the inverse problem for synthetic data $\mathbf{d}^{(k)}$ corresponding to a specific model parameter vector $\mathbf{s}^{(k)}$, one that is zero except for its kth element, which is unity (i.e., a unit spike at row k). This suggests a procedure for calculating the resolution: construct the desired $\mathbf{s}^{(k)}$, solve the forward problem to generate $\mathbf{d}^{(k)}$, solve the inverse problem, and then interpret the result as the kth column of the resolution matrix (Fig. 5.5A and B). The great advantage of this technique is that $\mathbf{R}$ in its entirety need not be constructed. Furthermore, the technique will work when the inverse problem is solved by an iterative method, such as the biconjugate gradient method, that does not explicitly construct $\mathbf{G}^{-g}$.

If the resolution of a problem is sufficiently good that the pattern for two well-separated model parameters does not overlap, or overlaps only minimally, then the calculation of two columns of the resolution matrix can be combined into one. One merely solves the inverse problem for synthetic data corresponding to a model parameter vector containing two unit spikes. Nor need one stop with two; a model parameter vector corresponding to a grid of spikes (i.e., a checkerboard) allows the resolution to be assessed throughout the model volume (Fig. 5.9C and D). If the problem has perfect resolution, this checkerboard pattern will be perfectly reproduced. If not, portions of the model volume with poor resolution will contain fuzzy spikes. The main limitation of this technique is that it makes the detection of unlocalized sidelobes very difficult, as an unlocalized sidelobe associated with a particular spike will tend to be confused with a localized sidelobe of another spike.

Rows of the resolution matrix and the covariance matrix also can be calculated individually (see Menke, 2014).

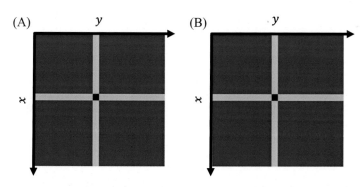

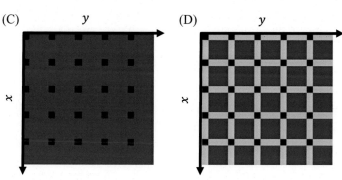

FIG. 5.9 Resolution of an acoustic tomography problem solved with the minimum-length method. The physical model space is a 20×20 grid of pixels on an (x, y) grid. Data are measured only along rows and columns, as in Fig. 3.2. (Top row) One column of the resolution matrix, for a model parameter near the center of the (x, y) grid, calculated using two methods, (A) by computing the complete model resolution matrix $\mathbf{R}$ and extracting one column and (B) by calculating the column separately. (Bottom row) Checkerboard resolution test showing (C) true checkerboard and (D) reconstructed checkerboard. Scripts gdama05_08 and 09 and gdapy05_08 and 09.

5.14 Resolution analysis without a data kernel

Suppose that the solution $\mathbf{m}^{est}$ to an inverse problem and its posterior covariance [cov $\mathbf{m}$] are known, but the data kernel $\mathbf{G}$ that led to them are not. As we will see in Chapter 12, such cases can arise when ensembles are considered the solution of an inverse problem. One can approximate $\mathbf{m}^{est}$ as the sample mean and [cov $\mathbf{m}$] as the sample covariance of the ensemble, but the data kernel $\mathbf{G}$ cannot be reconstructed. Nevertheless, one would like to say something about the resolution of the problem, and in particular, whether a localized average $\mathbf{m}'$ of the model parameters can be found whose variance is significantly better than that of $\mathbf{m}^{est}$.

One way to proceed is to consider the model parameters as data for a new inverse problem in which the data kernel is the identity matrix; that is, interpret the equation $\mathbf{m} = \mathbf{m}^{est}$ (with posterior covariance [cov $\mathbf{m}$]) as an inverse problem $\mathbf{m} = \mathbf{I}\mathbf{m}^{est}$ (with covariance [cov $\mathbf{m}$]) (Menke and Blatter, 2019). Here, $\mathbf{m}^{est}$ plays the role of the *data* and [cov $\mathbf{m}$] the role of the *data covariance*.

For the Dirichlet spread function, we set $\alpha_1 = 0$, $\alpha_2 = 1$, $\alpha_3 = \sigma_m^2 \varepsilon^2$, $\mathbf{G} = \mathbf{I}$, and $[\text{cov}_u\, \mathbf{d}] = \sigma_m^{-2}[\text{cov}\, \mathbf{m}]$ in Eq. (5.39), leading to the solution

$$\mathbf{m}' = \mathbf{G}'^{-g}\mathbf{m}^{est} \text{ with } \mathbf{G}'^{-g} = \left\{\mathbf{I} + \varepsilon^2[\text{cov}\, \mathbf{m}]\right\}^{-1}$$

$$\mathbf{R}' = \mathbf{G}'^{-g}\mathbf{G} = \mathbf{G}'^{-g} \text{ and } [\text{cov}_u\, \mathbf{m}'] = \mathbf{G}'^{-g}[\text{cov}_u\, \mathbf{m}]\mathbf{G}'^{-g} \quad (5.44)$$

Here, we have defined α_3 so that the scaling factors of σ_m^{-2} and σ_m^2 cancel. The new solution (or localized average) $\mathbf{m}'$ is different than $\mathbf{m}^{est}$ and is a function of the weight parameter ε^2. Points on the trade-off curve of resolution $\mathbf{R}'$ and variance [var$_u\mathbf{m}'$] can be explored by varying ε^2, and one chosen that represents a better trade-off of resolution and variance than is implicit in the original solution $\mathbf{m}^{est}$. The procedure for the Backus-Gilbert spread function is analogous; one merely substitutes the Backus-Gilbert generalized inverse.

Another way to analyze this problem is to write the covariance in terms of its eigenvalues and eigenvectors, $[\text{cov}_u\, \mathbf{m}] = \mathbf{V}\mathbf{\Lambda}\mathbf{V}^T$ (see Eqs. 1A.23 or 1B.23), where the eigenvalues λ_i in the main diagonal of $\mathbf{\Lambda}$ are sorted from smallest to largest. Now consider the new variable $\tilde{\mathbf{m}} = \mathbf{V}^T\mathbf{m}$. By the usual rules of error propagation, it has covariance $[\text{cov}_u\, \tilde{\mathbf{m}}] = \mathbf{V}^T\mathbf{V}\mathbf{\Lambda}\mathbf{V}^T\mathbf{V} = \mathbf{\Lambda}$; that is, the elements of $\tilde{\mathbf{m}}$ are uncorrelated, with variances equal to the eigenvalues λ_i, which increased with their position i in the vector. The challenge is to build a generalized inverse that utilizes mainly the low-variance $\tilde{m}$s, so that the resulting localized average has low variance. This is exactly what the Dirichlet generalized inverse $\mathbf{G}'^{-g}$ is doing, as can be seen by inserting the eigenvalue expansion into the formula for the generalized inverse

$$\mathbf{G}'^{-g} = \left\{\mathbf{I} + \varepsilon^2[\text{cov}_u\, \mathbf{m}]\right\}^{-1} = \left\{\mathbf{V}\mathbf{V}^T + \varepsilon^2\mathbf{V}\mathbf{\Lambda}\mathbf{V}^T\right\}^{-1} = \mathbf{V}\left\{\mathbf{I} + \varepsilon^2\mathbf{\Lambda}\right\}^{-1}\mathbf{V}^T$$

$$= \mathbf{V}\begin{bmatrix} \left(1 + \varepsilon^2\lambda_1\right)^{-1} & & \\ & \ddots & \\ & & \left(1 + \varepsilon^2\lambda_M\right)^{-1} \end{bmatrix}\mathbf{V}^T \quad (5.45)$$

Here, we have used the rule $\mathbf{V}\mathbf{V}^T = \mathbf{V}^T\mathbf{V} = \mathbf{I}$. For small eigenvalues, the factor $(1 + \varepsilon^2\lambda_i)^{-1} \approx 1$, but declines to zero as the size of the eigenvalue increases. Consequently, the generalized inverse tends to retain the effect of the smaller eigenvalues but suppress the effect of the larger ones.

5.15 Problems

5.1 Consider an underdetermined problem in which each datum is the sum of three neighboring model parameters, that is, $d_i^{obs} = m_{i-1} + m_i + m_{i+1}$ for $2 \le i \le (M-1)$ and $M = 100$. Compute and plot both the Dirichlet and Backus-Gilbert model resolution matrices. Use the standard Backus-Gilbert weight function $w(i, j) = (i - j)^2$. Interpret the results.

5.2 This problem builds upon Problem 5.1. How does the Backus-Gilbert result change if you use the weight function $w(i, j) = |i - j|^{1/2}$, which gives less weight to distant sidelobes?

5.3 This problem is especially difficult. Consider a two-dimensional acoustic tomography problem like the one discussed in Section 2.7, consisting of a 20×20 rectangular array of pixels, with observations only along rows and columns. (A) Design an appropriate Backus-Gilbert weight function that quantifies the spread of resolution. (B) Write a script that calculates the model resolution matrix $\mathbf{R}$. (C) Plot a few representative rows of $\mathbf{R}$, but where each row is reorganized into a two-dimensional image, using the same scheme that was applied to the model parameters. Interpret the results. (Hint: You will need to switch back and forth between a 20×20 rectangular array of model parameters and a length, $M = 400$, column vector of model parameters, as in Section 5.12).

References

Backus, G.E., Gilbert, J.F., 1967. Numerical application of a formalism for geophysical inverse problems. Geophys. J. R. Astron. Soc. 13, 247–276.

Backus, G.E., Gilbert, J.F., 1968. The resolving power of gross earth data. Geophys. J. R. Astron. Soc. 16, 169–205.

Backus, G.E., Gilbert, J.F., 1970. Uniqueness in the inversion of gross Earth data. Philos. Trans. R. Soc. Lond. A 266, 123–192.

Menke, W., 2014. Review of the generalized least squares method. Surv. Geophys. 36, 1–25.

Menke, W., Blatter, D., 2019. Trade-off of resolution and variance computed from ensembles of solutions, with application to Markov Chain Monte Carlo methods. Geophys. J. Int. 218, 1522–1536.

Minster, J.F., Jordan, T.J., Molnar, P., Haines, E., 1974. Numerical modelling of instantaneous plate tectonics. Geophys. J. R. Astron. Soc. 36, 541–576.

Wiggins, R.A., 1972. The general linear inverse problem: implication of surface waves and free oscillations for Earth structure. Rev. Geophys. Space Phys. 10, 251–285.

Solution of the linear, Normal inverse problem, viewpoint 3: Maximum likelihood methods

6.1 The mean of a group of measurements

Suppose that an experiment is performed N times and that each time a single datum d_i is collected. Suppose further that these data are all noisy measurements of the same model parameter m_1. In the view of probability theory, N realizations of random variables, all of which have the same probability density function, have been measured. If these random variables are Normal, their joint probability density function can be characterized in terms of a variance σ_d^2 and a mean m_1 (see Section 2.4) as

$$p(\mathbf{d}) = \sigma_d^{-N}(2\pi)^{-N/2} \exp\left\{ -\tfrac{1}{2}\sigma_d^{-2} \sum_{i=1}^{N} (d_i - m_1)^2 \right\} \tag{6.1}$$

The observed data $\mathbf{d}^{obs}$ can be represented graphically as a point in the N-dimensional space whose coordinate axes are $d_1, d_2, \cdots, d_N$ (Fig. 6.1). The probability density function for the data can also be graphed (Fig. 6.2). Note that the probability density function is centered about the line $d_1 = d_2 = \cdots = d_N$, since all the ds are supposed to have the same mean. Furthermore, the pdf is spherically symmetric, since all the ds have the same variance.

Suppose that we guess a value for the unknown mean and variance, thus fixing the center and diameter of the probability density function. We can then calculate its numerical value at the data $p(\mathbf{d}^{obs})$. If the guessed values of mean and variance are close to being correct, then $p(\mathbf{d}^{obs})$ should be a relatively large number. If the guessed values are incorrect, then the probability, or *likelihood*, of the observed data will be small. We can imagine sliding the cloud of probability in Fig. 6.2 up along the line and adjusting its diameter until its probability at the point $\mathbf{d}^{obs}$ is maximized.

This procedure defines a method of estimating the unknown parameters in the pdf, the *method of maximum likelihood*. It asserts that the optimum values of the parameters maximize the probability that the observed data are in fact observed. In other words, the value of the probability density function at the point $\mathbf{d}^{obs}$ is made as large as possible. For this reason, $p(\mathbf{d}^{obs})$ is called the *likelihood*. The maximum is located by differentiating $p(\mathbf{d}^{obs})$ with respect to mean and variance and setting the result to zero as

$$\frac{\partial}{\partial m_1} p\left(\mathbf{d}^{obs}\right) = 0 \ \text{ and } \ \frac{\partial}{\partial \sigma_d} p\left(\mathbf{d}^{obs}\right) = 0 \tag{6.2}$$

Maximizing $\log p(\mathbf{d}^{obs})$ gives the same result as maximizing $p(\mathbf{d}^{obs})$, as $\log(p)$ is a monotonic function of p. We therefore compute derivatives of the *log-likelihood function*, $L = \log p(\mathbf{d}^{obs})$ (Fig. 6.3). Ignoring the overall normalization of $(2\pi)^{-N/2}$, we have

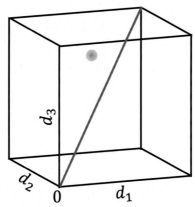

FIG. 6.1   The data are represented by a single point (gray) in a space whose dimensions equal the number of observations (in this case, 3). These data are realizations of random variables with the same mean and variance. Nevertheless, they do not necessarily fall on the line $d_1 = d_2 = d_3$ (blue). Scripts `gdama06_01` and `gdapy06_01`.

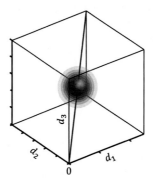

FIG. 6.2   If the data d_i are assumed to be uncorrelated with equal mean and uniform variance, their probability density function $p(d)$ is a spherical cloud (red) centered on the line $d_1 = d_2 = d_3$ (blue). Scripts `gdama06_02` and `gdapy06_02`.

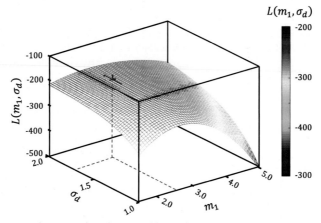

FIG. 6.3   Log-likelihood surface for 100 realizations of random variables with equal mean $m_1 = 2.5$ and uniform variance $\sigma_d^2 = 1.5$. Near the maximum, the curvature in the direction of m_1 is greater than in the direction of σ_d, indicating that the former can be determined to greater certainty. Scripts `gdama06_03` and `gdapy06_03`.

$$L \equiv \log p\left(\mathbf{d}^{obs}\right) = -N \log(\sigma_d) - \tfrac{1}{2}\sigma_d^{-2} \sum_{i=1}^{N} \left(d_i^{obs} - m_1\right)^2$$

$$\frac{\partial L}{\partial m_1} = 0 = \tfrac{1}{2}\sigma_d^{-2} 2 \sum_{i=1}^{N} \left(d_i^{obs} - m_1\right) \tag{6.3}$$

$$\frac{\partial L}{\partial \sigma_d} = 0 = -\frac{N}{\sigma_d} + \sigma_d^{-3} \sum_{i=1}^{N} \left(d_i^{obs} - m_1\right)^2$$

These equations can be solved for the estimated mean and variance as

$$m_1^{est} = \frac{1}{N} \sum_{i=1}^{N} d_i^{obs} \text{ and } \sigma_d^{est} = \left[\frac{1}{N} \sum_{i=1}^{N} \left(d_i^{obs} - m_1^{est}\right)^2\right]^{\frac{1}{2}} \tag{6.4}$$

The estimate for m_1 is just the usual formulas for the sample mean. The estimate for σ_d^{est} is the root mean squared error and also is almost the formula for the sample standard deviation, except that it has a leading factor of $1/N$, instead of $1/(N-1)$. We note that these estimates arise as a direct consequence of the assumption that the data possess a Normal pdf. If the data were not Normally distributed, then the arithmetic mean might not be an appropriate estimate of the mean of the pdf. (As we shall see in Section 9.2, the *sample median* is the maximum likelihood estimate of the mean of an exponential pdf.)

6.2   Maximum likelihood applied to inverse problems

In the simple case, the data in the linear inverse problem $\mathbf{d}^{obs} = \mathbf{Gm}$ have a multivariate Normal pdf, as given by

$$p(\mathbf{d}) \propto \exp\left\{-\tfrac{1}{2}(\mathbf{d} - \mathbf{Gm})^T [\text{cov } \mathbf{d}]^{-1}(\mathbf{d} - \mathbf{Gm})\right\} \tag{6.5}$$

We assume that the model parameters are unknown but (for the sake of simplicity) that the data covariance is known. We can then apply the method of maximum likelihood to estimate the model parameters. The optimum values for the model parameters are the ones that maximize the probability that the observed data are in fact observed. The maximum of $p(\mathbf{d}^{obs})$ occurs when the argument of the exponential is a maximum or when the quantity given by

$$\left(\mathbf{d}^{obs} - \mathbf{Gm}\right)^T [\text{cov } \mathbf{d}]^{-1} \left(\mathbf{d}^{obs} - \mathbf{Gm}\right) \tag{6.6}$$

is a minimum. But this expression is just a weighted measure of prediction error E. The maximum likelihood estimate of the model parameters is nothing but the weighted least-squares solution, where the weighting matrix is the inverse of the covariance matrix of the data (in the notation of Chapter 4, $\mathbf{W}_e = [\text{cov } \mathbf{d}]^{-1}$). If the data happen to be uncorrelated and all have equal variance, then $[\text{cov } \mathbf{d}]^{-1} = \sigma_d^2 \mathbf{I}$, and the maximum likelihood solution is the simple least-squares solution. If the data are uncorrelated but their variances are all different (say, $\sigma_{d_i}^2$), then the prediction error is given by

$$E \equiv \sum_{i=1}^{N} \sigma_{d_i}^{-2} e_i^2 \text{ with } e_i \equiv d_i^{obs} - d_i^{pre} \tag{6.7}$$

where e_i is the prediction error for each datum. Each individual error is weighted by the reciprocal of its standard deviation; the most *certain* data are weighted most.

We have justified the use of the L_2 norm through the application of probability theory. The least-squares procedure for minimizing the L_2 norm of the prediction error makes sense if the data are uncorrelated, have equal variance, and obey Normal statistics. If the data are not Normally distributed, then other measures of prediction error may be more appropriate.

6.3   Prior pdfs

The least-squares solution does not exist when the linear problem is underdetermined. From the standpoint of probability theory, the pdf of the data $p(\mathbf{d}^{obs})$ has no well-defined maximum with respect to variations of the model parameters. At best, it has a ridge of maximum probability (Fig. 6.4).

We must add prior information that causes the pdf to have a well-defined peak in order to solve an underdetermined problem. One way to accomplish this goal is to write the prior information about the model parameters as a pdf $p_A(\mathbf{m})$, where the subscript A indicates "*a priori*" or "*prior*." The mean of this probability density function is then the value we expect the model parameter vector to have, and its shape reflects the certainty of this expectation.

Prior pdfs of the model parameters can take a variety of forms. For instance, if we expected that the model parameters are close to $\langle \mathbf{m} \rangle$, we might use a Normal pdf with mean $\langle \mathbf{m} \rangle$ and variance that reflects the certainty of our knowledge (Fig. 6.5). If the prior value of one model parameter was more certain than another, we might use different variances for the different model parameters (Fig. 6.6). The general Normal case, with covariance $[\text{cov } \mathbf{m}]_A$, is

$$p_A(\mathbf{m}) \propto \exp\left\{ -\tfrac{1}{2}(\mathbf{m} - \langle \mathbf{m} \rangle)^T [\text{cov } \mathbf{m}]_A^{-1}(\mathbf{m} - \langle \mathbf{m} \rangle) \right\} \tag{6.8}$$

Equality constraints can be implemented with a pdf that contains a ridge (Fig. 6.7). This pdf is non-Normal, but might be approximated by a Normal pdf with nonzero covariance if the expected range of the model parameters was small. Inequality constraints can also be represented by a prior pdf but are inherently non-Normal (Fig. 6.8).

Similarly, one can summarize the state of knowledge about the measurements with a prior probability density function $p_A(\mathbf{d})$. It simply summarizes the observations, so its mean is $\mathbf{d}^{obs}$ and its covariance is the prior covariance $[\text{cov } \mathbf{d}]$.

$$p_A(\mathbf{d}) \propto \exp\left\{ -\tfrac{1}{2}\left(\mathbf{d} - \mathbf{d}^{obs}\right)^T [\text{cov } \mathbf{m}]^{-1}\left(\mathbf{d} - \mathbf{d}^{obs}\right) \right\} \tag{6.9}$$

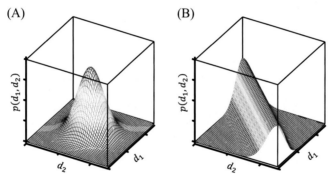

FIG. 6.4   (A) Probability density function $p(d_1, d_2)$ with a well-defined peak. (B) Probability density function with a ridge. Scripts gdama06_04 and gdapy06_04.

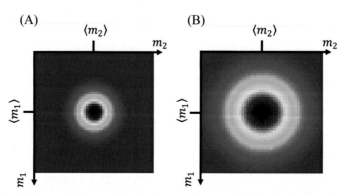

FIG. 6.5   Prior information about model parameters m_1 and m_2 is represented by a probability density function $p(m_1, m_2)$. Most probable values are given by means $\langle m_1 \rangle$ and $\langle m_2 \rangle$. Width of the pdf reflects certainty of knowledge: (A) certain, (B) uncertain. Scripts gdama06_05 and gdapy06_05.

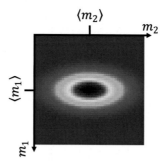

FIG. 6.6 Prior information about model parameters m_1 and m_2 represented with a probability density function $p(m_1, m_2)$. The model parameters are thought to be near $\langle m_1 \rangle$, with uncertainty in the former less than the uncertainty in the latter. Scripts gdama06_06 and gdapy06_06.

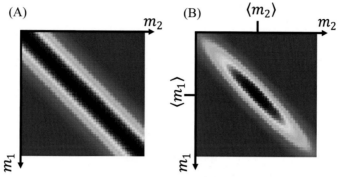

FIG. 6.7 Prior information about model parameters m_1 and m_2 represented with a probability density function $p(m_1, m_2)$. (A) Case when the values of m_1 and m_2 are unknown, but believed to be correlated. (B) Approximation of (A) with a Normal probability density function with finite variance. Scripts gdama06_07 and gdapy06_07.

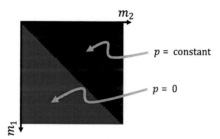

FIG. 6.8 Prior information about model parameters m_1 and m_2 represented with a probability density function $p(m_1, m_2)$. The values of the model parameters are unknown, but the relationship $m_1 \leq m_2$ is believed to hold exactly. This is a non-Normal pdf. Scripts gdama06_08 and gdapy06_08.

An important attribute of $p_A(\mathbf{m})$ and $p_A(\mathbf{d})$ is that they contain information about the model parameters $\mathbf{m}$ and the data $\mathbf{d}$, respectively. The amount of information can be quantified by the *information gain*, a scalar number S defined as

$$S[p_A(\mathbf{m})] = \int p_A(\mathbf{m}) \, \log\left(\frac{p_A(\mathbf{m})}{p_N(\mathbf{m})}\right) \mathrm{d}^M \mathbf{m}$$

$$S[p_A(\mathbf{d})] = \int p_A(\mathbf{d}) \, \log\left(\frac{p_A(\mathbf{d})}{p_N(\mathbf{d})}\right) \mathrm{d}^N \mathbf{d}$$

(6.10)

Here, the *null* pdfs $p_N(\mathbf{m})$ and $p_N(\mathbf{d})$ express the state of complete ignorance about the model parameters and data, respectively. When the ranges of $\mathbf{m}$ and $\mathbf{d}$ are bounded, the null probability density functions can be taken to be proportional to a constant, that is, $p_N(\mathbf{m}) \propto$ constant and $p_N(\mathbf{d}) \propto$ constant, meaning $\mathbf{m}$ and $\mathbf{d}$ "could be anything." However, when $\mathbf{m}$ and $\mathbf{d}$ are unbounded, the uniform pdf does not exist, and some other probability density function, such as a very wide Normal pdf, must be used, instead. The quantity $-S$ is sometimes called the *relative entropy* between the two pdfs. A wide pdf is "more random" than a narrow one; it has more *entropy*.

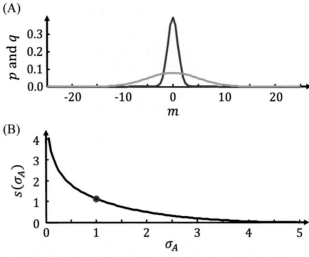

FIG. 6.9 (A) In this example, a wide Normal pdf (green, $\sigma_N = 5$) is used for the null probability density function $p_N(m)$ and a narrow Normal pdf (red, $\sigma_A = 1$) is used for the prior probability density function $p_A(m)$. (B) The information gain S decreases as the width of $p_A(m)$ is increased. The case in (A) is depicted with a red circle. Scripts gdama06_09 and gdapy06_09.

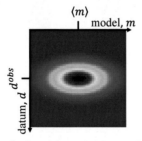

FIG. 6.10 Prior joint probability density function $p_A(m,d)$ of model parameter m and datum d. The pdf is peaked at mean values $\langle m \rangle$ and d^{obs}. Scripts gdama06_10 and gdapy06_10.

The information gain is always a nonnegative number and is only zero when $p_A(\mathbf{m}) = p_N(\mathbf{m})$ and $p_A(\mathbf{d}) = p_N(\mathbf{d})$ (Fig. 6.9). The information gain S has the following properties (Tarantola and Valette, 1982b): (1) the information gain of the null pdf is zero; (2) all pdfs except the null pdf have positive information gain; (3) the more sharply peaked the probability density function becomes, the larger its information gain; and (4) the information gain is invariant under reparameterizations.

We can summarize the state of knowledge about the inverse problem before it is solved by first defining a prior probability density function for the data $p_A(\mathbf{d})$ and then combining it with the prior probability density function for the model $p_A(\mathbf{m})$ (Fig. 6.10). The prior data probability density function simply summarizes the observations, so its mean is $\mathbf{d}^{obs}$ and its variance is equal to the prior variance of the data. Since the prior model probability density function is completely independent of the actual values of the data, we can form the joint prior probability density function simply by multiplying the two as

$$p_A(\mathbf{m}, \mathbf{d}) = p_A(\mathbf{m}) p_A(\mathbf{d}) \tag{6.11}$$

6.4 Maximum likelihood for an exact theory

Suppose that the model is the general explicit equation $\mathbf{d} = \mathbf{g}(\mathbf{m})$, which may or may not be linear. This equation defines a surface in the space of model parameters and data along which the solution must lie (Fig. 6.11). The maximum likelihood problem then translates into finding the maximum of the joint pdf $p_A(\mathbf{m}, \mathbf{d})$ (or, equivalently, its logarithm) on the surface $\mathbf{d} = \mathbf{g}(\mathbf{m})$ (Tarantola and Valette, 1982a):

$$\text{maximize } p_A(\mathbf{m}, \mathbf{d}) \text{ subject to the constraint } \mathbf{d} = \mathbf{g}(\mathbf{m}) \tag{6.12}$$

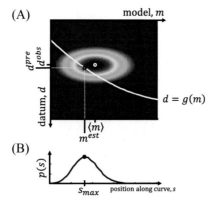

FIG. 6.11   (A) Prior joint probability density function $p_A(m,d)$ (colors) of model parameter m and datum d represents the idea that the model parameter is near its prior value $\langle m \rangle$ and the datum is near its observed value d^{obs} (white circle). The data and model parameters are believed to be related by an exact theory $d = g(m)$ (white curve). The estimated model parameter m^{est} and predicted datum d^{pre} fall on this curve at the point of maximum probability (black dot). (B) Probability density p evaluated along the curve. Scripts gdama06_11 and gdapy06_11.

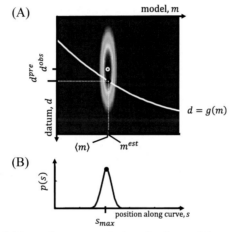

FIG. 6.12   (A) If the prior model parameter $\langle m \rangle$ is much more certain than the observed datum d^{obs}, the solution is close to $\langle m \rangle$ but may be far from d^{obs}. (B) The probability density function p evaluated along the curve. Scripts gdama06_12 and gdapy06_12.

If the prior probability density function for the model parameters is much more certain than that of the observed data (i.e., if $\sigma_m \ll \sigma_d$), then the estimate of the model parameters (the maximum likelihood point) tends to be close to the prior model parameters (Fig. 6.12). On the other hand, if the data are far more certain than the model parameters (i.e., if $\sigma_d \ll \sigma_m$), then the estimates of the model parameters primarily reflect information contained in the data (Fig. 6.13).

In the case of a Normal pdf, and the linear theory $\mathbf{d} = \mathbf{Gm}$, we need only to substitute $\mathbf{Gm}$ for $\mathbf{d}$ in the expression for $p_A(\mathbf{d})$ to obtain

$$
\begin{aligned}
\text{minimize } \ &\Phi(\mathbf{m}) = L(\mathbf{m}) + E(\mathbf{m}) \\
\text{with } \ L(\mathbf{m}) &\equiv (\langle \mathbf{m} \rangle - \mathbf{m})^T [\text{cov } \mathbf{m}]_A^{-1} (\langle \mathbf{m} \rangle - \mathbf{m}) \\
\text{and } \ E(\mathbf{m}) &\equiv \left(\mathbf{d}^{obs} - \mathbf{Gm} \right)^T [\text{cov } \mathbf{d}]^{-1} \left(\mathbf{d}^{obs} - \mathbf{Gm} \right)
\end{aligned}
\tag{6.13}
$$

Comparison with Section 4.15 indicates that this is the weighted damped least-squares problem, with

$$
\varepsilon^2 \mathbf{W}_m = [\text{cov } \mathbf{m}]_A^{-1} \quad \text{and} \quad \mathbf{W}_e = [\text{cov } \mathbf{d}]^{-1}
\tag{6.14}
$$

so its solution is the Generalized Least-Squares solution

$$
\mathbf{m}^{est} = \left[\mathbf{F}^T \mathbf{F} \right]^{-1} \mathbf{F}^T \text{ with}
$$
$$
\mathbf{F} \equiv \begin{bmatrix} [\text{cov } \mathbf{d}]^{-\frac{1}{2}} \mathbf{G} \\ [\text{cov } \mathbf{m}]_A^{-\frac{1}{2}} \mathbf{I} \end{bmatrix} \quad \text{and} \quad \mathbf{f} \equiv \begin{bmatrix} [\text{cov } \mathbf{d}]^{-\frac{1}{2}} \mathbf{d}^{obs} \\ [\text{cov } \mathbf{m}]_A^{-\frac{1}{2}} \langle \mathbf{m} \rangle \end{bmatrix}
\tag{6.15}
$$

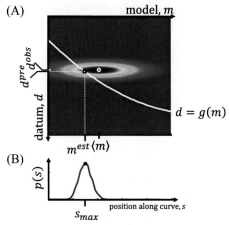

FIG. 6.13 (A) If the prior model parameter $\langle m \rangle$ is much less certain than the observed datum d^{obs}, the solution is close to d^{obs} but may be far from $\langle m \rangle$. (B) The probability density function p evaluated along the curve. Scripts gdama06_13 and gdapy06_13.

The matrices $[\text{cov } \mathbf{d}]^{-\frac{1}{2}}$ and $[\text{cov } \mathbf{m}]_A^{-\frac{1}{2}}$ can be interpreted as the *certainty* of $\mathbf{d}^{obs}$ and $\langle \mathbf{m} \rangle$, respectively, since they are numerically large when the uncertainty of these quantities is small. Thus the top part of the GLS equation $\mathbf{Fm} = \mathbf{f}$ is the data equation $\mathbf{Gm} = \mathbf{d}^{obs}$, weighted by its certainty, and the bottom part is the prior equation $\mathbf{m} = \langle \mathbf{m} \rangle$, weighted by its certainty. Thus the weight matrices $\mathbf{W}_m$ and $\mathbf{W}_e$ that we encountered in Chapter 4 have important probabilistic interpretations.

The vector $\mathbf{f}$ in Eq. (6.15) has unit covariance $[\text{cov } \mathbf{m}] = \mathbf{I}$, as its component quantities $[\text{cov } \mathbf{d}]^{-\frac{1}{2}}\mathbf{d}^{obs}$ and $[\text{cov } \mathbf{m}]_A^{-\frac{1}{2}}\langle \mathbf{m} \rangle$ each have unit covariance (e.g., by the usual rules of error propagation, $[\text{cov } \mathbf{d}]^{-\frac{1}{2}}[\text{cov } \mathbf{d}][\text{cov } \mathbf{d}]^{-\frac{1}{2}T} = \mathbf{I}$). Thus the covariance of the estimated model parameters is

$$[\text{cov } \mathbf{m}] = \left[\mathbf{F}^T\mathbf{F}\right]^{-1} \tag{6.16}$$

Somewhat incidentally, we note that, had the prior information involved a linear function $\mathbf{Hm} = \mathbf{h}^{pri}$ of the model parameters, with covariance $[\text{cov } \mathbf{h}]_A$, in contrast to the model parameters themselves, the appropriate form of Eq. (6.15) would be

$$\mathbf{F} = \begin{bmatrix} [\text{cov } \mathbf{d}]^{-\frac{1}{2}}\mathbf{G} \\ [\text{cov } \mathbf{h}]_A^{-\frac{1}{2}}\mathbf{H} \end{bmatrix} \text{ and } \mathbf{f} = \begin{bmatrix} [\text{cov } \mathbf{d}]^{-\frac{1}{2}}\mathbf{d}^{obs} \\ [\text{cov } \mathbf{h}]_A^{-\frac{1}{2}}\mathbf{h}^{pri} \end{bmatrix} \tag{6.17}$$

Many important types of prior information can be put into this form. For instance, the prior information that the model parameters sum to unity corresponds to $\mathbf{H} = [1, 1, \cdots, 1]$ and $\mathbf{h}^{pri} = [1]$.

When performing calculations, methods that solve the linear equation $[\mathbf{F}^T\mathbf{F}]\,\mathbf{m}^{est} = \mathbf{F}^T\mathbf{f}$, such as the biconjugate gradient method, are usually more efficient than those that explicitly compute $[\mathbf{F}^T\mathbf{F}]^{-1}$. Unfortunately, when the full covariance matrix is needed, this matrix inverse needs to be calculated anyway, as $[\text{cov } \mathbf{m}] = [\mathbf{F}^T\mathbf{F}]^{-1}$. However, variances of a few model parameters can be "spot-checked" by solving for individual columns $\mathbf{c}^{(i)}$ of $[\text{cov } \mathbf{m}]$ using the equation $[\mathbf{F}^T\mathbf{F}]\,\mathbf{c}^{(i)} = \mathbf{s}^{(i)}$, where $\mathbf{s}^{(r)}$ is the ith column of the identify matrix (in which case var $m_i^{est} = c_i^{(i)}$).

6.5 Inexact theories

Generalized Least Squares, as it is embodied in Eqs. (6.15)–(6.17), is extremely useful. Nevertheless, it is somewhat unsatisfying from the standpoint of a probabilistic analysis because the theory has been assumed to be exact. In many realistic problems, there are errors associated with the theory. Some of the assumptions that go into the theory may be unrealistic, or it may be an approximate form of a clumsier but exact theory.

In this case, the model equation $\mathbf{d} = \mathbf{g}(\mathbf{m})$ can no longer be represented by a simple surface. It has become "fuzzy" because there are now errors associated with it (Fig. 6.14B; Tarantola and Valette, 1982b). Instead of a surface, one might envision a pdf $p_g(\mathbf{m}, \mathbf{d})$ centered about $\mathbf{d} = \mathbf{g}(\mathbf{m})$, with width proportional to the uncertainty of the theory.

Rather than find the maximum likelihood point of $p_A(\mathbf{m}, \mathbf{d})$ on a surface, we should instead combine $p_A(\mathbf{m}, \mathbf{d})$ and $p_g(\mathbf{m}, \mathbf{d})$ into a single pdf and find the maximum likelihood point in the overall volume (Fig. 6.14C). To proceed, we need a way of combining two probability density functions, each of which contains information about the data and model parameters.

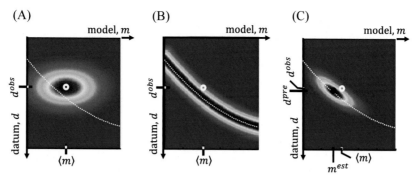

FIG. 6.14 (A) The prior probability density function $p_A(m,d)$ represents the state of knowledge before the theory is applied. Its mean (white circle) is the prior model parameter $\langle m \rangle$ and observed data d^{obs}. (B) An inexact theory is represented by the probability density function $p_g(m,d)$, which is centered about the exact theory (dotted white curve). (C) The product $p_T(m,d) = p_A(m,d)p_g(m,d)$ combines the prior information and theory. Its peak is at the estimated model m^{est} and predicted data d^{pre}. Scripts `gdama06_14` and `gdapy06_14`.

We have already encountered one special case of a combination in our discussion of Bayesian inference (Section 2.7). After adjusting the variable names to match the current discussion, Bayes theorem takes the form

$$p(\mathbf{m}|\mathbf{d}) = \frac{p(\mathbf{m}|\mathbf{d})p(\mathbf{m})}{p(\mathbf{d})} \quad \text{or} \quad p(\mathbf{m}|\mathbf{d}) \propto p(\mathbf{m}|\mathbf{d})p(\mathbf{m}) \tag{6.18}$$

Note that the second form omits the denominator, which is not a function of the model parameters and hence acts only as an overall normalization. This second form can be interpreted as *updating* the information in $p(\mathbf{m})$ (identified now as the prior information) with $p(\mathbf{m}|\mathbf{d})$ (identified now with the data and quantitative model). Thus, in Bayesian inference, probability density functions are combined by multiplication.

In the general case, we denote the process of combining two probability density functions as $p_3 = C(p_1, p_2)$, meaning that functions 1 and 2 are combined into function 3. Then, clearly, the process of combining must have the following properties (adapted from Tarantola and Valette, 1982b):

1. The order in which two probability density functions are combined should not matter; that is, $C(p_1, p_2)$ should be commutative: $C(p_1, p_2) = C(p_2, p_1)$.
2. The order in which three probability density functions are combined should not matter; that is, $C(p_1, p_2)$ should be associative: $C(p_1, C(p_2, p_3)) = C(C(p_1, p_2), p_3))$.
3. Combining a pdf with the null pdf p_N should return the same pdf, that is, $p_1 = C(p_1, p_N)$
4. The combination $C(p_1, p_2)$ should never be everywhere zero except if p_1 or p_2 is everywhere zero.
5. The combination $C(p_1, p_2)$ should be invariant under reparameterizations.

These conditions can be shown to be satisfied by the choice (Tarantola and Valette, 1982b):

$$C(p_1, p_2) = \frac{p_1 p_2}{p_N} \tag{6.19}$$

(at least up to an overall normalization). Note that if the null pdf p_N is constant (as we shall assume for the rest of this chapter), one combines pdfs simply by multiplying them:

$$p_T(\mathbf{m}, \mathbf{d}) = p_A(\mathbf{m}, \mathbf{d}) \, p_g(\mathbf{m}, \mathbf{d}) \tag{6.20}$$

Here, the subscript T means the combined or *total* pdf. Note that as the error associated with the theory increases, the maximum likelihood point moves back toward the prior values of model parameters and observed data (Fig. 6.15). The limiting case in which the theory is infinitely accurate is equivalent to the case in which the pdf is replaced by a distinct surface.

The maximum likelihood point of $p_T(\mathbf{m}, \mathbf{d})$ is specified by both a set of estimated model parameters $\mathbf{m}^{est}$ and a set of predicted data $\mathbf{d}^{pre}$. They are determined simultaneously. This result is different from the one obtained for an exact theory (Section 6.4). In that section, we maximized the probability density function with respect to the model parameters only to determine the most probable model parameters $\mathbf{m}^{est}$ and afterward can calculate the predicted data $\mathbf{d}^{pre}$ via $\mathbf{d}^{pre} = \mathbf{G}\mathbf{m}^{est}$.

FIG. 6.15 The rows of the figure have the same format as Fig. 6.14. If the theory is made more and more inexact (compare (A–C) with (D–F)), the solution (black circle) moves toward the maximum likelihood point of the prior pdf (white circle). Scripts `gdama06_15` and `gdapy06_15`.

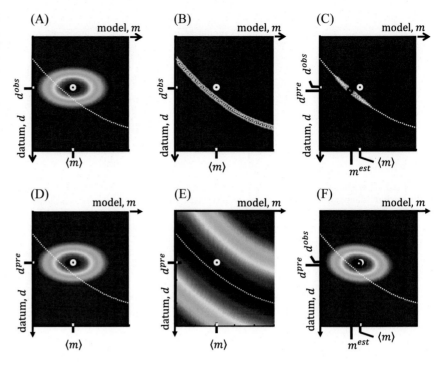

Furthermore, the most probable pair $(\mathbf{m}^{est}, \mathbf{d}^{pre})$ of model parameters and predicted data is *not* the same as the most probable $\mathbf{m}^{est}$, irrespective of the estimate of the data. As an analogy, consider that in the US in 2021, the most popular car model-color combination was the black Ford F150, but that the most popular car color was white. The most probable model, irrespective of the data, is the maximum likelihood point of

$$p(\mathbf{m}) = \int p_T(\mathbf{m}, \mathbf{d}) \, \mathrm{d}^N d \qquad (6.21)$$

Here, the integration is performed over the entire range of the ds. This distinction emphasizes that multivariate pdfs are complicated, and that the answers one obtains from them can be dependent on the nuances of the questions that are asked.

6.6   Exact theory as a limiting case of an inexact one

To illustrate this method, we rederive the GLS solution, with prior probability density function:

$$p_A(\mathbf{m}, \mathbf{d}) \propto$$
$$\exp\left\{ -\tfrac{1}{2}(\langle\mathbf{m}\rangle - \mathbf{m})^T[\text{cov } \mathbf{m}]_A^{-1}(\langle\mathbf{m}\rangle - \mathbf{m}) - \tfrac{1}{2}\left(\mathbf{d}^{obs} - \mathbf{d}\right)^T[\text{cov } \mathbf{d}]^{-1}\left(\mathbf{d}^{obs} - \mathbf{d}\right) \right\} \qquad (6.22)$$

If there are no errors in the theory, then its probability density function is "infinitely narrow" and can be represented by a Dirac delta function

$$p_g(\mathbf{m}, \mathbf{d}) = \delta\left(\mathbf{d}^{obs} - \mathbf{Gm}\right) \qquad (6.23)$$

Performing the projection "integrates away" the delta function

$$p(\mathbf{m}) = \int p_T(\mathbf{m}, \mathbf{d}) \, \mathrm{d}^N d = \int p_A(\mathbf{m}, \mathbf{d}) p_g(\mathbf{m}, \mathbf{d}) \, \mathrm{d}^N d$$
$$\propto \exp\left\{ -\tfrac{1}{2}(\langle\mathbf{m}\rangle - \mathbf{m})^T[\text{cov } \mathbf{m}]_A^{-1}(\langle\mathbf{m}\rangle - \mathbf{m}) - \tfrac{1}{2}\left(\mathbf{d}^{obs} - \mathbf{Gm}\right)^T[\text{cov } \mathbf{d}]^{-1}\left(\mathbf{d}^{obs} - \mathbf{Gm}\right) \right\} \qquad (6.24)$$

This pdf is exactly the one we encountered in the GLS problem (the logarithm of which is shown in Eq. (6.13)).

6.7   Inexact theory with a normal pdf

In the general linear, Normally distributed case, we assume that all the component probability density functions are Normal and that the theory is the linear equation $\mathbf{d} = \mathbf{Gm}$, so that

$$p_A(\mathbf{m}) \propto \exp\left\{-\tfrac{1}{2}(\langle\mathbf{m}\rangle - \mathbf{m})^T[\text{cov } \mathbf{m}]_A^{-1}(\langle\mathbf{m}\rangle - \mathbf{m})\right\}$$

$$p_A(\mathbf{d}) \propto \exp\left\{-\tfrac{1}{2}\left(\mathbf{d}^{obs} - \mathbf{d}\right)^T[\text{cov } \mathbf{d}]^{-1}\left(\mathbf{d}^{obs} - \mathbf{d}\right)\right\} \tag{6.25}$$

$$p_g(\mathbf{m}, \mathbf{d}) \propto \exp\left\{-\tfrac{1}{2}(\mathbf{d} - \mathbf{Gm})^T[\text{cov } \mathbf{g}]^{-1}(\mathbf{d} - \mathbf{Gm})\right\}$$

Here, the theory is represented by a Normal pdf with covariance $[\text{cov } \mathbf{g}]$. The total pdf $p_T(\mathbf{m}, \mathbf{d})$ is the product of these three pdfs. As we noted in Section 3.4, products of Normal pdfs are themselves Normal, so $p_T(\mathbf{m}, \mathbf{d})$ is Normal. We need to determine the mean of this pdf, as it is also the maximum likelihood point that defines $\mathbf{m}^{est}$ and $\mathbf{d}^{pre}$.

To simplify the algebra, we first define combined quantities

$$\mathbf{x} \equiv \begin{bmatrix} \mathbf{d} \\ \mathbf{m} \end{bmatrix} \text{ and } \langle\mathbf{x}\rangle \equiv \begin{bmatrix} \mathbf{d}^{obs} \\ \langle\mathbf{m}\rangle \end{bmatrix} \text{ and } [\text{cov } \mathbf{x}] \equiv \begin{bmatrix} [\text{cov } \mathbf{d}] & 0 \\ 0 & [\text{cov } \mathbf{m}]_A \end{bmatrix} \tag{6.26}$$

The first two products in the total pdf can then be combined into an exponential, with the argument given by

$$-\tfrac{1}{2}(\langle\mathbf{x}\rangle - \mathbf{x})^T[\text{cov } \mathbf{x}]^{-1}(\langle\mathbf{x}\rangle - \mathbf{x}) \tag{6.27}$$

To express the third product in terms of $\mathbf{x}$, we define a matrix $\mathbf{X} \equiv [\mathbf{I} \;\; -\mathbf{G}]$, such that $\mathbf{Xx} = \mathbf{d} - \mathbf{Gm} = 0$. The argument of the third product's exponential is then given by

$$-\tfrac{1}{2}(\mathbf{Xx})^T[\text{cov } \mathbf{g}]^{-1}(\mathbf{Xx}) \tag{6.28}$$

The total pdf is proportional to an exponential with argument

$$-\tfrac{1}{2}(\langle\mathbf{x}\rangle - \mathbf{x})^T[\text{cov } \mathbf{x}]^{-1}(\langle\mathbf{x}\rangle - \mathbf{x}) - \tfrac{1}{2}(\mathbf{Xx})^T[\text{cov } \mathbf{g}]^{-1}(\mathbf{Xx})$$

$$= -\tfrac{1}{2}\mathbf{x}^T\left[[\text{cov } \mathbf{x}]^{-1} + \mathbf{X}^T[\text{cov } \mathbf{g}]^{-1}\mathbf{X}\right]^{-1}\mathbf{x} - \mathbf{x}^T[\text{cov } \mathbf{x}]^{-1}\langle\mathbf{x}\rangle - \tfrac{1}{2}\langle\mathbf{x}\rangle^T[\text{cov } \mathbf{x}]^{-1}\langle\mathbf{x}\rangle \tag{6.29}$$

We would like to manipulate this expression into the standard form of the argument of a Normal pdf, that is, an expression involving just a single vector, say $\mathbf{x}^*$, and a single covariance matrix, say $[\text{cov } \mathbf{x}^*]$, related by

$$-\tfrac{1}{2}(\mathbf{x} - \mathbf{x}^*)^T[\text{cov } \mathbf{x}^*]^{-1}(\mathbf{x} - \mathbf{x}^*)$$

$$= -\tfrac{1}{2}\mathbf{x}^T[\text{cov } \mathbf{x}^*]^{-1}\mathbf{x} - \mathbf{x}^T[\text{cov } \mathbf{x}^*]^{-1}\mathbf{x}^* - \tfrac{1}{2}(\mathbf{x}^*)^T[\text{cov } \mathbf{x}^*]^{-1}\mathbf{x}^* \tag{6.30}$$

We can identify $\mathbf{x}^*$ and $[\text{cov } \mathbf{x}^*]$ by matching terms of equal powers of x with Eq. (6.29). Matching the quadratic terms yields

$$[\text{cov } \mathbf{x}^*]^{-1} = [\text{cov } \mathbf{x}]^{-1} + \mathbf{X}^T[\text{cov } \mathbf{g}]^{-1}\mathbf{X} \tag{6.31}$$

and matching the linear terms yields

$$\mathbf{x}^T[\text{cov } \mathbf{x}]^{-1}\langle\mathbf{x}\rangle = \mathbf{x}^T[\text{cov } \mathbf{x}^*]^{-1}\mathbf{x}^* \text{ or } \mathbf{x}^* = [\text{cov } \mathbf{x}^*][\text{cov } \mathbf{x}]^{-1}\langle\mathbf{x}\rangle \tag{6.32}$$

These choices do not match the constant term, but such a match is not necessary, because the constant term effects only the overall normalization of the probability density function $p_T(\mathbf{x})$. The vector $\mathbf{x}^*$ corresponds to the maximum likelihood point of $p_T(\mathbf{x})$ and so can be considered the solution to the inverse problem. This solution has covariance $[\text{cov } \mathbf{x}^*]$.

Eq. (6.31) for $[\text{cov } \mathbf{x}^*]^{-1}$ can be explicitly inverted to yield an expression for $[\text{cov } \mathbf{x}^*]$, using a matrix identity called the *Woodbury formula* (Woodbury, 1950). It relates square, invertible matrices $\mathbf{A}$ and $\mathbf{B}$ and rectangular matrices $\mathbf{U}$ and $\mathbf{V}$ by

$$[\mathbf{A} + \mathbf{UBV}]^{-1} = \left(\mathbf{I} - \mathbf{A}^{-1}\mathbf{U}[\mathbf{B}^{-1} + \mathbf{VA}^{-1}\mathbf{U}]^{-1'}\mathbf{V}\right)\mathbf{A}^{-1} \tag{6.33}$$

Equating $\mathbf{A} = [\text{cov } \mathbf{x}]^{-1}$, $\mathbf{B} = [\text{cov } \mathbf{g}]^{-1} \mathbf{U} = \mathbf{X}^T$ and $\mathbf{V} = \mathbf{X}$, we find

$$[\text{cov } \mathbf{x}^*] = \left[[\text{cov } \mathbf{x}]^{-1} + \mathbf{X}^T [\text{cov } \mathbf{g}]^{-1} \mathbf{X} \right]^{-1} = $$
$$\left(\mathbf{I} - [\text{cov } \mathbf{x}] \mathbf{X}^T \left[[\text{cov } \mathbf{g}] + \mathbf{X}[\text{cov } \mathbf{x}]\mathbf{X}^T \right]^{-1'} \mathbf{X} \right) [\text{cov } \mathbf{x}] \tag{6.34}$$

The solution $\mathbf{x}^*$ is given by

$$\mathbf{x}^* = [\text{cov } \mathbf{x}^*][\text{cov } \mathbf{x}]^{-1} \langle \mathbf{x} \rangle = $$
$$\left(\mathbf{I} - [\text{cov } \mathbf{x}] \mathbf{X}^T \left[[\text{cov } \mathbf{g}] + \mathbf{X}[\text{cov } \mathbf{x}]\mathbf{X}^T \right]^{-1'} \mathbf{X} \right) \langle \mathbf{x} \rangle \tag{6.35}$$

Nothing in this derivation requires the special forms of $\mathbf{X}$ and $[\text{cov } \mathbf{x}]$ assumed before that made $\mathbf{Xx} = 0$ separable into an explicit linear inverse problem. So, Eq. (6.35) is the solution to the completely general implicit linear inverse problem.

When $\mathbf{Xx} = 0$ is an explicit equation, the formula for $\mathbf{x}^*$ in Eq. (6.35) can be decomposed into its component vectors $\mathbf{d}^{pre}$ and $\mathbf{m}^{est}$ by substituting the definition of $\mathbf{X}$ and $[\text{cov } \mathbf{x}]$ (Eq. 6.26) into it and performing the matrix multiplications. An explicit formula for the estimated model parameters is then given by

$$\mathbf{m}^{est} = \langle \mathbf{m} \rangle + \mathbf{G}^{-g} \left(\mathbf{d}^{obs} - \mathbf{G} \langle \mathbf{m} \rangle \right) = \mathbf{G}^{-g} \mathbf{d}^{obs} + (\mathbf{I} - \mathbf{R}) \langle \mathbf{m} \rangle$$

with $\mathbf{G}^{-g} = [\text{cov } \mathbf{m}]_A \mathbf{G}^T \left([\text{cov } \mathbf{d}] + [\text{cov } \mathbf{g}] + \mathbf{G}[\text{cov } \mathbf{m}]_A \mathbf{G}^T \right)^{-1} = [\text{cov } \mathbf{m}]_A \mathbf{G}^T \mathbf{Q}^{-1} \tag{6.36}$

and $\mathbf{Q} \equiv [\text{cov } \mathbf{d}] + [\text{cov } \mathbf{g}] + \mathbf{G}[\text{cov } \mathbf{m}]_A \mathbf{G}^T$ and $\mathbf{R} \equiv \mathbf{G}^{-g} \mathbf{G}$

where we have used the generalized inverse $\mathbf{G}^{-g}$ and resolution matrix $\mathbf{R}$ notation for convenience. The corresponding predicted data are just $\mathbf{d}^{pre} = \mathbf{Gm}^{est}$ (see Eq. 11.39). The generalized inverse in Eq. (6.36) is reminiscent of the minimum-length inverse $\mathbf{G}^{-g} = \mathbf{G}^T(\mathbf{GG}^T)^{-1}$ (Eq. 4.34). The generalized inverse also can be written

$$\mathbf{G}^{-g} = \left(\mathbf{G}^T([\text{cov } \mathbf{d}] + [\text{cov } \mathbf{g}])^{-1} \mathbf{G} + [\text{cov } \mathbf{m}]_A^{-1} \right)^{-1} \mathbf{G}^T([\text{cov } \mathbf{d}] + [\text{cov } \mathbf{g}])^{-1}$$
$$= \mathbf{A}^{-1} \mathbf{G}^T([\text{cov } \mathbf{d}] + [\text{cov } \mathbf{g}])^{-1} \text{ with } \mathbf{A} \equiv \mathbf{G}^T([\text{cov } \mathbf{d}] + [\text{cov } \mathbf{g}])^{-1} \mathbf{G} + [\text{cov } \mathbf{m}]_A^{-1} \tag{6.37}$$

which is reminiscent of the minimum-length inverse $\mathbf{G}^{-g} = (\mathbf{G}^T \mathbf{G})^{-1} \mathbf{G}^T$ (Eq. 4.15). Both forms of the generalized inverse depend only upon the sum of the covariance of the data $[\text{cov } \mathbf{d}]$ and the covariance of the theory $[\text{cov } \mathbf{g}]$; that is, they make only a combined contribution. This is the most important insight gained from this problem, for the exact-theory case (Eq. 6.15) can be made identical to the inexact-theory case (Eqs. 5.36 and 5.37) with the substitution

$$[\text{cov } \mathbf{d}] \text{ is replaced by } [\text{cov } \mathbf{d}] + [\text{cov } \mathbf{g}] \tag{6.38}$$

and is therefore a form of Generalized Least Squares. The equivalence of the two forms of the generalized inverse can be verified by equating them, moving the matrix inverses to the opposite sides of the equation, and manipulating the result. Presuming that the covariance of the theory already has been moved into $[\text{cov } \mathbf{d}]$, we have

$$\left(\mathbf{G}^T[\text{cov } \mathbf{d}]^{-1} \mathbf{G} + [\text{cov } \mathbf{m}]_A^{-1} \right)^{-1} \mathbf{G}^T[\text{cov } \mathbf{d}]^{-1} \stackrel{?}{=} [\text{cov } \mathbf{m}]_A \mathbf{G}^T \left([\text{cov } \mathbf{d}] + \mathbf{G}[\text{cov } \mathbf{m}]_A \mathbf{G}^T \right)^{-1}$$
$$\mathbf{G}^T[\text{cov } \mathbf{d}]^{-1} \left([\text{cov } \mathbf{d}] + \mathbf{G}[\text{cov } \mathbf{m}]_A \mathbf{G}^T \right) \stackrel{?}{=} \left(\mathbf{G}^T[\text{cov } \mathbf{d}]^{-1} \mathbf{G} + [\text{cov } \mathbf{m}]_A^{-1} \right) [\text{cov } \mathbf{m}]_A \mathbf{G}^T \tag{6.39}$$
$$\mathbf{G}^T + \mathbf{G}^T[\text{cov } \mathbf{d}]^{-1} \mathbf{G}[\text{cov } \mathbf{m}]_A \mathbf{G}^T = \mathbf{G}^T[\text{cov } \mathbf{d}]^{-1} \mathbf{G}[\text{cov } \mathbf{m}]_A \mathbf{G}^T + \mathbf{G}^T$$

Hence, from the point of view of computations, one can continue to use the GLS formulation (Eq. 6.15) after adjusting the covariance using Eq. (6.38).

As the estimated model parameters are a linear combination of observed data and prior model parameters, we can calculate their covariance as

$$[\text{cov } \mathbf{m}^{est}] = \mathbf{G}^{-g}[\text{cov } \mathbf{d}] \mathbf{G}^{-g T} + [\mathbf{I} - \mathbf{R}][\text{cov } \mathbf{m}]_A [\mathbf{I} - \mathbf{R}]^T \tag{6.40}$$

This expression differs from those derived in Chapters 4 and 5 because it contains a term dependent on the prior model parameter covariance $[\text{cov } \mathbf{m}]_A$. This significance of this term is explored further in Section 7.4. An equivalent formula for the covariance is

$$[\text{cov } \mathbf{m}^{est}] = \left(\mathbf{I} - [\text{cov } \mathbf{m}]_A \mathbf{G}^T \mathbf{Q}^{-1} \mathbf{G} \right) [\text{cov } \mathbf{m}]_A \tag{6.41}$$

It is obtained by substituting the generalized inverse (Eq. 6.36) into Eq. (6.40)

$$[\text{cov }\mathbf{m}^{est}] = \mathbf{G}^{-g}[\text{cov }\mathbf{d}]\mathbf{G}^{-gT} + [\mathbf{I} - \mathbf{G}^{-g}\mathbf{G}][\text{cov }\mathbf{m}]_A[\mathbf{I} - \mathbf{G}^{-g}\mathbf{G}]^T$$

$$= \mathbf{G}^{-g}[\text{cov }\mathbf{d}]\mathbf{G}^{-gT} + [\text{cov }\mathbf{m}]_A - \mathbf{G}^{-g}\mathbf{G}[\text{cov }\mathbf{m}]_A - [\text{cov }\mathbf{m}]_A\mathbf{G}^T\mathbf{G}^{-gT} + \mathbf{G}^{-g}\mathbf{G}[\text{cov }\mathbf{m}]_A\mathbf{G}^T\mathbf{G}^{-gT}$$

$$= \mathbf{G}^{-g}\{[\text{cov }\mathbf{d}] + \mathbf{G}[\text{cov }\mathbf{m}]_A\mathbf{G}^T\}\mathbf{G}^{-gT} + [\text{cov }\mathbf{m}]_A - \mathbf{G}^{-g}\mathbf{G}[\text{cov }\mathbf{m}]_A - [\text{cov }\mathbf{m}]_A\mathbf{G}^T\mathbf{G}^{-gT} \quad (6.42)$$

$$= [\text{cov }\mathbf{m}]_A\mathbf{G}^T\mathbf{Q}^{-1}\mathbf{Q}\mathbf{Q}^{-1}\mathbf{G}[\text{cov }\mathbf{m}]_A + [\text{cov }\mathbf{m}]_A - [\text{cov }\mathbf{m}]_A\mathbf{G}^T\mathbf{Q}^{-1}\mathbf{G}[\text{cov }\mathbf{m}]_A - [\text{cov }\mathbf{m}]_A\mathbf{G}^T\mathbf{Q}^{-1}\mathbf{G}[\text{cov }\mathbf{m}]_A$$

$$= [\text{cov }\mathbf{m}]_A - [\text{cov }\mathbf{m}]_A\mathbf{G}^T\mathbf{Q}^{-1}\mathbf{G}[\text{cov }\mathbf{m}]_A$$

6.8 Limiting cases

We can examine a few interesting limiting cases of problems that have uncorrelated prior covariance $[\text{cov }\mathbf{m}]_A = \sigma_m^2\mathbf{I}$, $[\text{cov }\mathbf{d}] = \sigma_d^2\mathbf{I}$, and $[\text{cov }\mathbf{g}] = \sigma_g^2\mathbf{I}$.

Suppose that the data and theory are exactly known, so that $\sigma_d^2 \to 0$ and $\sigma_g^2 \to 0$ and that the prior model parameters are zero, so that $\langle\mathbf{m}\rangle = 0$. The solution is then given by

$$\mathbf{m}^{est} = \mathbf{G}^T(\mathbf{GG}^T)^{-1}\mathbf{d}^{obs} = (\mathbf{G}^T\mathbf{G})^{-1}\mathbf{G}^T\mathbf{d}^{obs} \quad (6.43)$$

Note that the solution does not depend on the prior model variance, since the data and theory are infinitely more accurate than the prior model parameters. These solutions are just the minimum-length and least-squares solutions, which (as we now see) are simply two different aspects of the same solution. The minimum-length form of the solution, however, exists only when the problem is purely underdetermined; the least-squares form exists only when the problem is purely overdetermined.

If the prior model parameters are not equal to zero, then a second term appears in the estimated solution:

$$\mathbf{m}^{est} = \mathbf{G}^{-g}\mathbf{d}^{obs} + (\mathbf{I} - \mathbf{R})\langle\mathbf{m}\rangle \quad (6.44)$$

The second term only contributes when the model resolution matrix $\mathbf{R} \neq \mathbf{I}$. As we have shown in Eq. (5.10), $\mathbf{R} = \mathbf{I}$ in a overdetermined simple least-squares problem, so the solution does not depend upon $\langle\mathbf{m}\rangle$. On the other hand, $\mathbf{R} = \mathbf{G}^T(\mathbf{GG}^T)^{-1}\mathbf{G} \neq \mathbf{I}$ in a underdetermined minimum-length problem, so in that case the solution does depend upon $\langle\mathbf{m}\rangle$. Adding prior information with finite error to an inverse problem that features exact data and theory only affects the underdetermined part of the solution.

In the case of indefinitely inexact data and theory, we have $\sigma_d^2 \to \infty$ or $\sigma_g^2 \to \infty$ (or both). The solution becomes

$$\mathbf{m}^{est} = \langle\mathbf{m}\rangle \quad (6.45)$$

As the data and theory contain no information, we simply recover the prior model parameters.

In the case of indefinitely inexact prior information, we have $\sigma_m^2 \to \infty$; that is, there is no prior knowledge of the model parameters, and

$$\mathbf{G}^{-g} = \mathbf{G}^T(\mathbf{GG}^T)^{-1} \quad (6.46)$$

Indefinitely weak prior information and finite-error data and theory produce the same results as finite-error prior information and error-free data and theory.

6.9 Model and data resolution in the presence of prior information

The Generalized Least-Squares solution in Eq. (6.17) does not distinguish the weighted data equation $[\text{cov }\mathbf{d}]^{-\frac{1}{2}}\mathbf{Gm} = [\text{cov }\mathbf{d}]^{-\frac{1}{2}}\mathbf{d}^{obs}$ from the weighted prior information equation $[\text{cov }\mathbf{h}]^{-\frac{1}{2}}\mathbf{Hm} = [\text{cov }\mathbf{h}]^{-\frac{1}{2}}\mathbf{h}^{pri}$; the latter is simply appended to the bottom of the former to create the combined equation $\mathbf{Fm} = \mathbf{f}$, which is then solved by simple least squares. Consequently, in analogy to simple least squares, we can define a generalized inverse $\mathbf{F}^{-g}$ and a model resolution matrix $\mathbf{R}^F$ as:

$$\mathbf{F}^{-g} = [\mathbf{F}^T\mathbf{F}]^{-1}\mathbf{F}^T \text{ so that } \mathbf{m}^{est} = \mathbf{F}^{-g}\mathbf{f}$$
$$\mathbf{R}^F = \mathbf{F}^{-g}\mathbf{F} \text{ so that } \mathbf{m}^{est} = \mathbf{R}^F\mathbf{m}^{true} \quad (6.47)$$

However, when defined in this way, the resolution is perfect, as:

$$\mathbf{R}^F = \mathbf{F}^{-g}\mathbf{F} = \left[\mathbf{F}^T\mathbf{F}\right]^{-1}\mathbf{F}^T\mathbf{F} = \mathbf{I} \tag{6.48}$$

Hence, $\mathbf{R}^F$ is not a useful quantity; we need to construct one that is more informative.

The model parameters depend upon both $\mathbf{d}^{obs}$ and $\mathbf{h}^{pri}$. This dependence is implicit in the equation $\mathbf{m}^{est} = \mathbf{F}^{-g}\mathbf{f}$, as $\mathbf{f}$ depends on $\mathbf{d}^{obs}$ and $\mathbf{h}$, but can be made explicit by rewriting this equation as:

$$\mathbf{m}^{est} = \mathbf{G}^{-g}\mathbf{d}^{obs} + \mathbf{H}^{-g}\mathbf{h}^{pri}$$

$$\text{with } \mathbf{G}^{-g} = \mathbf{A}^{-1}\mathbf{G}^T[\text{cov } \mathbf{d}]^{-1} \text{ and } \mathbf{H}^{-g} = \mathbf{A}^{-1}\mathbf{H}^T[\text{cov } \mathbf{h}]^{-1} \tag{6.49}$$

$$\text{and } \mathbf{A} = \mathbf{F}^T\mathbf{F} = \mathbf{G}^T[\text{cov } \mathbf{d}]^{-1}\mathbf{G} + \mathbf{H}^T[\text{cov } \mathbf{h}]^{-1}\mathbf{H}$$

Consider, for the moment, the special case of $\mathbf{h} = 0$; we will relax this requirement later. This case commonly arises in practice, for example, for the prior information of smoothness. The estimated model parameters depend only upon $\mathbf{d}^{obs}$, that is $\mathbf{m}^{est} = \mathbf{G}^{-g}\mathbf{d}^{obs} + 0$. We can use the forward equation to predict data associated with true model parameters $\mathbf{d}^{pre} = \mathbf{G}\mathbf{m}^{true}$ and then invert these predictions back to recovered model parameters $\mathbf{m}^{est} = \mathbf{G}^{-g}\mathbf{d}^{pre} + 0$. Hence, we obtain the usual formula for resolution:

$$\mathbf{m}^{est} = \mathbf{R}^G\mathbf{m}^{true} \text{ with } \mathbf{R}^G \equiv \mathbf{G}^{-g}\mathbf{G} \tag{6.50}$$

(except that the formula for $\mathbf{G}^{-g}$ in Eq. (6.49) depends on both $\mathbf{G}$ and $\mathbf{H}$). Unlike $\mathbf{R}^F$, $\mathbf{R}^G$ is not equal to $\mathbf{I}$ and is a useful measure of resolution.

We now relax that condition that $\mathbf{h}^{pri} = 0$. Suppose that prior information is complete, in the sense of that a unique *prior model* $\mathbf{m}^H$ can be obtained by a weighted least-squares solution of the prior information equation alone, that is, $\mathbf{m}^H = [\mathbf{H}^T[\text{cov } \mathbf{h}]^{-1}\mathbf{H}]^{-1}\mathbf{H}^T[\text{cov } \mathbf{h}]^{-1}\mathbf{h}^{pri}$. Not all prior information is complete in this sense, but it always can be approximated as complete by adding smallness information of the form $\mathbf{m} \approx 0$, but giving that information extremely large variance. We use this prior model $\mathbf{m}^H$ as a reference model, defining the deviation of a given model from it as $\Delta\mathbf{m} = \mathbf{m} - \mathbf{m}^H$. The generalized least-squares solution can be rewritten in terms of this deviation:

$$\Delta\mathbf{m}^{est} = \mathbf{m}^{est} - \mathbf{m}^H = \mathbf{A}^{-1}\left(\mathbf{G}^T[\text{cov } \mathbf{d}]^{-1}\mathbf{d}^{obs} + \mathbf{H}^T[\text{cov } \mathbf{h}]^{-1}\mathbf{h}^{pri}\right) - \mathbf{m}^H$$

$$= \mathbf{A}^{-1}\left(\mathbf{G}^T[\text{cov } \mathbf{d}]^{-1}\mathbf{d}^{obs} + \mathbf{H}^T[\text{cov } \mathbf{h}]^{-1}\mathbf{h}^{pri}\right) - \mathbf{A}^{-1}\mathbf{A}\mathbf{m}^H$$

$$= \mathbf{A}^{-1}\left(\mathbf{G}^T[\text{cov } \mathbf{d}]^{-1}\mathbf{d}^{obs} + \mathbf{H}^T[\text{cov } \mathbf{h}]^{-1}\mathbf{h} - \mathbf{A}\mathbf{m}^H\right)$$

$$= \mathbf{A}^{-1}\left(\mathbf{G}^T[\text{cov } \mathbf{d}]^{-1}\mathbf{d}^{obs} + \mathbf{H}^T[\text{cov } \mathbf{h}]^{-1}\mathbf{h} - \mathbf{G}^T[\text{cov } \mathbf{d}]^{-1}\mathbf{G}\mathbf{m}^H - \mathbf{H}^T[\text{cov } \mathbf{h}]^{-1}\mathbf{H}\left[\mathbf{H}^T[\text{cov } \mathbf{h}]^{-1}\mathbf{H}\right]^{-1}\mathbf{H}^T[\text{cov } \mathbf{h}]^{-1}\mathbf{h}^{pri}\right)$$

$$= \mathbf{A}^{-1}\left(\mathbf{G}^T[\text{cov } \mathbf{d}]^{-1}\mathbf{d}^{obs} + \mathbf{H}^T[\text{cov } \mathbf{h}]^{-1}\mathbf{h} - \mathbf{G}^T[\text{cov } \mathbf{d}]^{-1}\mathbf{G}\mathbf{m}^H - \mathbf{H}^T[\text{cov } \mathbf{h}]^{-1}\mathbf{h}^{pri}\right)$$

$$= \mathbf{G}^{-g}\left(\mathbf{d}^{obs} - \mathbf{G}\mathbf{m}^H\right) = \mathbf{G}^{-g}\left(\mathbf{d}^{obs} - \mathbf{d}^H\right)$$

$$\tag{6.51}$$

Thus the deviation of the model from $\mathbf{m}^H$ depends only on the deviation of the data from those predicted by $\mathbf{m}^H$:

$$\Delta\mathbf{m} = \mathbf{G}^{-g}\Delta\mathbf{d} \text{ with } \Delta\mathbf{m} \equiv \mathbf{m} - \mathbf{m}^H \text{ and } \Delta\mathbf{d} \equiv \mathbf{d} - \mathbf{d}^H \tag{6.52}$$

and furthermore

$$\mathbf{G}\Delta\mathbf{m} = \Delta\mathbf{d} \text{ since } \mathbf{G}\Delta\mathbf{m} = \mathbf{G}\left(\mathbf{m} - \mathbf{m}^H\right) = \mathbf{G}\mathbf{m} - \mathbf{G}\mathbf{m}^H = \mathbf{d} - \mathbf{d}^H = \Delta\mathbf{d} \tag{6.53}$$

We can now combine $\Delta\mathbf{d} = \mathbf{G}\Delta\mathbf{m}$ with $\Delta\mathbf{m} = \mathbf{G}^{-g}\Delta\mathbf{d}$ into the usual statements about resolution,

$$\Delta\mathbf{m}^{est} = \mathbf{R}^G\Delta\mathbf{m}^{true} \text{ with } \mathbf{R}^G \equiv \mathbf{G}^{-g}\mathbf{G}$$
$$\Delta\mathbf{d}^{pre} = \mathbf{N}^G\Delta\mathbf{d}^{obs} \text{ with } \mathbf{N}^G \equiv \mathbf{G}\mathbf{G}^{-g} \tag{6.54}$$

These results are due to Menke (2014) and Menke and Creel (2021).

Resolution matrices with unit row sum are attractive, because they imply that while the inversion process smooths out the solution, it preserves its overall amplitude. Defining a vector of ones as $\mathbf{w} = [1, 1, \cdots, 1]^T$, the condition

$\mathbf{R}^G \mathbf{w} = \mathbf{w}$ implies that the rows of the model resolution matrix $\mathbf{R}^G$ all sum to unity, and the condition $\mathbf{Hw} = 0$ implies that the rows of the prior information matrix $\mathbf{H}$ all sum to zero, and as has been pointed out by An (2022), the latter condition implies the former:

$$\mathbf{Rw} \overset{?}{=} \mathbf{w}$$

$$\left[\mathbf{G}^T[\text{cov } \mathbf{d}]^{-1}\mathbf{G} + \mathbf{H}^T[\text{cov } \mathbf{h}]^{-1}\mathbf{H}\right]^{-1}\mathbf{G}^T[\text{cov } \mathbf{d}]^{-1}\mathbf{Gw} \overset{?}{=} \mathbf{w}$$

$$\mathbf{G}^T[\text{cov } \mathbf{d}]^{-1}\mathbf{Gw} \overset{?}{=} \mathbf{G}^T[\text{cov } \mathbf{d}]^{-1}\mathbf{Gw} + \mathbf{H}^T[\text{cov } \mathbf{h}]^{-1}[\mathbf{Hw}]$$

$$\mathbf{G}^T[\text{cov } \mathbf{d}]^{-1}\mathbf{Gw} = \mathbf{G}^T[\text{cov } \mathbf{d}]^{-1}\mathbf{Gw} + 0$$

(6.55)

Except when $[\text{cov } \mathbf{h}]^{-1} = 0$, this is a necessary as well as sufficient condition, as $\mathbf{H}^T\mathbf{H}$ cannot be zero. We have encountered prior information matrices with unit row sum previously; they include the first difference operators $\mathbf{D}^{(1)}$ and $\mathbf{D}^{(2)}$.

The resolution matrix $\mathbf{R}^G$ is a good choice for quantifying resolution. However, the quantity being resolved is the deviation of the model from the prior model and not the model itself. The distinction, while of minor significance in cases where $\mathbf{m}^H$ has a simple shape, is more important when $\mathbf{m}^H$ is complicated.

6.10　Relative entropy as a guiding principle

In Section 6.3, we introduced the information gain, S (Eq. 6.10), as a way of quantifying the difference in information content between two probability density functions. It can be used as a guiding principle for constructing solutions to inverse problems. The idea is to find the probability density function $p_T(\mathbf{m})$—the solution to the inverse problem—that minimizes the information gain of $p_T(\mathbf{m})$ relative to the prior probability density function $p_A(\mathbf{m})$. Thus as little information as possible has been added to the prior information to create the solution. The quantity S is the *relative entropy* of the two probability density functions, so this method is called the *Maximum Relative Entropy* method and is often abbreviated MRE (Kapur, 1989). Some authors define the entropy as $+S$, in which case it is called the *Minimum Relative Entropy* method (also abbreviated MRE).

Constraints need to be added to the minimization of S, or else the solution simply would be $p_T(\mathbf{m}) = p_A(\mathbf{m})$. One of these constraints must be that the area beneath $p_T(\mathbf{m})$ is unity. The choice of the other constraints depends on the particular type of inverse problem, that is, whether it is under- or overdetermined.

As an example, consider the underdetermined problem, where the equation $\mathbf{d}^{obs} = \mathbf{Gm}$ can be assumed to hold in the mean. The minimization problem is

$$\text{minimize } S = \int p_T(\mathbf{m}) \log\left(\frac{p_T(\mathbf{m})}{p_A(\mathbf{m})}\right) d^M m \text{ with constraints}$$

$$\int p_T(\mathbf{m}) d^M m = 1 \text{ and } \int p_T(\mathbf{m})\left(\mathbf{d}^{obs} - \mathbf{Gm}\right) d^M m = 0$$

(6.56)

Here, $p_T(\mathbf{m})$ is unknown and the prior pdf $p_A(\mathbf{m})$ is prescribed. The second constraint indicates that the mean (expected) value of the prediction error $\mathbf{e} = \mathbf{d}^{obs} - \mathbf{Gm}$ is zero. This minimization problem can be solved by using the Euler-Lagrange method. It states that the integral $\int F[f(\mathbf{m}), \mathbf{m}] d^M m$ is minimized subject to the integral constraint $\int G[f(\mathbf{m}), \mathbf{m}] d^M m$ when $\Phi = F + \lambda G$ is minimiaxed with respect to f. Here, λ is a Lagrange multiplier. In our case, we introduce one Lagrange multiplier λ_0 associated with the first constraint and a vector $\boldsymbol{\lambda}$ of Lagrange multipliers associated with the second.

$$\Phi(\mathbf{m}) = p_T \log p_T - p_T \log p_A + \lambda_0 p_T + \boldsymbol{\lambda}^T\left(\mathbf{d}^{obs} - \mathbf{Gm}\right)p_T$$

(6.57)

Differentiating with respect to p_T, and setting the result to zero yields

$$\frac{\partial \Phi}{\partial p_T} = 0 = \log p_T + 1 - \log p_A + \lambda_0 + \boldsymbol{\lambda}^T\left(\mathbf{d}^{obs} - \mathbf{Gm}\right)$$

(6.58)

solving for p_T and exponentiating yields

$$p_T(\mathbf{m}) = p_A(\mathbf{m}) \exp\left\{-(1 + \lambda_0) - \boldsymbol{\lambda}^T\left(\mathbf{d}^{obs} - \mathbf{Gm}\right)\right\}$$

(6.59)

Now suppose that the prior pdf $p_A(\mathbf{m})$ is Normal, with prior value $\langle \mathbf{m} \rangle$ and prior covariance $[\text{cov } \mathbf{m}]_A$

$$p_A(\mathbf{m}) \propto \exp\left\{-\tfrac{1}{2}(\langle \mathbf{m} \rangle - \mathbf{m})^T [\text{cov } \mathbf{m}]_A^{-1}(\langle \mathbf{m} \rangle - \mathbf{m})\right\} \tag{6.60}$$

then

$$p_T(\mathbf{m}) \propto \exp\{-A(\mathbf{m})\}$$
$$\text{with } A(\mathbf{m}) \equiv \tfrac{1}{2}(\langle \mathbf{m} \rangle - \mathbf{m})^T [\text{cov } \mathbf{m}]_A^{-1}(\langle \mathbf{m} \rangle - \mathbf{m}) + (1 + \lambda_0) + \boldsymbol{\lambda}^T\left(\mathbf{d}^{obs} - \mathbf{Gm}\right) \tag{6.61}$$

We now assert that the best estimate $\mathbf{m}^{est}$ of the model parameter is the mean of this pdf, which is also its maximum likelihood point. This point occurs where $A(\mathbf{m})$ is minimum:

$$\frac{\partial A}{\partial m_q} = 0 \text{ or } 0 = -[\text{cov } \mathbf{m}]_A^{-1}(\langle \mathbf{m} \rangle - \mathbf{m}) - \mathbf{G}^T\boldsymbol{\lambda} \tag{6.62}$$

Premultiplying by $\mathbf{G}[\text{cov } \mathbf{m}]_A$ and substituting in the constraint equation $\mathbf{d}^{obs} = \mathbf{Gm}$ (which is assumed to hold in the mean) and rearranging yields

$$\boldsymbol{\lambda} = \left[\mathbf{G}[\text{cov } \mathbf{m}]_A\mathbf{G}^T\right]^{-1}\left(\mathbf{d}^{obs} - \mathbf{G}\langle \mathbf{m} \rangle\right) \tag{6.63}$$

Substituting this expression for $\boldsymbol{\lambda}$ into Eq. (6.61) and rearranging yields the solution

$$\mathbf{m}^{est} - \langle \mathbf{m} \rangle = [\text{cov } \mathbf{m}]_A\mathbf{G}^T\left[\mathbf{G}[\text{cov } \mathbf{m}]_A\mathbf{G}^T\right]^{-1}\left(\mathbf{d}^{obs} - \mathbf{G}\langle \mathbf{m} \rangle\right) \tag{6.64}$$

Thus the principle of maximum relative entropy, when applied to the underdetermined problem, yields the weighted minimum-length solution with $\mathbf{W}_m^{-1} = [\text{cov } \mathbf{m}]_A$ (compare Eq. (6.64) with Eq. (4.48)). Many of the other inverse theory solutions that were developed in this chapter using maximum likelihood techniques can also be derived using the MRE principle (Woodbury, 2011).

6.11   Equivalence of the three viewpoints

We can arrive at the same general solution to the linear inverse problem by three distinct routes.
Viewpoint 1. The solution is obtained by minimizing a weighted sum of L_2 prediction error and L_2 solution simplicity

$$\text{minimize} : \mathbf{e}^T\mathbf{W}_e\mathbf{e} + \varepsilon^2(\mathbf{m} - \langle \mathbf{m} \rangle)^T\mathbf{W}_m(\mathbf{m} - \langle \mathbf{m} \rangle)^T \tag{6.65}$$

where ε^2 is a weighting factor.
Viewpoint 2. The solution is obtained by minimizing a weighted sum of three terms: the Dirichlet spreads of model resolution and data resolution and the size of the model covariance.

$$\text{minimize} : \alpha_1 \text{ spread}(\mathbf{R}) + \alpha_2 \text{ spread}(\mathbf{N}) + \alpha_3 \text{ size}([\text{cov}_u \mathbf{m}]) \tag{6.66}$$

Here, the αs are weighting factors.
Viewpoint 3. The solution is obtained by maximizing the likelihood of the joint Normal pdf of data, prior model parameters, and theory.

$$\text{maximize} : L = \log\left(p_T(\mathbf{m}, \mathbf{d})\right) \tag{6.67}$$

These derivations emphasize the close relationship among the L_2 norm, the Dirichlet spread function, and the Normal probability density function.

6.12   Chi-square test for the compatibility of the prior and observed error

The Generalized Least Squares in Eq. (6.17) is based upon the minimization of the total error Φ (see Eq. 6.13), which is the sum of an error E associated with the observations and the error L associated with the a priori information:

$$\Phi = E + L$$
$$E = \mathbf{e}^T \mathbf{e} \text{ with } \mathbf{e} = [\text{cov } \mathbf{d}]^{-\frac{1}{2}} \left(\mathbf{d}^{obs} - \mathbf{Gm} \right)$$
$$L = \mathbf{l}^T \mathbf{l} \text{ with } \mathbf{l} = [\text{cov } \mathbf{h}]_A^{-\frac{1}{2}} \left(\mathbf{h}^{pri} - \mathbf{Hm} \right) \tag{6.68}$$

Each type of error is weighted by its *certainty*, as quantified by a function of its covariance matrix: $[\text{cov } \mathbf{d}]^{-\frac{1}{2}}$ in the case of the prediction error $\mathbf{e}$ and $[\text{cov } \mathbf{h}]_A^{-\frac{1}{2}}$ in the case of $\mathbf{l}$. These are prior covariances, meaning that they are established independently of the outcome of the inverse problem, say from a broad knowledge of the accuracy of the measurement technique and the quality of the prior information. As discussed in Section 6.5, the individual errors $\mathbf{e}$ and $\mathbf{l}$ are uncorrelated random variables with zero mean and unit variance; the weighting normalizes their amplitudes. Thus E, L, and Φ are chi-squared distributed random variables (see Eq. 3.28), with N, K, and $N+K$ degrees of freedom, respectively, where N is the number of data and K is the number of pieces of a priori information. Because they are chi-squared distributed, they have mean values of $\overline{E} = N$, $\overline{L} = K$, and $\overline{\Phi} = N + L$.

Once the inverse problem is solved, estimate of the errors E^{est}, L^{est}, and Φ^{est} can be determined by inserting $\mathbf{m}^{est}$ into Eq. (6.68). An important issue is whether these estimates are compatible with the means stated before; that is, whether their values are compatible with the *null hypothesis* that any difference between the estimated and expected value is due to random variation. If the null hypothesis is unlikely to be true, then error is not behaving as expected. The model might be incorrect, or the a priori covariances may be poorly estimated.

The estimated error decreases as more model parameters are added to the inverse problem, since they are able to fit increasingly more of the variability of the data. In the $N+K=M$ case, the estimated error can zero, even when the true error is greater than zero. Thus while Φ^{est} is a chi-squared distrbuted random variable, it has only $\nu_\Phi = N+K-M$ degrees of freedom, where M is the number of model pararameters. Consequently, it has mean ν_Φ and variance $2\nu_\Phi$. The null hypothesis is unlikely to be true when Φ^{est} falls outside the 95% confidence interval:

$$\nu_\Phi - 2(2\nu_\Phi)^{\frac{1}{2}} < \Phi < \nu_\Phi + 2(2\nu_\Phi)^{\frac{1}{2}} \tag{6.69}$$

Errors that are to the right of the interval correspond to models that fit the data more poorly than can be expected by random variation alone. This case can occur when the model is too simple; too few model parameters are available to capture the true variability of the data and prior information. Errors to the left of the interval correspond to models that *overfit* the data; that is, the error is smaller than can be expected by random variation alone. The overfit case can occur when the model is too complex. So many model parameters are available that even noise is being fit.

The individual compatibility of the errors E^{est} and of L^{est} with their respective prior variances can also be tested (although with a caveat described later). These errors are chi-squared distributed, but with fewer degrees of error than the total error Φ^{est}. An important issue is how to partition the loss of degrees of freedom associated with the M model parameters between E^{est} and L^{est}. The *Welch-Satterthwaite approximation* spreads it in proportion to the number of component errors, so that E^{est} is assigned $\nu_E = \nu_\Phi N/(N+K)$ degrees of freedom and L^{est} is assigned $\nu_L = \nu_\Phi K/(N+K)$ degrees of freedom. The 95% confidence intervals are approximately:

$$\nu_E - 2(2\nu_E)^{\frac{1}{2}} < E < \nu_E + 2(2\nu_E)^{\frac{1}{2}} \text{ and } \nu_L - 2(2\nu_L)^{\frac{1}{2}} < L < \nu_L + 2(2\nu_L)^{\frac{1}{2}} \tag{6.70}$$

The caveat is that the accuracy of the Welch-Satterthwaite approximation varies from one inverse problem to another and is not always sufficient for a correct conclusion to be drawn from the test. This problem can be avoided by generating empirical pdfs for E^{est} and L^{est} (and Φ^{est}, too), solving the inverse problem many times with noisy synthetic data, computing the error associated with every solution, and constructing an empirical probability density function (histogram) from them.

The test for the compatibility of the estimated error is illustrated in the following example. Suppose that an unknown time series $\mathbf{m}$, with samplng interval $\Delta t = 1$ and length $M = 101$, is measured by three different observers, so that

$$\mathbf{d} = \mathbf{Gm} = \begin{bmatrix} \mathbf{I} \\ \mathbf{I} \\ \mathbf{I} \end{bmatrix} \mathbf{m} \tag{6.71}$$

where $\mathbf{I}$ is an $M \times M$ identity matrix. The vector of observations $\mathbf{d}^{obs}$ is of length $N = 3M = 303$ and contains measurement noise with zero mean and a variance of, say, $\sigma_d^2 = (0.1)^2$. Furthermore, suppose that the time series is believed to be smoothly varying, corresponding to the a prior information $\mathbf{Hm} = \mathbf{h}$ with

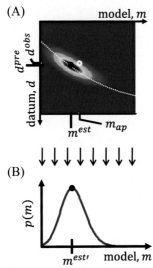

FIG. 6.16 (A) The total joint probability density function $p_T(m,d)$ can be considered the solution to the inverse problem. Its maximum likelihood point (black circle) gives an estimate of the model parameter m^{est} and a prediction of the data d^{pre}. (B) The function $p_T(m,d)$ is projected onto the m-axis, by integrating over d, to form the probability density function $p(m)$, of the model parameter irrespective of the datum. This function also has a maximum likelihood point $m^{est\prime}$ which in general can be different than m^{est}. The distinction points out the difficulty of defining a unique "solution" to an inverse problem. Scripts gdama06_16 and gdapy06_16.

$$\mathbf{H} = \frac{1}{(\Delta t)^2} \begin{bmatrix} 1 & -2 & 1 & & & \\ & 1 & -2 & 1 & & \\ & & & \ddots & & \\ & & & 1 & -2 & 1 \end{bmatrix} \text{ and } \mathbf{h} = 0 \qquad (6.72)$$

Here, $\mathbf{H}$ is a $K \times M$ matrix of second differences that approximates the second derivative, with $K = M - 2 = 99$. If we also know that $\mathbf{m}$ is approximately sinusoidal, say with an angular frequency of about $\omega_0 = 0.3$ radians/s and an amplitude of about $A = 2$, then a good choice for the variance of the a prior information is $\sigma_A^2 = (\omega_0^2 A)^2/2$, as the second derivative of $m(t) = A \sin \omega_0 t$ is $\ddot{m}(t) = -\omega_0^2 A \sin \omega_0 t$ and the average value of $[\ddot{m}(t)]^2$, which is a measure of its variance, is $(\omega_0^2 A)^2/2$. We start with an $\mathbf{m}^{true}$ (Fig. 6.16A) with the derived properties and create synthetic data $\mathbf{d}^{obs}$ by adding Normally distributed random noise to $\mathbf{d}^{true} = \mathbf{Gm}^{true}$. We then solve the inverse problem and compare the errors with the confidence ranges from Eqs. (6.69) and (6.70):

$$\begin{aligned} \Phi^{est} &= 303 \text{ with a 95\% confidence range of } 252 < \Phi < 350 \\ E^{est} &= 227 \text{ with a 95\% confidence range of } 184 < E < 269 \\ L^{est} &= 57 \text{ with a 95\% confidence range of } 50 < L < 98 \end{aligned} \qquad (6.73)$$

Because the errors are all within the 95% confidence ranges, the null hypothesis cannot be rejected. The differences between the observed values and expected values may be due to random variation. The results of the inversion are compatible with the initial assumptions about the a priori variances.

An alternative (and arguably better) set of confidence limits can be created by solving the inverse problem many times, each with a different realization of observational noise, tabulating the resulting errors and then constructing empirical probability density functions from them (Fig. 6.17B–D). These pdfs give 95% confidence ranges that are similar to, but not exactly the same, as those calculated from Eqs. (6.69) and (6.70). In this particular example, they lead to the same conclusion: that the null hypothesis cannot be rejected.

6.13   The F-test of the significance of the reduction of error

We sometimes have *two* candidate models for describing an inverse problem, one of which is more complicated than the other (in the sense that it possesses a greater number of model parameters). Suppose that Model A is more complicated than Model B and that the total estimated error Φ_B^{est} for Model A is less than the total estimated error Φ_A^{est} for Model B: $\Phi_B^{est} < \Phi_A^{est}$. Does Model A really fit the data and a priori information better than Model B?

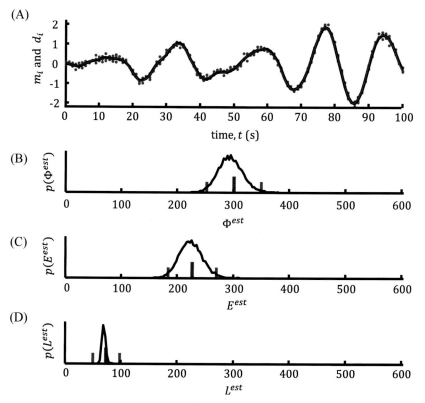

FIG. 6.17 Chi-squared test for the compatibility of the prior and posterior errors. (A) The smooth model (black curve) is measured by three different observers at a suite of closely spaced points (blue dots). The model is reconstructed (red) using the data together with the prior information that its second derivative is small. (B) Histogram (black curve) of the total error Φ^{est} for 10,000 realizations of the data, compared to mean (red bar) and 95% confidence interval (blue bars) of the corresponding chi-squared pdf. (C) Same as (B), but for the observational error E^{est}. (D) Same as (C), but for prior error L^{est}. Scripts gdama06_17 and gdapy06_17.

The answer to this question depends on the magnitude of the difference. Almost any complicated model will fit better than a less complicated one. The relevant question is whether the fit is *significantly* better, that is, whether the improvement is too large to be accounted for by random fluctuations in the data. For statistical reasons that will be cited, we pretend, in this case, that the two inverse problems are solved with two different realizations of the data.

As discussed in the previous section, Φ^{est} is chi-squared distributed with $\nu = N + K - M$ degrees of freedom. Consequently, Φ^{est} has a mean of ν and the quantity Φ^{est}/ν has a mean of unity. The ratio

$$F = \frac{\Phi_A^{est}/\nu_A}{\Phi_B^{est}/\nu_B} \tag{6.74}$$

will tend to be less than unity when model A is the better fit, and greater than unity when model B is the better fit. If the ratio is only slightly less than or greater than unity, the difference in fit may be entirely the result of random fluctuations in the data and therefore may not be significant. Values well outside of this range are significant in the sense that they are unlikely to have arisen from random variation. The null hypothesis is that deviation from the value $F = 1$ is due to random variation. One model is said to fit *significantly* better than the other only when the null hypothesis is unlikely to be true.

The probability density function $p(F, \nu_A, \nu_B)$ is called the *Fisher-Snedecor* pdf (or the F-pdf, for short). Its functional form is known, but cannot be written in terms of elementary functions, so we omit it here. It is unimodal, with mean and variance given by

$$\overline{F} = \frac{\nu_B}{\nu_B - 2} \quad \text{and} \quad \sigma_F^2 = \frac{2\nu_B^2(\nu_A + \nu_B - 2)}{\nu_A(\nu_B - 2)^2(\nu_B - 4)} \tag{6.75}$$

For large degrees of freedom, $\overline{F} \approx 1$. The quantities F^{est} and $1/F^{est}$ play symmetrical roles, in the sense that the first quantifies the improvement of fit of model A with respect to model B, and the latter, the improvement of fit of model B with respect to model A. Thus, in testing the null hypothesis that any difference between the two estimated variances is due to random variation, we should compute the probability that F is smaller than $1/F^{est}$ or larger than F^{est}:

$$P(F < 1/F^{est} \text{ or } F > F^{est}) \tag{6.76}$$

The null hypothesis can be rejected if this probability is less than 5%. This probability is computed as:

MATLAB®
| ```
Fobs = (EA/vA) / (EB/vB); % F-value
% probability value associated with F-value
if( Fobs<1 )
    Fobs=1/Fobs;
end
P = 1 - (fcdf(Fobs,vA,vB)-fcdf(1/Fobs,vA,vB));
``` |

<div align="right">[gdama06_18]</div>

Python
| ```
Fobs = (EA/vA) / (EB/vB); # F-value
# probability value associated with F-value
if( Fobs<1.0 ):
    Fobs=1.0/Fobs;
P = 1.0 - (st.f.cdf(Fobs,vA,vB)-st.f.cdf(1.0/Fobs,vA,vB));
``` |

[gdapy06_18]

Here, the MATLAB® function `fcdf()` and the Python method `st.f.cdf()` (where `st` is an abbreviation for `scipy.stats`) compute the cumulative probability of F.

In the special case of no prior information, $\Phi^{est}=E^{est}$, $K=0$, and $\nu_\Phi=\nu_E=N-M$. Furthermore, in the common case where the data are uncorrelated and with equal variance, $[\mathrm{cov}\ \boldsymbol{d}]=\sigma_d^2\boldsymbol{I}$ and the F-ratio simplifies to:

$$F = \left[\frac{\boldsymbol{e}_A^T\boldsymbol{e}_A}{\sigma_d^2(N-M_A)}\right] \bigg/ \left[\frac{\boldsymbol{e}_B^T\boldsymbol{e}_B}{\sigma_d^2(N-M_B)}\right] = \frac{\left(\sigma_{dA}^{est}\right)^2/\sigma_d^2}{\left(\sigma_{dB}^{est}\right)^2/\sigma_d^2} = \frac{\left(\sigma_{dA}^{est}\right)^2}{\left(\sigma_{dB}^{est}\right)^2} \tag{6.77}$$

In this case, the F-ratio is just the ratio of the estimated data variances, and the null hypothesis is that any departure of that ratio from unity is due to random variation. An example is shown in Fig. 6.18.

As an example of a case that includes prior information, we return to the time series reconstruction problem described at the end of the previous section and performed it two different ways, with densely sampled Case A with M model parameters (with M odd), and with sparsely sampled Case B with $(M+1)/2$ model parameters, formed by selecting every other Case A model parameter, and by linearly interpolating between them. The data kernels are then:

FIG. 6.18 Hypothetical data set (red circles), fit (blue curve) with (A) a straight line and (B) a cubic polynomial. Although the cubic fit appears superior, an F-test reveals that this level of improvement of fit will be obtained 6.4% of the time under the null hypothesis that it is due to random variation. Consequently, the improvement of fit is not significant at the 95% level. Scripts gdama06_18 and gdapy06_18.

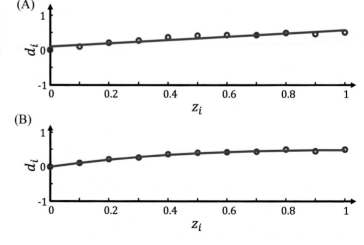

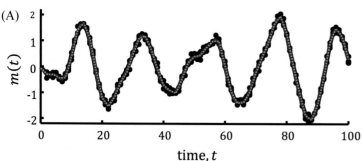

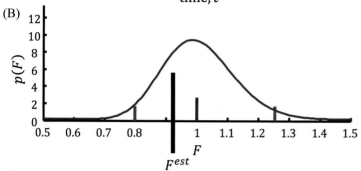

FIG. 6.19 (A) True time series, $m(t)$ (black curve), observed (dots), estimated sparse solution (red), estimated dense solution (green). (B) Probability density function $p(F)$ (red curve) showing median (red bar), mean (green bar), region of 95% probability (blue bars), and observed F-value for this experiment (black bar). See text for further discussion. Scripts gdama06_19 and gdapy06_19.

$$\mathbf{G}_A = \begin{bmatrix} \mathbf{I} \\ \mathbf{I} \\ \mathbf{I} \end{bmatrix} \text{ and } \mathbf{G}_B = \begin{bmatrix} \mathbf{C} \\ \mathbf{C} \\ \mathbf{C} \end{bmatrix} \text{ where } \mathbf{C} = \begin{bmatrix} 1 & 0 & 0 & 0 & 0 \\ \frac{1}{2} & \frac{1}{2} & 0 & 0 & 0 \\ 0 & 1 & 0 & 0 & 0 \\ 0 & \frac{1}{2} & \frac{1}{2} & 0 & 0 \\ 0 & 0 & 1 & 0 & 0 \\ 0 & 0 & \frac{1}{2} & \frac{1}{2} & 0 \\ & & \vdots & & \end{bmatrix} \cdots \tag{6.78}$$

Here $\mathbf{C}$ is a matrix that takes a sparsely sampled vector of length $(M+1)/2$ into a densely sampled vector of length M by linearly interpolating between values. The solution, for a particular realization of the noise, gives $F^{est} = 0.969$ (Fig. 6.19A), which corresponds to a case where Case A (the densely sampled case) fits better than case B (the sparsely sampled). However, the probability that $F^{est} < 1/0.969$ or $F^{est} > 0.969$ is 78.5%, much higher than the 5% needed to reject the null hypothesis. The estimated value of 0.969 is within the central part of the F-pdf and not near its tails (Fig. 6.19B).

6.14 Problems

6.1 Suppose that a random variable m is defined on the interval $[-1,1]$. A reasonable choice for the null probability density function is $p_N(m) = \frac{1}{2}$, meaning that m can be anywhere within the interval with equal probability.
(A) Calculate (either analytically or numerically) the information gain S of the prior probability density function $p_A(m) = \frac{1}{2} + cm$, where $0 < c < \frac{1}{2}$. Note that the larger the constant c, the higher the probability that m will fall in the positive half of the interval. (B) Make and interpret a plot of $S(c)$.

6.2 Suppose that a Normal probability density function with mean $\langle m \rangle$ and variance σ_m^2 is used to represent prior information about the following types of model parameters: (A) the density of sea water, (B) the shear velocity at 100-km depth in the Earth, (C) the O^{18} to O^{16} ratio in glacial ice. Propose plausible values of $\langle m \rangle$ and σ_m^2 in each case, citing references that justify your values.

6.3 Suppose that you are fitting a cubic polynomial to data $d_i = m_1 + m_2 z_i + m_3 z_i^2 + m_4 z_i^3$, but have exact prior information that $m_1 = 2m_2 = 4m_3 = 8m_4$. Write a script to solve this problem using Eq. (4.67). Set up the problem so that $N = 50$, $0 < z_i < 1$, and $m_1^{true} = 1$, and generate synthetic data with $\sigma_d^2 = (0.1)^2$. How well do the data alone (i.e., without the prior information) constrain the ratios between the ms?

6.4 This problem builds upon Problem 6.3. Suppose that you are fitting a cubic polynomial to data $d_i = m_1 + m_2 z_i + m_3 z_i^2 + m_4 z_i^3 3$, but have inexact prior information that $m_1 = 2m_2 = 4m_3 = 8m_4$. Write a script to solve this problem using Eq. (6.17). Use a range of values for the variance σ_m^2 of the prior information, from very uncertain to very certain. Set up the problem so that $N=50$, $0<z_i<1$, and $m_1^{true}=1$, and generate synthetic data with $\sigma_d^2=(0.1)^2$. How well do the data alone (i.e., without the prior information) constrain the ratios between the ms?

6.5 This problem builds upon Problems 6.3 and 6.4. Modify your script in Problem 6.4 in the case where the model of the data fitting a cubic polynomial is thought to be inexact, with a variance $\sigma_g^2=(0.05)^2$. How does this modification change your results?

6.6 Write a MATLAB® or Python script to empirically generate the $p(F)$ probability density function with $\nu_1=\nu_2=20$ by (A) using a random number generator to generate batches of 20 Normally distributed random numbers with zero mean and unit variance. (B) Calculating χ_{20}^2 by summing the squares of each batch. (C) Calculating F for pairs of batches. (D) Repeat many times, creating a histogram of the resulting Fs, and normalize to unit area to produce an empirical estimate of $p(F)$. (E) Compare your result with the result of MATLAB® or Python's $p(F)$ function.

6.7 This problem expands upon Problem 6.6. Suppose that the random numbers in step (A) are drawn from a uniform pdf with zero mean and unit variance, but that F is calculated the same as before (let us call it F_0). (A) How different is $p(F_0)$ from $p(F)$? (B) How would treating data with uniformly distributed error as if they were Normally distributed affect the results of an F-test? Hint: A pdf that is uniform between a and b has a variance of $(b-a)^2/12$.

References

An, M., 2022. On resolution matrices. Pure Appl. Geophys. 121. https://doi.org/10.1007/s00024-022-03211-9.

Kapur, J.N., 1989. Maximum Entropy Models in Science and Engineering. Wiley, New York. 636 pp.

Menke, W., 2014. Review of the generalized least squares method. Surv. Geophys. 36, 1–25.

Menke, W., Creel, R., 2021. Gaussian process regression reviewed in the context of inverse theory. Surv. Geophys. 42, 473–503. https://doi.org/10.1007/s10712-021-09640-w.

Tarantola, A., Valette, B., 1982a. Generalized non-linear inverse problems solved using the least squares criterion. Rev. Geophys. Space Phys. 20, 219–232.

Tarantola, A., Valette, B., 1982b. Inverse problems¼quest for information. J. Geophys. 50, 159–170.

Woodbury, M.A., 1950. Woodbury, Inverting Modified Matrices, Memorandum Rept. 42. Statistical Research Group, Princeton University, Princeton, NJ. 4 pp.

Woodbury, A.D., 2011. Minimum relative entropy, Bayes and Kapur. Geophys. J. Int. 185, 181–189.

7

Data assimilation methods including Gaussian process regression and Kalman filtering

7.1 Smoothness via the prior covariance matrix

The notion that a function is smooth implies correlation between its values at adjacent points. Consequently, correlation can be quantified by a covariance matrix, because its elements express the degree of correlation between pairs of points. Functions with different degrees of smoothness have different covariance matrices. Asserting a prior covariance matrix, as is done in Generalized Least Squares (GLS), is sometimes equivalent (depending on the covariance matrix) to imposing prior information of smoothness.

To see how this works, consider model parameters m_i and m_j that represent a material parameter, such as density $m(\mathbf{x})$, at two points $\mathbf{x}^{(i)}$ and $\mathbf{x}^{(j)}$ separated by a distance r_{ij}. Considering $m_i \equiv m(\mathbf{x}^{(i)})$ and $m_j \equiv m(\mathbf{x}^{(j)})$ as random variables, the degree of correlation between them is expressed by the (i, j) element of the covariance matrix. For a smooth function, we expect that the prior covariance $[\text{cov}_A \, \mathbf{m}]_{ij}$ slowly declines with the distance r_{ij}, because pairs of neighboring points are more correlated than well-separated ones. If the degree of correlation depends *only* upon the separation of the model parameters, and not their absolute location, then $[\text{cov}_A \, \mathbf{m}]_{ij} = \sigma_{mA}^2 c(r_{ij})$, where σ_{mA}^2 is a prior variance and $c(r)$ is an *auto-covariance function*. It obeys $c(0) = 1$ and $|c(r)| \leq 1$. In this case, the model function $m(\mathbf{x})$ is said to be *stationary*. As an example, the three-dimensional exponential covariance

$$[\text{cov}_A \, \mathbf{m}]_{ij} = \sigma_{mA}^2 \exp\{-sr_{ij}\} \text{ with } r_{ij} = |x_i - x_j| + |y_i - y_j| + |z_i - z_j| \tag{7.1}$$

and the three-dimensional Gaussian covariance

$$[\text{cov}_A \, \mathbf{m}]_{ij} = \sigma_{mA}^2 \exp\{-s^2 r_{ij}^2\} \text{ with } r_{ij}^2 = (x_i - x_j)^2 + (y_i - y_j)^2 + (z_i - z_j)^2 \tag{7.2}$$

both imply that the model function is smooth over a scale length s^{-1}. The smaller the scale factor, the smoother the medium. However, the two cases differ in the degree to which more distant points are correlated, with the degree of correlation being less in the Gaussian case, because it has the shorter tails.

7.2 Realizations of a medium with a specified covariance matrix

Images that represent a realization of a two-dimensional random function $m(x, y)$ can be very useful for providing a sense of how the properties of $m(x, y)$ correspond to the properties of $\sigma_{mA}^2 c(r)$ and for creating realistic synthetic data sets with which to test software. Two such images are shown in Fig. 7.1, corresponding to exponential and Gaussian covariance matrices, both with the same scale factor s. Note that their texture is very different, with the image associated with the exponential covariance having the rougher texture. This behavior is due to $c(r)$ having a nonzero slope at $r = 0$, which implies that the correlation declines with rapidly offset, even when the offset is very small.

Images like these are easy to compute, but require Fourier analysis, a subject covered only briefly in this book (in Section 15.9; see also Menke, 2022a). Nevertheless, we present the algorithm here for the benefit of those readers already familiar with the subject. The idea is to start with an image, say $n(x, y)$, that consists of uncorrelated, Normally

FIG. 7.1   Realizations of random media with (A) exponential and (B) Gaussian auto-covariance functions. Scripts gdama07_01 and gdapy07_01.

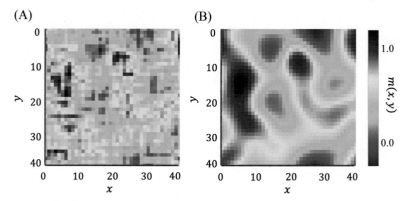

distributed random noise with zero mean and unit variance and to modify it so that it has the desired covariance. The modification process uses three key facts. First, that the amplitude spectral density $\left|\widetilde{n}(k_x, k_y)\right|$ of uncorrelated noise is approximately constant. Here, the tildes imply Fourier transformation from position (x, y) to wavenumber (k_x, k_y). Second, the auto-covariance $c(x, y)$ is just a normalized form of the autocorrelation function. Third, the amplitude spectral density of the autocorrelation is proportional to the square of the amplitude spectral density of the original function $m(x, y)$. These facts allow us to define a conversion factor $\left|\widetilde{\widetilde{c}}(k_x, k_y)\right|^{\frac{1}{2}}$ that shapes the amplitude spectral density of $n(x, y)$ into the one implied by $c(x, y)$

$$\widetilde{\widetilde{m}}(k_x, k_y) = \left|\widetilde{\widetilde{c}}(k_x, k_y)\right|^{\frac{1}{2}} \widetilde{\widetilde{n}}(k_x, k_y) \tag{7.3}$$

We apply this procedure by generating the image $n(x, y)$ of random noise and an image $c(x, y)$ of the desired auto-covariance, Fourier transforming them to $\widetilde{\widetilde{n}}(k_x, k_x)$ and $\widetilde{\widetilde{c}}(k_x, k_x)$, applying Eq. (7.3), and inverse Fourier transforming $\widetilde{\widetilde{m}}(k_x, k_x)$ to $m(x, y)$. The point $(x, y) = (0, 0)$ should be in the center of the $c(x, y)$ image, so that no parts of that function with significant amplitude are cut off. Finally, the image $m(x, y)$ needs to be normalized to the correct variance σ_m^2.

```
                          MATLAB®
% make m(x,y) with desired autocovariance, c using fact that
% a.s.d. of autocorrelation is power spectrum of data
sqrtabsctt = abs(fft2(c)).^0.5;  % target asd
n = random('Normal',0.0,1.0,Nx,Ny);  % random noise
m = ifft2(sqrtabsctt.*fft2(n)); % shape spectrum on n into d
sigmamest = std(m(:)); % sqrt variance
m = (sigmam/sigmamest)*m; % normalize to correct variance
                                                    [gdama07_01]
```

```
                          Python
# make m(x,y) with desired autocovariance, c using fact that
# a.s.d. of autocorrelation is power spectrum of data
sqrtabsctt=np.sqrt(np.abs(np.fft.fft2(c)));  # target asd
# random noise
n = np.random.normal(loc=0.0,scale=1.0,size=(Nx,Ny));
# shape spectrum on n into d
m = np.fft.ifft2(np.multiply(sqrtabsctt,np.fft.fft2(n)));
sigmamest = np.std(m);  # sqrt variance
m = np.real(m);
m = (sigmam/sigmamest)*m; # normalize to correct variance
                                                    [gdapy07_01]
```

Examples are shown in Figs. 7.1 and 7.2.

(A)　　　　　　(B)

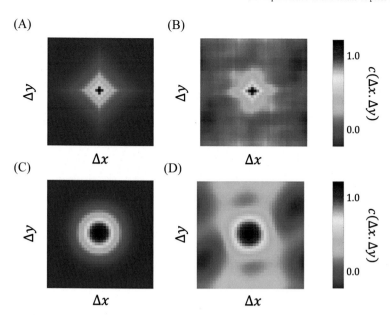

(C)　　　　　　(D)

FIG. 7.2 Auto-covariance functions c($\Delta x, \Delta y$). (A) Exact exponential auto-covariance. (B) Auto-covariance of one realization of a function with exponential auto-covariance. (C) Exact Gaussian auto-covariance. (D) Auto-covariance of one realization of a function with Gaussian auto-covariance. Scripts gdama07_01 and gdapy07_01.

7.3 Equivalence of two forms of prior information

The prior covariance matrix [cov_A **m**] enters into GLS through the measure of total error in prior information $L = \mathbf{l}^T[\text{cov}_A \mathbf{m}]^{-1}\mathbf{l}$, where the individual errors are $\mathbf{l} = \mathbf{m} - \langle\mathbf{m}\rangle$, with $\langle\mathbf{m}\rangle$ a prior value. In past chapters, we have mostly considered uncorrelated covariance matrices [$\text{cov}_A \mathbf{m}] = \sigma_{mA}^2\mathbf{I}$, but this discussion highlights the fact that covariance matrices with correlation can be used to impose smoothness.

The expression for total error in prior information can be simplified by defining a new vector $\tilde{\mathbf{l}} \equiv \mathbf{D}\mathbf{l}$, where $\mathbf{D}^T\mathbf{D} = [\text{cov}_A \mathbf{m}]^{-1}$, for then $\tilde{\mathbf{l}}^T\tilde{\mathbf{l}} = \mathbf{l}^T[\text{cov}_A \mathbf{m}]^{-1}\mathbf{l} = L$. One possible choice is $\mathbf{D} = [\text{cov}_A \mathbf{m}]^{-\frac{1}{2}}$ (the symmetric square root of [$\text{cov}_A \mathbf{m}$]). This equivalence highlights the fact that smoothness can be implemented through two different types of prior information

$$\mathbf{l} = \mathbf{m} - \langle\mathbf{m}\rangle \text{ with covariance } [\text{cov}_A \mathbf{m}]$$
$$\text{or} \tag{7.4}$$
$$\tilde{\mathbf{l}} = \mathbf{D}\mathbf{m} - \mathbf{D}\langle\mathbf{m}\rangle \text{ with covariance } \sigma_D^2$$

Here, the variance σ_D^2 expresses the accuracy to which $\tilde{\mathbf{l}} \approx \mathbf{0}$. For every prior covariance matrix [$\text{cov}_A \mathbf{m}$], there is an equivalent weight matrix $\mathbf{D} = [\text{cov}_A \mathbf{m}]^{-\frac{1}{2}}$, and for every weight matrix $\mathbf{D}$, there is an equivalent prior covariance matrix [$\text{cov}_A \mathbf{m}] = [\mathbf{D}^T\mathbf{D}]^{-1}$ (presuming the matrix inverses to exist).

We encountered the weight matrix $\mathbf{D}$ in Section 4.12, where we showed that the steepness of a one-dimensional model $m_i = m(x_i)$ is quantified by $\mathbf{D}^{(1)}\mathbf{m}$, where $\mathbf{D}^{(1)}$ is the first-difference matrix. Similarly, we showed that roughness is quantified by $\mathbf{D}^{(2)}\mathbf{m}$, where $\mathbf{D}^{(2)}$ is the second-difference matrix.

For one-dimensional problems, correspondences between some pairs of [$\text{cov}_A \mathbf{m}$] and $\mathbf{D}$ are known (Menke and Eilon, 2015; Menke and Creel, 2021a,b; Menke 2022a)

$$\text{exponential}$$
$$[\text{cov}_A \mathbf{m}]_{nm} = \sigma_{mA}^2 \exp\{-s\Delta x|n - m|\}$$
$$\mathbf{D}_E \propto s^{-1}\mathbf{D}^{(1)} + \mathbf{I}$$
$$\text{long tailed bell:}$$
$$[\text{cov}_A \mathbf{m}]_{nm} = \sigma_{mA}^2(1 + s\Delta x|n - m|)\exp(-s\Delta x|n - m|)$$
$$\mathbf{D}_L \propto s^{-2}\mathbf{D}^{(2)} - \mathbf{I} \tag{7.5}$$
$$\text{short} - \text{tailed bell (Gaussian):}$$
$$[\text{cov}_A \mathbf{m}]_{nm} = \sigma_{mA}^2 \exp\{-s^2(\Delta x)^2(n - m)^2\}$$
$$\mathbf{D}_G \propto \mathbf{I} - \tfrac{1}{4}s^{-2}\mathbf{D}^{(2)} + \left(\frac{1}{32}\right)s^{-4}\mathbf{D}^{(4)} + \cdots$$

FIG. 7.3 (A–C) Exponential, Long-tailed Bell and Gaussian covariance matrices, respectively, for the one-dimensional case with $M=100$. (D–E) Cross-sections along column 50 of these matrices (red) compared with approximations (green) based on differential operators. Scripts `gdama07_02` and `gdapy07_02`.

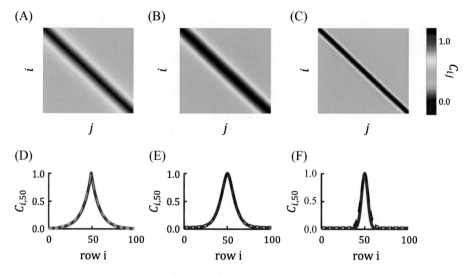

The expressions for $\mathbf{D}_E$ and $\mathbf{D}_L$ are exact, but the one for $\mathbf{D}_G$ is approximate (Fig. 7.3). For small-scale parameter $s \ll 1$, the exponential covariance is approximately equivalent to prior information of flatness ($\mathbf{D}_E \propto \mathbf{D}^{(1)}$) and the long-tailed bell covariance to smoothness ($\mathbf{D}_L \propto \mathbf{D}^{(2)}$). In all three of these cases, $\mathbf{D}$ is sparse, whereas $[\text{cov}_A \, \mathbf{m}]$ is nonsparse, implying that methods that use $\mathbf{D}$ are the more efficient solution method.

As was discussed in Section 6.4, GLS combines a linear data equation $\mathbf{Gm}=\mathbf{d}^{obs}$ (with prior covariance $[\text{cov} \, \mathbf{d}]$) and a linear prior information equation $\mathbf{Hm}=\mathbf{h}^{pri}$ (with prior covariance $[\text{cov} \, \mathbf{h}]$). Its solution, written in terms of deviations, is

$$\mathbf{m}^{est} = \mathbf{m}^{H} + \Delta\mathbf{m} \quad \text{with} \quad \Delta\mathbf{m} = \mathbf{G}^{-g}\Delta\mathbf{d} \quad \text{and} \quad \Delta\mathbf{d} = \left(\mathbf{d} - \mathbf{Gm}^{H}\right) \tag{7.6}$$

Here, $\mathbf{m}^{H}$ is the *prior solution*, which is the solution implied by the prior information, acting alone:

$$\mathbf{m}^{H} = \left[\mathbf{H}^{T}[\text{cov} \, \mathbf{h}]^{-1}\mathbf{H}\right]^{-1}\mathbf{H}^{T}[\text{cov} \, \mathbf{h}]^{-1}\mathbf{h}^{pri} \tag{7.7}$$

The two equivalent forms of the generalized inverse are

$$1^{st} \text{ form}: \mathbf{G}^{-g} = \left[\mathbf{G}^{T}[\text{cov} \, \mathbf{d}]^{-1}\mathbf{G} + \mathbf{H}^{T}[\text{cov} \, \mathbf{h}]^{-1}\mathbf{H}\right]^{-1}\mathbf{G}^{T}\left[\mathbf{C}^{(d)}\right]^{-1}$$

$$\text{and}$$

$$2^{nd} \text{ form}: \mathbf{G}^{-g} = \left[\mathbf{H}^{T}[\text{cov} \, \mathbf{h}]^{-1}\mathbf{H}\right]^{-1}\mathbf{G}^{T}\left[[\text{cov} \, \mathbf{d}] + \mathbf{G}\left[\mathbf{H}^{T}[\text{cov} \, \mathbf{h}]^{-1}\mathbf{H}\right]^{-1}\mathbf{G}^{T}\right]^{-1} \tag{7.8}$$

This result is due to Tarantola and Valette (1982), although the forms here have been generalized from the ones they develop. The first form is a generalization of damped least squares and the second form, of damped minimum length. The equivalence of these two forms can be demonstrated by equating them and moving each of the matrix inverses to the opposite side of the equation

$$\left[\mathbf{G}^{T}\left[\mathbf{C}^{(d)}\right]^{-1}\mathbf{G} + \mathbf{H}^{T}\left[\mathbf{C}^{(h)}\right]^{-1}\mathbf{H}\right]^{-1}\mathbf{G}^{T}\left[\mathbf{C}^{(d)}\right]^{-1} \stackrel{?}{=} \left[\mathbf{H}^{T}\left[\mathbf{C}^{(h)}\right]^{-1}\mathbf{H}\right]^{-1}\mathbf{G}^{T}\left[\left[\mathbf{C}^{(d)}\right] + \mathbf{G}\left[\mathbf{H}^{T}\left[\mathbf{C}^{(h)}\right]^{-1}\mathbf{H}\right]^{-1}\mathbf{G}^{T}\right]^{-1}$$

$$\mathbf{G}^{T}\left[\mathbf{C}^{(d)}\right]^{-1}\left[\left[\mathbf{C}^{(d)}\right] + \mathbf{G}\left[\mathbf{H}^{T}\left[\mathbf{C}^{(h)}\right]^{-1}\mathbf{H}\right]^{-1}\mathbf{G}^{T}\right] \stackrel{?}{=} \left[\mathbf{G}^{T}\left[\mathbf{C}^{(d)}\right]^{-1}\mathbf{G} + \mathbf{H}^{T}\left[\mathbf{C}^{(h)}\right]^{-1}\mathbf{H}\right]\left[\mathbf{H}^{T}\left[\mathbf{C}^{(h)}\right]^{-1}\mathbf{H}\right]^{-1}\mathbf{G}^{T} \tag{7.9}$$

$$\left[\mathbf{G}^{T} + \mathbf{G}^{T}\left[\mathbf{C}^{(d)}\right]^{-1}\mathbf{G}\left[\mathbf{H}^{T}\left[\mathbf{C}^{(h)}\right]^{-1}\mathbf{H}\right]^{-1}\mathbf{G}^{T}\right] = \left[\mathbf{G}^{T}\left[\mathbf{C}^{(d)}\right]^{-1}\mathbf{G}\left[\mathbf{H}^{T}\left[\mathbf{C}^{(h)}\right]^{-1}\mathbf{H}\right]^{-1}\mathbf{G}^{T} + \mathbf{G}^{T}\right]$$

When the prior covariance of the model parameters is imposed, then $\mathbf{H}=\mathbf{I}$, $[\text{cov} \, \mathbf{h}]=[\text{cov}_A \, \mathbf{m}]$, and $\mathbf{m}^{(h)}=\langle\mathbf{m}\rangle$. The second form of the generalized inverse

$$G^{-g} = [\text{cov}_A \, \mathbf{m}] \, \mathbf{G}^T \left[[\text{cov } \mathbf{d}] + \mathbf{G}[\text{cov}_A \, \mathbf{m}]\mathbf{G}^T\right]^{-1} \tag{7.10}$$

is simpler, because it involves the fewest matrix inverses. When the weight matrix $\mathbf{D}$ is imposed, then $\mathbf{H} = \mathbf{D}$, $[\text{cov } \mathbf{h}] = \sigma_D^2 \mathbf{I}$, and $\mathbf{m}^{(h)} = \mathbf{D}^{-1}\mathbf{h}^{pri}$. The first form of the generalized inverse

$$\mathbf{G}^{-g} = \left[\mathbf{G}^T[[\text{cov } \mathbf{d}]]^{-1}\mathbf{G} + \sigma_D^{-2}\mathbf{D}^T\mathbf{D}\right]^{-1}\mathbf{G}^T[[\text{cov } \mathbf{d}]]^{-1} \tag{7.11}$$

is the simpler.

7.4 Gaussian process regression

When GLS is used to estimate a function $m(\mathbf{x})$ based on noisy observations of the function itself, the process is called *Gaussian Process Regression* (GPR) (Rasmussen and Williams, 2006). In order to facilitate the bookkeeping, we divide the model parameters into two groups, *control* model parameters, $m_i^{(c)}$, $i = 1 \cdots N$, for which observations d_i^{obs} are available, and *target* model parameters, $m_i^{(t)}$ $i = 1 \cdots M$, for which estimates are needed because no observations are available. These model parameters are associated with positions $\mathbf{x}_i^{(c)}$ and $\mathbf{x}_i^{(t)}$, respectively. The data kernel is $\mathbf{G} = [\mathbf{I} \ \mathbf{0}]$ and the data equation is

$$\mathbf{Gm} = [\mathbf{I} \ \mathbf{0}] \begin{bmatrix} \mathbf{m}^{(c)} \\ \mathbf{m}^{(t)} \end{bmatrix} = \mathbf{m}^{(c)} = \mathbf{d}^{obs} \tag{7.12}$$

The prior information is that the model parameters have correlation expressed by the prior covariance matrix $[\text{cov}_A \, \mathbf{m}]$. This matrix can be divided into target and control blocks

$$[\text{cov}_A \, \mathbf{m}] = \begin{bmatrix} [\text{cov}_A \, \mathbf{m}]^{(cc)} & [\text{cov}_A \, \mathbf{m}]^{(ct)} \\ [\text{cov}_A \, \mathbf{m}]^{(tc)} & [\text{cov}_A \, \mathbf{m}]^{(tt)} \end{bmatrix} \tag{7.13}$$

The matrix is symmetric, so $[\text{cov}_A \, \mathbf{m}]^{(tc)} = [\text{cov}_A \, \mathbf{m}]^{(ct)}$. With the assumption of uncorrelated data $[\text{cov} \mathbf{d}] = \sigma_d^2\mathbf{I}$, Eq. (7.10) implies

$$\mathbf{G}^{-g} = \begin{bmatrix} [\text{cov}_A \, \mathbf{m}]^{(cc)} & [\text{cov}_A \, \mathbf{m}]^{(ct)} \\ [\text{cov}_A \, \mathbf{m}]^{(tc)} & [\text{cov}_A \, \mathbf{m}]^{(tt)} \end{bmatrix} [\mathbf{I} \ \mathbf{0}] \left(\begin{bmatrix} [\text{cov}_A \, \mathbf{m}]^{(cc)} & [\text{cov}_A \, \mathbf{m}]^{(ct)} \\ [\text{cov}_A \, \mathbf{m}]^{(tc)} & [\text{cov}_A \, \mathbf{m}]^{(tt)} \end{bmatrix} \begin{bmatrix} \mathbf{I} \\ \mathbf{0} \end{bmatrix} + \sigma_d^2\mathbf{I} \right)^{-1}$$

$$= \begin{bmatrix} [\text{cov}_A \, \mathbf{m}]^{(cc)} \\ [\text{cov}_A \, \mathbf{m}]^{(tc)} \end{bmatrix} \left[[\text{cov}_A \, \mathbf{m}]^{(cc)} + \sigma_d^2\mathbf{I}\right]^{-1} = [\text{cov}_A \, \mathbf{m}]^{(L)}\mathbf{Q}^{-1} \tag{7.14}$$

$$\text{with } \mathbf{Q} \equiv [\text{cov}_A \, \mathbf{m}]^{(cc)} + \sigma_d^2\mathbf{I} \text{ and } [\text{cov}_A \, \mathbf{m}]^{(L)} \equiv \begin{bmatrix} [\text{cov}_A \, \mathbf{m}]^{(cc)} \\ [\text{cov}_A \, \mathbf{m}]^{(tc)} \end{bmatrix}$$

Usually, only the solution for the target points is of interest

$$\mathbf{m}^{(t)est} = \left\langle \mathbf{m}^{(t)} \right\rangle + \Delta\mathbf{m}^{(t)} \text{ and } \Delta\mathbf{d} = \left(\mathbf{d}^{obs} - \left\langle \mathbf{m}^{(c)} \right\rangle \right)$$

$$\Delta\mathbf{m}^{(t)} = \mathbf{G}_{tc}^{-g}\Delta\mathbf{d} = [\text{cov}_A \, \mathbf{m}]^{(tc)}\mathbf{Q}^{-1}\Delta\mathbf{d} \tag{7.15}$$

An efficient technique for calculating the deviation $\Delta\mathbf{m}^{(t)}$ is to solve the linear system $\mathbf{Q}\boldsymbol{\lambda} = \Delta\mathbf{d}$ for the vector $\boldsymbol{\lambda}$ and then to perform the multiplication $\Delta\mathbf{m}^{(t)} = [\text{cov}_A\mathbf{m}]^{(tc)} \boldsymbol{\lambda}$. Another equivalent form of the solution is

$$\mathbf{m}^{est} = \langle \mathbf{m} \rangle + \mathbf{G}^{-g} \left(\mathbf{d}^{obs} - \mathbf{G}\langle \mathbf{m} \rangle \right) = \mathbf{G}^{-g}\mathbf{d}^{obs} + (\mathbf{I} - \mathbf{G}^{-g}\mathbf{G})\langle \mathbf{m} \rangle$$

$$\text{with } \mathbf{G} = [\mathbf{I} \ \mathbf{0}] \text{ and } \mathbf{G}^{-g} = [\text{cov}_A\mathbf{m}]\mathbf{G}^T\mathbf{Q}^{-1} = [\text{cov}_A \, \mathbf{m}]^{(L)}\mathbf{Q}^{-1} \tag{7.16}$$

It is useful for computing the posterior variance, because it separates the effects of random variation in $\mathbf{d}^{obs}$ from that in $\langle \mathbf{m} \rangle$. By the usual rules of error propagation, the posterior covariance is

$$[\text{cov}\mathbf{m}] = \sigma_d^2\mathbf{G}^{-g}\mathbf{G}^{-gT} + (\mathbf{I} - \mathbf{G}^{-g}\mathbf{G})[\text{cov}_A \, \mathbf{m}](\mathbf{I} - \mathbf{G}^{-g}\mathbf{G})^T \tag{7.17}$$

Note that the covariance of the estimated solution depends both on the variance σ_d^2 of the observed data and the prior covariance of the model parameters $[\text{cov}_A \, \mathbf{m}]$. The first term dominates when the data density is sufficient for the model to be faithfully reconstructed. When the data are too sparse, then the second term dominates and the estimated model can be *anywhere* between bounds controlled by σ_{mA}^2.

FIG. 7.4   (A) Random time series, $m(x)$ (black), with a Gaussian auto-covariance ($\sigma_{mA}^2=1$, $s^2=50$), irregularly sampled data, $d(x)$ (circles), with data gaps growing from left to right, together with Gaussian Process Regression estimate (red). (B) Standard deviation σ_m of an ensemble of 10,000 estimates, with random fluctuation due to observational error $\sigma_d=0.1$ (black), only, and both observational error and prior model error $\sigma_{mA}^2=1$ (red). The finely sampled left side of the plot is controlled by the observational error and the coarsely sampled right side by prior error. Scripts gdama07_03 and gdapy07_03.

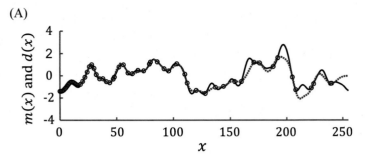

(A)

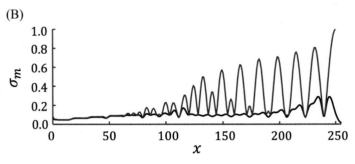

(B)

Were many experiments to be performed, identical except for observational noise, and an ensemble of solutions be computed, all using the same prior model, say $\langle \mathbf{m} \rangle = 0$, then the variability of solutions will be controlled by the first term in Eq. (7.17) only. The parts of the solutions that are unconstrained by data all will satisfy $\langle \mathbf{m} \rangle \approx 0$ and will have low variance. In order for the ensemble to include the uncertainty associated with the second term in Eq. (7.17), one must add random fluctuation to $\langle \mathbf{m} \rangle$ in Eq. (7.16) (Fig. 7.4).

The formula for covariance (Eq. 7.17) can be substantially simplified through algebraic manipulation

$$[\mathrm{cov}\ \mathbf{m}] =$$

$$\sigma_d^2 \mathbf{G}^{-g} \mathbf{G}^{-gT} + [\mathrm{cov}_A\ \mathbf{m}] - \mathbf{G}^{-g}\mathbf{G}[\mathrm{cov}_A\ \mathbf{m}] - [\mathrm{cov}_A\ \mathbf{m}]\mathbf{G}^T\mathbf{G}^{-gT} + \mathbf{G}^{-g}\mathbf{G}[\mathrm{cov}_A\ \mathbf{m}]\mathbf{G}^T\mathbf{G}^{-gT} =$$

$$[\mathrm{cov}_A\ \mathbf{m}] + \mathbf{G}^{-g}\left\{ \sigma_d^2\mathbf{I} + \mathbf{G}[\mathrm{cov}_A\ \mathbf{m}]\mathbf{G}^T \right\}\mathbf{G}^{-gT} - \mathbf{G}^{-g}\mathbf{G}[\mathrm{cov}_A\ \mathbf{m}] - [\mathrm{cov}_A\ \mathbf{m}]\mathbf{G}^T\mathbf{G}^{-gT} =$$

$$[\mathrm{cov}_A\ \mathbf{m}] + \mathbf{G}^{-g}\left\{ \sigma_d^2\mathbf{I} + [\mathrm{cov}_A\ \mathbf{m}]^{(cc)} \right\}\mathbf{G}^{-gT} - \mathbf{G}^{-g}\mathbf{G}[\mathrm{cov}_A\ \mathbf{m}] - [\mathrm{cov}_A\ \mathbf{m}]\mathbf{G}^T\mathbf{G}^{-gT} = \tag{7.18}$$

$$[\mathrm{cov}_A\ \mathbf{m}] + [\mathrm{cov}_A\ \mathbf{m}]\mathbf{G}^T\mathbf{Q}^{-1}\mathbf{Q}\mathbf{Q}^{-1}\mathbf{G}[\mathrm{cov}_A\ \mathbf{m}] - [\mathrm{cov}_A\ \mathbf{m}]\mathbf{G}^T\mathbf{Q}^{-1}\mathbf{G}[\mathrm{cov}_A\ \mathbf{m}] - [\mathrm{cov}_A\ \mathbf{m}]\mathbf{G}^T\mathbf{Q}^{-1}\mathbf{G}[\mathrm{cov}_A\ \mathbf{m}] =$$

$$[\mathrm{cov}_A\ \mathbf{m}] + [\mathrm{cov}_A\ \mathbf{m}]\mathbf{G}^T\mathbf{Q}^{-1}\mathbf{G}[\mathrm{cov}_A\ \mathbf{m}] - [\mathrm{cov}_A\ \mathbf{m}]\mathbf{G}^T\mathbf{Q}^{-1}\mathbf{G}[\mathrm{cov}_A\ \mathbf{m}] - [\mathrm{cov}_A\ \mathbf{m}]\mathbf{G}^T\mathbf{Q}^{-1}\mathbf{G}[\mathrm{cov}_A\ \mathbf{m}] =$$

$$[\mathrm{cov}_A\ \mathbf{m}] - [\mathrm{cov}_A\ \mathbf{m}]\mathbf{G}^T\mathbf{Q}^{-1}\mathbf{G}[\mathrm{cov}_A\ \mathbf{m}]$$

Noting that $[\mathrm{cov}_A\ \mathbf{m}]\mathbf{G}^T = [\mathrm{cov}_A\ \mathbf{m}]^{(L)}$, we have

$$[\mathrm{cov}\ \mathbf{m}] = [\mathrm{cov}_A\ \mathbf{m}] - [\mathrm{cov}_A\ \mathbf{m}]^{(L)}\mathbf{Q}^{-1}\ [\mathrm{cov}_A\ \mathbf{m}]^{(L)T} \tag{7.19}$$

Extracting the control-control and target-target blocks gives

$$[\mathrm{cov}\ \mathbf{m}]^{(cc)} = [\mathrm{cov}_A\ \mathbf{m}]^{(cc)} - [\mathrm{cov}_A\ \mathbf{m}]^{(cc)}\mathbf{Q}^{-1}[\mathrm{cov}_A\ \mathbf{m}]^{(cc)}$$
$$[\mathrm{cov}\ \mathbf{m}]^{(tt)} = [\mathrm{cov}_A\ \mathbf{m}]^{(tt)} - [\mathrm{cov}_A\ \mathbf{m}]^{(tc)}\mathbf{Q}^{-1}[\mathrm{cov}_A\ \mathbf{m}]^{(tc)T} \tag{7.20}$$

For very certain data, $\sigma_d^2 \to 0$ and $\mathbf{Q} \to [\mathrm{cov}_A\ \mathbf{m}]^{(cc)}$, so

$$\mathbf{G}^{-g} \to \begin{bmatrix} \mathbf{I} \\ [\mathrm{cov}_A\ \mathbf{m}]^{(tc)}\left[[\mathrm{cov}_A\ \mathbf{m}]^{(cc)}\right]^{-1} \end{bmatrix} \text{ and } [\mathrm{cov}\ \mathbf{m}]^{(cc)} \to 0 \text{ and}$$

$$[\mathrm{cov}\ \mathbf{m}]^{(tt)} = [\mathrm{cov}_A\ \mathbf{m}]^{(tt)} - [\mathrm{cov}_A\ \mathbf{m}]^{(tc)}\left[[\mathrm{cov}_A\ \mathbf{m}]^{(cc)}\right]^{-1}[\mathrm{cov}_A\ \mathbf{m}]^{(tc)T} \tag{7.21}$$

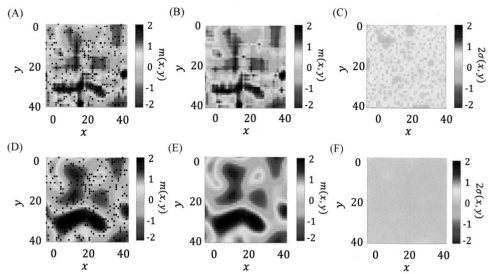

FIG. 7.5 Example of Gaussian Process Regression (GPR) for exponential auto-covariance. (A) True function, $m(x,y)$ (colors) with data (dots). (B) GPR estimate of the function. (C) Two sigma confidence limits. (D–F) Same as (A–C), except for Gaussian auto-covariance. Scripts `gdama07_04` and `gdapy07_04`.

The estimates of the control points exactly equal the data and have no error, and the estimates of the target points have a variance that is less than the prior variance. This limit is called *Kriging* (Krige, 1951) and is a form of true interpolation. For very uncertain data, $\sigma_d^2 \to \infty$, so

$$\mathbf{G}^{-g} \to 0 \quad \text{and} \quad [\text{cov } \mathbf{m}]^{(cc)} = [\text{cov}_A \, \mathbf{m}]^{(cc)} \quad \text{and} \quad [\text{cov } \mathbf{m}]^{(tt)} = [\text{cov}_A \, \mathbf{m}]^{(tt)} \tag{7.22}$$

The estimates $\mathbf{m}^{est}$ of the model parameters are just their prior values $\langle \mathbf{m} \rangle$ and the posterior variance is just the prior variance.

The GPR solution $\mathbf{m}^{est}$ fits the data up to its prior variance σ_d^2 and has the desired spatial correlation. Consequently, GPR is an extremely effective tool for "interpolating" irregularly spaced observations onto a regular grid (Fig. 7.5). Except for the Kriging limit, GPR is not true interpolation, for the data values are not preserved; that is, the estimated control points $\mathbf{m}^{(c)est}$ are only approximately equal to $\mathbf{d}^{obs}$. However, this behavior is desirable when the data are noisy, for it reduces the deleterious effect of data outliers.

A serious limitation of GPR for problems with a large amount of data is that the $N \times N$ prior covariance matrix $[\text{cov}_A \mathbf{m}]^{(cc)}$ that appears in the formula for $\mathbf{Q}$ is not sparse, so that the solution, which involves $\mathbf{Q}^{-1}$, is computationally expensive. As was discussed in Section 7.3, sometimes it may be possible to use the equivalent formulation that is sparse, provided that the matrix $\mathbf{D} = [\text{cov}_A \, \mathbf{m}]^{-\frac{1}{2}}$ can be deduced.

7.5 Prior information of dynamics

Prior information of flatness ($dm/dx = 0$, approximated as $\mathbf{D}^{(1)}\mathbf{m} = 0$) and prior information of smoothness ($d^2\mathbf{m}/dx^2 = 0$, approximated as $\mathbf{D}^{(2)}\mathbf{m} = 0$) are both differential equations, albeit very simple ones. The GLS formalism is not limited to these simple cases. It implements the general form of prior information $\mathbf{Hm} = \mathbf{h}^{pri}$ (with covariance $[\text{cov}\mathbf{h}]$) irrespective of the complexity of the matrix $\mathbf{H}$. Consequently, it can be used to implement the prior information *dynamics*; that is, that the model parameters obey some linear differential equation of the form

$$\mathcal{D} \, m(x,y,z,t) = s(x,y,z,t) \tag{7.23}$$

merely by setting $\mathbf{H} = \mathbf{D}$, where $\mathbf{D}$ is the finite difference version of the *differential operator* $\mathcal{D}$ and $\mathbf{h}^{pri} = \mathbf{s}$, where $\mathbf{s}$ is the discrete version of the *source* function s. The covariance matrix $[\text{cov}\mathbf{h}] = [\text{cov}\mathbf{s}]$ then expresses the accuracy to which $\mathbf{Dm} = \mathbf{s}$ holds. In order to reinforce the connection to GLS, we will continue to use the variables $\mathbf{H}$ and $\mathbf{h}^{pri}$.

Suppose that one is measuring temperature $m(x,t)$ in an one-dimensional rod, as a function of position x and time t (Fig. 7.6A). The sides of the rod are insulated, so that heat can only flow along its axis, and its ends are held at a constant temperature (say, zero). An often-encountered goal is to estimate the temperature on a uniformly sampled

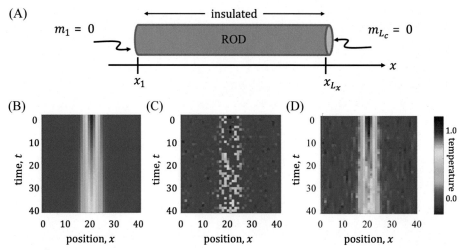

FIG. 7.6 Example of the use of GLS to reconstruct the temperature history $m(x,t)$ of a cooling rod. (A) Schematic diagram of the rod, showing insulated side and zero temperature ends at x_1 and x_{L_x}. (B) True solution. (C) Noisy observations $\mathbf{d}^{(i)}$. (D) Data assimilation solution in which GLS is used to combine the thermal diffusion equation and the observations to reconstruct temperature. Scripts gdama07_05 and 06 and gdapy07_05 and 06.

space-time grid (x_n, t_m), where $1 \leq n \leq L_x$, $1 \leq m \leq L_t$, on the basis of noisy measurements of temperature made at irregularly spaced positions and times.

Prior information that heat is conserved as it flows from hot to cold regions is expressed by the *heat diffusion equation*, which must be satisfied at all times and all positions within the rod

$$\mathcal{D} m(x,t) = 0 \quad \text{with} \quad \mathcal{D} = \frac{\partial}{\partial t} - \kappa \frac{\partial^2}{\partial x^2} \tag{7.24}$$

Here, κ is a material constant called the *thermal diffusivity*. The ends of the rod at x_1 and x_{L_x} are held at zero temperature, so the *boundary conditions* are

$$m(x_1) = m(x_{L_x}) = 0 \tag{7.25}$$

and when the temperature at t_1 is known, the *initial condition* is

$$m(x_n, t_1) = m_0(x_n) \tag{7.26}$$

Differential equation, boundary conditions, and initial condition all contribute prior information to the problem. At positions in the interior of the rod ($1 < n < L_x$) and at time greater than zero ($m > 1$), the heat diffusion equation is satisfied. Its discrete form is

$$\frac{m(x_n, t_m) - m(x_n, t_{m-1})}{\Delta t} - \kappa \frac{m(x_{n-1}, t_{m-1}) - 2m(x_n, t_{m-1}) + m(x_{n+1}, t_{m-1})}{(\Delta x)^2} = 0 \tag{7.27}$$

The heat flow equation contributes $(L_x - 2) \times (L_x - 1)$ pieces of prior information. At the two ends of the rod, $n = 1$ and $n = L_x$, and for all times, $m \geq 1$, the discrete form of the boundary conditions are satisfied

$$m(x_1, t_m) = 0 \quad \text{and} \quad m(x_{L_x}, t_m) = 0 \tag{7.28}$$

The boundary conditions contribute $2 \times L_t$ pieces of prior information. At time zero and for all interior positions, $1 < n < L_x$, the initial conditions are satisfied:

$$m(x_n, t_1) = m_0(x_n) \tag{7.29}$$

The initial conditions contribute $(L_x - 2) \times 1$ pieces of prior information.

As in other multidimensional problems that we have examined, the two-dimensional function $m(x,t)$ needs to be folded out into a column vector $\mathbf{m}$ that contains temperature at all positions and times. The matrix $\mathbf{H}$ is sparse, with rows corresponding to the heat flow equation (four nonzero elements per row), boundary conditions (two per row), and initial conditions (one per row). As is typical of many inverse problems, the bookkeeping is the most difficult part of the solution process.

The total number of pieces of prior information is $L_x \times L_t$, which is equal to the number M of model parameters. This is a case where the prior information matrix $\mathbf{H}$ is square and prior information alone is enough to uniquely specify a solution $\mathbf{m}^{pri} = \mathbf{H}^{-1}\mathbf{h}^{pri}$. Typically, the differential equation and boundary conditions are assigned very small prior variance, because they correspond to very robust physical principles. On the other hand, the initial conditions often are poorly known and assigned high variance, leading to $\mathbf{m}^{pri}$ being very uncertain. The temperature observations $\mathbf{d}^{obs}$ add additional information that reduce this variance. They satisfy the equation $\mathbf{Gm} = \mathbf{d}^{obs}$ (with prior covariance [cov$\mathbf{d}$]), in which each row of $\mathbf{G}$ is zero except for a single instance of 1, which matches up the temperature measurement with the corresponding model parameter (see Section 4.18). The temperature data enable a GLS solution $\mathbf{m}^{est}$ (Fig. 7.6D), which has lower covariance than $\mathbf{m}^{pri}$. This process is called *data assimilation*. As the initial conditions are just the first time step of $\mathbf{m}^{est}$, the data assimilation process estimates them, too.

7.6 Data assimilation in the case of first-order dynamics

The heat diffusion equation is first order in time, meaning that its only time derivative is the first derivative $\partial/\partial t$. Many other differential equations with geophysical relevance are also first order in time or can be made so by a simple mathematical manipulation. Consider Newton's Second Law for the motion of an object $f = md^2u/dt^2$ (with force f, mass m, and position u), where u is the model parameter. Although second order in time, it can be made first order by introducing a second model parameter, velocity, $v \equiv du/dt$. It then becomes the matrix equation

$$\frac{\mathrm{d}}{\mathrm{d}t}\begin{bmatrix} u \\ v \end{bmatrix} = \begin{bmatrix} 0 & 1 \\ 0 & 0 \end{bmatrix}\begin{bmatrix} u \\ v \end{bmatrix} + \begin{bmatrix} 0 \\ f/m \end{bmatrix} \tag{7.30}$$

A very useful generalization of this equation is

$$\frac{\mathrm{d}}{\mathrm{d}t}\mathbf{m}(t) = \mathbf{Z}\,\mathbf{m}(t) + \mathbf{q}(t) \tag{7.31}$$

Here, $\mathbf{m}(t)$ is the model parameter vector at time t, $\mathbf{Z}$ is the time-independent *dynamics matrix*, and $\mathbf{q}(t)$ is the *source vector*. In the case we will develop as follows, individual model parameters, $m_i(t)$, $i = 1, \cdots, L_x$, will refer to the values of some unknown function, like temperature, at a given position and time, that is $m_i(t) \equiv m(x_i, z_i, z_i, t)$. When time is discretized as $t_m = (m-1)\Delta t$, with $1 \le m \le L_t$, the equation can be approximated as

$$\frac{\mathbf{m}^{(m)} - \mathbf{m}^{(m-1)}}{\Delta t} = \mathbf{Z}\,\mathbf{m}^{(m-1)} + \Delta t\mathbf{q}^{(m-1)} \tag{7.32}$$

Here, $\mathbf{m}^{(m)} \equiv \mathbf{m}(t_m)$ and $\mathbf{q}^{(m)} \equiv \mathbf{q}(t_m)$. In the heat diffusion case, (Eq 7.24), $\mathbf{Z} = \kappa\mathbf{D}^{(2)}$ and $\mathbf{q}^{(m-1)} = 0$. The matrix $\mathbf{D}^{(2)}$ contains the boundary conditions as its top and bottom rows and has second differences in its interior. The equation can be rearranged as

$$\mathbf{m}^{(m)} = \mathbf{D}\mathbf{m}^{(m-1)} + \mathbf{s}^{(m-1)} \quad \text{or} \quad -\mathbf{D}\mathbf{m}^{(m-1)} + \mathbf{I}\mathbf{m}^{(m)} = \mathbf{s}^{(m-1)}$$
$$\text{with } \mathbf{D} \equiv \Delta t\mathbf{Z} + \mathbf{I} \text{ and } \mathbf{s}^{(m)} \equiv \Delta t\mathbf{q}^{(n)} \tag{7.33}$$

The initial condition can be written as $\mathbf{m}^{(1)} = \mathbf{m}_0$. The model parameters for sequential times can be concatenated together into a single model parameter vector $\mathbf{m}$

$$\mathbf{m} = \begin{bmatrix} \mathbf{m}^{(1)} \\ \mathbf{m}^{(2)} \\ \vdots \\ \mathbf{m}^{(L_t)} \end{bmatrix} \tag{7.34}$$

The differential equation, together with its boundary conditions and initial conditions, becomes $\mathbf{Hm} = \mathbf{h}^{pri}$, with

$$\mathbf{H} = \begin{bmatrix} \mathbf{I} & & & & \\ -\mathbf{D} & \mathbf{I} & & & \\ & -\mathbf{D} & \mathbf{I} & & \\ & & \ddots & \ddots & \\ & & & -\mathbf{D} & \mathbf{I} \end{bmatrix} \quad \text{and} \quad \mathbf{h}^{pri} = \begin{bmatrix} \mathbf{m}_0 \\ \mathbf{s}^{(1)} \\ \mathbf{s}^{(2)} \\ \vdots \\ \mathbf{s}^{(L_t-1)} \end{bmatrix} \tag{7.35}$$

Here, $\mathbf{D}$ is $L_x \times L_x$ and $\mathbf{H}$ is $M \times M$, with $M = L_x \times L_t$. This form of $\mathbf{H}$ is *block lower bidiagonal*, meaning that nonzero values appear on its main diagonal and the one below it. We will assume that its prior covariance $[\text{cov}\mathbf{h}]$ is *block diagonal*, meaning that it has nonzero values only on its main diagonal, with first block $[\text{cov}\mathbf{m}_0]$ and the rest $[\text{cov}\mathbf{s}]$. This assumption implies that prior information for different times is uncorrelated.

In the data assimilation problem, we introduce N observations that satisfy the data equation $\mathbf{Gm} = \mathbf{d}^{obs}$. This equation is block diagonal, as long as the data measured at a given time $\mathbf{d}^{(m)}$ depend only on the model parameters $\mathbf{m}^{(m)}$ at that same time.

$$\mathbf{G} = \begin{bmatrix} \mathbf{0} & \mathbf{G}^{(2)} & & & \\ & & \mathbf{G}^{(3)} & & \\ & & & \ddots & \\ & & & & \mathbf{G}^{(L_t)} \end{bmatrix} \quad \text{and} \quad \mathbf{d}^{obs} = \begin{bmatrix} \mathbf{d}^{(2)} \\ \mathbf{d}^{(3)} \\ \vdots \\ \mathbf{d}^{(L_t)} \end{bmatrix} \tag{7.36}$$

(We assume that no data are measured at the time of the initial condition.) We also assume that the prior covariance $[\text{cov}\,\mathbf{d}]$ is block diagonal, implying that the error of measurements made at one time do not correlate with those at another.

In practice, the dynamics matrix $\mathbf{Z}$ is sparse, and the covariance matrices usually are taken to be proportional to identity matrices, so they are sparse, too, and the overall sparseness of the data assimilation problem is controlled by the data kernels $\mathbf{G}^{(m)}$. In some cases, including direct observations of the model parameters and ray-based tomography, the data kernels are sparse. Then, even though the $(M+N) \times M$ GLS equation $\mathbf{Fm} = \mathbf{f}$ is very large, the total number of nonzero elements is only moderately large and the solution via the biconjugate gradient method is practical (as in the temperature example). However, when $\mathbf{G}^{(m)}$ is not sparse, this method is not tractable. Fortunately, owing to the simple structure of $\mathbf{F}$, an alternative solution method, described in the next section, is available.

7.7 Data assimilation using Thomas recursion

The matrix that appears in the GLS solution, $\mathbf{A} \equiv \mathbf{F}^T\mathbf{F} = \mathbf{H}^T[\text{covh}]^{-1}\mathbf{H} + \mathbf{G}^T[\text{cov}\,\mathbf{d}]^{-1}\mathbf{G}$, is a symmetric *block tri-diagonal* matrix (meaning a symmetric matrix with nonzero value only on its central three diagonals). It has the form

$$\mathbf{A} = \begin{bmatrix} \mathbf{A}_1 & \mathbf{B}^T & & \\ \mathbf{B} & \mathbf{A}_2 & \mathbf{B}^T & \\ & & \ddots & \mathbf{B}^T \\ & & \mathbf{B} & \mathbf{A}_{L_t} \end{bmatrix} \quad \text{with} \quad \mathbf{B} = -[\text{cov}\,\mathbf{s}]^{-1}\,\mathbf{D}$$

$$\text{and} \quad \mathbf{A}_i = \begin{cases} \mathbf{D}^T[\text{cov}\,\mathbf{s}]^{-1}\mathbf{D} + [\text{cov}\,\mathbf{m}_0]^{-1} & (i=1) \\ \mathbf{D}^T[\text{cov}\,\mathbf{s}]^{-1}\mathbf{D} + [\text{cov}\,\mathbf{s}]^{-1} + \mathbf{G}^{(i)T}[\text{cov}\,\mathbf{d}]^{-1}\mathbf{G}^{(i)} & (1 < i < L_T) \\ [\text{cov}\,\mathbf{s}]^{-1} + \mathbf{G}^{(L_T)T}[\text{cov}\,\mathbf{d}]^{-1}\mathbf{G}^{(L_T)} & (i = L_T) \end{cases} \tag{7.37}$$

The GLS right-hand side, $\mathbf{a} \equiv \mathbf{F}^T\mathbf{f} = \mathbf{H}^T[\text{cov}\,\mathbf{h}]^{-1}\mathbf{h}^{pri} + \mathbf{G}^T[\text{cov}\,\mathbf{d}]^{-1}\mathbf{d}^{obs}$, is

$$\mathbf{a}_i = \begin{cases} -\mathbf{D}^T[\text{cov}\,\mathbf{s}]^{-1}\mathbf{s}^{(1)} + [\text{cov}\,\mathbf{m}_0]^{-1}\mathbf{m}_0 & (i=1) \\ -\mathbf{D}^T[\text{cov}\,\mathbf{s}]^{-1}\mathbf{s}^{(i)} + [\text{cov}\,\mathbf{s}]^{-1}\mathbf{s}^{(i-1)} + \mathbf{G}^{(i)T}[\text{cov}\,\mathbf{d}]^{-1}\mathbf{d}^{(i)} & (1 < i < L_T) \\ [\text{cov}\,\mathbf{s}]^{-1}\mathbf{s}^{(L_T-1)} + \mathbf{G}^{(L_T)T}[\text{cov}\,\mathbf{d}]^{-1}\mathbf{d}^{(L_T)} & (i = L_T) \end{cases} \tag{7.38}$$

The extremely efficient *Thomas method* can be used to solving this block tri-diagonal system of equations. It computes the solution with two recursions, each of length L_t. In the *forward recursion*, a set of $L_x \times L_x$ matrices $\hat{\mathbf{A}}_i$ and $L_x \times 1$ column vectors $\hat{\mathbf{a}}_i$ are calculated

$$\hat{\mathbf{A}}_i = \begin{cases} \mathbf{A}_1 & (i=1) \\ \left[\mathbf{A}_i - \mathbf{B}\hat{\mathbf{A}}_{i-1}^{-1}\mathbf{B}^T\right] & (i>1) \end{cases} \quad \text{and} \quad \hat{\mathbf{a}}_i = \begin{cases} \mathbf{a}_1 & (i=1) \\ \left[\mathbf{a}_i - \mathbf{B}\hat{\mathbf{A}}_{i-1}^{-1}\hat{\mathbf{a}}_{i-1}\right] & (i>1) \end{cases} \tag{7.39}$$

In the *backward recursion*, these quantities are used to calculate the solution:

$$\mathbf{m}^{(i)} = \begin{cases} \hat{\mathbf{A}}_{L_T}^{-1}\hat{\mathbf{a}}_{L_T} & (i=L_T) \\ \hat{\mathbf{A}}_i^{-1}\left[\hat{\mathbf{a}}_i - \mathbf{B}^T\mathbf{m}^{(i+1)}\right] & (1 \le i < L_T) \end{cases} \tag{7.40}$$

```
                              MATLAB®
% forward recursion
Ahinvi = zeros(Lx,Lx,Lt);
ahi = zeros(Lx,Lt);
Ahinvi(:,:,1) = inv(squeeze(Ai(:,:,1)));
ahi(:,1) = ai(:,1);
for i=(2:Lt)
  Ahinvi(:,:,i)=inv(Ai(:,:,i)-B*squeeze(Ahinvi(:,:,i-1))*(B'));
  ahi(:,i)=ai(:,i)-B*squeeze(Ahinvi(:,:,i-1))*ahi(:,i-1);
end
% backward recursion
mi = zeros(Lx,Lt);
mi(:,Lt) = squeeze(Ahinvi(:,:,Lt))*ahi(:,Lt);
for i=(Lt-1:-1:1)
  mi(:,i)=squeeze(Ahinvi(:,:,i))*(ahi(:,i)-B'*mi(:,i+1));
end
```
[gdama07_07]

```
                              Python
# forward recursion
Ahinvi = np.zeros((Lx,Lx,Lt));
ahi = np.zeros((Lx,Lt));
Ahinvi[0:Lx,0:Lx,0] = la.inv(Ai[0:Lx,0:Lx,0]);
ahi[0:Lx,0] = ai[0:Lx,0];
for i in range(1,Lt):
    Ahinvi[0:Lx,0:Lx,i] = ( la.inv(Ai[0:Lx,0:Lx,i]
        np.matmul(B,np.matmul(Ahinvi[0:Lx,0:Lx,i-1],B.T))) );
    ahi[0:Lx,i] = ( ai[0:Lx,i]- np.matmul(B,
        np.matmul(Ahinvi[0:Lx,0:Lx,i-1],ahi[0:Lx,i-1])) );
# backward recursion
mi = np.zeros((Lx,Lt));
mi[0:Lx,Lt-1] = np.matmul(Ahinvi[0:Lx,0:Lx,Lt-1],
                          ahi[0:Lx,Lt-1]);
for i in reversed(range(Lt-1)):
    mi[0:Lx,i] = np.matmul(Ahinvi[0:Lx,0:Lx,i],
                  ahi[0:Lx,i]-np.matmul(B.T,mi[0:Lx,i+1]));
```
[gdapy07_07]

The Thomas method reduces the problem of solving one, monolithic $(L_xL_t) \times (L_xL_t)$ system of equations to a sequence of L_t, smaller $L_x \times L_x$ systems. However, the improvement in efficiency is not nearly as dramatic as it may initially appear. In general, the inverse of a sparse matrix is not sparse, so even when the **A**s are sparse, the $\widehat{\mathbf{A}}$s are not. Thus the solution of each $L_x \times L_x$ system can be computationally intensive.

Taken together, the recursions propagate information both forward and backward in time. The values of the estimated model parameters at a given time t_m are affected by data collected in their past $t < t_m$ and in their future $t > t_m$.

7.8 Present-time solutions

Many time-dependent data assimilation problems are focused on the *present time*. Although a sequence of solutions are being computed, $\mathbf{m}^{(1)}, \mathbf{m}^{(2)}, \cdots$, for times $t_1, t_2\cdots$, the only one that is of interest is the one for most recent time—*now*. Today's weather, for instance, is of much greater interest to most people than yesterday's. This is not to say that only today's meteorological observations are of interest. To the contrary, yesterday's (and earlier) data may help inform the estimation of today's weather through the data assimilation process. However, as the estimate is needed now, the process cannot wait for tomorrow's measurements, even though they, too, might improve today' estimate.

The simplest strategy for solving this problem is to add rows to $\mathbf{H}$, $\mathbf{h}^{pri}$, $\mathbf{G}$, and $\mathbf{d}^{obs}$ at every time step and to re-solve the complete GLS problem for both past and present $\mathbf{m}^{(i)}$s, either by biconjugate gradients or the Thomas method. However, a disadvantage of this approach is that tremendous effort is put into reestimating past $\mathbf{m}^{(i)}$s, whose values, even though improved, are not of interest.

Close scrutiny of the recursion in the Thomas method indicates that only the forward recursion is needed to calculate the last set of model parameters $\mathbf{m}^{(L_t)}$, because the formula $\mathbf{m}^{(L_t)} = \widehat{\mathbf{A}}_{L_T}^{-1} \widehat{\mathbf{a}}_{L_T}$ in Eq. (7.40) depends only on quantities calculated in the last L_T iteration of the forward recursion. By viewing L_T as a moving target that represents the present time, the forward recursion can just be extended by one iteration for every time step. However, one small adjustment needs to be made, as can be understood by the following scenario. Suppose that the present time is $L_T = 20$. Then, $\mathbf{A}_{20}$ and $\mathbf{a}_{20}$ are computed using the ($i = L_T$) options of Eqs. (7.37) and (7.38). However, when the present time advances to $L_T = 21$, then $\mathbf{A}_{20}$ and $\mathbf{a}_{20}$ must be calculated through using the ($1 < i < L_T$) option, which is to say that the terms $\mathbf{D}^T[\text{covs}]^{-1}\mathbf{D}$ and $-\mathbf{D}^T[\text{covs}]^{-1}\mathbf{s}^{(20)}$, respectively, must be added to them.

One concern surrounding the use of the present-time Thomas method is that the physical significance of the various matrix operations that comprise it is hard to identify. Furthermore, the calculation of ancillary quantities, such as the posterior covariance $[\text{cov}\mathbf{m}^{(i)}]$, would seem to be tedious, at best. Fortunately, a much more intuitive method, *Kalman filtering* (Grewal and Andrews, 2014), is equivalent to the present-time Thomas method (Menke, 2022b) and overcomes these concern.

7.9 Kalman filtering

Kalman filtering (Grewal and Andrews, 2014) breaks the present-time data assimilation problem into four intuitive steps:

Step 1. At time $i = 1$, the solution and its covariance are given by the initial condition and the covariance of the initial condition, respectively. The best estimate of the solution is the initial condition itself, because no data have yet been collected.

$$\mathbf{m}^{(1)} = \mathbf{m}_0 \quad \text{and} \quad \left[\text{cov } \mathbf{m}^{(1)}\right] = [\text{cov } \mathbf{m}_0] \tag{7.41}$$

Step 2. The solution and its covariance are propagated forward in time using the dynamical equation (Eq. 7.33), and the result is considered to be *prior information* at that new time:

$$\mathbf{m}_A^{(i)} = \mathbf{D}\mathbf{m}^{(i-1)} + \mathbf{s}^{(i-1)} \quad \text{and} \quad \left[\text{cov } \mathbf{m}_A^{(i)}\right] = \mathbf{D}\left[\text{cov } \mathbf{m}^{(i-1)}\right]\mathbf{D}^T + [\text{cov } \mathbf{s}] \tag{7.42}$$

Note that the covariance $[\text{cov}\mathbf{m}_A^{(i)}]$ of the prior value is being computed using the standard error propagation formula. This covariance evolves with time and is, in general, not diagonal.

Step 3. At this new time step, a GLS problem is formulated with prior and data equations,

$$\mathbf{m}^{(i)} = \mathbf{m}_A^{(i)} \quad \text{with covariance} \quad \left[\text{cov } \mathbf{m}_A^{(i)}\right]$$
$$\mathbf{G}^{(i)}\mathbf{m}^{(i)} = \mathbf{d}^{(i)} \quad \text{with covariance} \quad [\text{cov } \mathbf{d}] \tag{7.43}$$

and solved using any of the equivalent GLS formulations. One such formulation gives

$$\mathbf{m}^{(i)} = \mathbf{A}^{-1}\left\{\left[\text{cov } \mathbf{m}_A^{(i)}\right]^{-1}\mathbf{m}_A^{(i)} + \mathbf{G}^{(i)T}[\text{cov } \mathbf{d}]^{-1}\mathbf{d}^{(i)}\right\} \quad \text{and} \quad \left[\text{cov } \mathbf{m}^{(i)}\right] = \mathbf{A}^{-1}$$
$$\text{with } \mathbf{A} = \left[\text{cov } \mathbf{m}_A^{(i)}\right]^{-1} + \mathbf{G}^{(i)T}[\text{cov } \mathbf{d}]^{-1}\mathbf{G}^{(i)} \tag{7.44}$$

Note that an estimate of the posterior covariance $[\text{cov}\mathbf{m}^{(i)}]$ arises naturally from this process. Step 4. Time is incremented and the process returns to Step 2, creating a recursion.

```
                               MATLAB®
sim1 = zeros(Lx,1);  % source term is zero
mK = zeros(Lx,Lt); % Kalman solution
varm = zeros(Lx,Lt);
mK(:,1) = m0;   % first soln is just initial condition
covm1 = sigmam02*eye(Lx);
varm(:,1) = diag(covm1);
covmAKim1 = covm1;
for i=(2:Lt)
    mAi = D*mK(:,i-1)+sim1;
    covmAi = D*covmAKim1*D' + sigmas2*eye(Lx);
    Gs = squeeze(Gi(:,:,i));
    AK=inv(covmAi) + sigmadi2*(Gs'*Gs);
    mK(:,i) = AK\( covmAi\mAi + sigmadi2*Gs'*dobs(:,i) );
    AKinv=inv(AK);
    varm(:,i)=diag(AKinv);
    covmAKim1 = AKinv;
end
                                              [gdama07_10]
```

```
                               Python
covmAKim1 = covm1;
for i in range(1,Lt):
    mAi = np.matmul(D,mK[0:Lx,i-1:i])+sim1;
    covmAi = ( np.matmul(D,np.matmul(covmAKim1,D.T)) +
              sigmas2*np.identity(Lx) );
    Gs = np.squeeze(Gi[0:Nd,0:Lx,i]);
    AK=la.inv(covmAi) + sigmadi2*np.matmul(Gs.T,Gs);
    AKinv=la.inv(AK);
    mK[0:Lx,i:i+1] = ( np.matmul(AKinv,la.solve(covmAi,mAi)
            + sigmadi2*np.matmul(Gs.T,dobs[0:Nd,i:i+1])) );
    varm[0:Lx,i:i+1]=gda_cvec(np.diag(AKinv));
    covmAKim1 = AKinv;
                                              [gdapy07_10]
```

The Kalman solution of the exemplary temperature problem (Fig. 7.7D) is very similar to—but not identical to—the full GLS solution (Fig. 7.7B), reflecting the fact that the lack of future measurements leads to only a modest degradation in the solution. The difference (Fig. 7.8A) between the solutions has a speckled appearance, reflecting the fact that the effect of a future datum is most dramatic when it is in the immediate future and spatially close to a given present-time position. The difference between the present-time Thomas and Kalman solutions is zero (up to numerical noise) (Fig. 7.8B).

7.10 Case of exact dynamics

In the limiting case of [cov$\mathbf{s}$] = 0, the solution obeys the dynamical equation exactly, and the only effect of the data is to improve the estimate of the initial conditions. This variant of the data assimilation problem often is solved using a method different than GLS, as will be described in Section 14.15. The underlying idea is to reformulate the problem so that the initial conditions $m_0(x)$ are the only model parameter

$$\text{find the } m_0(x) \text{ that minimizes } E = \mathbf{e}^T\mathbf{e}$$
$$\text{where } m_0(x) = m(x, t = 0) \text{ and } e_i = d_i - m(x_i, t_i) \tag{7.45}$$
$$\text{with the constraint } \mathcal{D}\, m(x, t) = 0$$

This formulation leads to substantial simplification of the solution process, using an approach called *adjoint methods*.

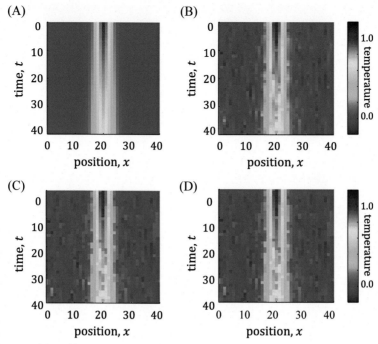

FIG. 7.7   Comparison of solutions of the cooling rod problem of Fig. 7.6. (A) True Solution. (B) GLS solution. (C) Present-time Thomas solution. (D) Kalman filtering solution. Scripts gdama07_10 and gdapy07_10.

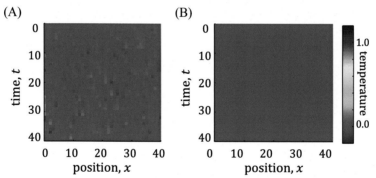

FIG. 7.8   Differences between solutions of the cooling rod problem of Fig. 7.7. (A) Kalman filtering solution minus GLS solution. (B) Kalman filtering solution minus present-time Thomas solution. Scripts gdama07_10 and gdapy07_10.

When $\mathcal{D}$ is first order in time, an initial condition is equivalent to a *source* acting at time $t=0$, because it is always possible to contrive an impulsive source $s(x,t) \equiv s_0(x)\delta(t)$, which, a moment after its application at $t=0$, leads to an $m(x, t=0+)$ that exactly equals any given initial condition $m_0(x)$. Hence, the minimization problem can be restated

$$\text{find the } s_0(x) \text{ that minimizes } E = \mathbf{e}^T\mathbf{e}$$
$$\text{where } m_0(x) = 0 \text{ and } e_i = d_i - m(x_i, t_i) \qquad (7.46)$$
$$\text{with the constraint } \mathcal{D}\, m(x,t) = s(x,t)$$

In the case of a second-order operator, $s(x,t)=s_0(x)\delta(t)+s_1(x)\partial\delta/\partial t$. We will return to this method in Section 14.13, which introduces and applies adjoint methods.

7.11 Problems

7.1 The file `Data/my_random_fields.txt` created by scripts `gdama07_01` and `gdapy07_01` contains $N=256$ data points $(x_i, y_i, d_i^{(E)}, d_i^{(G)})$ that sample two functions $d^{(E)}(x,y)$ and $d^{(G)}(x,y)$ on $0 < x < 41$ and $0 < y < 41$. The function $d^{(E)}(x,y)$ has an exponential auto-covariance and $d^{(E)}(x,y)$ has a Gaussian auto-covariance, both with scale factor $s=5$. The scripts `gdama07_03` and `gdapy07_03` grid these data using GPR, using the correct scale factor. Run test cases to evaluate the effect of using incorrect scale factors, in the range $0.5 < s < 50$.

7.2 An object moving horizontally in the u-direction is decelerating due to a resistive force $f = -cmv$ that is proportional to its mass m, velocity v, and a constant $c=0.01$. The car starts at position $u=0$ with initial velocity $v_0=1$. Newton's law implies $du/dt=v$ and $dv/dt=-cv$. (A) What are the dynamical quantities $\mathbf{Z}$, $\mathbf{q}$, $\mathbf{D}$, $\mathbf{s}$ in Eqs. (7.32) and (7.33)? (B) Create noise-free data by stepping Eq. (7.33) forward in time (say, with $\Delta t = 1$). (C) Create noisy data by adding Normally distributed random noise with zero mean and variance $\sigma_d^2 = (0.1)^2$ to the noise-free data. (D) Assuming that the dynamical equation (which is ultra-simplified) has variance $\sigma_s^2 = (0.01)^2$, use Kalman filtering to estimate $u(t)$. Comment upon the results.

7.3 This problem builds on Problem 7.2. How much better is an estimate of $u(t)$ when it is estimated by Generalized Least Squares (in which both past and future data influence the present)?

References

Grewal, M.S., Andrews, A.P., 2014. Kalman Filtering: Theory and Practice Using MATLAB, fourth ed. Wiley-IEEE Press, Hoboken, NJ.

Krige, D.G., 1951. A statistical approach to some basic mine valuation problems on the Witwatersrand. J. Chem. Metall. Min. Soc. South Afr. 52, 119–139.

Menke, W., 2022a. Environmental Data Analysis With MATLAB or Python, third ed. Elsevier, London, UK, ISBN: 978-0-323-95576-8.

Menke, W., 2022b. Links between Kalman filtering and data assimilation with generalized least squares. App. Math 13, 566–584. https://doi.org/10.4236/am.2022.136036.

Menke, W., Creel, R., 2021a. Gaussian process regression reviewed in the context of inverse theory. Surv. Geophys. 42, 473–503. https://doi.org/10.1007/s10712-021-09640-w.

Menke, W., Creel, R., 2021b. Why differential data work. Bull. Seismol. Soc. Am. https://doi.org/10.1785/0120210014.

Menke, W., Eilon, Z., 2015. Relationship between data smoothing and the regularization of inverse problems. Pure Appl. Geophys. 172, 2711–2726.

Rasmussen, C.E., Williams, C.K.I., 2006. Gaussian Processes for Machine Learning. The MIT Press, Cambridge, MA, ISBN: 026218253X. 272 pp.

Tarantola, A., Valette, B., 1982. Inverse problems=quest for information. J. Geophys. 50, 159–170.

8

Nonuniqueness and localized averages

8.1 Null vectors and nonuniqueness

In Chapters 4-6, we presented the basic method of finding estimates of the model parameters in a linear inverse problem. We showed that we could always obtain such estimates but that sometimes in order to do so we had to add prior information to the problem. We shall now consider the meaning and consequences of nonuniqueness in linear inverse problems and show that it is possible to devise solutions that do not depend at all on prior information.

As we shall show, however, these solutions are not estimates of the model parameters themselves but estimates of *weighted averages* (linear combinations) of the model parameters. When the linear inverse problem $\mathbf{Gm} = \mathbf{d}^{obs}$ has non-unique solutions, there exist nontrivial solutions (i.e., solutions with some nonzero m_i) to the *homogeneous* equation $\mathbf{Gm} = 0$. These solutions are called the *null vectors* of the inverse problem as premultiplying them by the data kernel yields zero. To see why nonuniqueness implies null vectors, suppose that the inverse problem has two distinct solutions $\mathbf{m}^{(1)}$ and $\mathbf{m}^{(2)}$ as

$$\begin{aligned} \mathbf{Gm}^{(1)} &= \mathbf{d}^{obs} \\ \mathbf{Gm}^{(2)} &= \mathbf{d}^{obs} \end{aligned} \tag{8.1}$$

Subtracting these two equations yields

$$\mathbf{G}\left(\mathbf{m}^{(1)} - \mathbf{m}^{(2)}\right) = 0 \tag{8.2}$$

Since the two solutions are by assumption distinct, their difference $\mathbf{m}^{null} = \mathbf{m}^{(1)} - \mathbf{m}^{(2)}$ is nonzero. The converse is also true; any linear inverse problem that has null vectors is nonunique. Note that the equation $\mathbf{Gm}^{null} = 0$ can be interpreted to mean that $\mathbf{m}^{null}$ is *perpendicular* to every row of $\mathbf{G}$ (as its dot product with every row is zero). Consequently, no linear combination of the rows of $\mathbf{G}$ can be a null vector. If $\mathbf{m}^{par}$ (where *par* stands for "particular") is any nonnull solution to

$$\mathbf{Gm}^{par} = \mathbf{d}^{obs} \tag{8.3}$$

(for instance, the minimum-length solution), then $\mathbf{m}^{par} + \alpha \mathbf{m}^{null}$ is also a solution with the same error for any choice of α. Note that as $\alpha \mathbf{m}^{null}$ is a null vector for any nonzero α, null vectors are only distinct if they are linearly independent. If a given inverse problem has q distinct null solutions, then the most general solution is

$$\mathbf{m}^{gen} = \mathbf{m}^{par} + \sum_{i=1}^{q} \alpha_i \mathbf{m}^{null(i)} \tag{8.4}$$

where *gen* stands for "general." We shall show in Section 9.6 that $0 \leq q \leq M$, that is, that there can be no more linearly independent null vectors than there are unknowns.

8.2   Null vectors of a simple inverse problem

As an example, consider the following very simple equations:

$$\begin{bmatrix} 1/4 & 1/4 & 1/4 & 1/4 \end{bmatrix} \begin{bmatrix} m_1 \\ m_2 \\ m_3 \\ m_4 \end{bmatrix} = \begin{bmatrix} d_1^{obs} \end{bmatrix} \tag{8.5}$$

This equation implies that only the mean value of a set of four model parameters has been measured. One obvious solution to this equation is $\mathbf{m} = d_1^{obs}[1,1,1,1]^T$ (in fact, this is the minimum-length solution). Three linearly independent null solutions can be determined by inspection as

$$\mathbf{m}^{null(1)} = \begin{bmatrix} 1 \\ -1 \\ 0 \\ 0 \end{bmatrix} \qquad \mathbf{m}^{null(2)} = \begin{bmatrix} 1 \\ 0 \\ -1 \\ 0 \end{bmatrix} \qquad \mathbf{m}^{null(3)} = \begin{bmatrix} 1 \\ 0 \\ 0 \\ -1 \end{bmatrix} \tag{8.6}$$

The most general solution is then

$$\mathbf{m}^{gen} = d_1^{obs} \begin{bmatrix} 1 \\ 1 \\ 1 \\ 1 \end{bmatrix} + \alpha_1 \begin{bmatrix} 1 \\ -1 \\ 0 \\ 0 \end{bmatrix} + \alpha_2 \begin{bmatrix} 1 \\ 0 \\ -1 \\ 0 \end{bmatrix} + \alpha_3 \begin{bmatrix} 1 \\ 0 \\ 0 \\ -1 \end{bmatrix} \tag{8.7}$$

where the αs are arbitrary parameters.

Finding a particular solution to this problem now consists of choosing values for the parameters α_i. If one chooses these parameters so that $\|\mathbf{m}\|_2^2$ is minimized, one obtains the minimum-length solution. Since the first vector is orthogonal to all the others, this minimum occurs when $\alpha_1 = \alpha_2 = \alpha_3 = 0$. We shall show in Section 9.6 that this is a general result: the minimum-length solution never contains any null vectors. Note, however, that if other definitions of solution simplicity are used (e.g., flatness or smoothness), those solutions will contain null vectors.

8.3   Localized averages of model parameters

In Chapters 4-6, we have sought an estimate of the solution vector $\mathbf{m}^{est}$. Another approach is to estimate some average of the model parameter $\langle \mathbf{m} \rangle = \mathbf{a}^T \mathbf{m}$, where $\mathbf{a}$ is some averaging vector. The average is said to be *localized* if this averaging vector consists mostly of zeros, except for some group of nonzero elements that multiplies model parameters centered about one particular model parameter. This definition makes particular sense when the model parameters possess some natural ordering in space and time, such as acoustic slowness as a function of depth in the Earth. For instance, if $M = 8$, the averaging vector $\mathbf{a} = [0,0,1/4,1/2,1/4,0,0,0]^T$ could be said to be localized about the fourth model parameter. The averaging vectors are usually normalized so that the sum of their elements is unity.

The advantage of estimating averages of the model parameters rather than the model parameters themselves is that very often it is possible to identify unique averages even when the model parameters themselves are not unique. To examine when uniqueness can occur, we compute the average of the general solution as

$$\langle \mathbf{m} \rangle = \mathbf{a}^T \mathbf{m}^{gen} = \mathbf{a}^T \mathbf{m}^{par} + \sum_{i=1}^{q} \alpha_i \mathbf{a}^T \mathbf{m}^{null(i)} \tag{8.8}$$

If $\mathbf{a}^T \mathbf{m}^{null(i)}$ is zero for all i, then $\langle \mathbf{m} \rangle$ is unique. The process of averaging has completely removed the nonuniqueness of the problem. Since $\mathbf{a}$ has M elements and there are $q < M$ constraints placed on $\mathbf{a}$, one can always find at least one vector that cancels (or "annihilates") the null vectors. One cannot, however, always guarantee that the averaging vector is localized around some particular model parameter. But, if $q \ll M$, one has some freedom in choosing $\mathbf{a}$ and there is some possibility of making the averaging vector at least somewhat localized. Whether this can be done depends on the structure of the null vectors, which, in turn, depends on the structure of the data kernel $\mathbf{G}$. Since the small-scale features of the model are unresolvable in many problems, unique localized averages can often be found.

8.4 Averages versus estimates

We can identify a type of dualism in inverse theory. Given a generalized inverse $\mathbf{G}^{-g}$ that in some sense solves $\mathbf{Gm} = \mathbf{d}^{obs}$, we can speak either of estimates of model parameters $\mathbf{m}^{est} = \mathbf{G}^{-g}\mathbf{d}^{obs}$ or of localized averages $\langle \mathbf{m} \rangle = \mathbf{G}^{-g}\mathbf{d}^{obs}$. The numerical values are the same but the interpretation is very different. When the solution is interpreted as a localized average, it can be viewed as a unique quantity that exists independently of any prior information applied to the inverse problem. Examination of the resolution matrix may reveal that the average is not especially localized and the solution may be difficult to interpret. When the solution is viewed as an estimate of a model parameter, the location of what is being solved for is clear. The estimate can be viewed as unique only if one accepts as appropriate whatever prior information was used to remove the inverse problem's underdeterminacy. In most instances, the choice of prior information is somewhat ad hoc, so the solution may still be difficult to interpret.

In the sample problem stated earlier, the data kernel has only one row. There is therefore only one averaging vector that will annihilate all the null vectors: one proportional to that row

$$\mathbf{a} = \begin{bmatrix} \frac{1}{4} & \frac{1}{4} & \frac{1}{4} & \frac{1}{4} \end{bmatrix}^T \tag{8.9}$$

This averaging vector is very poorly localized. In this problem, the structure of $\mathbf{G}$ is just too poor to form good averages. The generalized inverse to this problem is by inspection

$$\mathbf{G}^{-g} = \begin{bmatrix} 1 & 1 & 1 & 1 \end{bmatrix}^T \tag{8.10}$$

The model resolution matrix is therefore

$$\mathbf{R} = \mathbf{G}^{-g}\mathbf{G} = \frac{1}{4}\begin{bmatrix} 1 & 1 & 1 & 1 \\ 1 & 1 & 1 & 1 \\ 1 & 1 & 1 & 1 \\ 1 & 1 & 1 & 1 \end{bmatrix} \tag{8.11}$$

which is very poorly localized.

8.5 "Decoupling" localized averages from estimates

One way to think about a generalized inverse $\mathbf{G}^{-g}$ is as a set of "recipes" for creating unique averages—one recipe for each average. Suppose the ith row of the resolution matrix is denoted as $r_j^{(i)} \equiv R_{ij}$. For fixed i, the quantity $r_j^{(i)}$ specifies the elements $a_j \equiv r_j^{(i)}$ of an averaging vector $\mathbf{a}$. Similarly, suppose that the ith row of the generalized inverse and the data kernel are denoted as $\gamma_j^{(i)}$ and $g_j^{(i)}$, respectively. Then, the formula for the resolution matrix $\mathbf{R} = \mathbf{G}^{-g}\mathbf{G}$ implies

$$\mathbf{r}^{(i)} = \sum_{k=1}^{N} \gamma_k^{(i)} \mathbf{g}^{(k)} \text{ with } \mathbf{R} = \begin{bmatrix} \mathbf{r}^{(1)} \\ \mathbf{r}^{(2)} \\ \vdots \\ \mathbf{r}^{(M)} \end{bmatrix} \text{ and } \mathbf{G} = \begin{bmatrix} \mathbf{g}^{(1)} \\ \mathbf{g}^{(2)} \\ \vdots \\ \mathbf{G}^{(N)} \end{bmatrix} \tag{8.12}$$

This equation indicates that row $\mathbf{r}^{(i)}$ is built up from a weighted sum of the rows $\mathbf{g}^{(k)}$ of the data kernel (with weights $\gamma_k^{(i)}$), and consequently, defines a unique average. The ith localized average, say $\langle m^{(i)} \rangle$, is then

$$\left\langle m^{(i)} \right\rangle = \sum_{k=1}^{N} \gamma_k^{(i)} d_k^{obs} \tag{8.13}$$

Thus $\langle m^{(i)} \rangle$ is the value of a unique average of the model parameters, with the weights given by $\gamma_k^{(i)}$. In this interpretation, the generalized inverse has no purpose other than to supply the recipes for computing $\langle m^{(i)} \rangle$ and $\mathbf{r}^{(i)}$. Any connection between the generalized inverse and estimates of model parameters, or between it and minimization principles, is being ignored. The goodness of the average $\langle m^{(i)} \rangle$ is being judged solely by whether $r_j^{(i)}$ has desirable properties (such as being localized).

This viewpoint provides rationale for rescaling a generalized inverse so that the resulting model resolution matrix has unit row-sum. This is a desirable property, because it implies that the overall amplitude of the solution is being preserved, even though features are being smoothed out by the averaging. Suppose that $\mathbf{w} = [1, 1, 1, \cdots 1]^T$ is a vector of ones, so that $\mathbf{s} \equiv \mathbf{Rw}$ is a vector of row-sums of the resolution matrix. Then, defining $\mathbf{S} \equiv \text{diag}(\mathbf{s})$:

$$\mathbf{Rw} = \mathbf{G}^{-g}\mathbf{Gw} = \mathbf{s} = \mathbf{Sw} \text{ or } \left[\mathbf{S}^{-1}\mathbf{G}^{-g}\right]\mathbf{Gw} = \mathbf{w} \tag{8.14}$$

Thus we can define a *rescaled* generalized inverse $(\mathbf{G}^{-g})' \equiv \mathbf{S}^{-1}\mathbf{G}^{-g}$ that leads to a resolution matrix with unit row-sums. Of course, by changing the generalized inverse, it may no longer be one that its optimum according to some minimization principle. However, by supposition, our focus is not on those principles.

8.6   Nonunique averaging vectors and prior information

There are instances in which even nonunique averages of model parameters can be of value, especially when they are used in conjunction with prior knowledge of the nature of the solution (Wunsch and Minster, 1982). Suppose that one simply picks a localized averaging vector that does not necessarily annihilate all the null vectors and that, therefore, does not lead to a unique average. In the problem in Section 8.2, $\mathbf{a} = \frac{1}{3}[1,1,1,0]^T$ is such a vector. It is somewhat localized, being centered about the second model parameter. Note that it does not lead to a unique average, as

$$\langle m \rangle = \mathbf{a}^T \mathbf{m}^{gen} = d_1^{obs} + 0 + 0 + \frac{1}{3}\alpha_3 \tag{8.15}$$

is still a function of one of the arbitrary parameters α_i. Suppose, however, that there is prior knowledge that every m_i must satisfy $0 \leq m_i \leq 2d_1^{obs}$. Then from the equation for $\mathbf{m}^{gen}$, α_3 must be no greater than d_1^{obs} and no less than $-d_1^{obs}$. Since $-d_1^{obs} \leq \alpha_3 \leq d_1^{obs}$, the average has bounds $(2d_1^{obs}/3) \leq \langle m \rangle \leq (4d_1^{obs}/3)$. These constraints are considerably tighter than the prior bounds on m_i, which demonstrates that this technique has indeed produced some useful information. This approach works because even though the averaging vector does not annihilate all the null vectors, $\mathbf{a}^T \mathbf{m}^{null(i)}$ is small compared with the elements of the null vector. Localized averaging vectors often lead to small products since the null vectors often fluctuate rapidly about zero, indicating that small-scale features of the model are the most poorly resolved. A slightly more complicated example of this type is solved in Fig. 8.1.

This approach can be generalized as follows (Oldenburg, 1983):

$$\text{lower bound: } \text{minimize } \langle m \rangle = \mathbf{a}^T \mathbf{m} \text{ with respect to } \mathbf{m}$$
$$\text{upper bound: } \text{maximize } \langle m \rangle = \mathbf{a}^T \mathbf{m} \text{ with respect to } \mathbf{m} \tag{8.16}$$
$$\text{with the constraints } \mathbf{Gm} = \mathbf{d}^{obs} \text{ and } \mathbf{m}^{(l)} \leq \mathbf{m} \leq \mathbf{m}^{(u)}$$

Here, $\mathbf{m}^{(l)}$ and $\mathbf{m}^{(u)}$ are the prior lower and upper bounds on $\mathbf{m}$, respectively. Note that this formulation does not explicitly include null vectors; the constraint $\mathbf{Gm} = \mathbf{d}^{obs}$ is sufficient to ensure that the model parameters satisfy the data. This problem is a special case of the *linear programming problem*

$$\text{lower bound: } \text{find the } \mathbf{m} \text{ that minimizes } z = \mathbf{f}^T \mathbf{x}$$
$$\text{upper bound: } \text{find the } \mathbf{m} \text{ that maximizes } z = \mathbf{f}^T \mathbf{x} \tag{8.17}$$
$$\text{with the constraints } \mathbf{Ax} \leq \mathbf{b} \text{ and } \mathbf{Cx} = \mathbf{d} \text{ and } \mathbf{x}^{(l)} \leq \mathbf{x} \leq \mathbf{x}^{(u)}$$

Note that the minimization problem can be converted into a maximization problem by flipping the sign of $\mathbf{f}$. Furthermore, greater-than-or-equal inequality constraints can be converted to less-than-or-equal by multiplication by -1.

The linear programming problem was first studied by economists and business operations analysts. For example, z might represent the profit realized by a factory producing a product line, where the number of each product is given

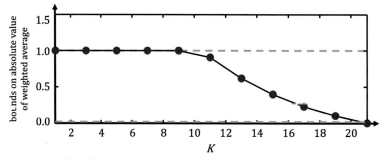

FIG. 8.1 Bounds on weighted averages of $M = 21$ model parameters, m_i, in a problem in which the only datum is that the sum of all model parameters is zero. When this observation is combined with the prior information that each model parameter must satisfy $0 \leq m_i \leq 1$ (blue lines), bounds can be placed on the weighted averages of the model parameter. The bounds shown here are for averages of K neighboring model parameters. Note that the bounds are tighter than the prior bounds only when $K > 10$. Scripts gdama_08_01 and gdapy_08_01.

by **x** and the profit on each item given by **f**. The problem is to maximize the total profit $\mathbf{f}^T\mathbf{x}$ without violating the constraint that one can produce only a positive amount of each product, or any other linear inequality constraints that might represent labor laws, union regulations, physical limitation of machines, etc.

Both MATLAB® and Python provide functions/methods for solving the linear programming problem. Eq. (8.17) is solved by calling it twice, once to minimize $\langle m \rangle$ and the other to maximize it.

MATLAB®

```
options = optimset('linprog');
options.Display = 'off';  % turns off annoying messages
[mest1,amin]=linprog( a,[],[],G,dobs,mlb,mub,options);
[mest2,amax]=linprog(-a,[],[],G,dobs,mlb,mub,options);
amax=-amax;
```
[gdama08_01]

Python

```
res = opt.linprog(c=a,A_eq=G,b_eq=dobs,
    bounds=np.concatenate((mlb,mub),axis=1));
x = gda_cvec(res.x);
amin[L,0] = res.fun;
status = res.success;
if( not status ):
    print("linear programming failed" );

res = opt.linprog(c=-a,A_eq=G,b_eq=dobs,
    bounds=np.concatenate((mlb,mub),axis=1));
x = gda_cvec(res.x);
amax[L,0] = -res.fun;
status = res.success;
if( not status ):
    print("linear programming failed" );
```
[gdapy08_01]

Here, `amin` and `amax` are lower and upper bounds on $\langle m \rangle$, respectively. In MATLAB®, the arguments of MATLAB's `linprog()` function and Python' `opt.linprog()` method include the averaging vector `a`, the data kernel `G`, the observed data `dobs`, and the prior upper and lower bounds `mlb` and `mub` on the model parameters.

An example using a Laplace transform-like data kernel is shown in Fig. 8.2.

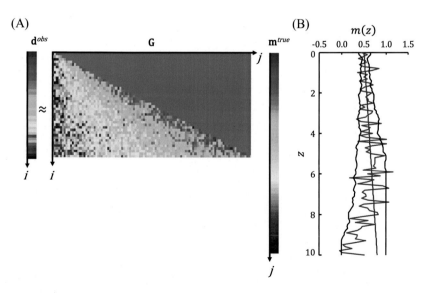

(A)

(B)

FIG. 8.2 (A) This underdetermined inverse problem, $\mathbf{d}^{obs} = \mathbf{Gm}$, has $M=100$ model parameters, m_i, $N=40$ data, d_i^{obs}. The data are weighted averages of the model parameters, from the surface down to a depth z_i, which increases with index i. The observed data include additive noise. (B) The true model parameters (red curve) increase linearly with depth z. The estimated model parameters (blue curve), computed using the minimum-length method, scatter about the true model at shallow depths ($z<6$) but decline toward zero at deeper depths due to poor resolution. Bounds (black lines) on localized averages of the model parameters, with an averaging width, $w=2$, are for prior information, $0 \leq m_i \leq 1$. Scripts gdama_08_02 and gdapy_08_02.

8.7   End-member solutions and squeezing

The nonuniqueness of an inverse problem can be explored by constructing *end-member solutions*, which satisfy the data to equal degrees, but which obey different kinds of prior information. For example, "shallow" and "deep" solutions might be useful end-members in a one-dimensional inversion for model parameters $m(z_i)$, where the auxiliary variable z represents depth, especially if $m(z_i)$ is thought to contain a single feature centered at unknown depth $z=z_0$. The shallow solution is one for which the feature is centered at the smallest possible depth (but without violating the data) and the deep solution is the one for which the feature is centered at the deepest possible depth (but without violating the data). A comparison of these two end-members gives a sense of the degree to which the data constrain the depth of the feature. This technique is sometimes referred to as *squeezing*, as the solution is alternately *squeezed* toward one prior notion of an end-member and then toward another.

The end-member solutions are constructed by starting with a solution that satisfies the data, and then adding to it just the right combination of null vectors to satisfy the prior information of deepness or shallowness as best as possible. This process can be implemented by using a weighted damped least-squares solution of the form (see Section 4.15)

$$\mathbf{m}^{est} = \left[\mathbf{G}^T\mathbf{G} + \varepsilon^2\mathbf{W}_m\right]^{-1}\mathbf{G}^T\mathbf{d}^{obs} \tag{8.18}$$

together with judicious choices of parameter ε^2 and diagonal weight matrix $\mathbf{W}_m$ chosen as a diagonal matrix with main diagonal, say $\mathbf{w}_m$. The shallow solution is found using a $\mathbf{w}_m$ that preferentially penalizes deep model parameters; for example

$$\mathbf{w}_m = \begin{bmatrix} & \text{smoothly} & \\ 1,1,1, & \cdots \text{ramping} \cdots w_0, w_0, w_0 \\ & \text{up to} & \end{bmatrix} \text{ with } w_0 \gg 1 \tag{8.19}$$

The deep solution is found using a $\mathbf{w}_m$ that preferentially penalizes shallow model parameters; for example

$$\mathbf{w}_m = \begin{bmatrix} & \text{smoothly} & \\ w_0, w_0, w_0 & \cdots \text{ramping} \cdots 1,1,1, \\ & \text{down to} & \end{bmatrix} \text{ with } w_0 \gg 1 \tag{8.20}$$

The two parameters ε^2 and w_0 are adjusted by trial and error, so that the corresponding solutions satisfy the data to an acceptable degree and are as different from one another as possible (Fig. 8.3).

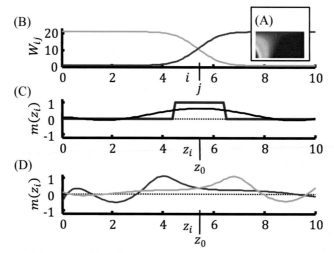

FIG. 8.3   Squeezed solutions for a simple inverse problem for model parameters $m(z_i)$, where z is depth. (A) The data kernel represents smooth averages of the data and has the form, $G_{ij} = c_i\exp(-c_iz_j)$, where the cs are constants. (B) Two weight functions W_{ij}, one that highly weights (and thus suppresses) shallow parts of the model (green), and the other that highly weights (and thus suppresses) deep parts of the model (red). (C) The true model (blue curve) is a boxcar centered at $z_0=5.4$. The ordinary damped least-squares solution (black curve) is smoother than the boxcar, but has negligible error. (D) The shallow solution (red curve) and deep solution (green curve) also have negligible error, but are concentrated at shallow and deep depths, respectively, than the ordinary damped least-squares solution. Taken together, they indicate that the data are sufficient to constrain the central part of the solution to the $4<z<7$ depth range. Scripts gdama08_03 and gdapy08_03.

Squeezing can be used to implement many other prior notion end-members, in addition to spatial localization. The general procedure is to represent the solution as a sum of patterns $\mathbf{p}^{(j)}$ with unknown coefficients $\widehat{m}_j$:

$$\mathbf{m} = \sum_{j=1}^{M} \widehat{m}_j \mathbf{p}^{(j)} = \mathbf{P}\widehat{\mathbf{m}} \quad \text{with} \quad P_{ij} = p_i^{(j)} \tag{8.21}$$

Two of these patterns, say $\mathbf{p}^{(A)}$ and $\mathbf{p}^{(B)}$), represent the end-members and the other patterns fill out the solution to make it complete. The inverse problem is then recast into one where the coefficients $\widehat{\mathbf{m}}$ are the model parameters with the substitution $\mathbf{m} = \mathbf{P}\widehat{\mathbf{m}}$, so that $\mathbf{Gm} = \mathbf{d}$ becomes $(\mathbf{GP})\widehat{\mathbf{m}} = \mathbf{d}$, and the solution of this new equation is squeezed toward large $\widehat{m}_A$ and small $\widehat{m}_B$ (and vice versa). For instance, the model parameters $m_i = m(z_i)$ (where z_i is an auxiliary variable) could be squeezed toward the end-members of polynomials of mostly low/high order by identifying Eq. (8.21) as a Taylor series (i.e., with $p^{(j)}(z_i) = z_i^{j-1}$) and by squeezing toward $\widehat{m}_1$ and away from $\widehat{m}_M$ (and vice versa). If the solution cannot be squeezed close to $\mathbf{p}^{(A)}$ without causing unacceptably large error, one is justified in concluding that, even though the data do not uniquely determine the solution, it requires that the solution cannot look like $\mathbf{p}^{(A)}$ (and similarly for $\mathbf{p}^{(B)}$).

8.8   Problems

8.1 What is the general solution to the problem $\mathbf{Gm} = \mathbf{d}$ with

$$\mathbf{G} = \begin{bmatrix} 1 & 1 & 0 & 0 \\ 0 & 0 & 1 & 1 \end{bmatrix}$$

8.2 Give some examples of physical problems where the model parameters can be assumed, with reasonable certainty, to fall between lower and upper bounds (and identify the bounds).

8.3 Suppose that $M = 21$, model parameters are known to have bounds $-1 \leq m_i \leq 1$. Suppose that the unweighted average of each three adjacent model parameters is observed so that the data kernel has the form

$$\mathbf{G} = \begin{bmatrix} 1 & 1 & 1 & 0 & 0 & 0 & 0 & 0 & 0 \\ 0 & 1 & 1 & 1 & 0 & 0 & 0 & 0 & 0 \\ 0 & 0 & 1 & 1 & 1 & 0 & 0 & 0 & 0 \\ & & & & \ddots & & & & \\ 0 & 0 & 0 & 0 & 0 & 0 & 1 & 1 & 1 \end{bmatrix}$$

Can useful bounds be placed on the weighted average of the three adjacent model parameters, where the weights are $[\frac{1}{4}, \frac{1}{2}, \frac{1}{4}]$? Adapt script gdama08_01 or gdapy08_01 to explore this problem.

8.4 Consider an inverse problem with $M = 8$ model parameters, $m_i = m(z_i) = \sin(\pi z_i)$, where the auxiliary depth variable z is in the range, $0 \leq z \leq 1$. (A) Construct a synthetic data set in which the observations are a triplicate set of measurements of m_i equally spaced depths, all made with an accuracy of $\sigma_d = 0.1$ (so that $N = 3M$). (B) Estimate a reference solution using simple least squares. (C) Use squeezing to develop end-member solutions for the case where the solution is represented as a Taylor series and where the end-members represent polynomials with mostly low-order or mostly high-order terms. (D) Show that the solution cannot be squeezed to one in which the low-order terms have negligible amplitude but for which the error is acceptably low.

References

Oldenburg, D.W., 1983. Funnel functions in linear and nonlinear appraisal. J. Geophys. Res. 88, 7387–7398.

Wunsch, C., Minster, J.F., 1982. Methods for box models and ocean circulation tracers: mathematical programming and non-linear inverse theory. J. Geophys. Res. 87, 5647–5662.

9

Applications of vector spaces

9.1   Model and data spaces

So far, we have used vector notation for the data $\mathbf{d}$ and model parameters $\mathbf{m}$ mainly because it facilitates the algebra. Additionally, the interpretations of vectors as *geometrical* quantities can be used to gain an insight into the properties of inverse problems. We therefore introduce the idea of vector spaces containing $\mathbf{d}$ and $\mathbf{m}$, which we shall denote $S(\mathbf{d})$ and $S(\mathbf{m})$, respectively. Any particular choice of $\mathbf{d}$ and $\mathbf{m}$ is then represented as a point in these spaces (or, equivalently, a vector connecting the origin to that point) (Fig. 9.1).

The linear equation $\mathbf{d} = \mathbf{Gm}$ can be interpreted as a *mapping* of vectors from $S(\mathbf{m})$ to $S(\mathbf{d})$ and its solution $\mathbf{m} = \mathbf{G}^{-g}\mathbf{d}$ as a mapping of vectors from $S(\mathbf{d})$ to $S(\mathbf{m})$.

One important property of a vector space, such as $S(\mathbf{m})$, is that its coordinate axes are arbitrary. Thus far, we have been using axes parallel to the individual model parameters, but we recognize that we are by no means required to do so. Any set of vectors that *spans* the space will serve as coordinate axes. The Mth dimensional space $S(\mathbf{m})$ is spanned by any M vectors, say, $\mathbf{m}^{(i)}$, as long as these vectors are *linearly independent*. An arbitrary vector lying in $S(\mathbf{m})$, say $\mathbf{m}^*$, can be expressed as a sum of these M *basis vectors*, written as

$$\mathbf{m}^* = \sum_{i=1}^{M} \alpha_i \mathbf{m}^{(i)} \tag{9.1}$$

where the αs are the *components* of the vector $\mathbf{m}^*$ in the new coordinate system. If the $\mathbf{m}^{(i)}$s are linearly dependent, then the vectors $\mathbf{m}^{(i)}$ lie in a subspace, or *hyperplane*, of $S(\mathbf{m})$ and Eq. (9.1) cannot be satisfied with any choice of the αs (Fig. 9.2).

We shall consider, therefore, transformations of the coordinate systems of the two spaces $S(\mathbf{d})$ and $S(\mathbf{m})$. Using $S(\mathbf{m})$ as an example, if $\mathbf{m}$ is the representation of a vector in one coordinate system and $\mathbf{m}'$ its representation in another, we can write the transformation as

$$\mathbf{m}' = \mathbf{Tm} \ \text{ and } \ \mathbf{m} = \mathbf{T}^{-1}\mathbf{m}' \tag{9.2}$$

where $\mathbf{T}$ is the *transformation* matrix. If the new basis vectors are still mutually orthogonal unit vectors, then $\mathbf{T}$ represents simple rotations or reflections of the coordinate axes. As we shall show later, however, it is sometimes convenient to choose a new set of basis vectors that are not unit vectors.

9.2   Householder transformations

In this section, we shall show that the minimum-length, least-squares, and constrained least-squares solutions can be found through simple transformations of the equation $\mathbf{d} = \mathbf{Gm}$. While the results are identical to the formulas derived in Chapter 4, the approach and emphasis are very different and will enable us to visualize these solutions in a new way. We shall begin by considering a purely underdetermined linear problem $\mathbf{d} = \mathbf{Gm}$, with $M > N$. Suppose we want to find the minimum-length solution (the one that minimizes $L = \mathbf{m}^T\mathbf{m}$). We shall show that it is easy to find

FIG. 9.1 (A) The model parameters represented as a vector **m** in the M-dimensional space $S(\mathbf{m})$ of all possible model parameters. (B) The data represented as a vector **d** in the N-dimensional space $S(\mathbf{d})$ of all possible data. Scripts gdama09_01 and gdapy09_01.

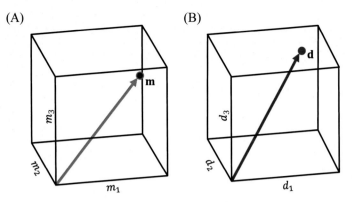

FIG. 9.2 (A) These three vectors span the three-dimensional space $S(\mathbf{m})$. (B) These three vectors do not span the space as they all lie on the same plane. Scripts gdama09_02 and gdapy09_02.

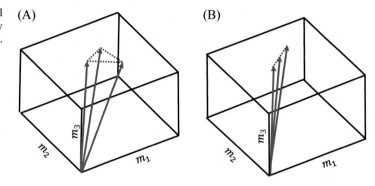

this solution by transforming the model parameters into a new, primed coordinate system $\mathbf{m}' = \mathbf{Tm}$. The inverse problem becomes

$$\mathbf{d} = \mathbf{GIm} = \mathbf{G}\left(\mathbf{T}^{-1}\mathbf{T}\right)\mathbf{m} = \left(\mathbf{GT}^{-1}\right)(\mathbf{Tm}) = \mathbf{G}'\mathbf{m}' \tag{9.3}$$

where $\mathbf{G}' = \mathbf{GT}^{-1}$ is the data kernel in the new coordinate system. The solution length becomes

$$L = \mathbf{m}^T\mathbf{m} = \left(\mathbf{T}^{-1}\mathbf{m}'\right)^T\left(\mathbf{T}^{-1}\mathbf{m}'\right) = \mathbf{m}'^T\left(\mathbf{T}^{-1T}\mathbf{T}^{-1}\right)\mathbf{m}' \tag{9.4}$$

Suppose that we could choose **T** so that $\mathbf{T}^{-1T}\mathbf{T}^{-1} = \mathbf{I}$. The solution length would then have the same form in both coordinate systems, namely, the sum of squares of the vector elements. Minimizing $\mathbf{m}'^T\mathbf{m}'$ would be equivalent to minimizing $\mathbf{m}^T\mathbf{m}$. Transformations of this type that do not change the length of the vector are called *unitary* transformations. They may be interpreted as rotations and reflections of the coordinate axes. We can see from Eq. (9.4) that unitary transformations satisfy $\mathbf{T}^{-1} = \mathbf{T}^T$.

Now suppose that we could also choose the transformation so that $\mathbf{G}'$ is the lower triangular matrix

$$\begin{bmatrix} G'_{11} & 0 & 0 & 0 & 0 & 0 & \cdots & 0 \\ G'_{21} & G'_{22} & 0 & 0 & 0 & 0 & \cdots & 0 \\ G'_{31} & G'_{32} & G'_{33} & 0 & 0 & 0 & \cdots & 0 \\ \vdots & \vdots & \vdots & \ddots & \vdots & \vdots & \cdots & 0 \\ G'_{N1} & G'_{N2} & G'_{N3} & \cdots & G'_{NN} & 0 & \cdots & 0 \end{bmatrix} \begin{bmatrix} m'_1 \\ m'_2 \\ m'_3 \\ \vdots \\ m'_M \end{bmatrix} = \begin{bmatrix} d_1 \\ d_2 \\ d_3 \\ \vdots \\ d_N \end{bmatrix} \tag{9.5}$$

Notice that no matter what values we pick for $m'_{N+1}..., m'_M$, we cannot change the value of $\mathbf{G}'\mathbf{m}'$ as the last $M - N$ columns of $\mathbf{G}'$ are all zero. On the other hand, we can solve for the first N elements of m'_i uniquely as

$$\begin{aligned} m'_1 &= [d_1]/G'_{11} \\ m'_2 &= \left[d_2 - G'_{21}m'_1\right]/G'_{22} \\ m'_3 &= \left[d_3 - G'_{31}m'_1 - G'_{32}m'_2\right]/G'_{33} \\ &\vdots \\ m'_N &= \left[d_N - G'_{N1}m'_1 - G'_{N2}m'_2 - \cdots - G'_{N,N-1}m'_{N-1}\right]/G'_{NN} \end{aligned} \tag{9.6}$$

This process is known as *back-solving*. As the first N elements of $\mathbf{m}'$ are thereby determined, $\mathbf{m}'^{T}\mathbf{m}'$ can be minimized by setting the remaining elements $m'_{N+1}...m'_M$ equal to zero. The solution in the original coordinate system is then $\mathbf{m}^{est} = \mathbf{T}^{-1}\mathbf{m}'$. As this solution satisfies the data exactly and minimizes the solution length, it is equal to the minimum-length solution (Section 4.10).

We have employed a transformation process that separates the determined and undetermined linear combinations of model parameters into two distinct groups so that we can deal with them separately. We can now easily determine the null vectors for the inverse problem. In the transformed coordinates, they are the set of vectors whose first N elements are zero and whose last $M - N$ elements are zero except for one element. There are clearly $M - N$ such vectors, so we have established that there are never more than M null vectors in a purely underdetermined problem. The null vectors can easily be transformed back into the original coordinate system by premultiplication by $\mathbf{T}^{-1}$. As all but one element of the transformed null vectors are zero, this operation just selects a column of $\mathbf{T}^{-1}$ (or, equivalently, a row of $\mathbf{T}$).

One transformation that can triangularize a matrix is called a *Householder transformation*. We shall discuss in Section 9.3 how it can be constructed. As we shall see later, Householder transformations have application to a wide variety of methods that employ the L_2 norm as a measure of size. These transformations provide an alternative method of solving such problems and additional insight into their structure.

The overdetermined linear inverse problem $\mathbf{d} = \mathbf{Gm}$ with $N > M$ can also be solved through the use of Householder transformations. In this case, we seek a solution that minimizes the prediction error $E = \mathbf{e}^T\mathbf{e}$. We seek a transformation with two properties: it must operate on the transformed prediction error $E = \mathbf{e}'^{T}\mathbf{e}'$, in such a way that minimizing $\mathbf{e}'^{T}\mathbf{e}'$ is the same as minimizing $\mathbf{e}^T\mathbf{e}$, and it must transform the data kernel into *upper-triangularized form*. The transformed prediction error is

$$
\begin{bmatrix} e'_1 \\ e'_2 \\ e'_3 \\ \vdots \\ e'_M \\ e'_{M+1} \\ \vdots \\ e'_N \end{bmatrix} = \begin{bmatrix} d'_1 \\ d'_2 \\ d'_3 \\ \vdots \\ d'_M \\ d'_{M+1} \\ \vdots \\ d'_N \end{bmatrix} - \begin{bmatrix} G'_{11} & G'_{12} & G'_{13} & \cdots & G'_{1M} \\ 0 & G'_{22} & G'_{23} & \cdots & G'_{2M} \\ 0 & 0 & G'_{33} & \cdots & G'_{3M} \\ \vdots & \vdots & \vdots & \ddots & \vdots \\ 0 & 0 & 0 & \cdots & G'_{MM} \\ 0 & 0 & 0 & \cdots & 0 \\ \vdots & \vdots & \vdots & \vdots & \vdots \\ 0 & 0 & 0 & 0 & 0 \end{bmatrix} \begin{bmatrix} m_1 \\ m_2 \\ m_3 \\ \vdots \\ m_M \end{bmatrix} \tag{9.7}
$$

No matter what values we choose for $\mathbf{m}$, we cannot alter the last $N - M$ elements of $\mathbf{e}'$ as the last $N - M$ rows of the transformed data kernel are zero. We can, however, set the first M elements of $\mathbf{e}'$ equal to zero by satisfying the first M equations, $\mathbf{e}' = \mathbf{d}' - \mathbf{G}'\mathbf{m}$, exactly. As the top part of $\mathbf{G}'$ is triangular, we can use the back-solving technique described earlier. The total error is then the length of the last $N - M$ elements of $\mathbf{e}'$, written as

$$
E = \sum_{i=M+1}^{N} e'^{2}_{i} \tag{9.8}
$$

As with the underdetermined case, we have used the Householder transformations to separate the problem into two parts: data that can be satisfied exactly and data that cannot be satisfied at all. The solution is chosen so that it minimizes the length of the prediction error, and the least square solution is thereby obtained.

Finally, we note that the constrained least-squares problem can also be solved with Householder transformations. Suppose that we want to solve $\mathbf{d} = \mathbf{Gm}$ in the least-squares sense except that we want the solution to obey p linear equality constraints of the form $\mathbf{h} = \mathbf{Hm}$. Because of the constraints, we do not have complete freedom in choosing the model parameters. We therefore employ Householder transformations to separate those linear combinations of $\mathbf{m}$ that are completely determined by the constraints from those that are completely undetermined. This process is precisely the same as the one used in the underdetermined problem and consists of finding a transformation, say, $\mathbf{T}$, that triangularizes $\mathbf{Hm} = \mathbf{h}$ as

$$
\mathbf{h} = \{\mathbf{HT}^{-1}\}\{\mathbf{Tm}\} = \mathbf{H}'\mathbf{m}' \tag{9.9}
$$

The first p elements of $\mathbf{m}'$ are now completely determined and can be computed by back-solving the triangular system. The same transformation can be applied to $\mathbf{d} = \mathbf{Gm}$ to yield the transformed inverse problem $\mathbf{d} = \{\mathbf{GT}^{-1}\}\{\mathbf{Gm}\} = \mathbf{G}'\mathbf{m}'$. But $\mathbf{G}'$ will not be triangular as the transformation was designed to triangularize $\mathbf{H}$, not $\mathbf{G}$. As the first p elements of $\mathbf{m}'$ have been determined by the constraints, we can partition $\mathbf{G}'$ into two submatrices $\mathbf{G}' = [\mathbf{G}'_1, \mathbf{G}'_2]$, where $\mathbf{G}'_1$ multiplies the p determined model parameters and $\mathbf{G}'_2$ multiplies the as yet unknown $(M - p)$ model parameters

$$\left[\mathbf{G}_1', \mathbf{G}_2'\right] \begin{bmatrix} \begin{bmatrix} m_1' \\ \vdots \\ m_p' \end{bmatrix} \\ \begin{bmatrix} m_{p+1}' \\ \vdots \\ m_M' \end{bmatrix} \end{bmatrix} = \mathbf{d} \tag{9.10}$$

The equation can be rearranged into standard form by subtracting the part involving the already-determined model parameters:

$$\left[\mathbf{G}_2'\right] \begin{bmatrix} m_{p+1}' \\ \vdots \\ m_M' \end{bmatrix} = \mathbf{d} - \mathbf{G}_1' \begin{bmatrix} m_1' \\ \vdots \\ m_p' \end{bmatrix} \tag{9.11}$$

The equation is now a completely overdetermined one in the $(M-p)$ unknown model parameters and can be solved as described earlier. Finally, the solution is transformed back into the original coordinate system by $\mathbf{m}^{est} = \mathbf{T}^{-1}\mathbf{m}'$.

9.3   Designing householder transformations

For a transformation to preserve length, it must be a unitary transformation (i.e., it must satisfy $\mathbf{T}^T = \mathbf{T}^{-1}$). Any transformation of the form

$$\mathbf{T} = \mathbf{I} - \frac{2\mathbf{v}\mathbf{v}^T}{\mathbf{v}^T\mathbf{v}} \tag{9.12}$$

(where $\mathbf{v}$ is any vector) is a unitary transformation as

$$\mathbf{T}^{-1}\mathbf{T} = \mathbf{T}^T\mathbf{T} = \left(\mathbf{I} - \frac{2\mathbf{v}\mathbf{v}^T}{\mathbf{v}^T\mathbf{v}}\right)\left(\mathbf{I} - \frac{2\mathbf{v}\mathbf{v}^T}{\mathbf{v}^T\mathbf{v}}\right) = \mathbf{I} - \frac{4\mathbf{v}\mathbf{v}^T}{\mathbf{v}^T\mathbf{v}} + \frac{4\mathbf{v}\mathbf{v}^T}{\mathbf{v}^T\mathbf{v}} = \mathbf{I} \tag{9.13}$$

Here, we have used the facts that $[\mathbf{v}\mathbf{v}^T]^T = \mathbf{v}\mathbf{v}^T$ and $\mathbf{v}\mathbf{v}^T\mathbf{v}\mathbf{v}^T = \mathbf{v}\mathbf{v}^T/\mathbf{v}^T\mathbf{v}$. Eq. (9.12) can be shown to be the most general form of a unitary transformation. The problem is to find the vector $\mathbf{v}$, such that the transformation triangularizes a given matrix. To do so, we shall begin by finding a sequence of transformations, each of which converts to zeros either the elements beneath the main diagonal of one column of the matrix (for premultiplication of the transformation) or the elements to the right of the main diagonal of one row of the matrix (for postmultiplication by the transformation). The first i columns are converted to zeros by

$$\mathbf{T} = \mathbf{T}^{(i)}\mathbf{T}^{(i-1)}\mathbf{T}^{(i-2)}\cdots\mathbf{T}^{(1)} \tag{9.14}$$

In the overdetermined problem, applying M of these transformations produces an upper-triangularized matrix. In the $M=3$, $N=4$ case, the transformations proceed as

$$\mathbf{G} \rightarrow \mathbf{T}^{(1)}\mathbf{G} \rightarrow \mathbf{T}^{(2)}\mathbf{T}^{(1)}\mathbf{G} \rightarrow \mathbf{T}^{(3)}\mathbf{T}^{(2)}\mathbf{T}^{(1)}\mathbf{G}$$

$$\begin{bmatrix} x & x & x \\ x & x & x \\ x & x & x \\ x & x & x \end{bmatrix} \rightarrow \begin{bmatrix} x & x & x \\ 0 & x & x \\ 0 & x & x \\ 0 & x & x \end{bmatrix} \rightarrow \begin{bmatrix} x & x & x \\ 0 & x & x \\ 0 & 0 & x \\ 0 & 0 & x \end{bmatrix} \rightarrow \begin{bmatrix} x & x & x \\ 0 & x & x \\ 0 & 0 & x \\ 0 & 0 & 0 \end{bmatrix} \tag{9.15}$$

where the xs symbolize nonzero matrix elements. Consider the ith column of $\mathbf{G}$, denoted by the vector $\mathbf{g} = [G_{1i}, G_{2i}, \cdots G_{Ni}]^T$. We want to construct a transformation such that

$$\mathbf{g}' = \mathbf{T}\mathbf{g} = \left[G_{1i}', G_{2i}', \cdots, G_{ii}', 0, \cdots 0\right]^T \tag{9.16}$$

Substituting the expression for the transformation yields

$$\mathbf{Ig} - \frac{2\mathbf{v}\mathbf{v}^T}{\mathbf{v}^T\mathbf{v}}\mathbf{g} = \begin{bmatrix} G_{1i} \\ G_{2i} \\ \vdots \\ G_{Ni} \end{bmatrix} - \frac{2\mathbf{v}\mathbf{g}^T}{\mathbf{v}^T\mathbf{v}} \begin{bmatrix} v_1 \\ v_2 \\ \vdots \\ v_N \end{bmatrix} = \begin{bmatrix} G_{1i}' \\ G_{2i}' \\ \vdots \\ G_{ii}' \\ 0 \\ \vdots \\ 0 \end{bmatrix} \tag{9.17}$$

Here, we have used $\mathbf{vv}^T\mathbf{g}=\mathbf{vg}^T\mathbf{v}$. As the term $2\mathbf{vg}^T/\mathbf{v}^T\mathbf{v}$ is a scalar, we can only zero the last $N-i$ elements of $\mathbf{g}'$ if we choose $\mathbf{v}$ such that $[2\mathbf{vg}^T/\mathbf{v}^T\mathbf{v}][v_{i+1},\cdots,v_N]^T=[G_{i+1,i},\cdots,G_{N,\ i}]^T$. We therefore set $[v_{i+1},\cdots,v_N]^T=[G_{i+1,i},\cdots,G_{N,i}]^T$ and choose the other i elements of $\mathbf{v}$ so that the normalization is correct. As this is but a single constraint on i elements of $\mathbf{v}$, we have considerable flexibility in making the choice. It is convenient to choose the first $i-1$ elements of $\mathbf{v}$ as zero (this choice is both simple and causes the transformation to leave the first $i-1$ elements of $\mathbf{g}$ unchanged). This leaves the ith element of $\mathbf{v}$ to be determined by the condition that $2\mathbf{vg}^T/\mathbf{v}^T\mathbf{v}=1$. It is easy to show that $v_i=G_{ii}-\alpha_i$, where

$$\alpha_i^2 = \sum_{j=i}^{N} G_{ji}^2 \tag{9.18}$$

Although the sign of α_i is arbitrary, choosing it to have a sign opposite that of G_{ii} is best, because it improves numerical stability of the algorithm. The Householder transformation is then defined by the vector

$$\mathbf{v} = [0\ 0\ \cdots\ 0\ (G_{ii} - \alpha_i)\ G_{i+1,i}\ G_{i+2,i}\ \cdots\ G_{N,i}]^T \tag{9.19}$$

Finally, we note that the $(i+1)$st Householder transformation does not destroy any of the zeros created by the ith, as long as we apply them in order of decreasing number of zeros. Thus we can apply a succession of these transformations to triangularize an arbitrary matrix. The inverse transformations are also trivial to construct, as they are just the transforms of the forward transformations. An example is shown in gdama09_03 and gdapy09_03.

9.4   Transformations that do not preserve length

Suppose we want to solve the linear inverse problem $\mathbf{d} = \mathbf{Gm}$ in the sense of finding a solution $\mathbf{m}^{est}$ that minimizes a weighted combination of prediction error and solution simplicity as

$$\text{minimize:}\ E + L = \mathbf{e}^T\mathbf{W}_e\mathbf{e} + \mathbf{m}^T\mathbf{W}_m\mathbf{m} \tag{9.20}$$

It is possible to find transformations $\mathbf{m}'=\mathbf{T}_m\mathbf{m}$ and $\mathbf{e}'=\mathbf{T}_e\mathbf{e}$ which, although they do not preserve the length, change it in precisely such a way that $E+L=\mathbf{e}'^T\mathbf{e}'+\mathbf{m}'^T\mathbf{m}'$ (Wiggins, 1972). The weighting factors are identity matrices in the new coordinate system.

Consider the weighted measure of length $L=\mathbf{m}^T\mathbf{W}_m\mathbf{m}$. If we could factor the weighting matrix into the product $\mathbf{W}_m=\mathbf{T}_m^T\mathbf{T}_m$, where $\mathbf{T}_m$ is some matrix, then we could identify $\mathbf{T}_m$ as a transformation of coordinates

$$L = \mathbf{m}^T\mathbf{W}_m\mathbf{m} = \mathbf{m}^T\mathbf{T}_m^T\mathbf{T}_m\mathbf{m} = [\mathbf{T}_m\mathbf{m}]^T[\mathbf{T}_m\mathbf{m}] = \mathbf{m}'^T\mathbf{m}' \tag{9.21}$$

This factorization can be accomplished in several ways. If we have built $\mathbf{W}_m$ up from the $\mathbf{D}$ matrix, then $\mathbf{W}_m=\mathbf{D}^T\mathbf{D}$ and $\mathbf{T}_m=\mathbf{D}$. If not, then we can rely upon the fact that $\mathbf{W}_m$, being symmetric, has a symmetric square root, so that $\mathbf{W}_m = \mathbf{W}_m^{\frac{1}{2}}\mathbf{W}_m^{\frac{1}{2}} = \mathbf{W}_m^{\frac{1}{2}T}\mathbf{W}_m^{\frac{1}{2}}$, in which case $\mathbf{T}_m=\mathbf{W}_m^{\frac{1}{2}}$. Then

$$\begin{array}{lll}
\mathbf{m}' = \mathbf{W}_m^{\frac{1}{2}}\mathbf{m}\text{ or }\mathbf{m}' = \mathbf{Dm} & & \mathbf{m} = \mathbf{W}_m^{-\frac{1}{2}}\mathbf{m}'\text{ or }\mathbf{m} = \mathbf{D}^{-1}\mathbf{m}' \\
\mathbf{d}' = \mathbf{W}_e^{\frac{1}{2}}\mathbf{d} & \text{and} & \mathbf{d} = \mathbf{W}_e^{-\frac{1}{2}}\mathbf{d}' \\
\mathbf{G}' = \mathbf{W}_e^{\frac{1}{2}}\mathbf{GW}_m^{-\frac{1}{2}}\text{ or }\mathbf{G}' = \mathbf{W}_e^{\frac{1}{2}}\mathbf{GD}^{-1} & & \mathbf{G} = \mathbf{W}_e^{-\frac{1}{2}}\mathbf{G}'\mathbf{W}_m^{\frac{1}{2}}\text{ or }\mathbf{G} = \mathbf{W}_e^{-\frac{1}{2}}\mathbf{G}'\mathbf{D}
\end{array} \tag{9.22}$$

We have encountered this weighting before, in Eqs. (4.49) and (4.50). In fact, the damped least-squares solution (Eq. 4.38)

$$\mathbf{m}' = \left[\mathbf{G}'^T\mathbf{G}' + \varepsilon^2\mathbf{I}\right]^{-1}\mathbf{G}'^T\mathbf{d}' \tag{9.23}$$

(written here in terms of the primed variables) transforms into the weighted damped least-squares solution (Eq. 4.49 with $\langle\mathbf{m}\rangle$ set to zero) under this transformation:

$$\begin{aligned}
\mathbf{W}_m^{-\frac{1}{2}}\mathbf{m}^{est} &= \left[\mathbf{W}_m^{-\frac{1}{2}}\mathbf{G}^T\mathbf{W}_e^{\frac{1}{2}}\mathbf{W}_e^{\frac{1}{2}}\mathbf{GW}_m^{-\frac{1}{2}} + \varepsilon^2\mathbf{I}\right]^{-1}\mathbf{W}_m^{-\frac{1}{2}}\mathbf{G}^T\mathbf{W}_e^{\frac{1}{2}}\mathbf{W}_e^{\frac{1}{2}}\mathbf{d}^{obs} \\
\mathbf{m}^{est} &= \left[\mathbf{G}^T\mathbf{W}_e\mathbf{G} + \varepsilon^2\mathbf{W}_m\right]^{-1}\mathbf{G}^T\mathbf{W}_e\mathbf{d}^{obs}
\end{aligned} \tag{9.24}$$

The square root of a symmetric matrix $\mathbf{W}$ is defined via its eigenvalue decomposition. Let $\mathbf{\Lambda}$ and $\mathbf{U}$ be the eigenvalues and eigenvectors, respectively, of $\mathbf{W}$ so that $\mathbf{W}=\mathbf{U\Lambda U}^T$. Then

$$\mathbf{T} = \mathbf{W}^{\frac{1}{2}} = \mathbf{U\Lambda}^{\frac{1}{2}}\mathbf{U}^T\ \text{ because }\ \mathbf{W}^{\frac{1}{2}}\mathbf{W}^{\frac{1}{2}} = \mathbf{U\Lambda}^{\frac{1}{2}}\mathbf{U}^T\mathbf{U\Lambda}^{\frac{1}{2}}\mathbf{U}^T = \mathbf{U\Lambda U}^T = \mathbf{W} \tag{9.25}$$

Here, we have used the facts that $\mathbf{U}^T\mathbf{U} = \mathbf{I}$ and $\mathbf{\Lambda}^{\frac{1}{2}}\mathbf{\Lambda}^{\frac{1}{2}} = \mathbf{\Lambda}$. The square root of a symmetric matrix is computed as

MATLAB®
`Wmsqrt = sqrtm(Wm);`
[gdama09_04]

Python
`Wmsqrt = la.sqrtm(Wm);`
[gdapy09_04]

so the eigenvalue decomposition need not be used explicitly. Yet another possible definition of the transformation $\mathbf{T}$ is

$$\mathbf{T} = \mathbf{\Lambda}^{\frac{1}{2}}\mathbf{U}^T \tag{9.26}$$

as then $\mathbf{T}^T\mathbf{T} = \mathbf{U}\mathbf{\Lambda}^{\frac{1}{2}}\mathbf{\Lambda}^{\frac{1}{2}}\mathbf{U}^T = \mathbf{U}\mathbf{\Lambda}\mathbf{U}^T = \mathbf{W}$. The transformed inverse problem is then $\mathbf{d}' = \mathbf{G}'\mathbf{m}'$, where

$$\begin{aligned}
\mathbf{m}' &= \left\{\mathbf{\Lambda}_m^{\frac{1}{2}}\mathbf{U}_m^T\right\}\mathbf{m} & \mathbf{m} &= \left\{\mathbf{U}_m\mathbf{\Lambda}_m^{-\frac{1}{2}}\right\}\mathbf{m}' \\
\mathbf{d}' &= \left\{\mathbf{\Lambda}_e^{\frac{1}{2}}\mathbf{U}_e^T\right\}\mathbf{d} & \text{and} \qquad \mathbf{d} &= \left\{\mathbf{U}_e\mathbf{\Lambda}_e^{-\frac{1}{2}}\right\}\mathbf{d}' \\
\mathbf{G}' &= \left\{\mathbf{\Lambda}_e^{\frac{1}{2}}\mathbf{U}_e^T\right\}\mathbf{G}\left\{\mathbf{U}_m\mathbf{\Lambda}_m^{-\frac{1}{2}}\right\} & \mathbf{G} &= \left\{\mathbf{U}_e\mathbf{\Lambda}_e^{-\frac{1}{2}}\right\}\mathbf{G}'\left\{\mathbf{\Lambda}_m^{\frac{1}{2}}\mathbf{U}_m^T\right\}
\end{aligned} \tag{9.27}$$

where $\mathbf{W}_e = \mathbf{U}_e\mathbf{\Lambda}_e\mathbf{U}_e^T$ and $\mathbf{W}_m = \mathbf{U}_m\mathbf{\Lambda}_m\mathbf{U}_m^T$. It is sometimes convenient to transform weighted L_2 problems into one or another of these two forms before proceeding with their solution.

9.5   The solution of the mixed-determined problem

The concept of vector spaces is particularly helpful in understanding the mixed-determined problem, in which some linear combinations of the model parameters are overdetermined and some are underdetermined.

If the problem is to some degree underdetermined, then the equation $\mathbf{d} = \mathbf{Gm}$ contains information about only some linear combinations of the model parameters. We can think of these combinations as lying in a subspace $S_p(\mathbf{m})$ of the model parameters space. No information is provided about the part of the solution that lies in the rest of the space, which we shall call the *null space* $S_0(\mathbf{m})$. The part of $\mathbf{m}$ that lies in the null space is completely "unilluminated" by $\mathbf{d} = \mathbf{Gm}$, as the equation contains no information about these linear combinations of the model parameters.

On the other hand, if the problem is to some degree overdetermined, the product $\mathbf{Gm}$ may not be able to span $S(\mathbf{d})$ no matter what one chooses for $\mathbf{m}$. At best $\mathbf{Gm}$ may span a subspace $S_p(\mathbf{d})$ of the data space. Then no part of the data lying outside of this space, say, in $S_0(\mathbf{d})$, can be satisfied for any choice of the model parameters.

If the model parameters and data are divided into parts with subscript p that lie in the p spaces and parts with subscript 0 that lie in the null spaces, we can write $\mathbf{d} = \mathbf{Gm}$ as

$$\mathbf{d}_p + \mathbf{d}_0 = \mathbf{G}(\mathbf{m}_p + \mathbf{m}_0) \tag{9.28}$$

The solution length is then

$$L = \mathbf{m}^T\mathbf{m} = (\mathbf{m}_p + \mathbf{m}_0)^T(\mathbf{m}_p + \mathbf{m}_0) = \mathbf{m}_p^T\mathbf{m}_p + \mathbf{m}_0^T\mathbf{m}_0 \tag{9.29}$$

The cross terms involving $\mathbf{m}_0^T\mathbf{m}_p$ are zero because vector lying in different subspaces are perpendicular to one another. We can now define precisely what we mean by a solution to the mixed-determined problem that minimizes prediction error while adding a minimum of prior information: prior information is added to specify only those linear combinations of the model parameters that reside in the null space $S_0(\mathbf{m})$ and the prediction error is reduced to only the portion in the null space $S_0(\mathbf{d})$ by satisfying $\mathbf{e}_p = \mathbf{d}_p - \mathbf{Gm}_p = \mathbf{0}$ exactly.

One possible choice of prior information is $\mathbf{m}_0^{est} = 0$, which is sometimes called the *natural solution* of the mixed-determined problem. We note that when $\mathbf{d} = \mathbf{Gm}$ is purely underdetermined the natural solution is just the minimum-length solution, and when $\mathbf{d} = \mathbf{Gm}$ is purely overdetermined it is just the least-squares solution.

One might be tempted to view the natural solution as better than, say, the generalized least-squares solution, as the prior information is applied only to the part of the solution that is in the null space and does not increase the prediction error E. However, such an assessment is not clear-cut. If the prior information is indeed accurate, it should be fully utilized, even if its use leads to a slightly larger prediction error. Anyway, given noisy measurements, two slightly different prediction errors will not be statistically distinguishable. This analysis emphasizes that the choice of the inversion method must be tailored carefully to the problem at hand.

9.6 Singular-value decomposition and the natural generalized inverse

The p and null subspaces of the linear problem easily can be identified through a type of eigenvalue decomposition of the data kernel that is called the *singular-value decomposition*. We shall derive this decomposition in Section 9.7, but first we shall state the result and demonstrate its usefulness.

Any $N \times M$ square matrix can be written as the product of three matrices (Penrose, 1955; Lanczos, 1961):

$$\mathbf{G} = \mathbf{U\Lambda V}^T \tag{9.30}$$

The matrix $\mathbf{U}$ is an $N \times N$ matrix of eigenvectors that span the data space $S(\mathbf{d})$:

$$\mathbf{U} = \begin{bmatrix} \mathbf{u}^{(1)} & \mathbf{u}^{(2)} & \mathbf{u}^{(3)} & \cdots & \mathbf{u}^{(N)} \end{bmatrix} \tag{9.31}$$

where the $\mathbf{u}^{(i)}$s are the individual vectors. The vectors are orthogonal to one another and can be chosen to be of unit length so that $\mathbf{UU}^T = \mathbf{U}^T\mathbf{U} = \mathbf{I}$ (where the identity matrix is $N \times N$). Similarly, $\mathbf{V}$ is an $M \times M$ matrix of eigenvectors that span the model parameter space $S(\mathbf{m})$:

$$\mathbf{V} = \begin{bmatrix} \mathbf{v}^{(1)} & \mathbf{v}^{(2)} & \mathbf{v}^{(3)} & \cdots & \mathbf{v}^{(M)} \end{bmatrix} \tag{9.32}$$

Here the $\mathbf{v}^{(i)}$s are the individual orthonormal vectors, satisfying $\mathbf{VV}^T = \mathbf{V}^T\mathbf{V} = \mathbf{I}$ (where the identity matrix is $M \times M$). The matrix $\mathbf{\Lambda}$ is an $N \times M$ diagonal eigenvalue matrix whose diagonal elements are nonnegative and are called *singular values*. In the $N=4$, $M=3$ case

$$\mathbf{\Lambda} = \begin{bmatrix} \lambda_1 & 0 & 0 \\ 0 & \lambda_2 & 0 \\ 0 & 0 & \lambda_3 \\ 0 & 0 & 0 \end{bmatrix} \tag{9.33}$$

The singular values are usually arranged in order of decreasing size. Some of the singular values may be zero. We therefore partition $\mathbf{\Lambda}$ into a submatrix $\mathbf{\Lambda}_p$ of p nonzero singular values and several zero matrices as

$$\mathbf{\Lambda} = \begin{bmatrix} \mathbf{\Lambda}_p & 0 \\ 0 & 0 \end{bmatrix} \tag{9.34}$$

where $\mathbf{\Lambda}_p$ is a $p \times p$ diagonal matrix. The decomposition then becomes $\mathbf{U\Lambda V}^T = \mathbf{U}_p \mathbf{\Lambda}_p \mathbf{V}_p^T$, where $\mathbf{U}_p$ and $\mathbf{V}_p$ consist of the first p columns of $\mathbf{U}$ and $\mathbf{V}$, respectively. The other portions of the eigenvector matrices are canceled by the zeros in $\mathbf{\Lambda}$. The matrix $\mathbf{G}$ contains no information about the subspaces spanned by these portions of the data and model eigenvectors, which we shall call $\mathbf{U}_0$ and $\mathbf{V}_0$, respectively. As we shall soon prove, these are precisely the same spaces as the p and null spaces defined in the previous section.

The data kernel is not a function of the null eigenvectors $\mathbf{U}_0$ and $\mathbf{V}_0$, The equation $\mathbf{d} = \mathbf{Gm} = \mathbf{U}_p\mathbf{\Lambda}_p\mathbf{V}_p^T\mathbf{m}$ contains no information about the part of the model parameters in the space spanned by $\mathbf{V}_0$ as the model parameters $\mathbf{m}$ are multiplied by $\mathbf{V}_p$ (which is orthogonal to everything in $\mathbf{V}_0$). The eigenvector $\mathbf{V}_p$, therefore, lies completely in $S_p(\mathbf{m})$, and $\mathbf{V}_0$ lies completely in $S_0(\mathbf{m})$. Similarly, no matter what value $\{\mathbf{\Lambda}_p\mathbf{V}_p^T\mathbf{m}\}$ attains, it can have no component in the space spanned by $\mathbf{U}_0$ as it is multiplied by $\mathbf{U}_p$ (and $\mathbf{U}_0$ and $\mathbf{U}_p$ are orthogonal). Therefore $\mathbf{U}_p$ lies completely in $S_p(\mathbf{d})$ and $\mathbf{U}_0$ lies completely in $S_0(\mathbf{d})$.

We have demonstrated that the p and null spaces can be identified through the singular-value decomposition of the data kernel. The full spaces $S(\mathbf{m})$ and $S(\mathbf{d})$ are spanned by $\mathbf{V}$ and $\mathbf{U}$, respectively. The p spaces are spanned by the parts of the eigenvector matrices that have nonzero eigenvalues: $S_p(\mathbf{m})$ is spanned by $\mathbf{V}_p$ and $S_p(\mathbf{d})$ is spanned by $\mathbf{U}_p$. The remaining eigenvectors $\mathbf{V}_0$ and $\mathbf{U}_0$ span the null spaces $S_0(\mathbf{m})$ and $S_p(\mathbf{m})$, respectively. The p and null matrices are orthogonal and are normalized in the sense that $\mathbf{V}_p^T\mathbf{V}_p = \mathbf{U}_p^T\mathbf{U}_p = \mathbf{I}$, where $\mathbf{I}$ is $p \times p$ in size. However, as these matrices do not in general span the complete data and model spaces $\mathbf{V}_p\mathbf{V}_p^T$ and $\mathbf{U}_p\mathbf{U}_p^T$ are not, in general, identity matrices.

The natural solution to the inverse problem can be constructed from the singular-value decomposition. This solution must have a solution $\mathbf{m}^{est}$ that has no component in $S_0(\mathbf{m})$ and a prediction error $\mathbf{e}$ that has no component in $S_p(\mathbf{d})$. We therefore consider the solution

$$\mathbf{m}^{est} = \mathbf{V}_p\mathbf{\Lambda}_p^{-1}\mathbf{U}_p^T\mathbf{d}^{obs} \tag{9.35}$$

which is picked in analogy to the square matrix case. To show that $\mathbf{m}^{est}$ has no component in $S_0(\mathbf{m})$, we take the dot product of the equation with $\mathbf{V}_0$, which lies completely in $S_0(\mathbf{m})$, as

$$\mathbf{V}_0^T\mathbf{m}^{est} = \left(\mathbf{V}_0^T\mathbf{V}_p\right)\mathbf{\Lambda}_p^{-1}\mathbf{U}_p^T\mathbf{d}^{obs} = 0 \tag{9.36}$$

as $\mathbf{V}_0$ and $\mathbf{V}_p$ are orthogonal. To show that $\mathbf{e}$ has no component in $S_p(\mathbf{d})$, we take the dot product with $\mathbf{U}_p$ as

$$\begin{aligned}\mathbf{U}_p^T\mathbf{e} = \mathbf{U}_p^T\left(\mathbf{d}^{obs} - \mathbf{Gm}^{est}\right) &= \mathbf{U}_p^T\mathbf{d}^{obs} - \mathbf{U}_p^T\mathbf{U}_p\mathbf{\Lambda}_p\mathbf{V}_p^T\mathbf{V}_p\mathbf{\Lambda}_p^{-1}\mathbf{U}_p^T\mathbf{d}^{obs} \\ &= \mathbf{U}_p^T\mathbf{d}^{obs} - \mathbf{U}_p^T\mathbf{d}^{obs} = 0\end{aligned} \tag{9.37}$$

Here, we have used the rules $\mathbf{U}_p^T\mathbf{U}_p = \mathbf{I}$, $\mathbf{V}_p^T\mathbf{V}_p = \mathbf{I}$, and $\mathbf{\Lambda}_p\mathbf{\Lambda}_p^{-1} = \mathbf{I}$. Thus

$$\mathbf{G}^{-g} = \mathbf{V}_p\mathbf{\Lambda}_p^{-1}\mathbf{U}_p^T \tag{9.38}$$

is the *natural generalized inverse*. This generalized inverse is so useful that it is sometimes referred to as *the* generalized inverse, although of course there are other generalized inverses that can be designed for the mixed-determined problem that embody other kinds of prior information.

The singular-value decomposition is computed as

MATLAB®
| ```
[U, L, V] = svd(G,0);
lam = diag(L);
```                                                        [gdama09_06] |

Python
| ```
U, lam, VT = la.svd(G, full_matrices=False);
V = VT.T;
```                                                        [gdapy09_06] |

In both cases, the *economy* version of svd() is used, signaled by the 0 in MATLAB® and the full_matrices= False in Python. This choice leads to $\mathbf{U}$ and $\mathbf{V}$ being $N \times K$ and $N \times M$, respectively, where $K = \min(N, M)$. In the MATLAB® version, the diagonal matrix L of singular values has been reduced to a column vector lambda and in the Python version, the returned matrix VT of eigenvectors has been transposed.

The submatrices Up and Vp (with p an integer) are extracted via:

MATLAB®
| ```
Up = U(:,1:p);
Vp = V(:,1:p);
lambdap = lambda(1:p);
```                                                        [gdama09_05] |

Python
| ```
Up=U[0:N,0:p];
Vp=V[0:M,0:p];
lambdap = gda_cvec(lam[0:p]);
```                                                        [gdapy09_05] |

Note that the Python versions of Up and Vp are *views* into U and V and are not independent copies. The model parameters are estimated as

MATLAB®
`mest = Vp*((Up'*dobs)./lambdap);`
[gdama09_05]

Python
`mest = np.matmul(Vp,np.divide(np.matmul(Up.T,dobs),lambdap));`
[gdapy09_05]

Note that the calculation is organized so that the matrices $\mathbf{U}_p^T$ and $\mathbf{V}_p$ multiply vectors (as contrasted to matrices); the generalized inverse is never explicitly formed. The value of the integer p must be chosen so that zero eigenvalues are excluded from the estimate; else a division-by-zero error will occur. As discussed later, one often excludes near-zero eigenvalues as well, as the resulting solution (while not quite the natural solution) is often better behaved (Fig. 9.3).

The natural generalized inverse has model resolution

$$\mathbf{R} = \mathbf{G}^{-g}\mathbf{G} = \mathbf{V}_p\mathbf{\Lambda}_p^{-1}\mathbf{U}_p^T\mathbf{U}_p\mathbf{\Lambda}_p\mathbf{V}_p^T = \mathbf{V}_p\mathbf{V}_p^T \tag{9.39}$$

The model parameters will be perfectly resolved only if $\mathbf{V}_p$ spans the complete space of model parameters, that is, if there are no zero eigenvalues and $p=M$. The data resolution matrix is

$$\mathbf{N} = \mathbf{G}\mathbf{G}^{-g} = \mathbf{U}_p\mathbf{\Lambda}_p\mathbf{V}_p^T\mathbf{V}_p\mathbf{\Lambda}_p^{-1}\mathbf{U}_p^T = \mathbf{U}_p\mathbf{U}_p^T \tag{9.40}$$

The data are only perfectly resolved if $\mathbf{U}_p$ spans the complete space of data and $p=N$. Finally, we note that if the data are uncorrelated with uniform variance σ_d^2, the posterior covariance of the model parameter is

$$[\text{cov } \mathbf{m}] = \sigma_d^2\mathbf{G}^{-g}\mathbf{G}^{-gT} = \sigma_d^2\mathbf{V}_p\mathbf{\Lambda}_p^{-1}\mathbf{U}_p^T\mathbf{U}_p\mathbf{\Lambda}_p^{-1}\mathbf{V}_p^T = \sigma_d^2\mathbf{V}_p\mathbf{\Lambda}_p^{-2}\mathbf{V}_p^T \tag{9.41}$$

The covariance of the estimated model parameters is very sensitive to the smallest nonzero eigenvalue.

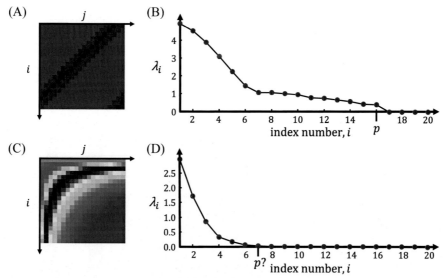

FIG. 9.3 (A) Hypothetical data kernel G_{ij} for an inverse problem with $M=20$ model parameters and $N=20$ observations. (B) Corresponding singular values λ_i have a clear cutoff at $p=16$. (C) Another hypothetical data kernel, also with $M=20$ and $N=20$. (D) Corresponding singular values do not have a clear cutoff, so the parameter p must be chosen in a more arbitrary fashion near $p=7$. Scripts gdama09_05 and gdapy09_05.

Eq. (9.41) for the posterior covariance omits any reference to the prior covariance $[\text{cov } \mathbf{m}]_A$ of the model parameters. The reason can be understood by writing the prior model parameters as $\langle \mathbf{m} \rangle = \langle \mathbf{m}_p \rangle + \langle \mathbf{m}_0 \rangle$, where $\langle \mathbf{m}_p \rangle$ and $\langle \mathbf{m}_0 \rangle$ are the parts of the prior model parameters in $S_p(\mathbf{m})$ and $S_0(\mathbf{m})$, respectively. The natural solution was created with the assumption that the error has been reduced to its minimum, and so is unaffected by the choice of $\langle \mathbf{m} \rangle$. It implies that the covariance of $\langle \mathbf{m}_p \rangle$ is infinite; that is, it has no influence on the solution. The natural solution was also created with the assumption that $\langle \mathbf{m}_0 \rangle$ is exactly zero, which implies that the covariance of $\langle \mathbf{m}_0 \rangle$ is zero. With these admittedly dubious assumptions, the solution is independent of the covariance of the prior information.

These assumptions can be relaxed. One could seek a solution that has minimum error but is close to some nonzero prior solution $\langle \mathbf{m} \rangle$ and having prior covariance $[\text{cov } \mathbf{m}]_A$. This can be achieved by adding null vectors to the natural solution, which change $\mathbf{m}^{est}$ without increasing the error $\mathbf{e}$. The general solution to the inverse problem (Eq. 8.4)

$$\mathbf{m}^{gen} = \mathbf{m}^{par} + \sum_{i=p+1}^{M} \alpha_i \mathbf{m}^{null(i)} \tag{9.42}$$

now takes the form

$$\mathbf{m}^{est} = \mathbf{G}_N^{-g} \mathbf{d}^{obs} + \mathbf{V}_0 \boldsymbol{\alpha} \tag{9.43}$$

where $\mathbf{G}_N^{-g} \equiv \mathbf{V}_p \boldsymbol{\Lambda}_p^{-1} \mathbf{U}_p^T$ is the natural generalized inverse and $\boldsymbol{\alpha}$ is a vector of coefficients (Wunsch and Minster, 1982). We then choose $\boldsymbol{\alpha}$ to minimize the L_2 distance between $\mathbf{m}^{est}$ and the prior solution $\langle \mathbf{m} \rangle$. The equation $\mathbf{m}^{est} \approx \langle \mathbf{m} \rangle$ implies

$$\mathbf{V}_0 \boldsymbol{\alpha} \approx \langle \mathbf{m}_0 \rangle \quad \text{with} \quad \langle \mathbf{m}_0 \rangle \equiv \langle \mathbf{m} \rangle - \mathbf{G}_N^{-g} \mathbf{d}^{obs} \tag{9.44}$$

This equation has least-squares solution, $\boldsymbol{\alpha} = [\mathbf{V}_0^T \mathbf{V}_0]^{-1} \mathbf{V}_0^T \langle \mathbf{m}_0 \rangle = \mathbf{V}_0^T \langle \mathbf{m}_0 \rangle$, implying that

$$\begin{aligned} \mathbf{m}^{est} = \mathbf{G}_N^{-g} \mathbf{d}^{obs} + \mathbf{V}_0 \mathbf{V}_0^T \langle \mathbf{m}_0 \rangle = \mathbf{G}_N^{-g} \mathbf{d}^{obs} + (\mathbf{I} - \mathbf{R}_N) \langle \mathbf{m}_0 \rangle = \\ \mathbf{G}_N^{-g} \mathbf{d}^{obs} - (\mathbf{I} - \mathbf{R}_N) \mathbf{G}_N^{-g} \mathbf{d}^{obs} + (\mathbf{I} - \mathbf{R}_N) \langle \mathbf{m} \rangle = \\ \mathbf{R}_N \mathbf{G}_N^{-g} \mathbf{d}^{obs} + (\mathbf{I} - \mathbf{R}_N) \langle \mathbf{m} \rangle = \mathbf{G}_N^{-g} \mathbf{d}^{obs} + (\mathbf{I} - \mathbf{R}_N) \langle \mathbf{m} \rangle \end{aligned} \tag{9.45}$$

Here we have used the rule $\mathbf{V}_p \mathbf{V}_p^T + \mathbf{V}_0 \mathbf{V}_0^T = \mathbf{I}$, which is derived by multiplying out the submatrices in $\mathbf{V}\mathbf{V}^T = \mathbf{I}$, the observation that $\mathbf{R}_N \equiv \mathbf{V}_p \mathbf{V}_p^T$ is the resolution matrix associated with the natural generalized inverse, and the rule $\mathbf{R}_N \mathbf{G}_N^{-g} = \mathbf{V}_p \mathbf{V}_p^T \mathbf{V}_p \boldsymbol{\Lambda}_p^{-1} \mathbf{U}_p^T = \mathbf{G}_N^{-g}$. By the usual rules of error propagation, the posterior covariance is

$$[\text{cov } \mathbf{m}] = \sigma_d^2 \mathbf{G}^{-g} \mathbf{G}^{-gT} + (\mathbf{I} - \mathbf{R}_N) [\text{cov } \mathbf{m}]_A (\mathbf{I} - \mathbf{R}_N)^T \tag{9.46}$$

We have encountered formulas similar to this one previously, for example, in Eq. (6.40). However, the statistical underpinnings of this equation are inconsistent: although $\mathbf{d}^{obs}$ contains noise, the prediction error is not allowed to increase, even slightly, from its minimum value in order to better fit the prior information (Fig. 9.4).

Instead of choosing a sharp cutoff for the singular values, all the singular values can be included in the generalized inverse, with the smaller ones being *damped*. We let $p = M$, but replace the reciprocals of all the singular values by $\lambda_i / (\varepsilon^2 + \lambda_i^2)$, where ε is a small number. This change has little effect on the larger singular values but prevents the smaller ones from leading to large variances. Of course, the solution is no longer the natural solution. While its variance is

FIG. 9.4 (A) The same linear inverse problem as in Fig. 9.3A. The data, where $\mathbf{d}^{obs}$, contain uncorrelated, Normally distributed noise. (B) Corresponding solutions, true solutions $\mathbf{m}^{true}$, natural solution $\mathbf{m}_N^{est}$, and damped minimum-length solution $\mathbf{m}_{DML}^{est}$. Scripts gdama09_05 and gdapy09_05.

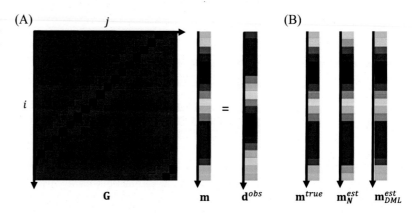

improved, its model and data resolution are degraded. In fact, this solution is precisely the damped least-squares solution discussed in Section 4.11

$$
\begin{aligned}
\left[\mathbf{G}^T\mathbf{G} + \varepsilon^2\mathbf{I}\right]^{-1}\mathbf{G}^T &= \left[\mathbf{V\Lambda U}^T\mathbf{U\Lambda V}^T + \varepsilon^2\mathbf{V}^T\mathbf{IV}\right]^{-1}\mathbf{V\Lambda U}^T \\
&= \left[\mathbf{V}^T\left(\mathbf{\Lambda}^2 + \varepsilon^2\mathbf{I}\right)\mathbf{V}\right]^{-1}\mathbf{V\Lambda U}^T = \mathbf{V}\left(\mathbf{\Lambda}^2 + \varepsilon^2\mathbf{I}\right)^{-1}\mathbf{V}^T\mathbf{V\Lambda U}^T \\
&= \mathbf{V}\left\{\left(\mathbf{\Lambda}^2 + \varepsilon^2\mathbf{I}\right)^{-1}\mathbf{\Lambda}\right\}\mathbf{U}^T
\end{aligned}
\tag{9.47}
$$

Here we have relied upon the relations $(\mathbf{ABC})^{-1} = \mathbf{C}^{-1}\mathbf{B}^{-1}\mathbf{A}^{-1}$, $\mathbf{V}^{-1} = \mathbf{V}^T$, and $\mathbf{U}^T\mathbf{U} = \mathbf{V}^T\mathbf{V} = \mathbf{I}$. The damping of the singular values corresponds to the addition of prior information that the model parameters are small. The precise value of the number used as the cutoff or damping parameter must be chosen by a trial-and-error process which weighs the relative merits of having a solution with small variance against those of having one that fits the data and is well resolved.

In Section 8.6, we discussed the problem of bounding nonunique averages of model parameters by incorporating prior inequality constraints into the solution of the inverse problem. We see that, through Eq. (9.43), singular-value decomposition provides a simple way of identifying the null vectors of $\mathbf{d} = \mathbf{Gm}$. In Section 8.6, upper and lower bounds on localized averages of the solution were found by determining the α that maximizes (for the upper bound) or minimizes (for the lower bound) the localized average $\langle m_i \rangle = \mathbf{a}^T\mathbf{m}$ with the constraint that $\mathbf{m}^l \leq \mathbf{m} \leq \mathbf{m}^u$, where $\mathbf{m}^l$ and $\mathbf{m}^u$ are prior bounds. The use of the natural solution guarantees that the prediction error is minimized in the L_2 sense.

The bounds on the localized average $\langle m \rangle = \mathbf{a}^T\mathbf{m}$ should be treated with some skepticism, as they depend on a particular choice of the solution (in this case one that minimizes the L_2 prediction error). If the total error E increases only slightly as this solution is perturbed and if this additional error can be considered negligible, then the true bounds of the localized average will be larger than those given earlier. In principle, one can handle this problem by forsaking the eigenvalue decomposition and simply determining the $\mathbf{m}$ that extremizes $\langle m \rangle$ with the constraints that $\mathbf{m}^l \leq \mathbf{m} \leq \mathbf{m}^u$ and that the total prediction error is less than some tolerable amount, say, E_M. For L_2 measures of the prediction error, this is a very difficult nonlinear problem. However, if the error is measured under the L_1 norm, it can be transformed into a linear programming problem.

9.7 Derivation of the singular-value decomposition

The singular-value decomposition can be derived in many ways. We follow here the treatment of Lanczos (1961). For an alternate derivation, see Menke and Menke (2011, Section 8.2). We first form an $(N+M) \times (N+M)$ square symmetric matrix $\mathbf{S}$ from $\mathbf{G}$ and $\mathbf{G}^T$ as

$$
\mathbf{S} = \begin{bmatrix} \mathbf{0} & \mathbf{G} \\ \mathbf{G}^T & \mathbf{0} \end{bmatrix}
\tag{9.48}
$$

Being symmetric, this matrix must have $(N+M)$ real eigenvalues λ_i and a complete set of eigenvectors $\mathbf{w}^{(i)}$, which solve $\mathbf{Sw}^{(i)} = \lambda_i\mathbf{w}^{(i)}$. Partitioning $\mathbf{w}^{(i)}$ into an upper part $\mathbf{u}^{(i)}$ of length N and a lower part $\mathbf{v}^{(i)}$ of length M, we obtain

$$
\begin{bmatrix} \mathbf{0} & \mathbf{G} \\ \mathbf{G}^T & \mathbf{0} \end{bmatrix} \begin{bmatrix} \mathbf{u}^{(i)} \\ \mathbf{v}^{(i)} \end{bmatrix} = \lambda_i \begin{bmatrix} \mathbf{u}^{(i)} \\ \mathbf{v}^{(i)} \end{bmatrix}
\tag{9.49}
$$

We shall now show $\mathbf{u}^{(i)}$ and $\mathbf{v}^{(i)}$ are the vectors, and λ_i the singular value, in the formula for the singular-value decomposition. Multiplying out the equation, we obtain $\mathbf{Gv}^{(i)} = \lambda_i\mathbf{u}^{(i)}$ and $\mathbf{G}^T\mathbf{u}^{(i)} = \lambda_i\mathbf{v}^{(i)}$. Suppose that there is a positive eigenvalue λ_i with eigenvector $[\mathbf{u}^{(i)}, \mathbf{v}^{(i)}]^T$. Then, $-\lambda_i$ is also an eigenvalue with eigenvector $[-\mathbf{u}^{(i)}, \mathbf{v}^{(i)}]^T$. If there are p positive eigenvalues, then there are p negative eigenvalues and $(N+M-2p)$ zero eigenvalues. Now by inserting the first equation into the second, and vice versa, we obtain

$$
\mathbf{G}^T\mathbf{Gv}^{(i)} = \lambda_i^2\mathbf{v}^{(i)} \quad \text{and} \quad \mathbf{GG}^T\mathbf{u}^{(i)} = \lambda_i^2\mathbf{u}^{(i)}
\tag{9.50a,b}
$$

These are eigenvalue equations involving square matrices $\mathbf{G}^T\mathbf{G}$ (which is $M \times M$) and $\mathbf{GG}^T$ (which is $N \times N$). Both matrices are nonnegative definite (see Sections 1A.4 and 1B.4), implying that their eigenvalue λ_i^2 is nonnegative, so that the λs are purely real. As a symmetric matrix can have no more distinct eigenvectors than its dimension $p \leq \min(N,M)$. However, there is an issue of the range of the index i, for $\mathbf{G}^T\mathbf{G}$ has M eigenvalues, whereas $\mathbf{GG}^T$ has N. We will resolve this issue later. As both matrices are square and symmetric, there are M vectors $\mathbf{v}^{(i)}$ that form a complete orthonormal set that spans $S(\mathbf{m})$, and there are N vectors $\mathbf{u}^{(i)}$ that form a complete orthonormal set spanning

$S(\mathbf{d})$. As they are complete, these sets can be arranged as matrices $\mathbf{V}$ and $\mathbf{U}$ that obey $\mathbf{V}^T\mathbf{V} = \mathbf{V}\mathbf{V}^T = \mathbf{I}$ and $\mathbf{U}^T\mathbf{U} = \mathbf{U}\mathbf{U}^T = \mathbf{I}$.

We now focus on the $N > M$ case (the $M > N$ case being similar). The original eigenvalue problem (Eq. 9.49) can be written as $\mathbf{S}\mathbf{W} = \mathbf{W}\Lambda$, where $\mathbf{w}^{(i)}$ are the columns of $\mathbf{W}$ and λ_i are the elements of the diagonal matrix, Λ. A valid choice of $\mathbf{W}$ is

$$\mathbf{W} = \frac{1}{\sqrt{2}}\begin{bmatrix} \begin{bmatrix} \mathbf{u}^{(1)}\cdots\mathbf{u}^{(p)} \end{bmatrix} & \begin{bmatrix} \mathbf{u}^{(p+1)}\cdots\mathbf{u}^{(M)} \end{bmatrix} & \sqrt{2}\begin{bmatrix} \mathbf{u}^{(M+1)}\cdots\mathbf{u}^{(N)} \end{bmatrix} & -\begin{bmatrix} \mathbf{u}^{(p+1)}\cdots\mathbf{u}^{(M)} \end{bmatrix} & -\begin{bmatrix} \mathbf{u}^{(p)}\cdots\mathbf{u}^{(1)} \end{bmatrix} \\ \begin{bmatrix} \mathbf{v}^{(1)}\cdots\mathbf{v}^{(p)} \end{bmatrix} & \begin{bmatrix} \mathbf{v}^{(p+1)}\cdots\mathbf{v}^{(M)} \end{bmatrix} & \mathbf{0} & \begin{bmatrix} \mathbf{v}^{(p+1)}\cdots\mathbf{v}^{(M)} \end{bmatrix} & \begin{bmatrix} \mathbf{v}^{(p)}\cdots\mathbf{v}^{(1)} \end{bmatrix} \end{bmatrix} \tag{9.51}$$

Here, the factors of $\sqrt{2}$ arise from normalizing $\mathbf{w}^{(i)}$ to unit length. This choice resolves the issue of the range of i; we have introduced ($N > M$) zero vectors, so that each of the N eigenvectors $\mathbf{u}^{(i)}$ is paired with some other vector, either $\mathbf{v}^{(i)}$ or a zero vector. The zero vectors are paired with zero eigenvalues, so that Eq. (9.50a) is still valid; they merely correspond to the equation $\mathbf{0} = \mathbf{0}$. It can be shown by inspection that $\mathbf{W}^T\mathbf{W} = \mathbf{W}\mathbf{W}^T = \mathbf{I}$; that is, the $\mathbf{w}$s are orthonormal. However, this choice of $\mathbf{W}$ does not guarantee that the eigenvalues in the matrix $\Lambda = \mathbf{W}^T\mathbf{S}\mathbf{W}$ are in standard descending order, as the members of some of the positive/negative pairs may be swapped. This problem can be corrected by examining each of the diagonal elements of Λ, and flipping its sign, and the sign of the corresponding $\mathbf{u}^{(i)}$, where necessary.

Eq. (9.50a) can be modified to $\mathbf{G}\mathbf{v}'^{(i)} = \lambda_i\mathbf{u}^{(i)}$, where $\mathbf{u}$s and $\mathbf{v}$s are drawn from the first N columns of $\mathbf{W}$ and the λs are the corresponding eigenvalues. This equation can be put into the form

$$\mathbf{G}\mathbf{V}' = \mathbf{U}\Lambda'$$

$$\text{with } \mathbf{U} = \begin{bmatrix} \mathbf{U}_p & \mathbf{U}_0 \end{bmatrix} \text{ and } \mathbf{V}' = \begin{bmatrix} \mathbf{V}_p & \mathbf{V}_0 & \mathbf{0} \end{bmatrix} \text{ and } \Lambda' = \begin{bmatrix} \Lambda_p & 0 & 0 \\ 0 & 0 & 0 \end{bmatrix} \tag{9.52}$$

Here, Λ_p are the positive eigenvalues. The modified matrix $\mathbf{V}'$ still obeys

$$\mathbf{V}'\mathbf{V}'^T = \begin{bmatrix} \mathbf{V} & 0 \end{bmatrix}\begin{bmatrix} \mathbf{V}^T \\ 0 \end{bmatrix} = \mathbf{V}\mathbf{V}^T = \mathbf{I} \tag{9.53}$$

Postmultiplying Eq. (9.52) by $\mathbf{V}'^T$ gives the singular-value decomposition

$$\mathbf{G} = \mathbf{U}\Lambda'\mathbf{V}'^T = \begin{bmatrix} \mathbf{U}_0 & \mathbf{U}_p \end{bmatrix}\begin{bmatrix} \Lambda_p & 0 & 0 \\ 0 & 0 & 0 \end{bmatrix}\begin{bmatrix} \mathbf{V}_p^T \\ \mathbf{V}_o^T \\ 0 \end{bmatrix} = \mathbf{U}_p\Lambda_p\mathbf{V}_p^T \tag{9.54}$$

This algorithm is demonstrated in scripts gdama09_05 and gdapy09_05; however, it is mostly of theoretical interest, as more efficient algorithms are available for computing the singular-value decomposition (as in gdama09_04 and gdapy09_04).

The $N < M$ case is handled in an analogous manner and leads to the same result. In that case, there are more $\mathbf{v}$s than $\mathbf{u}$s, and the excess $\mathbf{v}$s are the vectors paired with zero vectors.

9.8   Simplifying linear equality and inequality constraints

The singular-value decomposition can be used to simplify linear equality constraints. Consider the problem of solving $\mathbf{d} = \mathbf{G}\mathbf{m}$ in the sense of finding a solution that minimizes the L_2 prediction error subject to the $K < M$ constraints so that $\mathbf{H}\mathbf{m} = \mathbf{h}$. This problem can be reduced to the unconstrained problem $\mathbf{d}' = \mathbf{G}'\mathbf{m}'$ in $M' < M$ new model parameters. We first find the singular-value decomposition of the constraint matrix $\mathbf{H} = \mathbf{U}_p\Lambda_p\mathbf{V}_p^T$. If $p = K$, the constraints are consistent and determine p linear combinations of the unknowns. The general solution is then $\mathbf{m} = \mathbf{V}_p\Lambda_p^{-1}\mathbf{U}_p^T\mathbf{h} + \mathbf{V}_0\alpha$, where α is an arbitrary vector of length $M - p$ and is to be determined by minimizing the prediction error. Substituting this equation for $\mathbf{m}$ into $\mathbf{G}\mathbf{m} = \mathbf{d}$ and rearranging terms yields

$$\mathbf{G}\mathbf{V}_0\alpha = \mathbf{d} - \mathbf{G}\mathbf{V}_p\Lambda_p^{-1}\mathbf{U}_p^T\mathbf{h} \tag{9.55}$$

This equation in the unknown coefficients α now can be solved as an unconstrained least-squares problem. We note that we have encountered this problem in a somewhat different form during the discussion of Householder transformations (Section 9.2). The main advantage of using the singular-value decomposition is that it provides a test of the constraint's consistency.

The singular-value decomposition also can be used to simplify linear *inequality* constraints. Consider the L_2 problem

$$\text{minimize } \|\mathbf{d} - \mathbf{G}\mathbf{m}\|_2^2 \text{ subject to } \mathbf{H}\mathbf{m} \geq \mathbf{h} \tag{9.56}$$

We shall show that as long as $\mathbf{d} = \mathbf{Gm}$ is in fact overdetermined, this problem can be reduced to the simpler problem (Lawson and Hanson, 1974):

$$\text{minimize } \|\mathbf{m}'\|_2^2 \text{ subject to } \mathbf{H}'\mathbf{m}' \geq \mathbf{h}' \tag{9.57}$$

To demonstrate this transformation, we form the singular-value decomposition of the data kernel $\mathbf{G} = \mathbf{U}_p\boldsymbol{\Lambda}_p\mathbf{V}_p^T$. We first divide the data $\mathbf{d}$ into two parts $\mathbf{d} = [\mathbf{d}_p, \mathbf{d}_0]^T$, where $\mathbf{d}_p \equiv \mathbf{U}_p^T\mathbf{d}$ is in the $S_p(\mathbf{d})$ subspace and $\mathbf{d}_0 \equiv \mathbf{U}_0^T\mathbf{d}$ is in the $S_0(\mathbf{d})$ subspace. The prediction error is then $E = \|\mathbf{d}^{obs} - \mathbf{d}^{pre}\|_2^2$ or

$$E = \left\| \begin{bmatrix} \mathbf{U}_p^T\mathbf{d} \\ \mathbf{U}_0^T\mathbf{d} \end{bmatrix} - \begin{bmatrix} \mathbf{U}_p^T\mathbf{U}_p\boldsymbol{\Lambda}_p\mathbf{V}_p^T\mathbf{m} \\ 0 \end{bmatrix} \right\|_2^2 = \left\| \mathbf{d}_p - \boldsymbol{\Lambda}_p\mathbf{V}_p^T\mathbf{m} \right\|_2^2 + \|\mathbf{d}_0\|_2^2 = \|\mathbf{m}'\|_2^2 + \|\mathbf{d}_0\|_2^2 \tag{9.58}$$

where $\mathbf{m}' \equiv \mathbf{d}_p - \boldsymbol{\Lambda}_p\mathbf{V}_p^T\mathbf{m}$. Minimizing $\|\mathbf{m}'\|_2^2$ is the same as minimizing E, as the other term is a constant. Inverting this expression for the unprimed model parameters gives $\mathbf{m} = \mathbf{V}_p\boldsymbol{\Lambda}_p^{-1}[\mathbf{d}_p - \mathbf{m}'] = \mathbf{V}_p\boldsymbol{\Lambda}_p^{-1}[\mathbf{U}_p^T\mathbf{d} - \mathbf{m}']$. Substituting this expression into the constraint equation $\mathbf{Hm} \geq \mathbf{h}$ and rearranging terms yields

$$\left\{ -\mathbf{H}\mathbf{V}_p\boldsymbol{\Lambda}_p^{-1} \right\}\mathbf{m}' \geq \left\{ \mathbf{h} - \mathbf{H}\mathbf{V}_p\boldsymbol{\Lambda}_p^{-1}\mathbf{U}_p^T\mathbf{d} \right\} \tag{9.59}$$

which is in the desired form.

9.9 Inequality constraints

We shall now consider the solution of L_2 minimization problems with inequality constraints of the form

$$\text{minimize } \|\mathbf{d} - \mathbf{Gm}\|_2^2 \text{ subject to } \mathbf{Hm} \geq \mathbf{h} \tag{9.60}$$

We first note that problems involving $=$ and $\leq$ constraints can be reduced to this form. Equality constraints can be removed by the transformation described in Section 9.8, and $\leq$ constraints can be removed by multiplication by -1 to change them into $\geq$ constraints.

For this minimization problem to have any solution, the constraints $\mathbf{Hm} \geq \mathbf{h}$ must be consistent; there must be at least one $\mathbf{m}$ that satisfies all the constraints. We can view these constraints as defining a volume in $S(\mathbf{m})$. Each constraint defines a hyperplane that divides $S(\mathbf{m})$ into two half-spaces, one in which that constraint is satisfied (the *feasible* half-space) and the other in which it is violated (the *infeasible* half-space). The set of inequality constraints defines a volume in $S(\mathbf{m})$ which might be zero, finite, or infinite in extent. If the region has zero volume, then no feasible solution exists (Fig. 9.5B). If it has nonzero volume, then there is at least one feasible solution with the smallest prediction error (Fig. 9.5A). This volume has the shape of a polyhedron as its boundary surfaces are planes. It can be shown that the polyhedron must be convex: it can have no reentrant angles or grooves.

The starting point for solving the L_2 minimization problem with inequality constraints is the *Kuhn-Tucker theorem*, which describes the properties that any solution to this problem must possess. For any $\mathbf{m}$ that minimizes $\|\mathbf{d} - \mathbf{Gm}\|_2^2$ subject to p constraints $\mathbf{Hm} \geq \mathbf{h}$, it is possible to find a vector $\mathbf{y}$ of length p, such that

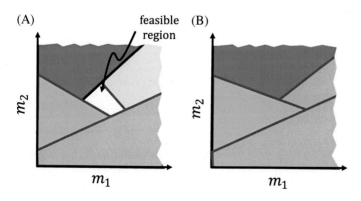

(A)   feasible (B)
          region
m_2

m_1 m_1

FIG. 9.5 (A) Each linear inequality constraint (lines) divides the space of model parameter $S(\mathbf{m})$ into two half-spaces, one infeasible (shaded) and the other feasible (white). Consistent constraints combine to form a convex, polygonal region within $S(\mathbf{m})$ (white) of feasible $\mathbf{m}$. (B) Inconsistent constraints do not form a feasible region.

$$\tfrac{1}{2}\nabla E = \nabla[\mathbf{Hm}] \cdot \mathbf{y} \quad \text{or} \quad -\mathbf{G}^T[\mathbf{d} - \mathbf{Gm}] = \mathbf{H}^T\mathbf{y} \quad \text{with } y_i > 0 \tag{9.61}$$

This equation states that the gradient of the error E can be written as a linear combination of hyperplane normal $\mathbf{H}^T$, with the combination specified by a vector $\mathbf{y}$ with nonnegative elements. In other words, $\mathbf{m}$ is only a solution when the direction that one must perturb it to decrease the error makes it infeasible. The elements of $\mathbf{y}$, which are nonnegative, can be partitioned into two parts $\mathbf{y}_E$ and $\mathbf{y}_S$ (possibly requiring reordering of the constraints) that satisfy

$$\mathbf{y} = \begin{bmatrix} \mathbf{y}_E \\ \mathbf{y}_S \end{bmatrix} \quad \text{and} \quad \begin{matrix} \mathbf{H}_E\mathbf{m} = \mathbf{h}_E \\ \mathbf{H}_S\mathbf{m} > \mathbf{h}_S \end{matrix} \quad \text{with} \quad \begin{matrix} [\mathbf{y}_E]_i = 0 \\ [\mathbf{y}_S]_i > 0 \end{matrix} \tag{9.62}$$

The first group of constraints is satisfied in the equality sense (thus the subscript E for *equality*). The rest are satisfied more loosely in the inequality sense (thus the subscript S for *slack*).

The theorem indicates that any feasible solution $\mathbf{m}$ is the minimum solution only if the direction in which one would have to perturb $\mathbf{m}$ to decrease the total error E causes the solution to cross some constraint hyperplane and become infeasible. The direction of decreasing error is $-\tfrac{1}{2}\nabla E = \mathbf{G}^T[\mathbf{d} - \mathbf{Gm}]$. The constraint hyperplanes have normal $+\nabla[\mathbf{Hm}] = \mathbf{H}^T$, which point into the feasible side. As $\mathbf{H}_E\mathbf{m} - \mathbf{h}_E = \mathbf{0}$, the solution lies exactly on the bounding hyperplanes of the $\mathbf{H}_E$ constraints. As $\mathbf{H}_S\mathbf{m} - \mathbf{h}_S > \mathbf{0}$, it lies within the feasible volume of the $\mathbf{H}_S$ constraints. Therefore an infinitesimal perturbation $\delta\mathbf{m}$ of the solution can only violate the $\mathbf{H}_E$ constraints. If it is not to violate these constraints, the perturbation must be made in the direction of feasibility, so that it must be expressible as a nonnegative combination of hyperplane normals, that is, $\delta\mathbf{m} \cdot \nabla[\mathbf{Hm}] \geq 0$. On the other hand, if it is to decrease the total prediction error it must satisfy $\delta\mathbf{m} \cdot \nabla E \leq 0$. These two conditions are incompatible with the Kuhn-Tucker theorem, as dotting Eq. (9.61) with $\delta\mathbf{m}$ yields $\delta\mathbf{m} \cdot \tfrac{1}{2}\nabla E = \delta\mathbf{m} \cdot \nabla[\mathbf{Hm}] \cdot \mathbf{y} \geq 0$ as both $\delta\mathbf{m} \cdot \nabla[\mathbf{Hm}]$ and $\mathbf{y}$ are positive. These solutions are indeed minimum solutions to the constrained problem (Fig. 9.6).

To demonstrate how the Kuhn-Tucker theorem can be used, we consider the simplified problem

$$\text{minimize } \|\mathbf{d} - \mathbf{Gm}\|_2^2 \text{ subject to } \mathbf{m} \geq \mathbf{0} \tag{9.63}$$

which is called *nonnegative least squares*. We find the solution using an iterative scheme of several steps (Lawson and Hanson, 1974):

Step 1. Start with an initial guess for $\mathbf{m}$. As $\mathbf{H} = \mathbf{I}$, each model parameter is associated with exactly one constraint. These model parameters can be separated into a set $\mathbf{m}_E$ that satisfies the constraints in the equality sense and a set $\mathbf{m}_S$ that satisfies the constraints in the inequality sense. The particular initial guess $\mathbf{m} = \mathbf{0}$ is clearly feasible and has all its elements in $\mathbf{m}_E$.

Step 2. Any model parameter m_i in $\mathbf{m}_E$ that has associated with it a negative gradient $[\nabla E]_i$ can be changed both to decrease the error and to remain feasible. Therefore if there is no such model parameter in $\mathbf{m}_E$, the Kuhn-Tucker theorem indicates that this $\mathbf{m}$ is the solution to the problem.

Step 3. If some model parameter m_i in $\mathbf{m}_E$ has a corresponding negative gradient, then the solution can be changed to decrease the prediction error. To change the solution, we select the model parameter corresponding to the most

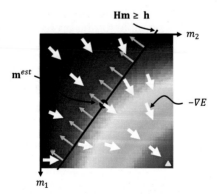

FIG. 9.6 The error $E(\mathbf{m})$ (colors) has a single minimum (yellow triangle). The linear equality constraint $\mathbf{Hm} \geq \mathbf{h}$ divides $S(\mathbf{m})$ into two half-spaces (black line, with gray arrows pointing into the feasible half-space). Solution (circle) lies on the boundary between the two half-spaces and therefore satisfies the constraint in the equality sense. At this point, the normal of the constraint hyperplane (gray arrow) is antiparallel to $-\nabla E$ (white arrows). Scripts gdama09_08 and gdapy09_08.

negative gradient and move it to the set $\mathbf{m}_S$. All the model parameters in $\mathbf{m}_S$ are now recomputed by solving the system $\mathbf{G}_S\mathbf{m}'_S = \mathbf{d}_S$ in the least-squares sense. The subscript S on the matrix indicates that only the columns multiplying the model parameters in $\mathbf{m}_S$ have been included in the calculation. All the $\mathbf{m}_E$s are still zero. If the new model parameters are all feasible, then we set $\mathbf{m} = \mathbf{m}'$ and return to Step 2.

Step 4. If some of the elements of $\mathbf{m}'_S$ are infeasible, we cannot use this vector as a new guess for the solution. Instead, we compute the change in the solution $\delta\mathbf{m} = \mathbf{m}'_S - \mathbf{m}_S$ and add as much of this vector as possible to the solution $\mathbf{m}_S$ without causing the solution to become infeasible. We therefore replace $\mathbf{m}_S$ with the new guess $\mathbf{m}_S + \alpha\delta\mathbf{m}$, where $\alpha = \min_i\{m_{Si}/(m_{Si} - m'_{Si})\}$ is the largest choice that can be made without some $\mathbf{m}_S$ becoming infeasible. At least one of the m_{Si}s has its constraint satisfied in the equality sense and must be moved back to $\mathbf{m}_E$. The process then returns to Step 3.

This algorithm contains two loops, one nested within the other. The outer loop successively moves model parameters from the group that is constrained to the group that minimizes the prediction error. The inner loop ensures that the addition of a variable to this latter group has not caused any of the constraints to be violated. Discussion of the convergence properties of this algorithm can be found in Lawson and Hanson (1974).

In MATLAB® and Python, the nonnegative least-squares solution is calculated as:

MATLAB®

```mest = lsqnonneg(G,dobs);```
```dpre = G*mest;```
 	
　　　　　　　　　　　　　　　　　　　　　　　　　　[gdama09_09]

Python

```vest, E = opt.nnls(G,dobs.ravel());```
```mest = gda_cvec(vest);```
 	
　　　　　　　　　　　　　　　　　　　　　　　　　[gdapy09_09]

Here, E is the prediction error and opt is an abbreviation for the `scipy.optimize` package. In a simple example, nonnegative least squares is applied to the problem of fitting a straight line to data (Fig. 9.7).

As a more realistic example, we analyze the problem of determining the mass distribution of an object from observations of its gravitational force (Fig. 9.8). Mass is an inherently positive quantity, so the positivity constraint constitutes indefinitely accurate prior information. The gravitational force d_i is measured at $N = 600$ points (x_i, y_i) above the object. The object (Fig. 9.8A) is subdivided into a grid of 20×20 pixels, each of mass m_j and position (x_j, y_j). According to Newton's inverse-square law, the data kernel is $G_{ij} = \gamma\cos(\theta_{ij})/R_{ij}^2$, where γ is the gravitational constant, θ_{ij} is the angle with respect to the vertical, and R_{ij} is the distance. Singular-value decomposition of $\mathbf{G}$ indicates that this mixed-determined problem is extremely nonunique, with only about 20 nonzero eigenvalues. Thus the solution that one obtains depends critically on the type of prior information that one employs. In this case, the natural solution (Fig. 9.8E), which contains 136 negative masses, is completely different from the nonnegative solution (Fig. 9.8F),

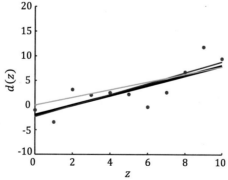

FIG. 9.7 Nonnegative least squares applied to a straight-line fit. The data (circles) scatter about the true line $d(z) = m_1 + m_2 z$ (black), with $\mathbf{m}^{true} = [-2, 1]^T$. The nonnegative least-squares solution $\mathbf{m}^{est} = [0.0, 0.8]^T$ (green), which has all nonnegative elements, can be compared to the ordinary least-squares solution $\mathbf{m}^{est} = [-2.3, 1.1]^T$ (red), which does not. Scripts gdama09_09 and gdapy09_09.

FIG. 9.8 (A) The vertical component of the force of gravity is measured on a horizontal line above a massive cube-shaped object. This cube contains a grid of 20×20 model parameters representing spatially varying density **m** (colors). The equation **Gm**=**d** embodies Newton's inverse-square law of gravity. (B) The $N=600$ gravitational force observations $\mathbf{d}^{obs}$ (black curve) and the gravitational force predicted by the natural solution (red curve, $p=15$) and nonnegative least squares (green curve). (C) True model. (D) Natural estimate of model, with $p=4$. (E) Natural estimate of model, with $p=15$. (F) Nonnegative estimate of model. Scripts gdama09_10 and gdapy09_10.

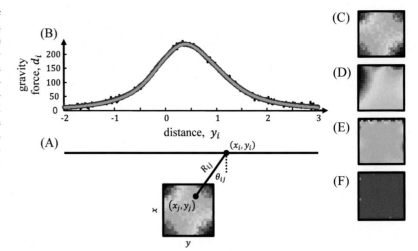

which contains none. Nevertheless, both satisfy the data almost exactly (Fig. 9.8A), with the prediction error E of the nonnegative solution about 1% larger than that of the natural solution. Ironically, neither the nonnegative solution nor the natural solution looks at all like the true solution (Fig. 9.8C), but a heavily damped solution (Fig. 9.8D) does.

The nonnegative least-squares algorithm can also be used to solve the problem (Lawson and Hanson, 1974)

$$\text{minimize } \|\mathbf{m}\|_2^2 \text{ subject to } \mathbf{Hm} \geq \mathbf{h} \tag{9.64}$$

and, by virtue of the transformation described in Section 9.8, the completely general problem. The method consists of forming the $(M+1) \times p$ equation

$$\mathbf{G'm'}=\mathbf{d'} \text{ with } \mathbf{G'} \equiv \begin{bmatrix} \mathbf{H}^T \\ \mathbf{h}^T \end{bmatrix} \text{ and } \mathbf{d'} \equiv \begin{bmatrix} \mathbf{0} \\ 1 \end{bmatrix} \tag{9.65}$$

and finding the $\mathbf{m'}$ that minimizes $\|\mathbf{G'm'} - \mathbf{d'}\|_2^2$ subject to $\mathbf{m'} \geq 0$ by the nonnegative least-squares algorithm described earlier. If the prediction error $\mathbf{e'} \equiv \mathbf{d'} - \mathbf{G'm'}$ is identically zero, then the constraints $\mathbf{Hm} \geq \mathbf{h}$ are inconsistent. Otherwise, the solution is $m_i = -e_i'/e_{M+1}'$ (which can also be written as $\mathbf{e'}=[-\mathbf{m},1]^T e_{M+1}'$). The solution is computed as

MATLAB®

``
Gp = [H, h]';
dp = [zeros(1,length(H(1,:))), 1]';
mp = lsqnonneg(Gp,dp);
ep = dp - Gp*mp;
m = -ep(1:end-1)/ep(end);
                                          [gdama09_12]
``

Python

``
Gp = np.concatenate((Hp,hp), axis=1).T
dp = gda_cvec(np.zeros((MHp,1)), 1.0);
t1, t2 = opt.nnls(Gp,dp.ravel());
mpp = gda_cvec(t1);
ep = dp - np.matmul(Gp,mpp);
Ne,1 = np.shape(ep);
mp = -ep[0:Ne-1,0:1]/ep[Ne-1,0];
                                          [gdapy09_12]
``

An example with $M=2$ model parameters and $N=3$ constraints is shown in Fig. 9.9. The code for converting a least squares with inequality constraints problem into a nonnegative least-squares problem using the procedure in Section 9.8 is

MATLAB®
| ```
[Up, Lp, Vp] = svd(G,0);
lambda = diag(Lp);
rlambda = 1./lambda;
Lpi = diag(rlambda);
% transformation 1
Hp = -H*Vp*Lpi;
hp = h + Hp*Up'*dobs;
% transformation 2
Gp = [Hp, hp]';
dp = [zeros(1,length(Hp(1,:))), 1]';
mpp = lsqnonneg(Gp,dp);
ep = dp - Gp*mpp;
mp = -ep(1:end-1)/ep(end);
% take mp back to m
mest = Vp*Lpi*(Up'*dobs-mp);
dpre = G*mest;
``` |

[gdama09_12]

Python
| ```
Up, lam, VpT = la.svd(G,full_matrices=False);
lam = gda_cvec(lam);
Vp = VpT.T;
rlambda = np.reciprocal(lam);
Lpi = np.diag(rlambda.ravel());
# transformation 1
Hp = -np.matmul(H,np.matmul(Vp,Lpi));
NHp, MHp = np.shape(Hp);
hp = h + np.matmul(Hp,np.matmul(Up.T,dobs));
# transformation 2
Gp = np.concatenate( (Hp,hp), axis=1).T
dp = gda_cvec( np.zeros((MHp,1)), 1.0 );
t1, t2 = opt.nnls(Gp,dp.ravel());
mpp = gda_cvec(t1);
ep = dp - np.matmul(Gp,mpp);
Ne,i = np.shape(ep);
mp = -ep[0:Ne-1,0:1]/ep[Ne-1,0];
# take mp back to m
mest = np.matmul(Vp,np.matmul(Lpi,(np.matmul(Up.T,dobs)-mp)));
dpre = np.matmul(G,mest);
``` |

[gdapy09_12]

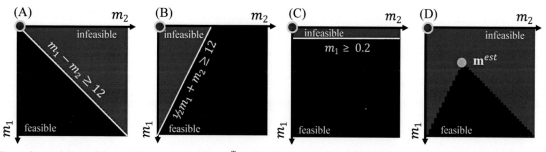

FIG. 9.9 Exemplary solution of the problem of minimizing $\mathbf{m}^T\mathbf{m}$ with $N=3$ inequality constraints $\mathbf{Hm} \geq \mathbf{h}$. (A–C) Each constraint divides the (m_1, m_2) plane into two half-places, one feasible and the other infeasible. (D) The intersection of the three feasible half-planes is polygonal in shape. The solution $\mathbf{m}^{est}$ (green circle) is the point in feasible area that is closest to the origin (red circle). Note that two of the three constrains are satisfied in the equality sense. Scripts gdama09_11 and gdapy09_11.

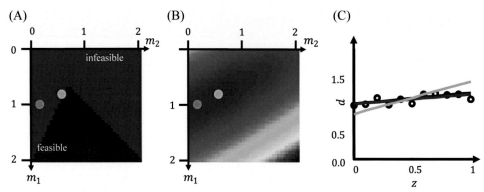

FIG. 9.10 Problem of minimizing the prediction error E subject to inequality constraints, applied to the straight-line problem. (A) Feasible region of the model parameters (the intercept m_1 and slope m_2 of the straight line). The unconstrained solution (orange circle) is outside the feasible region, but the constrained solution (green circle) is on its boundary. (B) The unconstrained solution (orange circle) is at the global minimum of the prediction error E, while the constrained solution is not. (C) Plot of the data $d(z)$ showing true data (black line), observed data (black circles), unconstrained prediction (red line), and constrained prediction (green line). Scripts gdama09_12 and gdapy09_12.

Here, the *economy* version of svd() is invoked by specifying a second argument of zero (in MATLAB®) or full_matrices=False (in Python). The economy version is more efficient, because it omits columns of $\mathbf{U}$ and $\mathbf{V}$ that do not appear in the singular-value decomposition. An exemplary problem is illustrated in Fig. 9.10.

We now demonstrate that this method does indeed solve the indicated problem (adapted from Lawson and Hanson, 1974). Step 1 is to show that e'_{M+1} is nonnegative. We first note that the gradient of the error satisfies $\tfrac{1}{2}\nabla E' = -\mathbf{G}'^T[\mathbf{d}' - \mathbf{G}'\mathbf{m}'] = -\mathbf{G}'^T\mathbf{e}'$, and that because of the Kuhn-Tucker theorem, $\mathbf{m}'$ and $\nabla E'$ satisfy

$$
\begin{aligned}
\mathbf{m}'_E = 0 \quad &\text{and} \quad [\nabla E']_E < 0 \\
\mathbf{m}'_s > 0 \quad &\text{and} \quad [\nabla E']_E = 0
\end{aligned}
\tag{9.66}
$$

These conditions imply $\mathbf{m}'^T\nabla E' = 0$. The error E' is therefore

$$
\begin{aligned}
E' = \mathbf{e}'^T\mathbf{e}' = \left[\mathbf{d}' - \mathbf{G}'\mathbf{m}'\right]^T\mathbf{e}' &= \mathbf{d}'^T\mathbf{e}' - \mathbf{m}'^T\mathbf{G}'^T\mathbf{e}' \\
= e'_{M+1} - \mathbf{m}'^T\mathbf{G}'^T\mathbf{e}' &= e'_{M+1} + \mathbf{m}'^T\tfrac{1}{2}\nabla E' = e'_{M+1}
\end{aligned}
\tag{9.67}
$$

as $\mathbf{d}'^T\mathbf{e}' = [0,1]^T\mathbf{e}' = e'_{M+1}$. The error E' is necessarily a nonnegative quantity, so if it is not identically zero, then e_{M+1}' must be greater than zero.

Step 2 is to show that the solution satisfies the inequality constraints $\mathbf{Hm} - \mathbf{h} \geq 0$. We start with the gradient of the error $\nabla E'$, which must have all nonnegative elements (or else the solution $\mathbf{m}'$ could be further minimized without violating the constraints $\mathbf{m}' \geq 0$):

$$
0 \leq \tfrac{1}{2}\nabla E' - \mathbf{G}'^T\mathbf{e}' = -[\mathbf{H} \quad \mathbf{h}]\begin{bmatrix} -\mathbf{m} \\ 1 \end{bmatrix} e'_{M+1} = (\mathbf{Hm} - \mathbf{h})e'_{M+1}
\tag{9.68}
$$

As e'_{M+1} is nonnegative, we have $\mathbf{Hm} - \mathbf{h} \geq 0$.

Step 3 is to show that the solution minimizes $\|\mathbf{m}\|_2^2$. This follows from the Kuhn-Tucker condition that, at a valid solution, the gradient of the error ∇E be represented as a nonnegative combination of the rows of $\mathbf{H}$:

$$
\nabla E = \nabla\|\mathbf{m}\|_2^2 = 2\mathbf{m} = -2[e'_1, \cdots, e'_M]^T/e'_{M+1} = 2\mathbf{H}^T[m'_1, \cdots, m'_M]^T/e'_{M+1}
\tag{9.69}
$$

Here we have used the fact that $\mathbf{e}' = \mathbf{d}' - \mathbf{G}'\mathbf{m}'$, together with the fact that the first M elements of $\mathbf{d}'$ are zero. Note that $\mathbf{m}'$ and e'_{M+1} are nonnegative, so the Kuhn-Tucker condition is satisfied.

Finally, we can also show that a feasible solution exists only when the error is not identically zero. Consider the contradiction that the error is identically zero but that a feasible solution exists. Then

$$
0 = \mathbf{e}'^T\mathbf{e}'/e'_{M+1} = [-\mathbf{m}, 1]^T[\mathbf{d}' - \mathbf{G}'\mathbf{m}'] = 1 + [\mathbf{Hm} - \mathbf{h}]^T\mathbf{m}'
\tag{9.70}
$$

As $\mathbf{m}' \geq 0$, the relationship $\mathbf{Hm} < \mathbf{h}$ is implied. This contradicts the constraint equations $\mathbf{Hm} \geq \mathbf{h}$, so that an identically zero error implies that no feasible solution exists and that the constraints are inconsistent.

9.10 Problems

9.1 This problem builds upon the discussion in Section 8.2. Write a MATLAB® or Python script that uses singular-value decomposition to compute the null vectors associated with the data kernel $\mathbf{G} = [1, 1, 1, 1]$. (A) The null vectors are different from that given in Eq. (8.7). Why? Show that the two sets are equivalent.

9.2 Modify the weighted damped least-squares "gap-filling" problem of Fig. 4.11 (scripts gdama04_12 and gdapy04_12) so that the prior information is applied only to the part of the solution in the null space. First, transform the weighted problem $\mathbf{Gm} = \mathbf{d}$ into an unweighted one $\mathbf{G'm'} = \mathbf{d'}$ using the transformation given by Eq. (9.22). Then, find the minimum-length solution (or the natural solution) $\mathbf{m'}^{est}$. Finally, transform back to obtain $\mathbf{m}^{est}$. How different is this solution from the one given in the figure? Compare the two prediction errors. In order to insure that $\mathbf{W}_e = \mathbf{D}^T\mathbf{D}$ has an inverse, which is required by the transformation, you should make $\mathbf{D}$ square by adding two rows, one at the top and the other at the bottom, that constrain the first and last model parameters to known values.

9.3 (A) Compute and plot the null vectors for the data kernel shown in Fig. 9.4A. (B) Suppose that the elements of $\mathbf{m}^{true}$ are drawn from a uniform pdf between 0 and 1. How large a contribution do the null vectors make to the true solution $\mathbf{m}^{true}$? (Hint: Write the model parameters as a linear combination of all the eigenvectors $\mathbf{V}$ by writing $\mathbf{m}^{true} = \mathbf{V}\boldsymbol{\alpha}$, where $\boldsymbol{\alpha}$ are coefficients, and then solving for $\boldsymbol{\alpha}$. Then examine the size of the elements of $\boldsymbol{\alpha}$.) You may wish to refer to scripts gdama09_05 and gdapy09_05 for the definition of $\mathbf{G}$.

9.4 Consider fitting a cubic polynomial $d_i = m_1 + m_2 z_i + m_3 z_i^2 + m_4 z_i^3$ to $N = 20$ data, d_i^{obs}, where the zs are evenly spaced on the interval, $(0, 1)$, where $m_2 = m_3 = m_4 = 1$, and where m_1 is varied from -1 to 1, as described later. (A) Write a script that generates synthetic observed data (including Normally distributed noise) for a particular choice of m_1, estimates the model parameters using both simple least squares and nonnegative least squares, plots the observed and predicted data, and outputs the total error. (B) Run a series of test cases for a range of values of m_1, and comment upon the results.

9.5 (A) Modify the script gdama09_11 (or gdapy09_11) so that the last constraint is $m_1 \geq 1.2$. How do the feasible region and the solution change? (B) Modify the script to add a constraint $m_2 \leq 0.2$. How do the feasible region and the solution change?

References

Lanczos, C., 1961. Linear Differential Operators. Van Nostrand-Reinhold, Princeton, NJ.

Lawson, C.L., Hanson, D.J., 1974. Solving Least Squares Problems. Prentice-Hall, Englewood Cliffs, NJ.

Menke, W., Menke, J., 2011. Environmental Data Analysis with MATLAB®. Academic Press, Elsevier Inc., Oxford, UK. 263 pp.

Penrose, R.A., 1955. A generalized inverse for matrices. Proc. Cambridge Philos. Soc. 51, 406–413.

Wiggins, R.A., 1972. The general linear inverse problem: implication of surface waves and free oscillations for Earth structure. Rev. Geophys. Space Phys. 10, 251–285.

Wunsch, C., Minster, J.F., 1982. Methods for box models and ocean circulation tracers: mathematical programming and non-linear inverse theory. J. Geophys. Res. 87, 5647–5662.

10

Linear inverse problems with non-Normal statistics

10.1 L_1 norms and exponential probability density functions

In Chapter 6, we showed that the method of least squares and the more general use of L_2 norms could be justified through the assumption that the data and prior model parameters followed Normal statistics. This assumption is not always appropriate; however, some data sets follow other probability density functions. The two-sided exponential pdf is one simple alternative. Here "two-sided" means that the function is defined on the interval $(-\infty, +\infty)$ and has tails in both directions. When Normal and exponential pdfs of the same mean $\langle d \rangle$ and variance σ_d^2 are compared, the exponential pdf is found to be much longer tailed (Fig. 10.1, Table 10.1):

$$\text{Normal} \qquad\qquad\qquad \text{exponential}$$
$$p(d) = \frac{1}{\sqrt{2\pi}\sigma_d} \exp\left\{ -\frac{(d - \langle d \rangle)^2}{2\sigma_d^2} \right\} \quad p(d) = \frac{1}{\sqrt{2}\sigma_d} \exp\left\{ -\sqrt{2}\frac{|d - \langle d \rangle|}{\sigma_d} \right\} \tag{10.1}$$

Note that the probability of realizing data far from $\langle d \rangle$ is much higher for the exponential pdf than for the Normal pdf. A few data, say, 4 standard deviations from the mean, are reasonably probable in a data set of, say, 1000 samples drawn from an exponential pdf, but very improbable for data drawn from a Normal pdf. We therefore expect that methods based on the exponential probability density function will be able to handle a data set with a few "bad" data (outliers) better than Normal methods. Methods that can tolerate a few outliers are said to be *robust* (Claerbout and Muir, 1973).

The exponential probability density function with mean `dbar` and variance `sd2` is calculated as

MATLAB®
```pE =(1/sqrt(2))*(1/sd)*exp(-sqrt(2)*abs((d-dbar))/sd);```
[gdama10_01]

Python
```pE = (1/sqrt(2)) * (1/sd)* np.exp(-sqrt(2)*np.abs(d-dbar)/sd);```
[gdapy10_01]

The two-sided exponential pdf is closely related to the one-sided exponential pdf $p(v) = \mu \exp(-v/\mu)$, where μ is the *scale parameter*. A realization d_r of the two-sided pdf can be created by sampling the one-sized pdf with the scale parameter set to $\mu = \sigma_d/\sqrt{2}$, to obtain a value v_r and using $d_r = \langle d \rangle + s_r v_r$. Here, s_r is a random sign (which can be generated by taking the sign of a random number sampled from any pdf that is symmetric about zero).

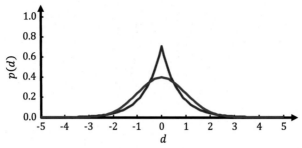

FIG. 10.1 Comparison of exponential probability density distribution (red) with a Normal probability density distribution (blue) of equal variance $\sigma_d^2 = 1$. The exponential probability density distribution is longer tailed. Scripts `gdama10_01` and `gdapy10_01`.

TABLE 10.1   The area beneath $\langle d \rangle \pm n\sigma_d$ for the normal and exponential pdf.

n	% Area of Normal pdf	% Area of exponential pdf
1	68.2	76
2	95.4	94
3	99.7	98.6
4	99.999+	99.7

MATLAB®
| ```
mu = sd/sqrt(2);                         % scale factor
rs = (2*(random('unid',2,Nr,1)-1)-1);    % random sign
re = random('exponential',mu,Nr,1);      % random 1-sided exp
dr = dbar + rs.*re;
                                              [gdama10_01]
``` |

Python
| ```
mu = sd/sqrt(2); # scale factor
# dr is random sign
dr = 2*np.random.randint(low=0,high=2,size=(Nr,1))-1.0;
drb = np.random.exponential(scale=mu,size=(Nr,1)); # exp pdf
dr = dbar + np.multiply(dra,drb);
                                              [gdapy10_01]
``` |

10.2   Maximum likelihood estimate of the mean of an exponential pdf

Exponential probability density functions bear the same relationship to L_1 norms as Normal pdfs bear to L_2 norms. To illustrate this relationship, we consider the joint distribution for N independent data, each with the same mean m_1 and variance σ_d^2. As the data are independent, the joint probability density function, or *likelihood*, is just the product of N univariate functions:

$$p(\mathbf{d}) = (2)^{-N/2}\sigma_d^{-N} \exp\left\{ -\frac{\sqrt{2}}{\sigma_d} \sum_{i=1}^{N} |d_i - m_1| \right\} \qquad (10.2)$$

To maximize the likelihood $p(\mathbf{d})$, we must maximize the argument of the exponential, which involves minimizing the sum of absolute residuals as

$$\text{minimize } E \equiv \sum_{i=1}^{N} |d_i - m_1| \qquad (10.3)$$

This is the L_1 norm of the prediction error in a linear inverse problem of the form $\mathbf{Gm} = \mathbf{d}$, where $M=1$, $\mathbf{G}=[1,1,...,1]^T$, and $\mathbf{m}=[m_1]$. Applying the principle of maximum likelihood, we obtain

$$\text{maximize}\ \ L = \log p(\mathbf{d}) = -\frac{N}{2}\log 2 - N\log\sigma_d - \frac{\sqrt{2}}{\sigma_d}\sum_{i=1}^{N}|d_i - m_1| \tag{10.4}$$

setting the derivatives to zero yields

$$\frac{\partial L}{\partial m_1} = 0 = -\frac{\sqrt{2}}{\sigma_d}\sum_{i=1}^{N}\text{sign}(d_i - m_1)$$

$$\frac{\partial L}{\partial \sigma_d} = 0 = -\frac{N}{\sigma_d} + \frac{\sqrt{2}}{\sigma_d^2}\sum_{i=1}^{N}|d_i - m_1| \tag{10.5}$$

The sign function $\text{sign}(x)$ equals $+1$ if $x>0$, -1 if $x<0$, and 0 if $x=0$. The derivative $d|x|/dx = \text{sign}(x)$, because if x is positive, then it is just equal to $dx/dx=1$ and if it is negative, then to -1. The first equation yields the implicit expression for $m_1 = \langle d \rangle^{est}$ for which $\sum_i \text{sign}(d_i - m_1)=0$. The second equation can then be solved for an estimate of the variance as

$$\sigma_d^{est} = \frac{\sqrt{2}}{N}\sum_{i=1}^{N}|d_i - m_1| \tag{10.6}$$

The formula for m_1^{est} is exactly the sample median; one finds an m_1 such that half the ds are less than m_1 and half are greater than m_1. There are then an equal number of negative and positive signs and the sum of the signs is zero. The median is a robust property of a set of data. Adding one outlier can at worst move the median from one central datum to another nearby central datum. While the maximum likelihood estimate of a Normal pdf's true mean is the sample arithmetic mean, the maximum likelihood estimate of an exponential pdf's true mean is the sample median.

The estimate of the variance also differs between the two distributions: in the Normal case, it is the square of the sample standard deviation, but in the exponential case, it is not. If there are an odd number of samples, then m_1^{est} equals the middle d_i^{obs}. If there is an even number of samples, any m_1^{est} between the two middle data maximizes the likelihood. In the odd case, the error E attains a minimum only at the middle sample, but in the even case, it is flat between the two middle samples (Fig. 10.2). We see, therefore, that L_1 problems of minimizing the prediction error of $\mathbf{Gm} = \mathbf{d}^{obs}$ can

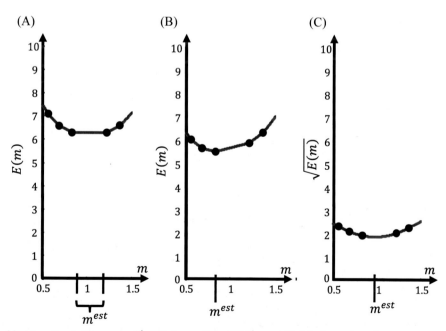

FIG. 10.2 Inverse problem to estimate the mean m^{est} of N observations. (A) L_1 error, $E(m)$ (red curve), with even N. The error has a flat minimum, bounded by two observations (circles), and the solution is nonunique. (B) L_1 error with odd N. The error is minimum at one of the observation points, and the solution is unique. (C) The L_2 error, both odd and even N. The error is minimum at a single point, which may not correspond to an observation point, and the solution is unique. Scripts gdama10_02 and gdapy10_02.

possess nonunique solutions that are distinct from the type of nonuniqueness encountered in the L_2 problems. The L_1 problems can still possess nonuniqueness owing to the existence of null solutions since the null solutions cannot change the prediction error under any norm. That kind of nonuniqueness leads to a completely unbounded range of estimates. The new type of nonuniqueness, on the other hand, permits the solution to take on any values between *finite bounds*.

We also note that regardless of whether N is even or odd, we can choose m_1^{est} so that one of the equations $\mathbf{Gm} = \mathbf{d}^{obs}$ is satisfied exactly (in this case, $m_1^{est} = d_k^{obs}$, where k is the index of the "middle" datum). This can be shown to be a general property of L_1 norm problems. Given N data and M unknowns related by $\mathbf{Gm} = \mathbf{d}^{obs}$, it is possible to choose $\mathbf{m}$ so that the L_1 prediction error is minimized and so that M of the equations are satisfied exactly.

The L_1 estimate of the mean $\langle d \rangle$ and square root of the variance σ_d are computed as

MATLAB®
| ```
dbarest = median(dr);
sdest = (sqrt(2)/Nr)*sum(abs(dr-dbarest));
```                                                   [gdama10_01] |

Python
| ```
dbarest = np.median(dr);
sdest = (sqrt(2)/Nr)*np.sum(np.abs(dr-dbarest));
```                                                  [gdapy10_01] |

10.3   The general linear problem

Consider the linear inverse problem $\mathbf{Gm} = \mathbf{d}$, in which the data $\mathbf{d}^{obs}$ are uncorrelated and with nonuniform but known variance $\sigma_{d_i}^2$, and where the model parameters have uncorrelated prior values $\langle m \rangle$, respectively, with nonuniform but known variance $\sigma_{m_i}^2$. Following ideas put forward in Chapter 6, the posterior pdf is then

$$p\left(\mathbf{m}|\mathbf{d}^{obs}\right) = (2)^{-(N+M)/2} \prod_{i=1}^{N} \sigma_{d_i}^{-1} \exp\left[-\sqrt{2}\frac{|e_i|}{\sigma_{d_i}}\right] \prod_{j=1}^{M} \sigma_{m_j}^{-1} \exp\left[-\sqrt{2}\frac{|l_j|}{\sigma_{m_j}}\right] \tag{10.7}$$

with $\mathbf{e} = \mathbf{d}^{obs} - \mathbf{Gm}$ and $\mathbf{l} = \langle \mathbf{m} \rangle - \mathbf{m}$

The maximum likelihood estimate of the model parameters occurs when the exponential is a minimum, that is, when the sum of the weighted L_1 prediction error and the weighted L_1 solution length is minimized:

$$\text{minimize } E + L \equiv \sum_{i=1}^{N} \frac{|e_i|}{\sigma_{d_i}} + \sum_{j=1}^{M} \frac{|l_j|}{\sigma_{m_j}} \tag{10.8}$$

In this case, the weighting factors are the reciprocals of the square root of the variance, in contrast to the Normal case, in which they are the reciprocals of the variances. Note that linear combinations of exponentially distributed random variables are not themselves exponential (unlike Normally distributed variables, which give rise to Normally distributed combinations). The covariance matrix of the estimated model parameters is, therefore, difficult both to calculate and to interpret since the manner in which it is related to confidence intervals varies from case to case. We shall focus our discussion on estimating only the model parameters themselves. However, we note that bootstrap analysis can be an effective means of estimating the confidence limits of the solution (see Section 11.12).

10.4   Solving L_1 norm problems by transformation to a linear programming problem

We shall show that the general linear problem can be transformed into a *linear programming* problem (see Section 8.6)

$$\text{find } \mathbf{x} \text{ that minimizes } z = \mathbf{f}^T\mathbf{x}$$
$$\text{with constraints } \mathbf{Ax} \leq \mathbf{b} \text{ and } \mathbf{Cx} = \mathbf{d} \text{ and } \mathbf{x}^{(l)} \leq \mathbf{x} \leq \mathbf{x}^{(u)} \tag{10.9}$$

We first define the L_1 versions of the weighted prediction error E and the weighted solution length L

$$E \equiv \sum_{i=1}^{N} \frac{|e_i|}{\sigma_{d_i}} \quad \text{and} \quad L \equiv \sum_{j=1}^{M} \frac{|\langle m_i \rangle - m_i|}{\sigma_{m_j}} \quad \text{with} \quad \mathbf{e} = \mathbf{d}^{obs} - \mathbf{Gm} \tag{10.10}$$

Note that these formulas are weighted by the square root of the variances of the measurement error and prior model parameters (σ_{d_i} and σ_{m_j}, respectively).

First, we shall consider the completely underdetermined linear problem in which the prior model parameters have mean $\langle m_i \rangle$ and square root of variance σ_{m_j}. The problem is to minimize the weighted solution length L subject to the constraint $\mathbf{Gm} = \mathbf{d}^{obs}$. We first introduce $5M$ new variables m_i', m_i'', α_i, x_i, and x_i', where $1 \leq i \leq M$. The linear programming problem may be stated as follows (Cuer and Bayer, 1980)

$$\text{minimize } z = \sum_{i=1}^{M} \frac{\alpha_i}{\sigma_{m_i}} \text{ subject to the constraints}$$

$$\mathbf{G}(\mathbf{m}' - \mathbf{m}'') = \mathbf{d}^{obs} \text{ and } \mathbf{m}' - \mathbf{m}'' + \mathbf{x} - \boldsymbol{\alpha} = \langle \mathbf{m} \rangle \text{ and } \mathbf{m}' - \mathbf{m}'' - \mathbf{x}' + \boldsymbol{\alpha} = \langle \mathbf{m} \rangle$$
$$\text{and } \mathbf{m}' \geq 0 \text{ and } \mathbf{m}'' \geq 0 \text{ and } \boldsymbol{\alpha} \geq 0 \text{ and } \mathbf{x} \geq 0 \text{ and } \mathbf{x}' \geq 0 \tag{10.11}$$

This linear programming problem has $5M$ unknowns, $N+2M$ equality constraints, and $5M$ inequality constraints. If one makes the identification $\mathbf{m} \equiv \mathbf{m}' - \mathbf{m}''$, the first equality constraint is equivalent to $\mathbf{Gm} = \mathbf{d}^{obs}$. The signs of the elements of $\mathbf{m}$ are not constrained even though those of $\mathbf{m}'$ and $\mathbf{m}''$ are. Defining the error in prior information as $\mathbf{l} \equiv \langle \mathbf{m} \rangle - \mathbf{m}$, the remaining equality constraints can be rewritten as

$$\boldsymbol{\alpha} - \mathbf{x} = -\mathbf{l} \text{ and } \boldsymbol{\alpha} - \mathbf{x}' = \mathbf{l} \tag{10.12}$$

where the α_i, x_i, and x_i' are nonnegative. Now if l_i is positive, the second equation requires $\alpha_i \geq l_i$, as the x_i' cannot be negative. The first equation can always be satisfied by choosing some appropriately large x_i. On the other hand, if l_i is negative, then the second constraint can always be satisfied by choosing some appropriately large x_i', but the first equation requires that $\alpha_i \geq -l_i$, as x_i cannot be negative. Taken together, these two constraints imply that $\alpha_i > |l_i|$. Minimizing $\sum_i^M \alpha_i / \sigma_{m_i}$ is therefore equivalent to minimizing the weighted solution length $L = \sum_i^M l_i / \sigma_{m_i}$. The L minimization problem has been converted to a linear programming problem.

The completely overdetermined problem of minimizing E with no prior information can be converted into a linear programming problem in a similar manner. We introduce $2M+3N$ new variables, m_i', m_i'', α_j, x_j, and x_j', where $1 \leq i \leq M$ and $1 \leq j \leq N$, that are related through $2N$ equality constraints and $2M+3N$ inequality constraints

$$\text{minimize } z = \sum_{i=1}^{M} \frac{\alpha_i}{\sigma_{d_i}} \text{ subject to the constraints}$$

$$\mathbf{G}(\mathbf{m}' - \mathbf{m}'') + \mathbf{x} - \boldsymbol{\alpha} = \mathbf{d}^{obs} \text{ and } \mathbf{G}(\mathbf{m}' - \mathbf{m}'') - \mathbf{x}' + \boldsymbol{\alpha} = \mathbf{d}^{obs}$$
$$\text{and } \mathbf{m}' \geq 0 \text{ and } \mathbf{m}'' \geq 0 \text{ and } \boldsymbol{\alpha} \geq 0 \text{ and } \mathbf{x} \geq 0 \text{ and } \mathbf{x}' \geq 0 \tag{10.13}$$

Defining the solution as $\mathbf{m} \equiv \mathbf{m}' - \mathbf{m}''$ and the prediction error as $\mathbf{e} \equiv \mathbf{d}^{obs} - \mathbf{Gm}$, the equality constraints can be rearranged to

$$\boldsymbol{\alpha} - \mathbf{x} = -\mathbf{e} \text{ and } \boldsymbol{\alpha} - \mathbf{x}' = \mathbf{e} \tag{10.14}$$

Following a similar argument as before, if e_i is positive, then the second equation implies that $\alpha_i \geq e_i$, as x_i' cannot be negative. The first equation can always be satisfied by choosing an appropriately large x_i. On the other hand, if e_i is negative, then the second equation can always be satisfied by choosing an appropriately large x_i', but the first equation implies $\alpha_i \geq -e_i$, as x_i cannot be negative. Taken together, these two equations imply $\alpha_i \geq |e_i|$. Consequently, minimizing $\sum_i^N \alpha_i / \sigma_{d_i}$ is the same as minimizing the weighted L_1 prediction error $E = \sum_i^N e_i / \sigma_{d_i}$.

The solution is computed as

MATLAB®

```
L = 2*M+3*N;     % number of unknowns
x = zeros(L,1); % unknowns in vector x
% f is length L
% minimize sum aplha(i)/sd(i)
f = zeros(L,1);
f(2*M+1:2*M+N)=1./sd;
% equality constraints
Aeq = zeros(2*N,L);
beq = zeros(2*N,1);
% Part 1: Equality constraints
% first equation G(mp-mpp)+x-alpha=d
Aeq(1:N,1:M)                = G;
Aeq(1:N,M+1:2*M)            = -G;
Aeq(1:N,2*M+1:2*M+N)        = -eye(N,N);
Aeq(1:N,2*M+N+1:2*M+2*N)    =  eye(N,N);
beq(1:N)                    =  dobs;
% second equation G(mp-mpp)-xp+alpha=d
Aeq(N+1:2*N,1:M)               =  G;
Aeq(N+1:2*N,M+1:2*M)           = -G;
Aeq(N+1:2*N,2*M+1:2*M+N)       =  eye(N,N);
Aeq(N+1:2*N,2*M+2*N+1:2*M+3*N) = -eye(N,N);
beq(N+1:2*N)                   =  dobs;
% Part 2: inequality constraints A x <= b
% everything positive
A = zeros(L+2*M,L);
b = zeros(L+2*M,1);
A(1:L,:) = -eye(L,L);
b(1:L) = zeros(L,1);
% mp and mpp have an upper bound (to prevent runaway soln.
% bound might need to be adjusted
A(L+1:L+2*M,:) = eye(2*M,L);
mupperbound=10*max(abs(mls)); % might need to be adjusted
b(L+1:L+2*M) = mupperbound;
% solve linear programming problem
[x, fmin] = linprog(f,A,b,Aeq,beq);
fmin=-fmin;
mest = x(1:M) - x(M+1:2*M);
dpre = G*mest
```

[gdama10_03]

Python

```python
f = np.zeros((L,1));
f[2*M:2*M+N,0:1]=np.reciprocal(sd);
# equality constraints
Aeq = np.zeros((2*N,L));
beq = np.zeros((2*N,1));
# first equation G(mp-mpp)+x-alpha=d
Aeq[0:N,0:M]                 =  G;
Aeq[0:N,M:2*M]               = -G;
Aeq[0:N,2*M:2*M+N]           = -np.identity(N);
Aeq[0:N,2*M+N:2*M+2*N]       =  np.identity(N);
beq[0:N,0:1]                 =  dobs;
# second equation G(mp-mpp)-xp+alpha=d
Aeq[N:2*N,0:M]               =  G;
Aeq[N:2*N,M:2*M]             = -G;
Aeq[N:2*N,2*M:2*M+N]         =  np.identity(N);
Aeq[N:2*N,2*M+2*N:2*M+3*N]   = -np.identity(N);
beq[N:2*N,0:1]               =  dobs;
# inequality constraints A x <= b
# part 1: everything positive
A = np.zeros((L+2*M,L));
b = np.zeros((L+2*M,1));
A[0:L,0:L] = -np.identity(L);
b[0:L,0:1] = np.zeros((L,1));
# part 2; mp and mpp have an upper bound (to prevent runaway
# soln. This bound might need to be adjusted
A[L:L+2*M,0:L] = np.eye(N=2*M,M=L);
mupperbound=10.0*max(abs(mls)); # adjust for problem at hand
b[L:L+2*M,0:1] = mupperbound;
# solve linear programming problem
res = opt.linprog(c=f,A_ub=A,b_ub=b,A_eq=Aeq,b_eq=beq);
x = gda_cvec(res.x);
fmin = -res.fun;
status = res.success;
if( status ):
    print("linear programming succeeded" );
else:
    print("linear programming failed" );
mest = x[0:M,0:1] - x[M:2*M,0:1];
dpre = np.matmul(G,mest);
```

[gdapy10_03]

These implementations add upper bounds for $\mathbf{m}'$ and $\mathbf{m}''$, ten times the largest element of the least-squares solution (an amount that might need to be adjusted for the problem at hand). Without this constraint, the algorithm could possibly return very large values for these variables, leading to a solution $\mathbf{m} = \mathbf{m}' - \mathbf{m}''$ that is inaccurate because of round-off error. An example of the L_1 problem applied to curve fitting is shown in Fig. 10.3.

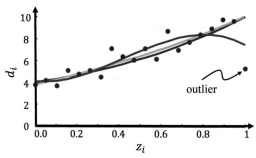

FIG. 10.3   Curve fitting using the L_1 norm. The true data (red curve) follow a cubic polynomial in an auxiliary variable z. Observations d_i^{obs} (red circles) have additive noise with zero mean and variance $\sigma_d^2 = 1$ drawn from an exponential probability density function. Note the outlier at $z = 1$. The L_1 fit (green curve) is not as affected by the outlier as a standard L_2 (least-squares) fit (blue curve). Scripts gdama10_03 and gdapy10_03.

The mixed-determined problem can be solved by any of several methods. By analogy to the L_2 methods described in Chapters 4 and 8, we could either pick some prior model parameters and minimize $E + L$ or separate the overdetermined model parameters from the underdetermined ones and apply prior information to the underdetermined ones only. The first method leads to a linear programming problem similar to the two cases stated earlier but with even more variables ($5M + 3N$), equality constraints ($2M + 2N$), and inequality constraints ($5M + 2N$):

$$\text{minimize } z = \sum_{i=1}^{M} \frac{\alpha_i}{\sigma_{m_i}} + \sum_{i=1}^{N} \frac{\alpha_i'}{\sigma_{m_i}} \text{ subject to the constraints}$$

$$\mathbf{m}' - \mathbf{m}'' + \mathbf{x} - \boldsymbol{\alpha} = \langle \mathbf{m} \rangle \text{ and } \mathbf{m}' - \mathbf{m}'' - \mathbf{x}' + \boldsymbol{\alpha} = \langle \mathbf{m} \rangle$$

$$\mathbf{G}(\mathbf{m}' - \mathbf{m}'') + \mathbf{x}'' - \boldsymbol{\alpha}' = \mathbf{d}^{obs} \text{ and } \mathbf{G}(\mathbf{m}' - \mathbf{m}'') - \mathbf{x}''' + \boldsymbol{\alpha}' = \mathbf{d}^{obs} \quad (10.15)$$

$$\text{and } \mathbf{m}' \geq 0 \text{ and } \mathbf{m}'' \geq 0 \text{ and } \boldsymbol{\alpha} \geq 0 \text{ and } \boldsymbol{\alpha}' \geq 0$$

$$\text{and } \mathbf{x} \geq 0 \text{ and } \mathbf{x}' \geq 0 \text{ and } \mathbf{x}'' \geq 0 \text{ and } \mathbf{x}''' \geq 0$$

The second method is more interesting. First, we use the singular-value decomposition to identify the null space of $\mathbf{G}$. The solution then has the form

$$\mathbf{m}^{est} = \sum_{i=1}^{p} a_i \mathbf{v}_p^{(i)} + \sum_{i=p+1}^{M} b_i \mathbf{v}_0^{(i)} = \mathbf{V}_p \mathbf{a} + \mathbf{V}_0 \mathbf{b} \quad (10.16)$$

where the $\mathbf{v}$s are eigenvectors and $\mathbf{a}$ and $\mathbf{b}$ are vectors of unknown coefficients. Only the vector $\mathbf{a}$ can affect the prediction error, so one uses the overdetermined algorithm to determine it

$$\text{find the } \mathbf{a} \text{ that minimizes } E = \left\| \mathbf{d}^{obs} - \mathbf{Gm} \right\|_1 = \sum_{i=1}^{N} \frac{\left| \mathbf{d}^{obs} - \mathbf{U}_p \boldsymbol{\Lambda}_p \mathbf{a} \right|_i}{\sigma_{d_i}} \quad (10.17)$$

where the singular-value decomposition is $\mathbf{G} = \mathbf{U}_p \boldsymbol{\Lambda}_p \mathbf{V}_p^T$ and we have used the fact that $\mathbf{Gm} = \mathbf{U}_p \boldsymbol{\Lambda}_p \mathbf{V}_p^T (\mathbf{V}_p \mathbf{a} + \mathbf{V}_0 \mathbf{b}) = \mathbf{U}_p \boldsymbol{\Lambda}_p \mathbf{a}$. Next, one uses the underdetermined algorithm to determine $\mathbf{b}$:

$$\text{find the } \mathbf{b} \text{ that minimizes } L = \left\| \langle \mathbf{m} \rangle - \mathbf{m} \right\|_1 = \sum_{i=1}^{M} \frac{\left| [\mathbf{V}_0 \mathbf{b}]_i - (\langle m_i \rangle - [\boldsymbol{\Lambda}_p \mathbf{a}]) \right|_i}{\sigma_{m_i}} \quad (10.18)$$

It is also possible to implement the basic underdetermined and overdetermined L_1 algorithms in such a manner that the many extra variables are never explicitly calculated (Cuer and Bayer, 1980). This procedure vastly decreases the storage and computation time required, making these algorithms practical for solving moderately large (say, $M = 1000$) inverse problems.

10.5   Solving L_1 norm problems by reweighted L_2 minimization

In the previous section, we solved the L_1 minimization problem by transforming it to an equivalent linear programming problem. The transformation is exact, but the formulation is cumbersome because many new variables are introduced (which, among other things, increases computer memory requirements). In this section, we show that the problem can also be transformed into an equivalent L_2 problem and solved with the standard least-squares methods. The transformation is only approximate and the method requires iteration to achieve an accurate solution, but is usually much faster than the linear programming transformation introduced in the previous section. (We will return to the issue of choosing between methods at the end of this section.)

We begin by considering the L_n norm; we will later focus on the L_1 norm by setting $n = 1$. Our goal is to manipulate $\|\mathbf{v}\|_n^n$, where $\mathbf{v}$ is an arbitrary vector, so that it looks like the square of a *weighted* L_2 norm $\|\mathbf{W}^{\frac{1}{2}}\mathbf{v}\|_2^2$. The former is defined as

$$\|\mathbf{v}\|_n^n \equiv \sum_k |v_k|^n \tag{10.19}$$

and the latter is defined as

$$\left\|\mathbf{W}^{\frac{1}{2}}\mathbf{v}\right\|_2^2 = \mathbf{v}^T \mathbf{W} \mathbf{v} = \sum_k w_k |v_k|^2 \tag{10.20}$$

Here, we have assumed a diagonal weight matrix, that is, $\mathbf{W} = \mathrm{diag}(\mathbf{w})$. Consider the choice:

$$w_k = (|v_k|^\gamma + \delta^\gamma)^{(n-2)/\gamma} \quad \text{with } 0 < \delta \ll 1 \text{ and } 1 \leq \gamma \leq 2 \tag{10.21}$$

Here, δ is a small positive number that prevents w_k from being singular should v_k^γ ever be zero and the factor γ is added for numerical stability. Frommlet and Nuel (2016) suggest values of $\delta = 10^{-5}$ and $\gamma = 2$. Inserting Eq. (10.21) into Eq. (10.20) yields:

$$\left\|\mathbf{W}^{\frac{1}{2}}\mathbf{v}\right\|_2^2 = \mathbf{v}^T \mathbf{W} \mathbf{v} = \sum_k \frac{|v_k|^n}{(|v_k|^\gamma + \delta^\gamma)^{-(n-2)/\gamma}} \approx \sum_k \frac{|v_k|^2}{|v_k|^{-(n-2)}} \approx \sum_k |v_k|^n = \|\mathbf{v}\|_n^n \tag{10.22}$$

Thus this choice of weighting makes the two norms equal. Frommlet and Nuel (2016) point out that significant round-off error can occur when Eq. (8.20) is used to evaluate the weights, and recommend instead the numerically more stable (and algebraically equivalent) formula:

$$w_k = \begin{cases} \delta^{n-2} \exp\left[\dfrac{n-2}{\gamma} \log 1\mathrm{p}\left(\left|\dfrac{v_k}{\delta}\right|^\gamma\right)\right] & \text{if } |v_k| \leq \delta \\[3mm] |v_k|^{n-2} \exp\left[\dfrac{n-2}{\gamma} \log 1\mathrm{p}\left(\left|\dfrac{\delta}{v_k}\right|^\gamma\right)\right] & \text{if } |v_k| > \delta \end{cases} \tag{10.23}$$

Here $\log 1\mathrm{p}(x) \equiv \log(1 + x)$; both MATLAB and Python implement `log1p(x)` as a function distinct from, and in cases, more accurate than, `log(x)`.

We now consider the special case of the L_1 norm, that is, $n = 1$. While the procedure has made $\|\mathbf{v}\|_1$ look like $\|\mathbf{W}^{\frac{1}{2}}\mathbf{v}\|_2^2$, it has introduced the undesirable complication of a weight matrix that is a function of $\mathbf{v}$. This behavior is problematical for a minimization problem where one seeks the model parameters $\mathbf{m}^{est}$ that minimize $\|\mathbf{W}^{\frac{1}{2}}\mathbf{v}\|_2^2$, for both $\mathbf{W}(\mathbf{m})$ and $\mathbf{v}(\mathbf{m})$ are functions of $\mathbf{m}$, and the standard least-squares formula, which requires $\mathbf{W}$ to be constant, do not apply. This problem can be handled by successive approximation. Starting with a weight matrix approximated as $\mathbf{W}^{(0)} = \mathbf{I}$, the standard least-squares formula is used to determine an initial approximation $\mathbf{m}^{(0)}$ for the model parameters. This solution is then iteratively improved, by first updating the weight matrix $\mathbf{W}^{(j)} = \mathbf{W}(\mathbf{m}^{(j-1)})$ and then using the least-squares formula to update the solution $\mathbf{m}^{(j)}$. This *reweighting* process is terminated when the solution no longer changes between successive iterations, whence $\mathbf{m}^{est} = \mathbf{m}^{(j)}$. Candes et al. (2008) analyze the properties of this algorithm (but only for the $n = 1$ case) and conclude that it will always converge to the correct solution.

As an example, consider the problem of finding the single model parameter m_1 that minimizes the error $\|\mathbf{e}\|_1$, where $\mathbf{e} = \mathbf{d}^{obs} - \mathbf{Gm}$. Here $\mathbf{d}^{obs} = \mathbf{Gm}^{true} + \mathbf{n}$ are the N observed data, $\mathbf{n}$ is exponentially distributed noise, and $\mathbf{G} = [1, \cdots, 1]$, that is, each datum equals the model parameter plus noise. We have previously shown that the solution of this L_1 problem is $m_1^{est} = m_1^{median} = \mathrm{median}(\mathbf{d}^{obs})$. The L_1 problem is transformed into the equivalent L_2 problem of minimizing $\|\mathbf{W}^{\frac{1}{2}}\mathbf{v}\|_2^2$ by making the correspondence $\mathbf{v} = \mathbf{e}$. The iterative form of the standard least-squares formula is

$$\mathbf{m}^{(j)} = \left[\mathbf{G}^T \mathbf{W}^{(j-1)} \mathbf{G}\right]^{-1} \mathbf{G}^T \mathbf{W}^{(j-1)} \mathbf{d}^{obs} \tag{10.24}$$

In the example, the solution is initialized to $m_1^{(0)} = m_1^{mean} = \mathrm{mean}(\mathbf{d}^{obs})$. The iterations quickly converge—typically in about 25 iterations—to the expected solution of $m_1^{est} = m_1^{mean}$ (Fig. 10.4A). The quality of the solution can be assessed using the function:

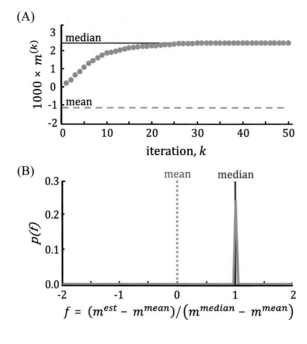

FIG. 10.4   (A) When applied to the first exemplary problem, the iterative algorithm rapidly converges to the median of the data (red line, the correct value). (B) An empirical probability density function $p(f)$ for the solution quality factor f is strongly peaked around $f=1$ (red line, the correct value). Scripts `gdama10_04` and `gdapy10_04`.

$$f = \frac{m_1^{est} - m_1^{mean}}{m_1^{median} - m_1^{mean}} \tag{10.25}$$

as f takes on a value of unity when the solution is correct. An empirical pdf, created by solving the problem *1000* times for different realizations of the noise $\mathbf{n}$, is very strongly peaked at $f=1$ (Fig. 10.4B), which verifies that the method very reliably converges to the correct answer.

The iterative algorithm can also be used to solve problems with prior information of the form $\mathbf{Hm}=\mathbf{h}^{pri}$. Consider, for instance, the problem of finding the $\mathbf{m}^{est}$ that minimizes:

$$\|\mathbf{e}\|_2^2 + \mu\|\mathbf{h}^{pri} - \mathbf{Hm}\|_1 \tag{10.26}$$

where the parameter μ quantifies the strength of the information. The weight matrix is calculated using the correspondence $\mathbf{v}=\mathbf{h}^{pri}-\mathbf{Hm}$, and the solution is calculated using the iterative formula (see Eq. 4.49)

$$\mathbf{m}^{(j)} = \left[\mathbf{G}^T\mathbf{G} + \mu\mathbf{H}^T\mathbf{W}^{(j-1)}\mathbf{H}\right]^{-1}\left[\mathbf{G}^T\mathbf{d}^{obs} + \mu\mathbf{H}^T\mathbf{W}^{(j-1)}\mathbf{h}^{pri}\right] \tag{10.27}$$

Prior information of *flatness* can be implemented by making the correspondences $\mathbf{h}^{pri}=0$ and $\mathbf{H}=\mathbf{D}$, where $\mathbf{D}$ is the first difference operator (Eq. 4.41). As we will discuss in Section 10.7, solutions that minimize $\|\mathbf{Dm}\|_1$ are significantly different in character from those that minimize $\|\mathbf{Dm}\|_2^2$; the former have a blocky (piecewise-constant) appearance while the latter tend to be smooth as well as flat. The L_1 minimization of slope, which is sometimes referred to as *Total Variation Regularization* (or sometimes just *TV*), is a qualitatively new type prior information that has wide application.

In most instances, the reweighting algorithm is to be preferred over the linear programming algorithm presented in Section 10.3. The main exception is when prior information in the form of inequality constraints is available. The linear programming algorithm can handle such constraints very naturally.

10.6   Solving L_∞ norm problems by transformation to a linear programming problem

While the L_1 norm weights "bad" data less than the L_2 norm, the L_∞ norm weights it more:

$$\text{minimize } E + L \equiv \|\mathbf{e}\|_\infty + \|\mathbf{l}\|_\infty = \max_i \frac{|e_i|}{\sigma_{d_i}} + \max_i \frac{|l_i|}{\sigma_{m_i}} \tag{10.28}$$

The prediction error and solution length are weighted by the reciprocal of their square root of their respective prior variances. Normally, one does not want to emphasize outliers, so the L_∞ form is useful mainly in that it can provide a "worst-case" estimate of the model parameters for comparison with estimates derived on the basis of other norms. If the estimates are in fact close to one another, one can be confident that the data are highly consistent. Since the L_∞ estimate is controlled only by the worst error or length, it is usually nonunique.

The general linear equation $\mathbf{Gm} = \mathbf{d}$ can be solved in the L_∞ sense by transformation into a linear programming problem using a variant of the method used to solve the L_1 problem. In the underdetermined problem, we again introduce new variables, m_i', m_i'', x_i, and x_i', each of length M and a single parameter α (for a total of $4M+1$ variables). Also, we again define $\mathbf{m} \equiv \mathbf{m}' - \mathbf{m}''$. The linear programming problem is

minimize α subject to the constraints

$$\mathbf{G}[\mathbf{m}' - \mathbf{m}''] = \mathbf{d}^{obs} \text{ and}$$
$$m_i' - m_i'' + x_i - \alpha\sigma_{m_i} = \langle m_i \rangle \text{ and } m_i' - m_i'' - x_i' + \alpha\sigma_{m_i} = \langle m_i \rangle \tag{10.29}$$
$$\text{and } \mathbf{m}' \geq 0 \text{ and } \mathbf{m}'' \geq 0 \text{ and } \alpha \geq 0 \text{ and } \mathbf{x} \geq 0 \text{ and } \mathbf{x}' \geq 0$$

We note that the second and third constraints can be written as

$$\alpha - \frac{x_i}{\sigma_{m_i}} = -\frac{\langle m_i \rangle - m_i}{\sigma_{m_i}} \equiv -\frac{l_i}{\sigma_{m_i}} \text{ and } \alpha - \frac{x_i'}{\sigma_{m_i}} = \frac{\langle m_i \rangle - m_i}{\sigma_{m_i}} \equiv \frac{l_l}{\sigma_{m_i}} \tag{10.30}$$

where the error in prior information is given by $l_i \equiv \langle m_i \rangle - m_i$, and where α, x_i, and x_i' are nonnegative. Using the same argument as was applied in the L_1 case, we conclude that these constraints require that $\alpha \geq |l_i|/\sigma_{m_i}$ for all i. As this problem has but a single parameter α, it must satisfy

$$\alpha \geq \max_i \frac{|l_i|}{\sigma_{m_i}} \tag{10.31}$$

Consequently, minimizing α yields the L_∞ solution.

The linear programming problem for the overdetermined case is

minimize α subject to the constraints

$$\sum_{j=1}^{M} G_{ij}\left[m_j' - m_j''\right] + x_i - \alpha\sigma_{d_i} = d_i^{obs} \text{ and } \sum_{j=1}^{M} G_{ij}\left[m_j' - m_j''\right] - x_i' + \alpha\sigma_{d_i} = d_i^{obs} \tag{10.32}$$
$$\text{and } \mathbf{m}' \geq 0 \text{ and } \mathbf{m}'' \geq 0 \text{ and } \alpha \geq 0 \text{ and } \mathbf{x} \geq 0 \text{ and } \mathbf{x}' \geq 0$$

Here, $\mathbf{m} \equiv \mathbf{m}' - \mathbf{m}''$, where $\mathbf{m}'$ and $\mathbf{m}''$ are of length M and $\mathbf{x}$ and $\mathbf{x}'$ are of length N. The code implementation is

MATLAB®

```
% variables
% m = mp - mpp
% x = [mp', mpp', alpha', x', xp']
% with mp, mpp length M; alpha length 1, x, xp, length N
L = 2*M+1+2*N;   % number of unknowns
L = 2*M+1+2*N;   % number of unknowns
x = zeros(L,1);  % unknowns in vector x
% f is length L
% minimize alpha
f = zeros(L,1);
f(2*M+1:2*M+1)=1;
% : Part 1equality constraints
Aeq = zeros(2*N,L);
beq = zeros(2*N,1);
% first equation G(mp-mpp)+x-alpha=d
Aeq(1:N,1:M)              =  G;
Aeq(1:N,M+1:2*M)          = -G;
Aeq(1:N,2*M+1:2*M+1)      = -1./sd;
Aeq(1:N,2*M+1+1:2*M+1+N)  =  eye(N,N);
beq(1:N)                  =  dobs;
% second equation G(mp-mpp)-xp+alpha=d
Aeq(N+1:2*N,1:M)              =  G;
Aeq(N+1:2*N,M+1:2*M)          = -G;
Aeq(N+1:2*N,2*M+1:2*M+1)      =  1./sd;
Aeq(N+1:2*N,2*M+1+N+1:2*M+1+2*N) = -eye(N,N);
beq(N+1:2*N)                  =  dobs;
% Part 2: inequality constraints A x <= b
% everything positive
A = zeros(L+2*M,L);
b = zeros(L+2*M,1);
A(1:L,:) = -eye(L,L);
b(1:L) = zeros(L,1);
% mp and mpp have an upper bound (to prevent runaway soln)
% might need to be adjusted for problem at hand
A(L+1:L+2*M,:) = eye(2*M,L);
mupperbound=10*max(abs(mls)); %
b(L+1:L+2*M) = mupperbound;
% solve linear programming problem
[x, fmin] = linprog(f,A,b,Aeq,beq);
fmin=-fmin;
mest = x(1:M) - x(M+1:2*M);
dpre = G*mest;
```

                                                                    [gdama10_05]

```
                              Python
f = np.zeros((L,1));
f[2*M,0]=1.0;
# equality constraints
Aeq = np.zeros((2*N,L));
beq = np.zeros((2*N,1));
# first equation G(mp-mpp)+x-alpha=d
Aeq[0:N,0:M]                  =  G;
Aeq[0:N,M:2*M]                =  -G;
Aeq[0:N,2*M:2*M+1]            =  -1./sd;
Aeq[0:N,2*M+1:2*M+1+N]        =  np.identity(N);
beq[0:N,0:1]                  =  dobs;
# second equation G(mp-mpp)-xp+alpha=d
Aeq[N:2*N,0:M]                    =  G;
Aeq[N:2*N,M:2*M]                  =  -G;
Aeq[N:2*N,2*M:2*M+1]              =  1./sd;
Aeq[N:2*N,2*M+1+N:2*M+1+2*N]      =  -np.identity(N);
beq[N:2*N,0:1]                    =  dobs;
# inequality constraints A x <= b
# part 1: everything positive
A = np.zeros((L+2*M,L));
b = np.zeros((L+2*M,1));
A[0:L,0:L] = -np.identity(L);
b[0:L,0:1] = np.zeros((L,1));
# part 2; mp and mpp have an upper bound (to prevent runaway
# soln. This bound might need to be adjusted
A[L:L+2*M,0:L] = np.eye(N=2*M,M=L);
mupperbound=10.0*np.max(np.abs(mls)); # adjust as needed
b[L:L+2*M,0:L] = mupperbound*np.ones((2*M,1));
  # solve linear programming problem
res = opt.linprog(c=f,A_ub=A,b_ub=b,A_eq=Aeq,b_eq=beq);
x = gda_cvec(res.x);
fmin = -res.fun;
status = res.success;
if( status ):
    print("linear programming succeeded" );
else:
    print("linear programming failed" );
mest = x[0:M,0:1] - x[M:2*M,0:1];
dpre = np.matmul(G,mest);
                                              [gdapy10_05]
```

As in the L_1 case, this implementation adds upper bounds on the variables $\mathbf{m}'$ and $\mathbf{m}''$. An example of the L_1 problem applied to curve fitting is shown in Fig. 10.5.

The mixed-determined problem can be solved by applying these algorithms and either of the two methods described for the L_1 problem.

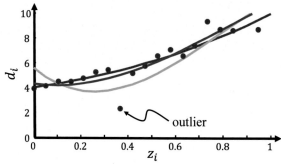

FIG. 10.5   Curve fitting using the L_∞ norm. The true data (red curve) follow a cubic polynomial in an auxiliary variable z. Observations d_i^{obs} (red circles) have additive noise with zero mean and variance $\sigma_d^2 = 1$ drawn from an exponential probability density function. Note the outlier at $z = 0.37$. The L_∞ fit (green curve) is more affected by the outlier than is a standard L_2 (least-squares) fit (blue curve). Scripts gdama10_05 and gdapy10_05.

10.7   The L_0 norm and sparsity

As discussed previously in Section 4.2, the pseudo-norm:

$$\|\mathbf{m}\|_0^0 \equiv \lim_{n \to 0} \|\mathbf{m}\|_n^n = \lim_{n \to 0} |m_i|^n = \text{number of nonzero elements of } \mathbf{m} \tag{10.33}$$

(where $\lim_{n \to 0} 0^n$ is taken to be zero) is a measure of the *sparsity* of $\mathbf{m}$, that is, the number of its nonzero elements.

Sparsity is a form of prior information that has many applications. Consider, for instance, the problem of fitting spectral peaks to data that was discussed in Section 2.9. One problem with implementing this inversion is that, initially, one does not know the number q of spectral peaks. One strategy, based on already-presented techniques, is to solve a sequence of problems, with $q = 1, 2, 3, \cdots$ peaks and to select the smallest q for which the prediction error $E(q)$ is acceptably small. An alternative strategy is to choose q to be a large number, but then to impose sparsity as prior information, that is, to minimize a weighted sum of the prediction error $\|\mathbf{e}\|_2^2$ and sparsity $\|\mathbf{m}\|_0^0$. This strategy suppresses the amplitude of any peak that is not needed to fit the data.

In another example, consider the CT tomography problem discussed in Section 2.9, where measurements of X-ray attenuation are used to determine an image of the opacity of body tissue $c(x, y)$. A useful type of prior information is that different tissues each tend to have its own distinct opacity, so that, to first approximation, the CT image is composed of patches (organs) of more-or-less uniform opacity, separated by sharp boundaries. Mathematically, this idea implies that the first differences of model parameters, measured in the x and y directions, are sparse and suggests the strategy of minimizing:

$$\|\mathbf{e}\|_2^2 + \mu \|\mathbf{D}_x \mathbf{m}\|_0^0 + \mu \|\mathbf{D}_y \mathbf{m}\|_0^0 \tag{10.34}$$

Here, μ is a positive constant that quantifies the strength of the prior information and $\mathbf{D}_x$ and $\mathbf{D}_y$ are matrices of first differences in the x and y directions, respectively. This strategy selects a *blocky* model.

While the reweighting algorithm presented in Section 10.4 cannot be used to solve a L_0 problem, it usually works for a fractional n that is close to zero (say, $n = 0.1$) and a solution obtained for that value can be considered an approximation for the L_0 solution. The word "usually' is used here, because the convergence properties of the reweighting algorithm are not well understood for $n < 1$ and may depend upon the matrices involved and the initial guess of the solution. That being said, the algorithm is one of the few available for solving L_0 problems. It should be used, but cautiously. The solution $\mathbf{m}^{est}$ should be compared to the corresponding L_1 solution to verify that it is at least as sparse, and its prediction error should be examined to verify that it is acceptably small.

When applied to the minimum-length problem:

$$\|\mathbf{e}\|_2^2 + \mu \|\mathbf{m}\|_0^0 \tag{10.35}$$

the reweighting algorithm is known to suffer from a solution *shrinkage* problem, meaning that the solution is always a bit smaller than if the zero-valued model parameters were to be removed and the problem resolved. This behavior suggests that value of μ in the iterative formula (call it μ'):

$$\mathbf{m}^{(j)} = \left[\mathbf{G}^T \mathbf{G} + \mu' \mathbf{W}^{(j-1)} \right]^{-1} \mathbf{G}^T \mathbf{d}^{obs} \tag{10.36}$$

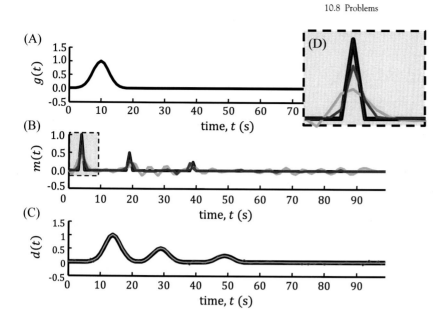

FIG. 10.6 The filter problem $\mathbf{d}^{obs} = \mathbf{g} * \mathbf{m}$ with a spiky filter $\mathbf{m}$. (A) The time series $\mathbf{g}$ is a smooth pulse. (B) The true filter $\mathbf{m}^{true}$ (black) is sparse, with three widely separated spikes. The estimated L_2 (green), L_1 (blue), and L_0 (red) filters have increasing spikiness. (C) All three filters (green) fit data (black). (D) Enlargement of the shaded region in C. Scripts `gdama10_06` and `gdapy10_06`.

should be smaller than the μ in Eq. (10.35). An analysis by Frommlet and Nuel (2016) indicates that $\mu' \approx \mu/4$. However, the issue is moot in many practical applications, where the value of μ is chosen by trial and error, as contrasted to being guided by a quantitative estimate of the certainty of the prior information.

The following example illustrates the process of finding a sparse solution. We seek the filter $\mathbf{m}^{est}$ that minimizes a weighted sum of prediction error and solution length $\|\mathbf{e}\|_2^2 + \mu\|\mathbf{m}\|_0^0$. Here the prediction error is $\mathbf{e} = \mathbf{d}^{obs} - \mathbf{Gm}$, the N observed data are $\mathbf{d}^{obs} = \mathbf{Gm}^{true} + \mathbf{n}$, and the Normally distributed noise is $\mathbf{n}$. The data kernel $\mathbf{G}$ is a Toeplitz matrix that implements convolution by a filter $\mathbf{g}$ (see Section 2.11). We choose $\mathbf{g}$ to consist of a single smooth pulse (Fig. 10.6A) and $\mathbf{m}^{true}$ to be sparse and to consist of three widely separated spikes (Fig. 10.6B). We compare the L_0, L_1, L_2 solutions, where the L_0 and L_1 solutions are computed using the reweighting algorithm with the correspondence $\mathbf{v} = \mathbf{m}$ and with the least-squares formula

$$\mathbf{m}^{(j)} = \left[\mathbf{G}^T\mathbf{G} + \mu\mathbf{W}^{(j-1)}\right]^{-1}\mathbf{G}^T\mathbf{d}^{obs} \tag{10.37}$$

All three solutions (Fig. 10.6B) fit the data (Fig. 10.6C) to within similar error. The L_0 solution matches the true solution extremely well, recovering the three spikes almost exactly. The L_1 solution is also fairly spiky, although not nearly as sparse as the L_0 solution. These two solutions can be contrasted to the L_2 solution, which is extremely ringy.

Had the true model been a boxcar function (Fig. 10.7B), as contrasted to a spike, then a solution $\mathbf{m}^{est}$, which minimizes a weighted sum of prediction error and solution slope $\|\mathbf{e}\|_2^2 + \mu\|\mathbf{Dm}\|_0^0$, would have been preferable. In this case, the L_0 and L_1 solutions are computed using the reweighting algorithm with the correspondence $\mathbf{v} = \mathbf{Dm}$ and with the least-squares formula:

$$\mathbf{m}^{(j)} = \left[\mathbf{G}^T\mathbf{G} + \mu\mathbf{D}^T\mathbf{W}^{(j-1)}\mathbf{D}\right]^{-1}\mathbf{G}^T\mathbf{d}^{obs} \tag{10.38}$$

As before, all three solutions (Fig. 10.7B) fit the data (Fig. 10.7C) to within similar error. The L_0 solution matches the true solution extremely well, recovering the boxcar function almost exactly.

The L_1 solution also does fairly well, except that it bevels the step slightly. As before, the L_2 solution is extremely ringy.

10.8 Problems

10.1 Solve the best-fitting plane problem of Section 4.7 under the L_1 norm and compare the estimated model parameters with those determined under the L_2 norm. How much does the estimated *strike* and *dip* of the plane change?

10.2 Solve the constrained best-fitting line problem of Section 4.21 under the L_1 norm, by these two methods: (A) by considering the point (z_0, d_0) as normal data with very small variance and (B) by explicitly including the constraint

FIG. 10.7 The filter problem $\mathbf{d}^{obs}=\mathbf{g}*\mathbf{m}$ with a piecewise constant filter $\mathbf{m}$. (A) The time series $\mathbf{g}$ is a smooth pulse. (B) The true filter $\mathbf{m}^{true}$ (black) consists of a single boxcar function. The estimated L_2 (green), L_1 (blue), and L_0 (red) filters have increasingly sharp steps. (C) All three filters (green) fit data (black). (D) Enlargement of the shaded region in C. Scripts gdama10_07 and gdapy10_07.

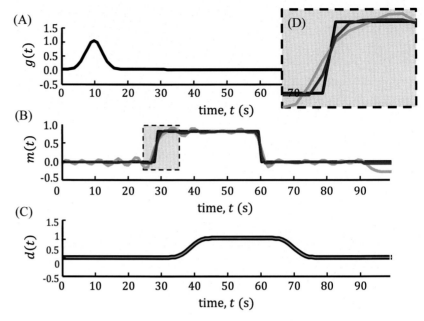

as a linear equality constraint within the linear programming problem. Compare the estimated model parameters with those determined under the L_2 norm.

10.3 This problem builds upon Problem 9.4. Consider fitting a cubic polynomial $d_i=m_1+m_2z_i+m_3z_i^2+m_4z_i^3$ to $N=20$ data, d_i^{obs}, where the zs are evenly spaced on the interval $(0,1)$, where $m_2=m_3=m_4=1$, and where m_1 is varied from -1 to 1, as described as follows. (A) Write a script that generates synthetic observed data (including exponentially distributed noise) for a particular choice of m_1 and estimates the model parameters under the L_1 norm, with the extra inequality constraint that $\mathbf{m} \geq 0$. (B) Run a series of test cases for a range of values of m_1 and comment upon the results.

10.4 Consider the problem of finding $M=100$ model parameters $\mathbf{m}$ that minimize $\|\mathbf{e}\|_2^2+\mu\|\mathbf{Hm}\|_n^n$, where μ is a small positive number, $\mathbf{e}=\mathbf{d}^{obs}-\mathbf{Gm}$ is a vector of errors, $\mathbf{d}^{obs}=\mathbf{Gm}^{true}+\mathbf{n}$ are the observed data, $\mathbf{n}$ is Normally distributed noise, and:

$$\mathbf{m}^{true} = \left[[1,\cdots,1]_{M/2}\ [2,\cdots,2]_{M/2}\right]^T$$

$$\mathbf{G} = \left[\begin{array}{cc}[1,\cdots,1]_{M/2} & [0,\cdots,0]_{M/2}\\ [0,\cdots,0]_{M/2} & [1,\cdots,1]_{M/2}\end{array}\right]^T$$

$$\mathbf{H} = \mathrm{diag}(\mathbf{a})\ \text{with}\ a_k = 1 + 0.001k\ \text{and}\ k = 1 \cdots M$$

The true model parameters $\mathbf{m}^{true}$ are divided into two groups, each of size $M/2$, with equal values within each group. The data kernel $\mathbf{G}$ sums each group separately, to yield $N=2$ data. The problem is very underdetermined, as $N \ll M$. The matrix $\mathbf{H}$ is almost the identity matrix, but weights the model parameters near the top of $\mathbf{m}$ slightly less than those at the bottom, so that, all other things being equal, the minimization of $\|\mathbf{Hm}\|_n^n$ selects a model that is squeezed toward the top. (A) Show that the true solution $\mathbf{m}^{true}$ is also the L_2 minimum-length solution $\mathbf{m}_{ML}=\mathbf{G}^T[\mathbf{GG}^T]^{-1}\mathbf{d}^{obs}$. (B) Compute the solution for $n=0, 1, 2$ and plot and interpret the result. (C) Repeat the solution in B but for $a_k=1+0.001(M-k+1)$. In what respect is the solution different, and why?

10.5 Solve the previous problem, but for $\mathbf{H}=\mathbf{D}$, where $\mathbf{D}$ is the $(M-2) \times M$ first difference operator:

$$\mathbf{D} = \begin{bmatrix} -1 & 1 & 0 & 0 & 0 & 0\\ 0 & -1 & 1 & 0 & 0 & 0\\ \ddots & \ddots & \ddots & \ddots & \ddots & \ddots\\ 0 & 0 & 0 & 0 & -1 & 1 \end{bmatrix}$$

Show that the model parameter vector becomes piecewise constant as $n \rightarrow 0$.

References

Candes, E.J., Wakin, M.B., Boyd, S.P., 2008. Enhancing sparsity by reweighted L1 minimization. J. Fourier Anal. Appl. 14, 877–905.

Claerbout, J.F., Muir, F., 1973. Robust modelling with erratic data. Geophysics 38, 826–844.

Cuer, M., Bayer, R., 1980. Fortran routines for linear inverse problems. Geophysics 45, 1706–1719.

Frommlet, F., Nuel, G., 2016. An adaptive ridge procedure for L0 regularization. PLoS ONE 11, e0148620.

11

Nonlinear inverse problems

11.1   Parameterizations

Variables representing the data and model parameters—the parameterization—must be selected at the start of any inverse problem. In many instances, this selection is rather ad hoc; there might be no strong reasons for selecting one parameterization over another. This can become a substantial issue in nonlinear inverse problems because the answer obtained by solving it is dependent on the parameterization. In other words, the solution is not invariant under nonlinear transformations of the variables. This is in contrast to the linear inverse problem with Normal statistics, in which solutions are invariant for any linear reparameterization of the data and model parameters.

As an illustration of this difficulty, consider the problem of fitting a straight line to the data pairs (1,1), (2,2), (3,3), (4,5). Suppose that we regard these data as (z, d) pairs, where z is an auxiliary variable. The least-squares fit is $d = 0.500 + 1.3z$. On the other hand, we might regard them as (d', z') pairs, where z' is the auxiliary variable. Least squares then gives $d' = 0.457 + 0.743z'$, which can be rearranged as $z' = 0.615 + 1.346d'$. These two straight lines have intercepts that differ by 20% and slopes that differ by 4% (see script gdama11_01 and gdapy11_01).

This discrepancy arises from two sources. The first is an inconsistent application of probability theory. In the earlier example, we alternately assumed that z was exactly known but d contained Normally distributed noise and that d was exactly known but z contained Normally distributed noise. These are two radically different assumptions about the distribution of errors, so it is no wonder that the solutions are different.

In theory, this first source of discrepancy can be avoided completely by recognizing and taking into account the fact that reparameterizing a problem changes the associated probability density functions. For instance, consider an inverse problem in which there is a model parameter m that is known to possess a uniform pdf $p(m)$ on the interval $(0,1)$ (Fig. 11.1A). If the inverse problem is reparameterized in terms of a new model parameter $m' = m^2$, then the transformed pdf $p(m')$ can be calculated using Eq. (3.13) as

$$p(m') = p[m(m')] \frac{\mathrm{d}m}{\mathrm{d}m'} = \tfrac{1}{2}(m')^{-\frac{1}{2}} \tag{11.1}$$

The pdf of m' is not uniform (Fig. 11.1B), and any inverse method developed to solve the problem under this parameterization must account for this fact.

The second source of discrepancy is more serious. Suppose that we could use some inverse theory to calculate the distribution of the model parameters under a particular parameterization. We could then use Eq. (11.1) to find their joint pdf under any arbitrary parameterization. Insofar as the pdf of the model parameters is the *answer* to the inverse problem, we would have the correct answer in the new parameterization. Probability density functions are *invariant* under changes of variables. However, a pdf is not always the answer for which we are looking. More typically, we need an estimate (a single number) based on a pdf (for instance, its maximum likelihood point or mean value).

Estimates are not invariant under changes in the parameterization. For example, suppose $p(m)$ has a uniform pdf, as before. Then, if $m' = m^2$, $p(m') = \tfrac{1}{2}(m')^{-\frac{1}{2}}$. The distribution in m has no maximum likelihood point, whereas the distribution in m' has one at $m' = m = 0$. The pdfs also have different means (expectations):

$$\langle m \rangle = \int_0^1 m \, p(m) \, \mathrm{d}m = \int_0^1 m \, \mathrm{d}m = \left. \frac{m^2}{2} \right|_0^1 = \tfrac{1}{2}$$

$$\langle m' \rangle = \int_0^1 m' \, p(m') \, \mathrm{d}m' = \int_0^1 \tfrac{1}{2}(m')^{\frac{1}{2}} \, \mathrm{d}m' = \left. \frac{2}{6}(m')^{3/2} \right|_0^1 = \frac{1}{3} \tag{11.2}$$

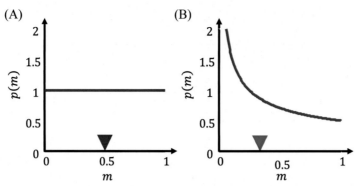

FIG. 11.1   (A) A probability density function $p(m)$ that is uniform on the interval $(0,1)$. (B) The corresponding probability density function $p(m')$ for the transformation $m' = m^2$. The mean (expectation) of each probability density function is indicated by a triangle. Scripts `gdama11_02` and `gdapy11_02`.

Even though m' equals the square of the model parameter m, the mean (expectation) of m' is not equal to the square of the expectation of m, that is, $\langle m \rangle^2 = \frac{1}{4}$ and not $\frac{1}{3}$. There is some advantage, therefore, in working explicitly with pdfs as long as possible, forming estimates only at the last step. If m and m' are two different parameterizations of model parameters, we want to avoid as much as possible sequences like

$$p(\mathbf{m}) \text{ leads to estimate } \mathbf{m}^{est} \text{ transformed to } (\mathbf{m}')^{est} \tag{11.3}$$

in favor of sequences like

$$p(\mathbf{m}) \text{ transformed to } p(\mathbf{m}') \text{ leads to estimate } (\mathbf{m}')^{est} \tag{11.4}$$

Note, however, that the mathematics for this second sequence is typically much more difficult than that for the first.

There are objective criteria for the "goodness" of a particular estimate of a model parameter. Suppose that we are interested in the value of a model parameter m. Suppose further that this parameter is either deterministic with a true value or (if it is a random variable) has a well-defined pdf from which the true expectation could be calculated if the distribution was known. Of course, we cannot know the true value; we can only perform experiments and then apply inverse theory to derive an estimate of the model parameter. As any one experiment contains noise, the estimate we derive will not coincide with the true value of the model parameter. But we can at least expect that if we perform the experiment enough times, the estimated values will scatter about the true value. If they do, then the method of estimating is said to be *unbiased*. Estimating model parameters by taking nonlinear combinations of estimates of other model parameters almost always leads to *bias*.

11.2   Linearizing transformations

One of the reasons for changing parameterizations is that an inverse problem can sometimes be transformed into a simpler form that can be solved by a known method. The problems that most commonly benefit from such transformations involve fitting exponential and power functions to data. Consider a set of (z_i, d_i) data pairs that are thought to obey the model $d_i = m_1 \exp(m_2 z_i)$. By making the transformation

$$m_1' = \log(m_1) \text{ and } m_2' = m_2 \text{ and } d_i' = \log(d_i) \tag{11.5}$$

we can write the model as the linear equation $d_i' = m_1' + m_2' z_i$, which can be solved by simple least-squares techniques. However, we must assume that the d_i' are independent Normally distributed random variables with uniform variance in order to justify rigorously the application of least-squares techniques to this problem. The pdf of the data in their original parameterization must therefore be non-Normal. (It must have a *log-normal* probability density function.)

The process of taking a logarithm amplifies the scattering of the near-zero points. Consequently, for a decaying exponential, the scatter of the transformed data about $(d_i')^{true}$ increases with z. The assumption that the d_i' have uniform variance, therefore, implies that the untransformed data d_i were measured with an accuracy that increases with z (Fig. 11.2). This assumption may well be inconsistent with the facts of the experiment. Linearizing transformations must be used with caution.

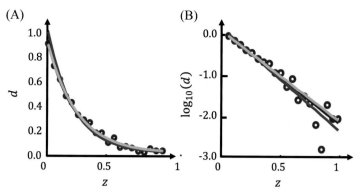

FIG. 11.2 (A) Least-squares fit to the exponential function $d_i = m_1 \exp(m_2 z_i)$ (Red curve). The true function $d(z)$ for $(m_1, m_2) = (0.7, -4.0)$. (Red circles) Data that include Normally distributed random noise with zero mean and variance $\sigma_d^2 = (0.2)^2$. (Green curve) Nonlinear least-squares solution using Newton's method. (Blue curve) Least-squares solution using the linearizing transformation $\log(d_i) = \log(m_1) + m_2 z_i$. Note that the two solutions are different. (B) Log-linear version of the graph in (A). Note that the scatter of the data increases with z, and that the solution based on the linearizing transformation is strongly affected by outliers associated with taking the logarithm of the data. Scripts gdama11_03 and gdapy11_03.

11.3 Error and log-likelihood in nonlinear inverse problems

Suppose that the data $\mathbf{d}$ in an inverse problem have a possibly non-Normal pdf $p(\mathbf{d}; \langle \mathbf{d} \rangle)$, where $\langle \mathbf{d} \rangle$ is the mean (or expected) value and the semicolon is used to indicate that $\langle \mathbf{d} \rangle$ is just a parameter in the pdf for $\mathbf{d}$ (as contrasted to a random variable). The principle of maximum likelihood—that the observed data are the most probable data—holds regardless of the form of $p(\mathbf{d}^{obs}; \langle \mathbf{d} \rangle)$. As long as the theory is explicit, we can assume that the theory predicts the mean of the data, that is, $\langle \mathbf{d} \rangle = g(\mathbf{m})$ and write the log-likelihood $L(\mathbf{m})$ as

$$L(\mathbf{m}) = \log p\left(\mathbf{d}^{obs}; g(\mathbf{m})\right) \equiv c - \tfrac{1}{2} E(\mathbf{m}) \tag{11.6}$$

Here, c is some constant, $E(\mathbf{m})$ is some function, and the factor of $1/2$ has been introduced so that $E(\mathbf{m})$ equals the L_2 weighted prediction error in the Normal case, that is, $E(\mathbf{m}) = [\mathbf{d}^{obs} - g(\mathbf{m})]^T [\text{cov } \mathbf{d}]^{-1} [\mathbf{d}^{obs} - g(\mathbf{m})]$. Thus it is always possible to define an "error" $E(\mathbf{m})$, such that minimizing the error $E(\mathbf{m})$ is equivalent to maximizing the log-likelihood $L(\mathbf{m})$. However, although $E(\mathbf{m})$ is the weighted prediction error in the Normal case, it does not necessarily have all the properties of L_2 case. For instance, it may not be zero when $\mathbf{d}^{obs} = g(\mathbf{m})$.

There is no simple means for determining whether a nonlinear inverse problem has a unique solution. Consider the very simple nonlinear model $d_i = m_1^2 + m_1 m_2 z_i$, with N Normally distributed data. This problem can be linearized by the transformation of variables $m_1' = m_1^2$ and $m_2' = m_1 m_2$ and can therefore be solved by the least-squares method if $N \geq 2$. Nevertheless, even if the primed parameters are unique, the unprimed ones are not: if $\mathbf{m}^{est}$ is a solution that minimizes the error, then $-\mathbf{m}^{est}$ is a solution with the same error. In this instance, although $E(\mathbf{m}')$ has a single minimum, $E(\mathbf{m})$ has two minima of equal depth.

We must examine the global properties of the prediction error $E(\mathbf{m})$ in order to determine whether the inverse problem is unique. If the function has but a single minimum point $\mathbf{m}^{est}$, then the solution is unique. If it has several minima of equal depth, the solution is nonunique, and prior information must be added to resolve the indeterminacy.

The error surface of a linear problem is always a paraboloid (Fig. 11.3), which can have only a simple range of shapes. An arbitrarily complex nonlinear inverse problem can have an arbitrarily complicated error. If $M = 2$ or 3, it may be possible to investigate the shape of the surface by graphical techniques (Fig. 11.4). For most realistic problems this is infeasible.

11.4 The grid search

One strategy for solving a nonlinear inverse problem is to exhaustively consider "every possible" solution and pick the one with the smallest error $E(\mathbf{m})$. Of course, it is impossible to examine "every possible" solution; but it is possible to examine a large set of trial solutions. When the trial solutions are drawn from a regular grid in model space, this procedure is called a *grid search*. Grid searches are most practical when:

(1) The total number of model parameters is small, say $M < 7$. The grid is M-dimensional, so the number of trial solutions is proportional to L^M, where L is the number of trial solutions along each dimension of the grid.

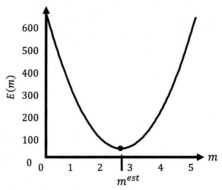

FIG. 11.3 L_2 prediction error $E(m)$ as a function of model parameter m for a typical linear inverse problem. The solution m^{est} minimizes the error. In the linear case, $E(m)$ is a paraboloid. Scripts `gdama11_04` and `gdapy11_04`.

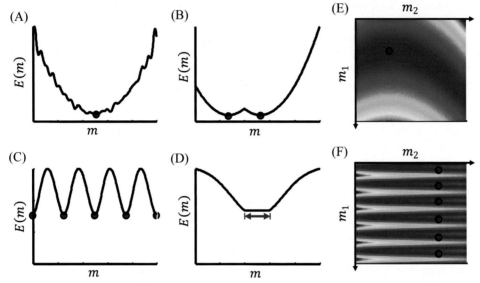

FIG. 11.4 (A–D) Prediction error E as a function of a single model parameter m. (A) A single minimum (red dot) corresponds to an inverse problem with a unique solution. (B) Two solutions. (C) Many well-separated solutions. (D) Finite range of solutions (red arrow). (E and F) Error (colors) as a function of two model parameters m_1 and m_2. (E) A single solution, with the minimum occurring within a nearly flat valley. (F) Many well-separated solutions. Scripts `gdama11_05` and `06` and `gdapy11_05` and `06`.

(2) The solution is known to lie within a specific range of values, which can be used to define the limits of the grid.

(3) The forward problem $\mathbf{d}^{pre} = \mathbf{g}(\mathbf{m})$ can be computed rapidly enough that the time needed to compute L^M of them is not prohibitive.

(4) The error $E(\mathbf{m})$ is smooth over the scale of the grid spacing $\Delta \mathbf{m}$, so that the minimum is not missed through the grid spacing being too coarse.

As an example, we use a grid search to solve the nonlinear problem

$$d_i = \sin(\omega m_1 x_i) + m_1 m_2 \tag{11.7}$$

The data are assumed to be Normally distributed and uncorrelated with uniform variance so that the appropriate error is

$$E(m_1, m_2) = \sum_{i=1}^{N} \left(d_i^{obs} - d_i^{pre} \right)^2 \tag{11.8}$$

A grid search script has three sections: the first section defines the grid of trial $\mathbf{m}$ values, the second evaluates $E(\mathbf{m})$ for every trial $\mathbf{m}$ and stores the results in a matrix $\mathbf{E}$, and the third searches the matrix for its smallest element and calculates $\mathbf{m}^{est}$ on the basis of its row and column indices. In the following exemplary code, Step 3 is being performed by the function `gda_Esurface()`, whose operation will be described later. A useful by-product of the grid search is a tabulation of E on the grid, which can be turned into an informative plot (Fig. 11.5).

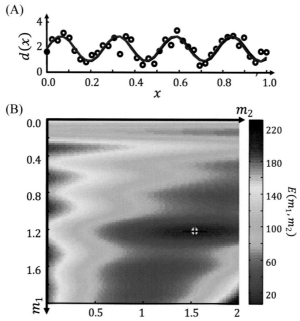

FIG. 11.5 A grid search is used to solve the nonlinear curve-fitting problem $d(x_i) = \sin(\omega_0 m_1 x_i) + m_1 m_2$. (A) The true data (black curve) are for $(m_1, m_2)^{true} = (1.21, 1.54)$. The observed data (black circles) have additive noise with variance $\sigma_d^2 = (0.4)^2$. The predicted data (red curve) are based on results of the grid search. (B) Error surface (colors), showing true solution (green circle), estimated solution (white circle), and refined estimated solution with 95% confidence intervals (red bars). Scripts gdama11_07 and gdapy11_07.

MATLAB®

```
% Define 2D grid
L = 101;
Dm = 0.02;
m1min=0;
m2min=0;
m1a = m1min+Dm*[0:L-1]';
m2a = m2min+Dm*[0:L-1]';
m1max = m1a(L);
m2max = m2a(L);

% tabulate error E on grid.
% Note m1 varies along rows of E and m2 along columns
E = zeros(L,L);
for j = [1:L]
for k = [1:L]
    dpre = sin(w0*m1a(j)*x) + m1a(j)*m2a(k);
    E(j,k) = (dobs-dpre)'*(dobs-dpre);
end
end

% Example of use of Esurface() function
% It refines the minimum using a quadratic approximation
% and provides an error estimate using the 2nd derivative
% note that the "column" variable is sent and returned
% before the "row" variable in this function
[m2est,m1est,E0,cov21,status]=gda_Esurface(m2a,m1a,E,sd^2);
sigmam1 = sqrt( cov21(2,2) );
sigmam2 = sqrt( cov21(1,1) );
                                              [gdama11_07]
```

```
                              Python
# Define 2D grid
L = 101;
Dm = 0.02;
m1min=0.0;
m2min=0.0;
m1a = gda_cvec(m1min+Dm*np.linspace(0,L-1,L));
m2a = gda_cvec(m2min+Dm*np.linspace(0,L-1,L));
m1max = m1a[L-1,0];
m2max = m2a[L-1,0];

# tabulate error E on grid.
# Note m1 varies along rows of E and m2 along columns
E = np.zeros((L,L));
for j in range(L):
    for k in range(L):
        dpre = np.sin(w0*m1a[j,0]*x) + m1a[j,0]*m2a[k,0];
        E[j,k] = np.matmul((dobs-dpre).T,(dobs-dpre));

# Example of use of Esurface() function
# It refines the minimum using a quadratic approximation
# and provides an error estimate using the 2nd derivative
# note that the "column" variable is sent and returned
# before the "row" variable in this function
m2est,m1est,E0,cov21,status=gda_Esurface(m2a,m1a,E,sd**2);
sigmam1 = sqrt( cov21[1,1] );
sigmam2 = sqrt( cov21[0,0] );
                                                [gdapy11_07]
```

In the vicinity of its minimum, the error surface can be approximated as a quadratic function of the model parameters

$$E(m_1, m_2) = E_0 + \mathbf{b}^T \Delta\mathbf{m} + \frac{1}{2}[\Delta\mathbf{m}]^T \mathbf{B}\Delta\mathbf{m} \tag{11.9}$$

Here, $\Delta\mathbf{m} = \mathbf{m} - \mathbf{m_0}$, where $\mathbf{m_0}$ is the location of the smallest value of error on the grid, which occurs, say, at row i_0 and column j_0. The parameters $\mathbf{b}$ and $\mathbf{B}$, which represent the first and second derivative of the error near its minimum, respectively, can be determined by using least squares to fit the parabolic function to a 3×3 portion of the error grid, centered on the minimum value. The 9×6 matrix equation for the unknowns is

$$\begin{bmatrix} E_{i_0-1,\,j_0-1} \\ E_{i_0,\,j_0-1} \\ E_{i_0+1,\,j_0-1} \\ E_{i_0-1,\,j_0} \\ E_{i_0,\,j_0} \\ E_{i_0+1,\,j_0} \\ E_{i_0-1,\,j_0+1} \\ E_{i_0,\,j_0+1} \\ E_{i_0+1,\,j_0+1} \end{bmatrix} = \begin{bmatrix} 1 & \Delta m_{i_0-1} & \Delta m_{j_0-1} & \frac{1}{2}\Delta m_{i_0-1}^2 & \Delta m_{i_0-1}\Delta m_{j_0-1} & \frac{1}{2}\Delta m_{j_0-1}^2 \\ 1 & \Delta m_{i_0} & \Delta m_{j_0-1} & \frac{1}{2}\Delta m_{i_0}^2 & \Delta m_{i_0}\Delta m_{j_0-1} & \frac{1}{2}\Delta m_{j_0-1}^2 \\ 1 & \Delta m_{i_0+1} & \Delta m_{j_0-1} & \frac{1}{2}\Delta m_{i_0+1}^2 & \Delta m_{i_0+1}\Delta m_{j_0-1} & \frac{1}{2}\Delta m_{j_0-1}^2 \\ 1 & \Delta m_{i_0-1} & \Delta m_{j_0} & \frac{1}{2}\Delta m_{i_0-1}^2 & \Delta m_{i_0-1}\Delta m_{j_0} & \frac{1}{2}\Delta m_{j_0}^2 \\ 1 & \Delta m_{i_0} & \Delta m_{j_0} & \frac{1}{2}\Delta m_{i_0}^2 & \Delta m_{i_0}\Delta m_{j_0} & \frac{1}{2}\Delta m_{j_0}^2 \\ 1 & \Delta m_{i_0+1} & \Delta m_{j_0} & \frac{1}{2}\Delta m_{i_0+1}^2 & \Delta m_{i_0+1}\Delta m_{j_0} & \frac{1}{2}\Delta m_{j_0}^2 \\ 1 & \Delta m_{i_0-1} & \Delta m_{j_0+1} & \frac{1}{2}\Delta m_{i_0-1}^2 & \Delta m_{i_0-1}\Delta m_{j_0+1} & \frac{1}{2}\Delta m_{j_0+1}^2 \\ 1 & \Delta m_{i_0} & \Delta m_{j_0+1} & \frac{1}{2}\Delta m_{i_0}^2 & \Delta m_{i_0}\Delta m_{j_0+1} & \frac{1}{2}\Delta m_{j_0+1}^2 \\ 1 & \Delta m_{i_0+1} & \Delta m_{j_0+1} & \frac{1}{2}\Delta m_{i_0+1}^2 & \Delta m_{i_0+1}\Delta m_{j_0+1} & \frac{1}{2}\Delta m_{j_0+1}^2 \end{bmatrix} \begin{bmatrix} E_0 \\ b_1 \\ b_2 \\ B_{11} \\ B_{12} \\ B_{22} \end{bmatrix} \tag{11.10}$$

(with $B_{21} = B_{12}$). The location of the minimum can be improved upon using the principle that the first derivative of the error is zero there

$$\left.\frac{\partial E}{\partial \mathbf{m}}\right|_{m^{est}} = 0 = \mathbf{b} + \mathbf{B}\left[\mathbf{m}^{est} - \mathbf{m_0}\right] \quad \text{so} \quad \mathbf{m}^{est} = \mathbf{m_0} - \mathbf{B}^{-1}\mathbf{b} \tag{11.11}$$

An estimate of the second derivative is

$$\frac{\partial^2 E}{\partial m_i \partial m_j}\bigg|_{m^{est}} \approx B_{ij} \tag{11.12}$$

Eq. (4.76) can be used to estimate the covariance of the model parameters. We provide MATLAB and Python functions, both named gda_Esurface(), that calculate the refined solution (see scripts gdama11_07 and gdapy11_07 for examples of their use). However, our implementation is limited to the $M=2$ case.

11.5 Newton's method

A completely different approach to solving the nonlinear inverse problem is to use information about the shape of the error $E(\mathbf{m})$ in the vicinity of a trial solution $\mathbf{m}^{(p)}$ to devise a better solution $\mathbf{m}^{(p+1)}$. One source of shape information is the derivatives of $E(\mathbf{m})$ at a trial solution $\mathbf{m}^{(p)}$, as Taylor's theorem indicates that the entire function can be built up from them. Expanding $E(\mathbf{m})$ in a Taylor series about the trial solution and keeping the first three terms, we obtain the parabolic approximation

$$E(\mathbf{m}) = E\left(\mathbf{m}^{(p)}\right) + \sum_{i=1}^{M} b_i \left(m - m_i^{(p)}\right) + \frac{1}{2} \sum_{i=1}^{M} \sum_{j=1}^{M} B_{ij} \left(m - m_i^{(p)}\right)\left(m - m_j^{(p)}\right)$$

$$\text{with } b_i \equiv \frac{\partial E}{\partial m_i}\bigg|_{\mathbf{m}^{(p)}} \text{ and } B_{ij} \equiv \frac{\partial^2 E}{\partial m_i \partial m_j}\bigg|_{\mathbf{m}^{(p)}} \tag{11.13}$$

Here, $\mathbf{b}$ is a column vector of first derivatives and $\mathbf{B}$ is a symmetric matrix of second derivatives of the error $E(\mathbf{m})$ evaluated at the trial solution $\mathbf{m}^{(p)}$. These derivatives can be calculated either by analytically differentiating $E(\mathbf{m})$, if its functional form is known, or by approximating it with finite differences

$$\frac{\partial E}{\partial m_i}\bigg|_{\mathbf{m}^{(p)}} \approx \frac{1}{\Delta m}\left\{E\left(\mathbf{m}^{(p)} + \Delta \mathbf{m}^{(i)}\right) - E\left(\mathbf{m}^{(p)}\right)\right\}$$

$$\frac{\partial^2 E}{\partial m_i \partial m_j}\bigg|_{\mathbf{m}^{(p)}} = \frac{1}{4(\Delta m)^2} \times \begin{cases} 4\left\{E\left(\mathbf{m}^{(p)} + \Delta \mathbf{m}^{(i)}\right) - 2E\left(\mathbf{m}^{(p)}\right) + E\left(\mathbf{m}^{(p)} - \Delta \mathbf{m}^{(i)}\right)\right\} & (i=j) \\ \left\{\begin{array}{l} E\left(\mathbf{m}^{(p)} + \Delta \mathbf{m}^{(i)} + \Delta \mathbf{m}^{(j)}\right) - E\left(\mathbf{m}^{(p)} - \Delta \mathbf{m}^{(i)} + \Delta \mathbf{m}^{(j)}\right) \\ -E\left(\mathbf{m}^{(p)} + \Delta \mathbf{m}^{(i)} - \Delta \mathbf{m}^{(j)}\right) + E\left(\mathbf{m}^{(p)} - \Delta \mathbf{m}^{(i)} - \Delta \mathbf{m}^{(j)}\right) \end{array}\right\} & (i \neq j) \end{cases} \tag{11.14}$$

Here, $\Delta \mathbf{m}^{(i)}$ is a small increment Δm in the ith direction, that is, $\Delta \mathbf{m}^{(i)} = \Delta m[0, \cdots 0, 1, 0, \cdots, 0]^T$, where the ith element is unity and the rest are zero. Note that these approximations are computationally expensive, in the sense that E must be evaluated many times for each instance of $\mathbf{b}$ and $\mathbf{B}$.

We can now find the minimum by differentiating this approximate form of $E(\mathbf{m})$ with respect to m_q and setting the result to zero:

$$\frac{\partial E}{\partial m_q} = 0 = b_q + \sum_{j=1}^{M} B_{qj}\left(m_j^{(p)} - m_j\right) \text{ or } \mathbf{m} = -\mathbf{B}^{-1}\mathbf{b} + \mathbf{m}^{(p)} \tag{11.15}$$

In the case of uncorrelated Normally distributed data with uniform variance and the linear theory, $\mathbf{d}^{obs} = \mathbf{Gm}$, the error is $E(\mathbf{m}) = (\mathbf{d}^{obs} - \mathbf{Gm})^T(\mathbf{d}^{obs} - \mathbf{Gm})$, from whence we find that $\mathbf{b} = -2\mathbf{G}^T(\mathbf{d}^{obs} - \mathbf{Gm}^{(p)})$, $\mathbf{G} = 2\mathbf{G}^T\mathbf{G}$ and

$$\mathbf{m} = -\mathbf{B}^{-1}\mathbf{b} + \mathbf{m}^{(p)} = \left[\mathbf{G}^T\mathbf{G}\right]^{-1}\mathbf{G}^T\left(\mathbf{d}^{obs} - \mathbf{Gm}^{(p)}\right) + \mathbf{m}^{(p)} = \left[\mathbf{G}^T\mathbf{G}\right]^{-1}\mathbf{G}^T\mathbf{d}^{obs} \tag{11.16}$$

which is the familiar least-squares solution. In this case, the result is independent of the trial solution and is exact.

When the theory is nonlinear, the error is $E(\mathbf{m}) = [\mathbf{d}^{obs} - \mathbf{g}(\mathbf{m})]^T[\mathbf{d}^{obs} - \mathbf{g}(\mathbf{m})]$, from which we conclude

$$\mathbf{b} = 2\mathbf{G}^{(p)T}\left[\mathbf{d}^{obs} - \mathbf{g}\left(\mathbf{m}^{(p)}\right)\right] \text{ and } \mathbf{B} \approx 2\mathbf{G}^{(p)T}\mathbf{G}^{(p)} \text{ with } G_{ij}^{(p)} \equiv \frac{\partial g_i}{\partial m_j}\bigg|_{\mathbf{m}^{(p)}} \tag{11.17}$$

so

$$\mathbf{m} - \mathbf{m}^{(p)} \approx \left[\mathbf{G}^{(p)T}\mathbf{G}^{(p)}\right]^{-1}\mathbf{G}^{(p)T}\left[\mathbf{d}^{obs} - \mathbf{g}\left(\mathbf{m}^{(p)}\right)\right] \tag{11.18}$$

Note that we have omitted the term involving the gradient of $\mathbf{G}^{(p)}$ in the formula for $\mathbf{B}$. In a linear theory, $\mathbf{G}$ is constant and its gradient is zero. We assume that the theory is sufficiently close to linear that here, too, it is insignificant.

The form of Eq. (11.18) is very similar to simple least squares. The matrix $\mathbf{G}^{(p)}$, which is the gradient of the model $\mathbf{g}(\mathbf{m})$ at the trial solution, acts as a data kernel. It can be calculated analytically, if its functional form is known or approximated by finite differences (see Eq. 11.14). The generalized inverse $\mathbf{G}_{(p)}^{-g} \equiv [\mathbf{G}^{(p)T}\mathbf{G}^{(p)}]^{-1}\mathbf{G}^{(p)T}$ relates the deviation of the data $\Delta\mathbf{d}$ from what is predicted by the trial solution to the deviation of the solution $\Delta\mathbf{m}$ from the trial solution, that is

$$\Delta\mathbf{m} = \mathbf{G}_{(p)}^{-g}\Delta\mathbf{d} \text{ with } \Delta\mathbf{m} \equiv \mathbf{m} - \mathbf{m}^{(p)} \text{ and } \Delta\mathbf{d} \equiv \mathbf{d}^{obs} - \mathbf{g}\left(\mathbf{m}^{(p)}\right) \tag{11.19}$$

However, because it is based on a truncated Taylor series, the solution is only approximate; it yields a solution that is improved over the trial solution, but not the exact solution. However, it can be iterated to yield a succession of improvements

$$\Delta\mathbf{m}^{(p+1)} = \mathbf{G}_{(p)}^{-g}\Delta\mathbf{d}^{(p)} \text{ with } \Delta\mathbf{m}^{(p+1)} \equiv \mathbf{m}^{(p+1)} - \mathbf{m}^{(p)} \text{ and } \Delta\mathbf{d}^{(p)} \equiv \mathbf{d}^{obs} - \mathbf{g}\left(\mathbf{m}^{(p)}\right) \tag{11.20}$$

until the error declines to an acceptably low level (Fig. 11.6). Unfortunately, while convergence of $\mathbf{m}^{(p)}$ to the value that globally minimizes $E(\mathbf{m})$ is often very rapid, it is not guaranteed. One common problem is the solution converging to a local minimum instead of the global minimum (Fig. 11.7). A critical part of the algorithm is picking the initial trial solution $\mathbf{m}^{(1)}$. The method may not find the global minimum if the trial solution is too far from it.

In Section 11.8, we will show that approximations for the covariance and resolution matrices can be achieved simply by replacing $\mathbf{G}$ and $\mathbf{G}^{-g}$ with $\mathbf{G}^{(p)}$ and $\mathbf{G}_{(p)}^{-g}$, respectively, in the usual formula.

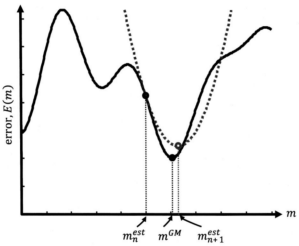

FIG. 11.6   The iterative method locates the global minimum m^{GM} of the error $E(m)$ (black curve) by determining the paraboloid (red curve) that is tangent to E at the trial solution m_n^{est}. The improved solution m_{n+1}^{est} is at the minimum (red circle) of this paraboloid and, under favorable conditions, can be closer to the solution corresponding to the global minimum than is the trial solution. Scripts gdama11_08 and gdapy11_08.

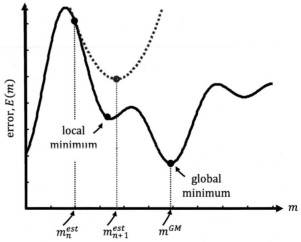

FIG. 11.7   If the trial solution m_n^{est} is too far from the global minimum m^{GM}, the method may converge to a local minimum. Scripts gdama11_09 and gdapy11_09.

As an example, we consider the exemplary problem from Section 11.4, in which $g_i(\mathbf{m}) = \sin(\omega_0 m_1 x_i) + m_1 m_2$. The matrix of partial derivatives is

$$\mathbf{G}^{(p)} = \begin{bmatrix} \omega_0 x_1 \cos\left(\omega_0 m_1^{(p)} x_1\right) + m_2^{(p)} & m_1^{(p)} \\ \omega_0 x_2 \cos\left(\omega_0 m_1^{(p)} x_2\right) + m_2^{(p)} & m_1^{(p)} \\ \vdots & \vdots \\ \omega_0 x_N \cos\left(\omega_0 m_1^{(p)} x_N\right) + m_2^{(p)} & m_1^{(p)} \end{bmatrix} \tag{11.21}$$

The main part of the following exemplary script is a loop that iteratively updates the trial solution. First, the deviation $\Delta\mathbf{d}$ (dd in the script) and the data kernel $\mathbf{G}^{(p)}$ are calculated using the trial solution $\mathbf{m}^{(p)}$. Then the deviation $\Delta\mathbf{m}$ is calculated using least squares. Finally, the trial solution is updated as $\mathbf{m}^{(p+1)} = \mathbf{m}^{(p)} + \Delta\mathbf{m}$.

```matlab
                              MATLAB®
% initial guess
mg=[1.1,0.95]';

Niter=10;
% iterate to improve initial guess
for k = (2:Niter)
    % predict data at current guess for solution
    dg = sin( w0*mg(1)*x) + mg(1)*mg(2);
    dd = dobs-dg;
    Eg=dd'*dd; % current error

    % build linearized data kernel
    G = zeros(N,2);
    G(:,1) = w0 * x .* cos( w0 * mg(1) * x ) + mg(2);
    G(:,2) = mg(2)*ones(N,1);

    % least squares solution for improvement to m
    dm = (G'*G)\(G'*dd);

    % update solution
    mg = mg+dm;
end
                                                    [gdama11_10]
```

```python
                               Python
# initial guess
mg=gda_cvec(1.1,0.95);
Niter=10;
# iterate to improve initial guess
for k in range(1,Niter):
    # predict data at current guess for solution
    dg = np.sin( w0*mg[0,0]*x) + mg[0,0]*mg[1,0];
    dd = dobs-dg;
    Eg=np.matmul(dd.T,dd); # current error

    # build linearized data kernel
    G = np.zeros((N,2));
    G[0:N,0:1]=w0*np.multiply(x,np.cos(w0*mg[0,0]*x))+mg[1,0];
    G[0:N,1:2]=mg[1,0]*np.ones((N,1));

    # least squares solution for improvement to m
    GTG = np.matmul(G.T,G);
    GTd = np.matmul(G.T,dd);
    dm = la.solve(GTG,GTd);

    # update solution
    mg = mg+dm;
                                                    [gdapy11_10]
```

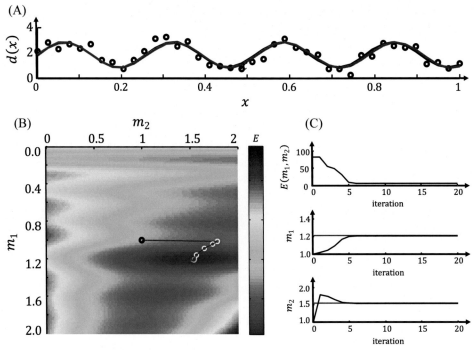

FIG. 11.8 Newton's method (linearized least squares) is used to solve the same nonlinear curve-fitting problem as in Fig. 11.5. (A) The observed data (black circles) are computed from the true data (black curve) by adding random noise. The predicted data (red curve) are based on the results of the method. (B) Error surface (colors), showing true solution (green dot), and a series of improved solutions (white circles connected by red lines) determined by the method. (C) Plot of error E and model parameters m_1 and m_2 as a function of iteration number. Scripts `gdama11_10` and `gdapy11_10`.

Results for the exemplary problem are shown in Fig. 11.8. This exemplary script iterates a fixed number of times, as specified by the variable `Niter`. A more elegant approach would be to perform a test that terminates the iterations when the error declines to an acceptable level or when the solution no longer changes significantly from iteration to iteration.

11.6   The implicit nonlinear inverse problem with Normally distributed data

We now generalize the iterative method of the previous section to the general case of the implicit theory $\mathbf{f}(\mathbf{d}, \mathbf{m}) = 0$, where $\mathbf{f}$ is of length $L \leq N + M$. We assume that the data $\mathbf{d}$ and prior model parameter $\langle \mathbf{m} \rangle$ are Normally distributed with covariance $[\text{cov } \mathbf{d}]$ and $[\text{cov } \mathbf{m}]_A$, respectively. If we let $\mathbf{x} = [\mathbf{d}, \mathbf{m}]^T$, we can think of the prior pdf of the data and model it as a cloud in the space $S(\mathbf{x})$ centered about the observed data and mean prior model parameters $\langle \mathbf{x} \rangle = [\mathbf{d}^{obs}, \langle \mathbf{m} \rangle]^T$, with a shape determined by the covariance matrix $[\text{covx}]$ (Fig. 11.9). This matrix contains $[\text{cov } \mathbf{d}]$ and $[\text{cov } \mathbf{m}]_A$ on its diagonal blocks, as in Eq. (6.26). In principle, the off-diagonal blocks could be made nonzero, indicating correlation between observed data and prior model parameters. However, specifying prior constraints is typically an ad hoc procedure for which one can seldom find motivation for introducing such a correlation.

The prior pdf is therefore

$$p_A(\mathbf{x}) \propto \exp\left\{ -\tfrac{1}{2}[\mathbf{x} - \langle \mathbf{x} \rangle]^T [\text{cov } \mathbf{x}]^{-1} [\mathbf{x} - \langle \mathbf{x} \rangle] \right\} \tag{11.22}$$

The theory $\mathbf{f}(\mathbf{x}) = 0$ defines a surface in $S(\mathbf{x})$ on which the predicted data and estimated model parameters $\mathbf{x}^{est} = [\mathbf{d}^{pre}, \mathbf{m}^{est}]^T$ must lie. Therefore the pdf for $\mathbf{x}^{est}$ is $p_A(\mathbf{x})$ evaluated on this surface (Fig. 11.10). When the surface is planar, this case is just the linear one described in Section 6.7 and the final pdf is Normal. On the other hand, if the surface is very "bumpy," the pdf on the surface will be very non-Normal and may be multimodal (Fig. 11.11).

One approach to estimating the solution is to find the maximum likelihood point of $p_A(\mathbf{x})$ on the surface $\mathbf{f}(\mathbf{x}) = 0$ (Fig. 11.10). This point can be found without explicitly determining the distribution on the surface. One just maximizes $p_A(\mathbf{x})$ with the constraint that $\mathbf{f}(\mathbf{x}) = 0$. One should keep in mind, however, that the maximum likelihood point of a non-Normal pdf may not be the most sensible estimate that can be made from it. Normal pdfs are symmetric, so their

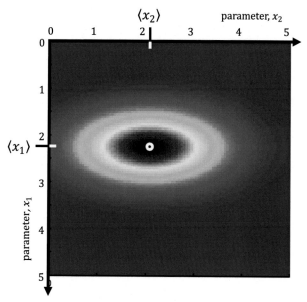

FIG. 11.9 The data and model parameters are grouped together in a vector $\mathbf{x}$. The prior information for $\mathbf{x}$ is then represented as a pdf (colors) in the $(M+N)$-dimensional space $S(\mathbf{x})$. Scripts gdama11_11 and gdapy11_11.

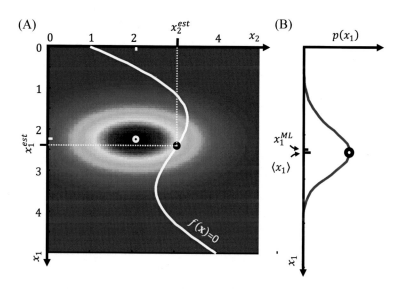

FIG. 11.10 (A) The estimated solution $\mathbf{x}^{est}$ (black circle) is at the point on the surface $\mathbf{f}(\mathbf{x})=0$ (white curve) where the prior pdf (colors) attains its largest value. (B) The probability density function $p(x_1)$ evaluated along the surface, as a function of position x_1. As the function is non-Normal, its mean $\langle x_1 \rangle$ may be distinct from its mode x_1^{ML} (although in this case they are similar). Scripts gdama11_12 and gdapy11_12.

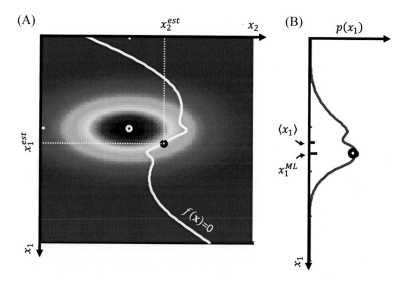

FIG. 11.11 (A) A highly nonlinear inverse problem corresponds to a complicated surface $\mathbf{f}(\mathbf{x})=0$ (white curve). (B) The probability density function $p(x_1)$ evaluated along the surface, as a function of position x_1. It may have several peaks, and as it is non-Normal, its mean $\langle x_1 \rangle$ may be distinct from its mode x_1^{ML}. Scripts gdama11_13 and gdapy11_13.

maximum likelihood point always coincides with their mean value. In contrast, the maximum likelihood point can be arbitrarily far from the mean in the non-Normal case (Fig. 11.11). Computing the mean, however, requires one to compute explicitly the distribution on the surface and then take its expectation (a much more difficult procedure).

These caveats aside, we proceed with the calculation of the maximum likelihood point by minimizing the argument of the exponential in $p_A(\mathbf{x})$ with the constraint that $\mathbf{f}(\mathbf{x})=0$ (adapted from Tarantola and Valette, 1982):

$$\text{minimize} \quad \Psi = [\mathbf{x} - \langle\mathbf{x}\rangle]^T[\text{cov } \mathbf{x}]^{-1}[\mathbf{x} - \langle\mathbf{x}\rangle] \quad \text{subject to} \quad \mathbf{f}(\mathbf{x}) = 0 \tag{11.23}$$

The Lagrange multiplier equations are obtained by minimizing $\Psi - 2\boldsymbol{\lambda}^T\mathbf{f}$ with respect to $\mathbf{x}$, where $-2\boldsymbol{\lambda}$ is a vector of Lagrange multipliers

$$\frac{\partial \Psi}{\partial x_i} - \sum_{j=1}^{L} 2\lambda_j \frac{\partial f_j}{\partial x_i} = 0 \quad \text{or} \quad [\text{cov } \mathbf{x}]^{-1}[\mathbf{x} - \langle\mathbf{x}\rangle] = \mathbf{F}^T\boldsymbol{\lambda} \tag{11.24}$$

Here, $\mathbf{F}$ is the matrix of partial derivatives $F_{ij} = \partial f_i/\partial x_j$. The Lagrange multipliers can be obtained by the following steps: moving the covariance to the right-hand side of the equation, multiplying by $\mathbf{F}$, and multiplying by $\{\mathbf{F}[\text{cov}\mathbf{x}]\mathbf{F}^T\}^{-1}$:

$$\boldsymbol{\lambda} = \left\{\mathbf{F}[\text{cov } \mathbf{x}]\mathbf{F}^T\right\}^{-1}\mathbf{F}[\mathbf{x} - \langle\mathbf{x}\rangle] \tag{11.25}$$

Substitution into Eq. (11.24) yields

$$[\mathbf{x} - \langle\mathbf{x}\rangle] = [\text{cov } \mathbf{x}]\mathbf{F}^T\left\{\mathbf{F}[\text{cov } \mathbf{x}]\mathbf{F}^T\right\}^{-1}\mathbf{F}[\mathbf{x} - \langle\mathbf{x}\rangle] \tag{11.26}$$

which must be solved simultaneously with the constraint equation $\mathbf{f}(\mathbf{x})=0$. These two equations are equivalent to the single equation

$$[\mathbf{x} - \langle\mathbf{x}\rangle] = [\text{cov } \mathbf{x}]\mathbf{F}^T\left\{\mathbf{F}[\text{cov } \mathbf{x}]\mathbf{F}^T\right\}^{-1}\left\{\mathbf{F}[\mathbf{x} - \langle\mathbf{x}\rangle] - \mathbf{f}(\mathbf{x})\right\} \tag{11.27}$$

as the original two equations can be recovered by premultiplying this equation by $\mathbf{F}$. The form of this equation is very similar to the linear solution of Eq. (6.35); in fact, it reduces to it in the case of the linear theory $\mathbf{f}(\mathbf{x})=\mathbf{X}\mathbf{x}$ (in which $\mathbf{F}=\mathbf{X}$ is a constant). As the unknown $\mathbf{x}$ appears on both sides of Eq. (11.27) and $\mathbf{f}(\mathbf{x})$ and $\mathbf{F}$ are functions of $\mathbf{x}$, this equation may be difficult to solve explicitly.

We now examine an iterative method of solving Eq. (11.27). This method consists of starting with some initial trial solution, say $\mathbf{x}^{(p)}$, where $p=1$, and then generating successive approximations as

$$\mathbf{x}^{(p+1)} = \langle\mathbf{x}\rangle + \mathbf{F}_{(p)}^{-g}\left\{\mathbf{F}^{(p)}\left[\mathbf{x}^{(p)} - \langle\mathbf{x}\rangle\right] - \mathbf{f}\left(\mathbf{x}^{(p)}\right)\right\}$$

$$\text{with} \quad \mathbf{F}_{(p)}^{-g} \equiv [\text{cov } \mathbf{x}]\mathbf{F}^{(p)T}\left\{\mathbf{F}^{(p)}[\text{cov } \mathbf{x}]\mathbf{F}^{(p)T}\right\}^{-1} \tag{11.28}$$

The superscript on $\mathbf{F}^{(p)}$ implies that it is evaluated at $\mathbf{x}^{(p)}$ and the quantity $\mathbf{F}_{(p)}^{-g}$ can be interpreted as a generalized inverse. If the initial guess is close enough to the maximum likelihood point, the successive approximations will converge to the true solution $\mathbf{x}^{est}$; else it may converge to a local minimum or even diverge.

11.7   The explicit nonlinear inverse problem with Normally distributed data

For the explicit theory $\mathbf{f}(\mathbf{d},\mathbf{m})=\mathbf{d}-\mathbf{g}(\mathbf{m})$ and $\mathbf{F}=[\mathbf{I} \quad -\mathbf{G}^{(p)}]$, where $G_{ij} \equiv \partial g_i/\partial m_j$ is the usual linearized data kernel. In this case, separate equations for $\mathbf{m}^{(p+1)}$ and $\mathbf{d}^{(p+1)}$ can be derived. We start by examining the generalized inverse $\mathbf{F}_{(p)}^{-g}$

$$\mathbf{F}_{(p)}^{-g} = \begin{bmatrix} [\text{cov } \mathbf{d}] & 0 \\ 0 & [\text{cov } \mathbf{m}]_A \end{bmatrix}\begin{bmatrix} \mathbf{I} \\ -\mathbf{G}^{(p)T} \end{bmatrix}\left\{\begin{bmatrix} \mathbf{I} & -\mathbf{G}^{(p)} \end{bmatrix}\begin{bmatrix} [\text{cov } \mathbf{d}] & 0 \\ 0 & [\text{cov } \mathbf{m}]_A \end{bmatrix}\begin{bmatrix} \mathbf{I} \\ -\mathbf{G}^{(p)T} \end{bmatrix}\right\}^{-1} =$$

$$= \begin{bmatrix} [\text{cov } \mathbf{d}]\mathbf{Q}^{-1} \\ [\text{cov } \mathbf{m}]_A\mathbf{G}^{(p)T}\mathbf{Q}^{-1} \end{bmatrix} \quad \text{where} \quad \mathbf{Q} \equiv \mathbf{G}^{(p)}[\text{cov } \mathbf{m}]_A\mathbf{G}^{(p)T} + [\text{cov } \mathbf{d}] \tag{11.29}$$

The following relationship holds between the two elements of $\mathbf{F}_{(p)}^{-g}$

$$\begin{aligned}
\left[\mathbf{F}_{(p)}^{-g}\right]_2 + \mathbf{G}^{(p)}\left[\mathbf{F}_{(p)}^{-g}\right]_1 &= [\text{cov }\mathbf{d}]\mathbf{Q}^{-1} + \mathbf{G}^{(p)}[\text{cov }\mathbf{m}]_A\mathbf{G}^{(p)T}\mathbf{Q}^{-1} \\
&= \left\{[\text{cov }\mathbf{d}] + \mathbf{G}^{(p)}[\text{cov }\mathbf{m}]_A\mathbf{G}^{(p)T}\right\}\mathbf{Q}^{-1} = \mathbf{Q}\mathbf{Q}^{-1} = \mathbf{I}
\end{aligned} \tag{11.30}$$

Consequently, the generalized inverse can be written as

$$\mathbf{F}_{(p)}^{-g} = \begin{bmatrix} \mathbf{I} - \mathbf{G}^{(p)}\mathbf{G}_{(p)}^{-g} \\ -\mathbf{G}_{(p)}^{-g} \end{bmatrix} \quad \text{where} \quad \mathbf{G}_{(p)}^{-g} \equiv [\text{cov }\mathbf{m}]_A\mathbf{G}^{(p)T}\mathbf{Q}^{-1} \tag{11.31}$$

Note that the term $\mathbf{G}^{(p)}\mathbf{G}_{(p)}^{-g}$ is a linearized data resolution matrix. We have encountered this form of the generalized inverse before, in Section 6.7 and Eq. (6.37), where it was shown to have the alternate form,

$$\mathbf{G}_{(p)}^{-g} = \mathbf{A}^{-1}[\text{cov }\mathbf{d}]^{-1}\mathbf{G}^{(p)T} \quad \text{where} \quad \mathbf{A} \equiv \mathbf{G}^{(p)T}[\text{cov }\mathbf{d}]^{-1}\mathbf{G}^{(p)} + [\text{cov }\mathbf{m}]_A^{-1} \tag{11.32}$$

The factor $\{\mathbf{F}^{(p)}[\mathbf{x}^{(p)} - \langle\mathbf{x}\rangle] - \mathbf{f}(\mathbf{x}^{(p)})\}$ in Eq. (11.28) can be written element-wise as

$$\begin{aligned}
\left\{\begin{bmatrix} \mathbf{I} & -\mathbf{G}^{(p)} \end{bmatrix}\left[\begin{bmatrix} \mathbf{d} \\ \mathbf{m} \end{bmatrix}^{(p)} - \begin{bmatrix} \mathbf{d}^{obs} \\ \langle\mathbf{m}\rangle \end{bmatrix}\right] - \left(\mathbf{d}^{(p)} - \mathbf{g}(\mathbf{m}^{(p)})\right)\right\} &= \\
\left(\mathbf{d}^{(p)} - \mathbf{G}^{(p)}\mathbf{m}^{(p)}\right) - \left(\mathbf{d}^{obs} - \mathbf{G}^{(p)}\langle\mathbf{m}\rangle\right) - \left(\mathbf{d}^{(p)} - \mathbf{g}(\mathbf{m}^{(p)})\right) &= \\
\mathbf{d}^{(p)} - \mathbf{G}^{(p)}\mathbf{m}^{(p)} - \mathbf{d}^{obs} + \mathbf{G}^{(p)}\langle\mathbf{m}\rangle - \mathbf{d}^{(p)} + \mathbf{g}(\mathbf{m}^{(p)}) &= \\
-\mathbf{d}^{obs} + \mathbf{g}(\mathbf{m}^{(p)}) + \mathbf{G}^{(p)}\left(\langle\mathbf{m}\rangle - \mathbf{m}^{(p)}\right)
\end{aligned} \tag{11.33}$$

Putting these pieces together, the "solution" part of Eq. (11.28) is

$$\mathbf{m}^{(p+1)} = \langle\mathbf{m}\rangle + \mathbf{G}_{(p)}^{-g}\left\{\mathbf{d}^{obs} - \mathbf{g}\left(\mathbf{m}^{(p)}\right) + \mathbf{G}^{(p)}\left(\mathbf{m}^{(p)} - \langle\mathbf{m}\rangle\right)\right\} \tag{11.34}$$

By Taylor's theorem, $\mathbf{g}(\mathbf{m}_2) \approx \mathbf{g}(\mathbf{m}_1) + \mathbf{G}^{(p)}(\mathbf{m}_2 - \mathbf{m}_1)$, where $\mathbf{m}_1$ and $\mathbf{m}_2$ are two solutions that are close to one another. Thus

$$\mathbf{g}\left(\mathbf{m}^{(p)}\right) + \mathbf{G}^{(p)}\left(\langle\mathbf{m}\rangle - \mathbf{m}^{(p)}\right) \approx \mathbf{g}(\langle\mathbf{m}\rangle) \tag{11.35}$$

and Eq. (11.34) becomes

$$\mathbf{m}^{(p+1)} = \langle\mathbf{m}\rangle + \mathbf{G}_{(p)}^{-g}\left\{\mathbf{d}^{obs} - \mathbf{g}(\langle\mathbf{m}\rangle)\right\} \tag{11.36}$$

By defining the deviation $\Delta\mathbf{m}^{(p+1)}$ of the updated solution with respect to the prior solution, and the deviation $\Delta\mathbf{d}$ of the observed data from that predicted by the prior solution, Eq. (11.36) takes the very simple form

$$\Delta\mathbf{m}^{(p+1)} = \mathbf{G}_{(p)}^{-g}\Delta\mathbf{d} \quad \text{with} \quad \Delta\mathbf{m}^{(p+1)} \equiv \mathbf{m}^{(p+1)} - \langle\mathbf{m}\rangle \quad \text{and} \quad \Delta\mathbf{d} = \mathbf{d}^{obs} - \mathbf{g}(\langle\mathbf{m}\rangle) \tag{11.37}$$

Superficially, this solution looks like the Newton method solution (Eq. 11.20), but the definitions of $\Delta\mathbf{m}$, $\Delta\mathbf{d}$, and $\mathbf{G}_{(p)}^{-g}$ differ between the two cases. One important difference is that deviations here are being calculated with respect to the prior solution, whereas they are being calculated with respect to the "current" solution in Newton's method. Another difference is that the generalized inverse here is GLS-like, in that it includes the effect of prior information, whereas in Newton's method it is simple least squares like.

The "current" solution $\mathbf{m}^{(p)}$ only influences the "updated" solution $\mathbf{m}^{(p+1)}$ through its effect on the generalized inverse. And it only influences the generalized inverse by specifying the point at which the partial derivative matrix $\mathbf{G}^{(p)}$ is to be evaluated.

The "predicted data" part of Eq. (11.28) is

$$\mathbf{d}^{(p+1)} = \mathbf{d}^{obs} + \left(\mathbf{I} - \mathbf{G}^{(p)}\mathbf{G}_{(p)}^{-g}\right)\left\{-\mathbf{d}^{obs} + \mathbf{g}\left(\mathbf{m}^{(p)}\right) + \mathbf{G}^{(p)}\left(\langle\mathbf{m}\rangle - \mathbf{m}^{(p)}\right)\right\} \tag{11.38}$$

It can be simplified by inserting the result for $\mathbf{m}^{(p+1)}$ and applying the Taylor series procedure:

$$
\begin{aligned}
\mathbf{d}^{(p+1)} &= \mathbf{d}^{obs} + \left\{ -\mathbf{d}^{obs} + \mathbf{g}\left(\mathbf{m}^{(p)}\right) + \mathbf{G}^{(p)}\left(\langle\mathbf{m}\rangle - \mathbf{m}^{(p)}\right) \right\} - \mathbf{G}^{(p)}\left(\mathbf{m}^{(p+1)} - \langle\mathbf{m}\rangle\right) = \\
&= \mathbf{g}\left(\mathbf{m}^{(p)}\right) + \left\{ \mathbf{G}^{(p)}\left(\langle\mathbf{m}\rangle - \mathbf{m}^{(p)}\right) \right\} - \mathbf{G}^{(p)}\left(\langle\mathbf{m}\rangle - \mathbf{m}^{(p+1)}\right) = \\
&= \mathbf{g}\left(\mathbf{m}^{(p)}\right) + \mathbf{G}^{(p)}\left(\mathbf{m}^{(p+1)} - \mathbf{m}^{(p)}\right) \approx \mathbf{g}\left(\mathbf{m}^{(p+1)}\right)
\end{aligned}
\tag{11.39}
$$

Thus the predicted data $\mathbf{d}^{(p+1)}$ are equal to the data predicted by the updated model $\mathbf{m}^{(p+1)}$.

11.8   Covariance and resolution in nonlinear problems

The iterative solution for the explicit problem $\mathbf{d}^{obs} = \mathbf{g}(\mathbf{m})$ supplemented with prior information $\mathbf{m} = \langle\mathbf{m}\rangle$ was derived in the previous section (see Eq. 11.36). A slightly rearranged version is

$$
\mathbf{m}^{(p+1)} = \mathbf{G}_{(p)}^{-g}\mathbf{d}^{obs} + \langle\mathbf{m}\rangle - \mathbf{G}_{(p)}^{-g}\mathbf{g}(\langle\mathbf{m}\rangle)
\tag{11.40}
$$

The only impediment to applying the standard rule of error propagation to this equation is our ignorance of the covariance of $\mathbf{g}(\langle\mathbf{m}\rangle)$. Nevertheless, random variation in $\langle\mathbf{m}\rangle$, which is quantified by $[\text{cov } \mathbf{m}]_A$, must propagate into random variation of $\mathbf{g}(\langle\mathbf{m}\rangle)$, so that the quantity must have a covariance, too. We approach the problem by approximating $\mathbf{g}(\langle\mathbf{m}\rangle)$ with the first two terms of its Taylor series. To first order, $\mathbf{g}(\langle\mathbf{m}\rangle) \approx \mathbf{g}(\mathbf{m}_0) + \mathbf{G}_0[\langle\mathbf{m}\rangle - \mathbf{m}_0]$, where $\mathbf{m}_0$ is a reference point close to $\langle\mathbf{m}\rangle$ and $\mathbf{G}_0$ is the gradient of $\mathbf{g}(\mathbf{m})$ evaluated at $\mathbf{m}_0$. Inserting into Eq. (11.40) and rearranging yields

$$
\mathbf{m}^{(p+1)} \approx \mathbf{G}_0^{-g}\mathbf{d}^{obs} + \left\{\mathbf{I} - \mathbf{G}_0^{-g}\mathbf{G}_0\right\}\langle\mathbf{m}\rangle - \mathbf{G}_0^{-g}\{\mathbf{g}(\mathbf{m}_0) - \mathbf{G}_0\mathbf{m}_0\}
\tag{11.41}
$$

As $\mathbf{m}_0$ is not a random variable, the third term on the right-hand side does not contribute to the covariance $\mathbf{m}^{(p+1)}$. Actually, we can set $\mathbf{m}_0$ to $\mathbf{m}^{(p)}$, as long as we understand that it is just a reference point, not a random variable. Then, standard error propagation gives

$$
\left[\text{cov } \mathbf{m}^{(p+1)}\right] \approx \mathbf{G}_{(p)}^{-g}[\text{cov } \mathbf{d}]\mathbf{G}_{(p)}^{-gT} + \left(\mathbf{I} - \mathbf{G}_{(p)}^{-g}\mathbf{G}^{(p)}\right)^T[\text{cov } \mathbf{m}]_A\left(\mathbf{I} - \mathbf{G}_{(p)}^{-g}\mathbf{G}^{(p)}\right)
\tag{11.42}
$$

Except for the index p, this is that same formula that was previously encountered in the context of the linear problem (see Eq. 6.40). Thus the covariance of a nonlinear problem approximately is the usual linear formula applied to the last iteration. The accuracy of the formula depends on the validity of the Taylor series approximation; it is most accurate when $\mathbf{g}(\mathbf{m})$ behaves linearly in the vicinity of both $\mathbf{m}^{(p)}$ and $\langle\mathbf{m}\rangle$, or, equivalently, when the shape of the generalized error surface is approximately a paraboloid.

Before working out formulas for resolution, we need to clarify just what is meant by the "resolution" of a nonlinear problem. We focus here on the model resolution of an underdetermined problem (that is, with $M > N$) by contrasting two very simple problems, one linear and one nonlinear, both with $N = 2$ and $M = 3$. The linear problem is

$$
d_1 = m_1 + m_2 \quad \text{and} \quad d_2 = m_2 + m_3
\tag{11.43}
$$

and the nonlinear problem is

$$
d_1 = \left[m_1^2 + m_2^2\right]^{\frac{1}{2}} \quad \text{and} \quad d_2 = \left[m_2^2 + m_3^2\right]^{\frac{1}{2}}
\tag{11.44}
$$

In the linear problem, each equation defines a plane in the space of model parameters (Fig. 11.12A) which intersect along a line, say $\overline{AB}$. The solution is nonunique and can lie anywhere along $\overline{AB}$. Now consider primed coordinates (m_1', m_2', m_3'), where $m_3' \propto [1, -1, 1]^T$ is parallel to $\overline{AB}$. The data only constrain the values of (m_1', m_2') and not m_3', although prior information may require that the solution be close to some point, say $\langle\mathbf{m}\rangle = [1, 1, 1]^T$ (Fig. 11.12B, yellow star). The direction of the m_3'-axis is independent of the choice of $\langle\mathbf{m}\rangle$. An average is only unique if it does not use the value of m_3', and such averages do not depend upon $\langle\mathbf{m}\rangle$. Thus, for instance, the localized averages defined by the rows of the resolution matrix

$$
\mathbf{R} = \begin{bmatrix} \frac{1}{2} & \frac{1}{2} & 0 \\ \frac{1}{4} & \frac{1}{2} & \frac{1}{4} \\ 0 & \frac{1}{2} & \frac{1}{2} \end{bmatrix}
\tag{11.45}
$$

are unique, because they all average $[1, -1, 1]^T$ to zero. Furthermore, they are somewhat localized.

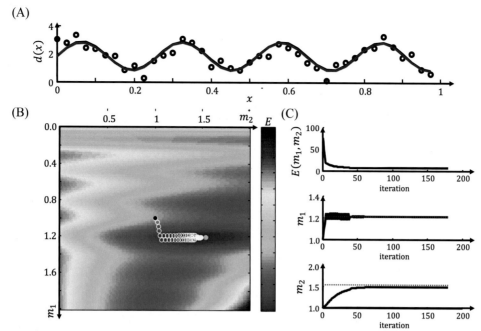

FIG. 11.12 The gradient method is used to solve the same nonlinear curve-fitting problem as in Fig. 11.5. (A) The observed data (black circles) are computed from the true data (black curve) by adding random noise. The predicted data (red curve) are based on the results of the method. (B) Error surface (colors), showing true solution (green dot), and a series of improved solutions (white circles) determined by the method. (C) Plot of error E and model parameters m_1 and m_2 as a function of iteration number. Scripts `gdama11_14` and `gdapy11_14`.

Now consider the nonlinear problem. Each equation defines a cylinder in the space of model parameters (Fig. 11.12C) that intersect along a curve, say $\widehat{AB}$. If we were to write $\mathbf{m} = \langle \mathbf{m} \rangle + \Delta \mathbf{m}$, then we can use Taylor's theorem to linearize the equations as

$$\Delta d_1 = \Delta m_1 + \Delta m_2 \quad \text{with} \quad \Delta d_1 \equiv \sqrt{2}d_1 - 2$$
$$\Delta d_2 = \Delta m_2 + \Delta m_3 \quad \text{with} \quad \Delta d_2 \equiv \sqrt{2}d_2 - 2 \tag{11.46}$$

Near $\langle \mathbf{m} \rangle$, the linearized equations have the same form as the linear ones. Consequently, the same resolution matrix $\mathbf{R}^{(p)} = \mathbf{R}$ defines unique averages of $\Delta \mathbf{m}$. However, because the direction of $\widehat{AB}$ changes along its length, these averages are unique only in the vicinity of the specified point $\langle \mathbf{m} \rangle = [1,1,1]^T$. Were we to choose another value for $\langle \mathbf{m} \rangle$, $\mathbf{R}^{(p)}$ would change accordingly.

This analysis brings out the important point that model resolution is *local* in nonlinear problems. It embodies averages that are unique only for small perturbations of the model parameters.

In a linear problem, the model resolution matrix is derived by inserting the data equation $\mathbf{d} = \mathbf{G}\mathbf{m}^{true}$ into solution $\mathbf{m}^{est} = \mathbf{G}^{-g}\mathbf{d}$ to obtain

$$\mathbf{m}^{est} = \mathbf{G}^{-g}\mathbf{G}\mathbf{m}^{true} = \mathbf{R}\mathbf{m}^{true} \quad \text{with} \quad \mathbf{R} \equiv \mathbf{G}^{-g}\mathbf{G} \tag{11.47}$$

When we try to do the same for a nonlinear problem $\mathbf{d} = \mathbf{g}(\mathbf{m}^{true})$, we can perform the first step $\mathbf{m}^{est} = \mathbf{G}_{(p)}^{-g}\mathbf{g}(\mathbf{m}^{true})$, but the result is not a matrix multiplying $\mathbf{m}^{true}$, so we cannot identify a resolution matrix. However, we can use Taylor's theorem to linearize $\mathbf{g}(\mathbf{m}^{true})$ about some reference value, say $\mathbf{m}^{(p)}$

$$\mathbf{g}\left(\mathbf{m}^{true}\right) \approx \mathbf{g}\left(\mathbf{m}^{(p)}\right) + \mathbf{G}^{(p)}\left[\mathbf{m}^{true} - \mathbf{m}^{(p)}\right] \tag{11.48}$$

from which it follows that

$$\Delta \mathbf{m}^{est} = \mathbf{R}^{(p)}\Delta \mathbf{m}^{true} \quad \text{with} \quad \mathbf{R}^{(p)} \equiv \mathbf{G}_{(p)}^{-g}\mathbf{G}^{(p)} \quad \text{and} \quad \Delta \mathbf{m}^{true} = \mathbf{m}^{true} - \mathbf{m}^{(p)}$$
$$\text{and} \quad \Delta \mathbf{m}^{est} = \mathbf{m}^{est} - \mathbf{m}^{est(p)} \quad \text{and} \quad \mathbf{m}^{est(p)} \equiv \mathbf{G}_{(p)}^{-g}\mathbf{g}\left(\mathbf{m}^{(p)}\right) \tag{11.49}$$

The linearized model resolution matrix $\mathbf{R}^{(p)}$ acts in a similar manner to the usual resolution matrix $\mathbf{R}$, except that it connects deviatoric quantities that represent small perturbations of the model parameters from reference values. Although limited by this proviso, the concept of resolution is still a useful one.

Our practice is to use the standard formulas for model and data resolution, after replacing $\mathbf{G}$ with $\mathbf{G}^{(p)}$ and $\mathbf{G}^{-g}$ with $\mathbf{G}_{(p)}^{-g}$. In problems with prior information, we use the forms of the resolution matrices that are described in Section 6.9.

11.9 Gradient-descent method

Occasionally, one encounters an inverse problem in which the error $E(\mathbf{m})$ and its gradient $[\nabla E]_i \equiv \partial E / \partial m_i$, are especially easy to calculate. It is possible to solve the inverse problem using this information alone, as the unit vector

$$\boldsymbol{\nu} \equiv -\left. \frac{\nabla E}{\lceil \nabla E \rceil} \right|_{\mathbf{m}^{(p)}} \tag{11.50}$$

points in the direction along which E becomes smaller. The current solution $\mathbf{m}^{(p)}$ can be improved to $\mathbf{m}^{(p+1)} = \mathbf{m}^{(p)} + \alpha \boldsymbol{\nu}$, where α is a positive number.

The only problem is that one does not immediately know an appropriate value for α. Too large and the minimum may be skipped over; too small and the convergence will be very slow. *Armijo's rule* provides an acceptance criterion for α:

$$E\left(\mathbf{m}^{(p+1)}\right) \leq E\left(\mathbf{m}^{(p)}\right) - c\alpha\boldsymbol{\nu}^T\nabla E \tag{11.51}$$

Here, c is an empirical constant in the range $(0,1)$ that is usually chosen to be about 10^{-4}. The rule says that a "good" α is one that leads not only to a decrease in error, but to a decrease that is in some sense "adequate"—but the details are complicated and we omit further discussion of them, here. One strategy for finding a "good" α is to start the iterative process with a "largish" value of α and to use it as long as it passes the test, but to decrease it whenever it fails, say using the rule α is replaced by $\alpha/2$. An example is shown in Fig. 11.13.

This *gradient-descent* method can be applied to problems with prior information simply by replacing the error $E(\mathbf{m})$ (and its gradient) with the generalized error $\Psi(\mathbf{m}) = E(\mathbf{m}) + L(\mathbf{m})$ (and its gradient). Here, $L(\mathbf{m})$ is the weighted error in prior information.

FIG. 11.13 Space of model parameters (m_1, m_2, m_3). (A) Depiction of the linear inverse problem with $N=2$ data, with each data equation graphing as a plane. The solution is nonunique and must lie on the line $\overline{AB}$ formed by the intersection of the two planes. (B) Enlargement of portion of (A) near the prior solution $\langle \mathbf{m} \rangle$ showing the transformed coordinate system (m_1', m_2', m_3') with m_3'-axis parallel to $\overline{AB}$. (C) Depiction of the nonlinear inverse problem with $N=2$ data, with each data equation graphing as a nonplanar surface. The solution is nonunique and must lie on the curve $\widehat{AB}$ formed by the intersection of the two surfaces.

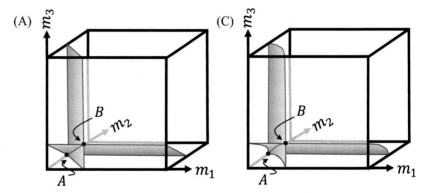

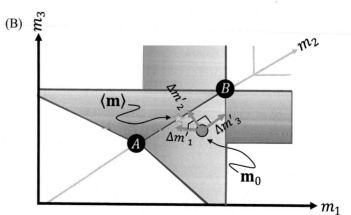

Both Newton's method and the gradient-descent method are iterative. In Newton's method, convergence is usually very rapid, so that only a few iterations are required. However, each iteration is computationally expensive, for $N \times M$ derivatives must be calculated to obtain the linearized data kernel $\mathbf{G}^{(p)}$ and then an $M \times M$ least-squares matrix equation must be solved. In the gradient-descent method, the number of required iterations can be large. However, each iteration is computationally inexpensive, for only M derivatives are calculated and no matrix equation needs to be solved. Our experience is that Newton's method is often the best choice for "small" inverse problems (say $M < 1000$), but that gradient descent is better for larger ones. Theoretically, gradient descent is the faster method in the limit $M \rightarrow \infty$.

A limitation of the gradient-descent method is that it provides no direct method for estimating the covariance of the solution. However, once the solution is found, the second derivative of error in its vicinity can be estimated using finite differences, as in Eq. (11.14), and the covariance inferred from the derivative using Eq. (4.76).

11.10 Choosing the null distribution for inexact non-Normal nonlinear theories

As we found in Section 6.5, the starting place for analyzing inexact theories is the rule for combining probability density functions to yield the total pdf p_T. It is built up by combining three pdfs p_T (prior), p_f or p_g (theory), and p_N (null).

$$p_T(\mathbf{d}, \mathbf{m}) = \frac{p_A(\mathbf{d}) p_A(\mathbf{m}) p_g(\mathbf{d}, \mathbf{m})}{p_N(\mathbf{d}) p_N(\mathbf{m})} \quad \text{or} \quad p_T(\mathbf{x}) = \frac{p_A(\mathbf{x}) p_g(\mathbf{x})}{p_N(\mathbf{x})} \tag{11.52}$$

In the Normal case, we have been assuming $p_N \propto$ constant. However, this definition is sometimes inadequate. For instance, if $\mathbf{x}$ is the Cartesian coordinate of an object in three-dimensional space, then $p_N \propto$ constant means that the object could be anywhere with equal probability. This is an adequate definition of the null distribution. On the other hand, if the position of the object is specified by the spherical coordinates $\mathbf{x} = [r, \theta, \varphi]^T$, then the statement $p_N \propto$ constant actually implies that the object is near the origin. The statement that the object could be anywhere is $p_N \propto r^2 \sin \theta$. The null distribution must be chosen with the physical significance of the vector $\mathbf{x}$ in mind.

Unfortunately, it is sometimes difficult to find a guiding principle with which to choose the null distribution. Consider the case of an acoustics problem in which a model parameter is the acoustic velocity v. At first sight it may seem that a reasonable choice for the null distribution is $p_N \propto$ constant. Acousticians, however, often work with the acoustic slowness $s = 1/v$. According to the transformation rule for pdfs (Eq. 3.13), the pdf $p_N(v) \propto$ constant implies $p_N(s) \propto s^2$. This is somewhat unsatisfactory, as one could, with equal plausibility, argue that $p_N(s) \propto$ constant, in which case $p_N(v) \propto v^2$. One possible solution to this dilemma is to choose a null pdf whose *form* is invariant under the reparameterization. The pdf that works in this case is $p_N(v) \propto 1/v$, as this implies $p_N(s) \propto 1/s$.

Thus, while a non-Normal inexact implicit theory can be handled with the same machinery as was applied to the Normal case of Section 11.7, more care must be taken when choosing the parameterization and defining the null distribution.

11.11 The genetic algorithm

Evolution is a natural process that drives a population of living organisms to the optimal condition of biological fitness; that is, evolving life is continuously solving the inverse problem of survival. Mathematical analogs of biological evolution can be harnessed to solve other inverse problems, too. Called *genetic algorithms*, they evolve a population of candidate solutions toward a best-estimate solution.

Biological evolution operates on a population of organisms that is composed of a large number of individuals and that reproduces for a large number of generations. The inheritable attributes of an individual are specified by its genome, which is composed of sequence of genes; most typically, a given attribute is influenced by just a few genes. The genome of a child differs from that of its parent (or parents), because of mutation, which randomly modifies the genome as it is copied from parent to child, and sexual reproduction, which causes a child to obtain some of its genes from each of two parents. The process of natural selection leads to the preferential survival of individuals who, by virtue of their particular genes, are better fitted to their environment. Selection leads to the average fitness of a population of organisms increasing from generation to generation.

In the genetic algorithm, the analog to the genome of an individual k is a model vector $\mathbf{m}^{(k)}$; the analog to a population of size K is a set of the model vectors $\mathbf{m}^{(1)}, \mathbf{m}^{(2)}, \cdots, \mathbf{m}^{(K)}$; and the analog to fitness is a function $F(E_k)$ that declines with error $E_k \equiv E(\mathbf{m}^{(k)})$. The genomes of individuals within the population are initialized to a set of trial values, and the population is allowed to evolve for a large number L of generations. The best estimate of the solution is the one, in the final generation, with the highest fitness (and lowest error).

A key issue is how to encode the model vector $\mathbf{m}^{(k)}$ into a genome. One possibility is to concatenate the binary representation of each element of the vector to form a single bit string $\mathbf{s}^{(k)}$ (where each element of $\mathbf{s}$ is either 0 or 1). The binary floating point representation of an element m_i could be used in the genome, but then mutations of the exponent part of the representation would lead to very large changes in value, which is undesirable. Usually better is representing m_i by a nonnegative integer p_i with a fixed number of bits (e.g., 4 bits, which can represent the numbers $0-15$; or 12 bits, which can represent the numbers $0-4095$; or 16 bits, which can represent the numbers $0-65{,}535$) and then linearly scaling it to the expected range of the model parameter, that is, $m_i = a_i + b_i p_i$, where a_i and b_i are prescribed constants. This leads to a bit string $\mathbf{s}$ that is the concatenation of M sets of bits, with none of the bits representing exponents. However, a disadvantage is that the model parameters are always within a fixed range and with "quantized" values.

In the following example, the 16-bit genome represents four 4-bit integer model parameters, each in the range $0-15$:

	$\mathbf{m}^{(k)}$	$\mathbf{s}^{(k)}$
Individual, k	$[2,6,5,4]^T$	0010011001010100

A mutation is accomplished by randomly choosing a bit—say bit 10—highlighted in bold in the following example—and flipping its value as the genome is copied from parent to child:

	$\mathbf{m}^{(k)}$	$\mathbf{s}^{(k)}$
Parent	$[2,6,5,4]^T$	0010011001010100
Child	$[2,6,1,4]^T$	0010011000010100

Sexual reproduction, or rather the simplified version of it called *recombination* (or *crossover*), is accomplished in the following steps: (1) two parents, called i and j in the following example, are randomly selected; (2) their genomes are cut into two pieces at the same randomly chosen point (between bits 6 and 7 in the following example, with one piece highlighted in bold); and (3) the pieces are swapped as the genomes are copied from the two parents to their two children:

	$\mathbf{m}^{(k)}$	$\mathbf{s}^{(k)}$		$\mathbf{m}^{(k)}$	$\mathbf{s}^{(k)}$
Parent, i	$[2,6,5,4]^T$	**001001**1001010100	parent, j	$[1,3,0,15]^T$	**000100**1100001111
Child 1	$[2,7,0,15]^T$	**001001**1100001111	child 2	$[1,2,5,4]^T$	**000100**1001010100

Note that the point at which the genomes were cut is not a multiple of 4 bits. Consequently, one pair of integer values is altered (an effect similar to mutation).

Mutation and crossover have somewhat different effects on a genome. Most single-bit mutations change an integer by only a small amount (Fig. 11.14). They allow the population to undergo, from generation to generation, a directed random walk toward greater fitness—a process akin to the one used in simulated annealing. Such a process will rapidly hone in on the nearest minima in error (though it may be only a local minima). Some single-bit mutations change an integer by a large amount. They allow individuals to explore wildly different regions of the space of model parameters and hence provide a way—when those individuals are especially fit—for the population to jump out of a local minimum into the global one. Except in very homogeneous populations, recombination can also lead to a large change in the genome and hence provides another mechanism for the population to jump out of a local minimum. Additionally, in cases where the inverse problem is approximately separable into two or more approximately independent subproblems, they provide an effective means of bringing together fit subsolutions into a single, extremely fit overall solution (discussed further later).

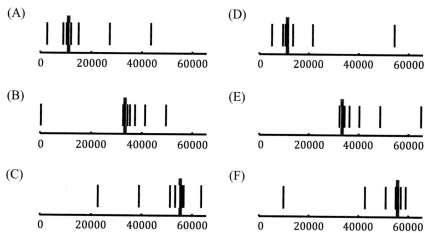

FIG. 11.14 Effect on one-bit mutations of the 16-bit standard binary code for the numbers (A) 11,111, (B) 33,333, and (C) 55,555. (C)–(E) Same as (A)–(C), except for the 16-bit Gray code. In each case, the original number is shown in red and the mutations in black. Scripts `gdama11_15` and `gdapy11_15`.

TABLE 11.1 Standard and Gray binary codes for integers in the $0-15$ range.

Decimal	Binary	Gray	Decimal	Binary	Gray
0	0000	0000	8	1000	1100
1	0001	0001	9	1001	1101
2	0010	0011	10	1010	1111
3	0011	0010	11	1011	1110
4	0100	0110	12	1100	1010
5	0101	0111	13	1101	1011
6	0110	0101	14	1110	1001
7	0111	0100	15	1111	1000

One idiosyncrasy of the standard binary representation of integers (Table 11.1) is that some adjacent integers are separated by only one mutation, but others are separated by many mutations. The integers 2 and 3 (0010 and 0011) differ by only one bit so that a single mutation can change a 2 to a 3, and vice versa. On the other hand, 7 and 8 (0111 and 1000) differ by four bits, so four mutations are required to change a 7 to an 8, and vice versa. Thus individuals of one generation may be very close to the exact solution, and yet the probability may be low that an individual in the next generation will mutate to it—an undesirable characteristic called a *Hamming wall*. It can be avoided by using *Gray coding* (Table 11.1), a nonstandard binary representation in which all adjacent integers differ by precisely one bit. In the earlier example, the Gray codes for 2 and 3 are 0011 and 0010, respectively, and the codes for 7 and 8 are 1000 and 1001, respectively. Both pairs differ by exactly one bit.

Choices encountered when implementing a genetic algorithm include the size of the population and whether or not its size varies between generations, the values to which the set of genomes are initialized, the rates of mutation and conjugation, the formula used to assess fitness, and the criteria used to decide when an acceptable best-estimate solution has been reached. Unfortunately, little theoretical guidance is available for making these choices; they usually need to be made by trial and error.

In our exemplary implementation of the nonlinear curve-fitting problem (as in Fig. 11.5), we use 16-bit Gray codes for each of the $M=2$ model parameters and scale their $0-65{,}535$ range onto the $(0,2)$ interval. We implement two functions `s=int2gray(i)` and `i=gray2int(s)` that translate between an integer `i` and a character string `s`. The string is of length-16 and represents the Gray code as a sequence of "0" and "1" characters. The use of character strings, as contrasted to bit strings, leads to some computational inefficiency, but greatly simplifies the manipulation and display of the codes.

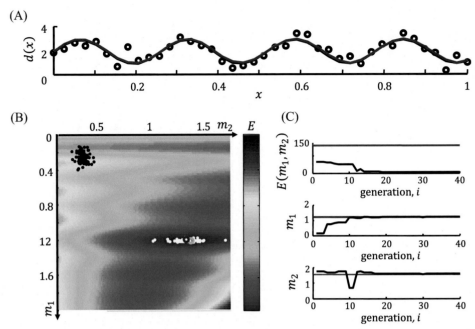

FIG. 11.15 The genetic algorithm is used to solve the same nonlinear curve-fitting problem as in Fig. 11.5. (A) The observed data (black circles) are computed from the true data (black curve) by adding random noise. The predicted data (red curve) are based on the results of the method. (B) Error surface (colors), showing true solution (green circle), the initial population of solutions (black dots), the population of solutions after 100 generations, and the best-estimate solution (red circle). (C) Plot of error E and model parameters m_1 and m_2 as a function of generation. Scripts gdama11_16 and gdapy11_16.

Both MATLAB® and Python versions need to be initialized by placing code at the beginning of the script.

MATLAB®
| ```
clear all;
global gray1 gray1index pow2;
load('gray1table.mat');
``` |
| [gdama11_16] |

Python
| ```
# setup for uing gray numbers
graydict=scipyio.loadmat("..\Data\gda_gray1table.mat");
gray1=graydict["gray1"];
gray1index=graydict["gray1index"];
graypow2=graydict["pow2"];
``` |
| [gdapy11_16] |

A population of $K=100$ candidate solutions (black dots in Fig. 11.15A) is initialized so that individuals scatter around $\mathbf{m}=[0.25,0.30]^T$, a point that is far from the true solution $\mathbf{m}^{true}=[1.21,1.54]^T$ (green circle in Fig. 11.15B). This population is then evolved for 100 generations.

New generations, all of size K, are formed by duplicating each individual of the previous generation to create a population of size $2K$, performing mutations and recombinations, determining each individual's fitness, and accepting only the K most fit individuals.

MATLAB®
``
Kold = K;
genes = [genes;genes];
K = 2*K;
```
                                                                [gdama11_16] |

| Python |
|---|
| ```
Kold = K;
genes = np.concatenate((genes,genes),axis=0);
K = 2*K;
```
                                                                [gdapy11_16] |

Single-bit mutations are performed at randomly chosen bit locations in half of individuals in each generation:

MATLAB®
``
k = unidrnd(16*M,1); % one random mutation
if(genes(i,k)=='0')
    genes(i,k)='1';
else
    genes(i,k)='0';
end
```
                                                                [gdama11_16] |

| Python |
|---|
| ```
k = np.random.randint(low=0,high=16*M);  # one random mutation
if( genes[i,k]=="0" ):
    genes[i,k]="1";
else:
    genes[i,k]="0";
```
                                                                [gdapy11_16] |

Here genes is a $K \times 16M$ character array, whose rows represent the genome of different individuals. A recombination is performed on one randomly selected pair of individuals in each generation:

```
j1 = unidrnd(K,1); % random individual
j2 = unidrnd(K,1); % another random individual
k = unidrnd(M*16-2,1)+1; % genome cut a random 2<=k<=16M-1
g1(1,1:16*M) = genes(j1,:);
g1(1,k+1:16*M) = genes(j2,k+1:16*M);
g2(1,1:16*M) = genes(j2,:);
g2(1,k+1:16*M) = genes(j1,k+1:16*M);
genes(j1,:) = g1;
genes(j2,:) = g2;
```
                                                                    [gdama11_16]

Python

```
j1 = np.random.randint(low=0,high=K); # random individual
j2 = np.random.randint(low=0,high=K); # another individual
k = np.random.randint(low=2,high=M*16-2); # split location
# be sure to  copy!
g1c = np.copy(genes[j1,0:M*16]);
g2c = np.copy(genes[j2,0:M*16]);
g1c[k:M*16] = genes[j2,k:M*16];
g2c[k:M*16] = genes[j1,k:M*16];
genes[j1,0:16*M]=g1c;
genes[j2,0:16*M]=g2c;
```
                                                                    [gdapy11_16]

The fitness F_k of individual k is assessed as:

$$F_k = r_k \exp\{-cE_k/E_{max}\} \text{ with } c = 5 \tag{11.53}$$

Here, E_{max} is the maximum error of any individual in that generation and r_k is a uniformly distributed random number on the interval $(0,1)$. The exponential implies that, on average, fitness declines rapidly with error. The multiplicative random number implies that some less-fit organisms survive—a strategy that prevents the population from becoming too homogeneous (and hence getting stuck in a local minimum).

The K individuals with the largest fitness survive into the next generation. They are identified by sorting the fitness vector, so that the column vector k gives their indices in increasing order of fitness. Only the individuals associated with the last half of k survive.

MATLAB®

```
[F_sorted, k] = sort(F);
k = k(Kold+1:2*Kold,1);
genes = genes(k,1:16*M);
K = Kold;
```
                                                                    [gdama11_16]

Python

```
isort = np.flipud(np.argsort(F,axis=0)).ravel();
isort = isort[0:Kold];
genes = genes[isort,0:16*M];
K=Kold;
```
                                                                    [gdapy11_16]

In the exemplary inverse problem, about 20 generations are needed for the population to evolve to a state that is close to the true solution (Fig. 11.15C). Little variation in the lowest-error individual is observed thereafter. Individuals of the final generation (white dots in the figure) scatter about the elliptical region of low error, with the most fit individual (red circle in the figure) being very close to the true solution (green circle).

In biology, different parts of an organism's genome code for different cellular components, such as proteins, and the function of these components are to some degree independent, that is, they operate separately to good approximation. Hypothetically, when present individually, a mutation-improved protein involved in the detection of food may lead to a modest gain in an individual's overall fitness, as might also a mutation-improved protein involved in digestion. The gain might be much more substantial, however, when both mutated genes are present in the same individual (as the organism can then digest the extra food that it finds). Recombination provides an efficient mechanism for bringing together two favorable versions of genes.

Many inverse problems share this trait. Their model parameters can be separated into several groups, each of which affect one group of data and leave the rest approximately unaffected. Recombination can be especially effective in such cases, since it brings together several groups of model parameters, each of which fit their subset of data well, into a solution with low overall error. However, the model parameters must be ordered in such a way that the members of a group are contiguous on the genome; else the recombination will not tend to swap groups as a whole. (Alternatively, the recombination algorithm can be redesigned to swap groups even though they are not contiguous.)

As an example, consider an inverse problem in which each of $N=M=20$ data is the arithmetic mean of three adjacent model parameters:

$$d_i = \frac{m_{i-2} + m_{i-1} + m_i}{3} \quad \text{with} \quad m_i \equiv 0 \text{ if } i < 1 \tag{11.54}$$

This problem is separable, in the sense that the error for a group of data at the left-hand side of the $(0,1)$ interval depends mostly on the values of the model parameters at the left-hand side of the interval, and the data on the right-hand side depend mostly on the model parameters on the right-hand side. Fig. 11.16 shows the minimum error as a function of generation, in each of two sets of five runs of the genetic algorithm, all with a population of $K=100$ candidate solutions. The first set (black curves) is for a genetic algorithm that implements a 50% mutation rate per generation but omits recombination. The second set (red curves) is for a genetic algorithm that implements both mutation and recombination. It has a reduced mutation rate, compared with the mutation-only set, of 12.5%, and a 10% recombination rate. The tendency of recombination to create random variation by combining partial genes has been suppressed by cutting the genome only at 16-bit gene boundaries. The algorithm that includes recombination converges, on average, faster, even though it has a much lower mutation rate.

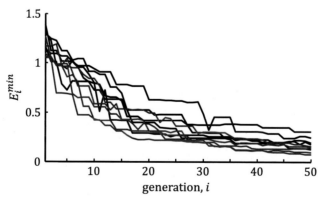

FIG. 11.16 Genetic algorithm applied to the approximately separable localized average problem discussed in the text. The minimum error E_i^{min} in a population of $K=100$ individuals declines with generation i. The rate of decline is faster for a combination of evolutionary parameters that includes recombination (red curves, 12.5% mutation rate, 5% recombination rate) than for one that omits it (black curves, 50% mutation, 0% recombination). Histories of five populations are shown for each parameter set. Scripts gdama11_17 and gdapy11_17.

11.12   Bootstrap confidence intervals

The pdf of the solution of a nonlinear problem is non-Normal, even when the data have Normally distributed error. The simple formulas that we developed for error propagation are not accurate in such cases, except perhaps when the nonlinearity is very weak. We describe here an alternative method of computing confidence intervals for the model parameters that performs the error propagation in an alternative way.

If many repeat data sets were available, the problem of estimating confidence intervals could be approached empirically. If we had repeated the experiment 1000 times, each time with the exact same experimental conditions, we would have a group of 1000 data sets, all similar to one another, but each containing a different pattern of observational noise.

We could then solve the inverse problem 1000 times, once for each repeat data set, make histograms of the resulting estimates of the model parameters and infer confidence intervals from them.

Repeat data sets are rarely available. However, it is possible to construct an approximate repeat data set by the random resampling with duplication of a single data set. The idea is to treat a set of N data as a pool of hypothetical observations and randomly draw N realized observations, which together constitute one repeat data set, from them. Even though the pool and the repeat data set are both of length N, they are not the same because the latter contains duplications. It can be shown that this process leads to a group of data sets that approximately have the probability density function of the original data.

As an example, suppose that a data set consists of N pairs of (x_i, d_i) observations, where x_i is an auxiliary variable. In a script, the resampling is performed as

MATLAB®

```
rowindex = unidrnd(N,N,1);
xresampled = x(rowindex);
dresampled = dobs(rowindex);
```
                                                                    [gdama11_18]

Python

```
rowindex = np.random.randint(low=0,high=N,size=(N,));
xresampled = x[rowindex,0:1];
dresampled = dobs[rowindex,0:1];
```
                                                                    [gdapy11_18]

Here, MATLAB's unidrnd() function and Python's np.random.randint() method return column vectors of N uniformly distributed integers. The inverse problem is then solved for many such repeat data sets, and the resulting estimates of the model parameters are saved. Confidence intervals are estimated by creating a histogram of each estimated model parameter m_i, normalizing it into an empirical pdf $p(m_i)$, integrating it to the cumulative distribution $P(m_i)$, and deriving confidence intervals from it.

<table>
<tr><th colspan="2">MATLAB®</th></tr>
</table>

```
Nbins=50;
m1hmin=min(m1save);
m1hmax=max(m1save);
Dm1bins = (m1hmax-m1hmin)/(Nbins-1);
m1bins=m1hmin+Dm1bins*[0:Nbins-1]';
m1hist = hist(m1save,m1bins);
pm1 = m1hist/(Dm1bins*sum(m1hist)); % empirical pdf p(m1)
Pm1 = Dm1bins*cumsum(pm1); % cumulative distribution
                                                  [gdama11_18]
```

<table>
<tr><th>Python</th></tr>
</table>

```
Lb1=50;
m1hmin=np.min(m1save);
m1hmax=np.max(m1save);
h1, e1 = np.histogram(m1save,Lb1,(m1hmin,m1hmax));  # create
histogram
Nb1 = len(h1);  # lengths of histogram
Ne1 = len(e1);  # length of edges
edges1 =  gda_cvec( e1[0:Ne1-1] );   # vector of edges
bins1 = edges1 + 0.5*(edges1[1,0]-edges1[0,0]); # centers
Db1 = bins1[1,0]-bins1[0,0];
pm1 = gda_cvec(h1)/(Db1*np.sum(h1)); # empirical pdf p(m1)
Pm1 = Db1*np.cumsum(pm1,axis=0); # cumulative distribution

                                                  [gdapy11_18]
```

Here, estimates of model parameter m_1 for all the repeat data sets have been saved in the column vector m1save. A histogram m1hist is computed from them using MATLAB's hist() function and PYTHON's np.histogram() method, after determining a reasonable range of m_1 values for its bins. The histogram is converted to an empirical pdf $p(m_1)$ by scaling it so that its integral is unity, and the MATLAB's cumsum() function and PYTHON's np.cumsum() method are used to integrate it to a cumulative distribution $P(m_1)$. MATLAB's find() function and Python's np.where() method are then used to determine the values of m_1 that enclose 95% of the area, defining 95% confidence limits for m_1. An example is shown in Fig. 11.17.

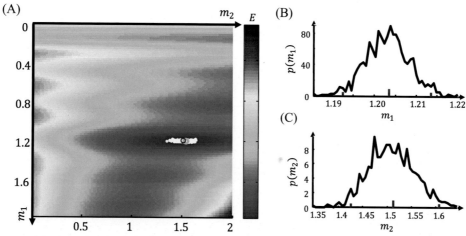

FIG. 11.17 Bootstrap confidence intervals for model parameters estimated using Newton's method for the same problem as in Fig. 11.5. (A) Error surface (colors), showing true solution (red circle), estimated solution (green circle), and bootstrap solutions (white dots). (B) Empirically derived pdf $p(m_1)$ with m_1^{est} (large red tick) and 95% confidence limits (small red ticks). (C) Same as (B), but for $p(m_2)$. Scripts gdama11_18 and gdapy11_18.

11.13　Problems

11.1 Use the principle of maximum likelihood to estimate the mean μ of N uncorrelated data, each of which is drawn from the same one-sided exponential probability density function $p(d_i) = \mu^{-1}\exp(-d_i/\mu)$ on the interval $(0, \infty)$. Suppose all the N data are equal to unity. What is the variance?

11.2 Experiment with the Newton's method example of Fig. 11.8, by changing the initial guess mg of the solution. Map out the region of the (m_1, m_2) plane for which the solution converges to the global minimum.

11.3 Solve the nonlinear inverse problem $d_i = m_1 z_i^{m_2}$, with z_i as an auxiliary variable on the interval $(1, 2)$ and with $\mathbf{m}^{true} = [3, 2]^T$, using both a linearizing transformation based on taking the logarithm of the equation and by Newton's method and compare the result. Use noisy synthetic data to test your scripts.

11.4 Suppose that the zs in the straight-line problem $d_i = m_1 + m_2 z_i$ are considered data, not auxiliary variables. (A) How should the inverse problem be classified? (B) Write a script that solves a test case using an appropriate method.

11.5 A column vector $\mathbf{d}$ is constructed by adding together scaled and shifted versions of vectors $\mathbf{a}$, $\mathbf{b}$, and $\mathbf{c}$. Noisy data $\mathbf{d}^{obs}$ of length N is observed. The vectors $\mathbf{a}$, $\mathbf{b}$, and $\mathbf{c}$, also of length N, are auxiliary variables. They are related by $d_i = Aa_{i+p} + Bb_{i+q} + Cc_{i+r}$, where A, B, and C are unknown constants and p, q, and r are unknown positive integers (assume $a_{i-p} = 0$ if $i + p > N$ and similarly for $\mathbf{b}$ and $\mathbf{c}$). This problem can be solved by using a three-dimensional grid search over p, q, and r only, as A, B, and C can be found using least squares once p, q, and r are specified. Write a script that solves a test case for $N = 50$. Assume that the error is $E = \mathbf{e}^T\mathbf{e}$ with $e_i = d_i^{obs} - (Aa_{i+p} + Bb_{i+q} + Cc_{i+r})$.

Reference

Tarantola, A., Valette, B., 1982. Generalized non-linear inverse problems solved using the least squares criterion. Rev. Geophys. Space Phys. 20, 219–232.

12

Monte Carlo methods

The term *Monte Carlo* is an allusion to Casino de Monte-Carlo (a gambling and entertainment complex in the Principality of Monaco, on the northern coast of the Mediterranean Sea) and refers to methods that employ randomness as part of their solution strategy. The term was coined by mathematician Nicholas Metropolis as a code word to disguise top secret research on the subject being done in the United States' nuclear weapons program during the 1940s.

12.1   The Monte Carlo search

The Monte Carlo search is a modification of the grid search (see Section 11.4), in which trial solutions are randomly generated, in contrast to being drawn from a regular grid in model space.

A typical script that implements the algorithm has two sections: the first section generates an initial solution $\mathbf{m}^{est}$ and its corresponding error; the second is a loop that randomly generates a trial solution $\mathbf{m}'$, calculates its corresponding weighted error E (or in problems with prior information, the generalized error $\Psi = E + L$, where L is the weighted error in prior information), and accepts it if the error is less than the previously accepted solution, that is, $\mathbf{m}^{est} = \mathbf{m}'$.

In its pure form, where each trial solution is generated independently of previous ones, the Monte Carlo search has only minor advantages over the grid search. In a later section, we will show that it can be improved by the introduction of correlation between successive trial solutions.

As in the grid search, bounds on the likely values of the solution need to be established, and the process of randomly generating trial solutions need to take them into account, to ensure that the process searches everywhere between the bounds. The Monte Carlo search has no analog to grid spacing, which limits the fidelity of an estimate determined by grid search. Given enough trials, the Monte Carlo estimate will become arbitrary close to the maximum likelihood point.

MATLAB®

```
for myiter = (2:Niter)
    % randomly select trial solution
    m1trial = random('Uniform',m1min,m1max);
    m2trial = random('Uniform',m2min,m2max);
    mtrial = [m1trial; m2trial];
    % predict data
    dtrial = sin(w0*m1trial*x) + m1trial*m2trial;
    % prediction error
    etrial = dobs-dtrial;
    Etrial = etrial'*etrial/sd2;
    % accept trial solution if error decreases
    if( Etrial<E0 )
        E0 = Etrial; % reset lowest error to this one
        m0 = mtrial; % reset best solution to this one
    end
end
```
[gdama12_01]

```
Python
for myiter in range(1,Niter):
    # randomly select trial solution
    m1trial = np.random.uniform(low=m1min,high=m1max);
    m2trial = np.random.uniform(low=m2min,high=m2max);
    mtrial = gda_cvec( m1trial, m2trial );
    # predict data
    dtrial = np.sin(w0*m1trial*x) + m1trial*m2trial;
    # prediction error
    etrial = dobs-dtrial;
    Etrial = np.matmul(etrial.T,etrial)/sd2;
    # accept trial solution if error decreases
    if( Etrial<E0 ):
        # reset best-solution so far to this one
        E0=Etrial;
        m0 = mtrial;
                                              [gdapy12_01]
```

FIG. 12.1   A Monte Carlo search is used to solve the simple nonlinear curve-fitting problem that was introduced in Chapter 11 (see Fig. 11.5). (A) The observed data (black circles) are computed from the true data (black curve) by adding random noise. The predicted data (red curve) are based on the results of the method. (B) Error surface (colors), showing true solution (green circle), and a series of improved solutions (white circles connected by red lines) determined by the method. (C) Plot of error E and model parameters m_1 and m_2 as a function of iteration number. Scripts gdama12_01 and gdapy12_01.

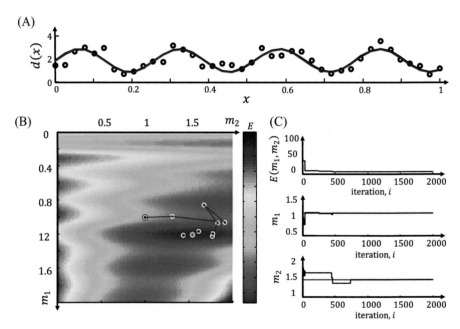

As an example, we apply the Monte Carlo search to the same simple nonlinear inverse problem that we have been studying previously. After a few thousand iterations of the method, the solution $\mathbf{m}^{est}$ is close to the point of minimum error (Fig. 12.1).

12.2   Simulated annealing

The Monte Carlo search (Section 12.1) is completely undirected. The whole model space is sampled randomly so that a point arbitrarily close to the global minimum of the error eventually is found, provided that enough trial solutions are examined. Unfortunately, the number of trial solutions that need to be computed may be very large—perhaps millions.

In contrast, Newton's method (Section 11.5) is completely directed. Local properties of the error—its slope and curvature—are used to determine the optimal improvement to a trial solution. Convergence to the global minimum of the error, when it occurs, is rapid and just a few trial solutions need be computed—perhaps just 10. Unfortunately, the method may only find a local minimum of the error that corresponds to an incorrect estimate of the model parameters.

The *simulated annealing* method combines the best features of these two methods. Like the Monte Carlo method, it samples the whole model space and so can avoid getting stuck in local minima. And like Newton's method, it uses local information to direct the sequence of trial solutions. Its development was inspired by the physical annealing of metals (Kirkpatrick et al., 1983), where an orderly minimum-energy crystal structure develops within the metal as it is slowly cooled from a red hot state. Initially, when the metal is hot, atomic motions are completely dominated by random thermal fluctuations, but as the temperature is slowly lowered, interatomic forces become more and more important. In the end, the atoms become a crystal lattice that represents a minimum-energy configuration.

In the simulated annealing algorithm, a parameter T is the analog to *temperature* and the error E is the analog to *energy*. Large values of T cause the algorithm to behave like a Monte Carlo search; small values cause it to act in a more directed way. As in physical annealing, one starts with a large T and then slowly decreases it as more and more trial solutions are examined. Initially, a very large volume of model space is randomly sampled, but the search becomes increasingly directed as it progresses.

The simulated annealing algorithm starts with an initial solution $\mathbf{m}^{(p)}$ with corresponding error $E(\mathbf{m}^{(p)})$, with $p=1$. A trial solution $\mathbf{m}'$ with corresponding error $E(\mathbf{m}')$ is then generated that is in the neighborhood of $\mathbf{m}^{(p)}$, say by adding to $\mathbf{m}^{(p)}$ an increment $\Delta\mathbf{m}$ drawn from a Normal pdf. The trial solution is always accepted as the solution $\mathbf{m}^{(p+1)}=\mathbf{m}'$ when $E(\mathbf{m}')\leq E(\mathbf{m}^{(p)})$, but it is *sometimes* accepted even when $E(\mathbf{m}')>E(\mathbf{m}^{(p)})$. To decide the latter case, a test parameter

$$\alpha = \frac{\exp\{-E(\mathbf{m}')/T\}}{\exp\{-E(\mathbf{m}^{(p)})/T\}} = \exp\left\{-\frac{E(\mathbf{m}') - E(\mathbf{m}^{(p)})}{T}\right\} \tag{12.1}$$

is computed, and the test solution is accepted with probability α. This can be accomplished by comparing α to a uniformly distributed random number r on the interval $(0,1)$.

An astute reader will realize that this procedure is just the Metropolis-Hastings algorithm applied to the pdf $p(\mathbf{m})\propto\exp\{-E(\mathbf{m})/T\}$. Furthermore, for $T=2$, it is the equal to the usual likelihood $p(\mathbf{m}|\mathbf{d}^{obs})\propto\exp\{-\frac{1}{2}E(\mathbf{m})\}$. The maximum likelihood point of $p(\mathbf{m})$ corresponds to the point of minimum error, regardless of the value of the parameter T. However, T controls the width of the minimum, with a larger T corresponding to a wider minimum. At first, T is high and the pdf is wide. The sequence of realizations samples a broad region of model space that includes the global minimum and, possibly, local minima as well. As T is decreased, the probability density function becomes increasingly peaked at the global minimum. In the limit of $T\rightarrow0$, all the realizations are at the maximum likelihood point, that is, the sought-after estimate $\mathbf{m}^{est}$ that minimizes the error.

In a script, the main part of the Metropolis-Hastings algorithm is a loop that computes the current value of T, randomly computes $\Delta\mathbf{m}$ to produce the trial solution $\mathbf{m}'$ and a corresponding error $E(\mathbf{m}')$, and applies the Metropolis-Hastings rules to either accept or reject $\mathbf{m}'$.

As an example, we apply the simulated annealing algorithm to the same simple nonlinear inverse problem that we have been studying previously. Initially, the algorithm visits local minima, but as the temperature falls, converges on the global minimum (Fig. 12.2). A few hundred iterations are needed to achieve a reasonably accurate estimate of solution $\mathbf{m}^{est}$, a number that is intermediate between the Monte Carlo search and Newton's method.

12.3   Advantages and disadvantages of ensemble solutions

Although the Monte Carlo search and Simulated Annealing utilize random processes, the objective of both is finding a single quantity—the estimated solution $\mathbf{m}^{est}$. An alternative solution strategy is to utilize randomness to *sample* the pdf $p(\mathbf{m})$ to produce an *ensemble* with a large number (say, millions) of members. Each member $\mathbf{m}^{(i)}$ of the ensemble is a possible solution, with the probability of that member being proportional to the percentage of that solution (and solutions close to it) in the ensemble.

The main disadvantage of an ensemble is that it can never fully capture the properties of $p(\mathbf{m})$. This is especially problematic when the number of model parameters is large. A million-member ensemble samples an $M=2$ model parameter space very well (as in Fig. 12.3), a $M=6$ model space sparsely, and a $M=20$ model space hardly at all. Furthermore, as two ensembles are different from one another due to statistical variation, different conclusions can be reached by examining them (especially when they undersample the space).

Ensemble solutions offer little advantage when the solution follows a simple, unimodal pdf, such as a Normal pdf. Although one can readily compute the mean and covariance of the ensemble (yielding $\mathbf{m}^{est}$ and the posterior [cov $\mathbf{m}$], respectively), these parameters usually can be estimated using faster methods, such as Newton's method.

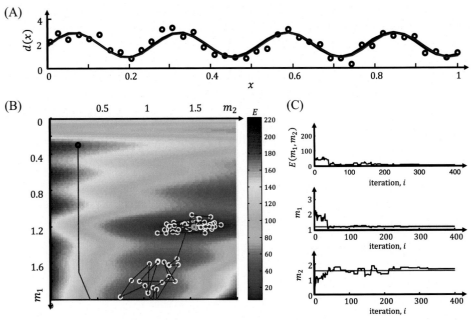

FIG. 12.2 Simulated annealing is used to solve the same nonlinear curve-fitting problem as in Fig. 12.1. (A) The observed data (black circles) are computed from the true data (black curve) by adding random noise. The predicted data (red curve) are based on the results of the method. (B) Error surface (colors), showing true solution (green circle), and a series of solutions (white circles connected by red lines) determined by the method. (C) Plot of error E and model parameters m_1 and m_2 as a function of iteration number. Scripts gdama12_02 and gdapy12_02.

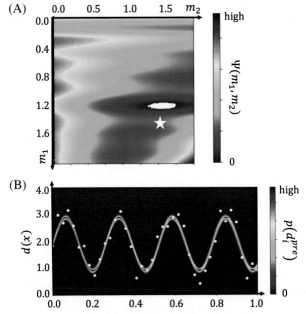

FIG. 12.3 Ensemble solution of the same nonlinear curve-fitting problem as in Fig. 12.1, modified through the introduction of weak prior information $\mathbf{m} = \langle \mathbf{m} \rangle$. (A) Generalized error surface $\Psi = E + L$ (colors), prior information (star), and million-member ensemble solution $\mathbf{m}^{(i)}$ (white dots) computed with the Metropolis-Hastings algorithm. (B) Observed data $\mathbf{d}^{obs}$ (white dots) superimposed on the empirical histogram $p(d_i^{pre})$ (colors) Scripts gdama12_03 and gdapy12_03.

Ensemble solutions offer advantages when the underlying pdf $p(\mathbf{m})$ is more complicated. For instance, a bimodal pdf corresponds to two radically different models, say $\mathbf{m}^{(1)}$ and $\mathbf{m}^{(2)}$, fitting the data to similar degree but having radically different properties. Given a sufficiently large ensemble, some members will scatter around $\mathbf{m}^{(1)}$ and some around $\mathbf{m}^{(2)}$. Once can scan through the ensemble and create an empirical pdf of an interesting property, which may turn out to be bimodal.

An example that we will return to subsequently concerns model parameters that represent a discretized continuous function, that is, $m_i = m(z_i)$, where z is depth. An interesting property, such as the depth z^{max} of the maximum, can easily be determined for a member $\mathbf{m}^{(i)}$, and the results for the entire ensemble binned to produce an empirical pdf $p(z^{max})$. Such a pdf would be extremely difficult, if not impossible, to compute by other methods.

12.4 The Metropolis-Hastings algorithm

The Metropolis-Hastings algorithm (Metropolis et al., 1953; Hastings, 1970) was introduced in Section 3.8; we discuss it here in greater detail. The algorithm generates a sequence $\mathbf{m}^{(1)}$, $\mathbf{m}^{(2)}$, $\mathbf{m}^{(3)}$ ⋯ of *states*, when in the limit of an indefinitely long sequence, samples the pdf, $p(\mathbf{m})$. The sequence is called a *Markov chain*, because the process of creating a member $\mathbf{m}^{(i)}$ depends only upon its immediate predecessor $\mathbf{m}^{(i-1)}$. Consequently, it is often referred to as a *Markov Chain Monte Carlo (MCMC)* method. Roughly speaking, an infinitely long chain winds its way through the entire space of $\mathbf{m}$s, visiting each point in the space indefinitely many times over. There will be instances in which two adjacent links in the chain go from $\mathbf{m}$ to $\mathbf{m}'$, and others where they go from $\mathbf{m}$ and $\mathbf{m}'$. If principle, one could count up the number of times the chain transitions from $\mathbf{m}$ and $\mathbf{m}'$ and the number of times it transitions from $\mathbf{m}'$ to $\mathbf{m}$. The *condition of detailed balance* asserts that these two numbers must be equal, that is $p(\mathbf{m}'|\mathbf{m})p(\mathbf{m}) = p(\mathbf{m}|\mathbf{m}')p(\mathbf{m}')$, where $p(\mathbf{m})$ and $p(\mathbf{m}')$ are probabilities of each of the states and $p(\mathbf{m}'|\mathbf{m})$ and $p(\mathbf{m}|\mathbf{m}')$ are the probabilities of transitioning from one state to another.

The Metropolis-Hastings algorithm views the transitional probabilities as the product of a proposed state q and an acceptance probability A, that is, $p(\mathbf{m}'|\mathbf{m}) = q(\mathbf{m}'|\mathbf{m})A(\mathbf{m}',\mathbf{m})$ and $p(\mathbf{m}|\mathbf{m}') = q(\mathbf{m}|\mathbf{m}')A(\mathbf{m},\mathbf{m}')$. Here, $q(\mathbf{m}'|\mathbf{m})$ is the probability that one *proposes* the state $\mathbf{m}'$, given that the current state is $\mathbf{m}$, and $A(\mathbf{m}',\mathbf{m})$ is the probability that one *accepts* the state $\mathbf{m}'$, given that the current state is $\mathbf{m}$. Similarly, $p(\mathbf{m}|\mathbf{m}')$ is the probability that one *proposes* the state $\mathbf{m}$, given that the current state is $\mathbf{m}'$, and $A(\mathbf{m},\mathbf{m}')$ is the probability that one *accepts* the state $\mathbf{m}$, given that the current state is $\mathbf{m}'$. Thus:

$$q(\mathbf{m}'|\mathbf{m})A(\mathbf{m}',\mathbf{m})p(\mathbf{m}) = q(\mathbf{m}|\mathbf{m}')A(\mathbf{m},\mathbf{m}')p(\mathbf{m}')$$
$$\frac{A(\mathbf{m}',\mathbf{m})}{A(\mathbf{m},\mathbf{m}')} = \alpha \quad \text{with} \quad \alpha \equiv \frac{p(\mathbf{m}')q(\mathbf{m}|\mathbf{m}')}{p(\mathbf{m})q(\mathbf{m}'|\mathbf{m})} \tag{12.2}$$

Here, the qs as considered known and As are constructed so that the equation holds. One possible choice is

$$A(\mathbf{m}',\mathbf{m}) = \min(1,\alpha) \quad \text{and} \quad A(\mathbf{m}|\mathbf{m}') = \min\left(1,\frac{1}{\alpha}\right) \tag{12.3}$$

since when $\alpha > 1$, $A(\mathbf{m}',\mathbf{m}) = 1$ and $A(\mathbf{m}|\mathbf{m}') = 1/\alpha$, in which case Eq. (12.2) is satisfied, and when $\alpha < 1$, $A(\mathbf{m}',\mathbf{m}) = \alpha$ and $A(\mathbf{m}|m') = 1$, in which case Eq. (12.2) again is satisfied.

In calculations, greater numerical stability is achieved by working with $\gamma = \ln(\alpha)$ as contrasted to α itself. Then, the acceptance tests become

$$A(\mathbf{m}',\mathbf{m}) = \begin{cases} 1 & (\gamma \geq 0) \\ \exp(\gamma) & (\text{otherwise}) \end{cases} \text{and} \; A(\mathbf{m}|\mathbf{m}') = \begin{cases} 1 & (\gamma \leq 0) \\ \exp(-\gamma) & (\text{otherwise}) \end{cases} \tag{12.4}$$

The process of creating an length-K ensemble of $\mathbf{m}$s that samples an arbitrary pdf $p(\mathbf{m})$ reduces to creating the Markov chain $\mathbf{m}^{(1)}$, $\mathbf{m}^{(2)}$, $\mathbf{m}^{(3)}$, ⋯, $\mathbf{m}^{(K)}$. The steps are as follows: (A) deciding upon the forms of $q(\mathbf{m}|\mathbf{m}')$ and $q(\mathbf{m}'|\mathbf{m})$; (B) setting $i=1$ and assigning an initial state $\mathbf{m}^{(i)}$; (C) sampling $q(\mathbf{m}^*|\mathbf{m}^{(i)})$ to create a proposed state $\mathbf{m}^*$; (D) computing the acceptance parameter α; (E) based on α, either accepting the proposed state $\mathbf{m}^*$ (in which case $\mathbf{m}^{(i+1)} = \mathbf{m}^*$) or rejecting it (in which case $\mathbf{m}^{(i+1)} = \mathbf{m}^{(i)}$), and incrementing i; and (F) repeating C-E many times, to generate the chain. The process is independent of the choice of $q(\mathbf{m}|\mathbf{m}')$, $q(\mathbf{m}'|\mathbf{m})$, and $\mathbf{m}^{(1)}$ in the limit of an indefinitely long chain. However, the choice is important for chains of practical length. The initial state $\mathbf{m}^{(1)}$ must be chosen so that $p(\mathbf{m}^{(1)})$ is nonnegligible, and $q(\mathbf{m}^*|\mathbf{m}^{(i)})$ must produce, with reasonably high probability, an $\mathbf{m}^*$ that is not so far from $\mathbf{m}^{(i)}$ that $p(\mathbf{m}^{(i)})$ is negligible, yet not so close that exploring the space of $\mathbf{m}$s takes a prohibitively long chain. A good practice is to discard the beginning portion of the chain (with the remainder said to be *burned in*), in order to reduce dependence on $\mathbf{m}^{(1)}$. An alternative (or complementary) strategy is to generate several chains with different starting $\mathbf{m}$s, and then concatenate them.

In order to apply the Metropolis-Hastings algorithm to an inverse problem with data and prior information, we invoke Bayes theorem, and write

$$p\left(\mathbf{m}|\mathbf{d}^{obs}\right) = \frac{p\left(\mathbf{d}^{obs}|\mathbf{m}\right)p_A(\mathbf{m})}{p\left(\mathbf{d}^{obs}\right)} \tag{12.5}$$

Here, $p_A(\mathbf{m})$ is the *prior*, which expressed prior expectations on the value of $\mathbf{m}$; $p(\mathbf{d}^{obs}|\mathbf{m})$ is the *likelihood*, which expresses the probability of the data given a particular value of the model; and $p(\mathbf{m}) \equiv p(\mathbf{m}|\mathbf{d}^{obs})$ is the *posterior*, that is, the solution. The factor of $p(\mathbf{d}^{obs})$ cancels from the Metropolis-Hastings formulas, so we will not discuss it further. The prior and likelihood are often assumed to be Normal pdfs, such as

$$p\left(\mathbf{d}^{obs}|\mathbf{m}\right) = (2\pi)^{-N/2}(\det[\text{cov } \mathbf{d}])^{-\frac{1}{2}}\exp(-\tfrac{1}{2}E(\mathbf{m}))$$
$$\text{with } E(\mathbf{m}) = \mathbf{e}^T[\text{cov } \mathbf{d}]^{-1}\mathbf{e} \text{ and } \mathbf{e} = \mathbf{d}^{obs} - \mathbf{g}(\mathbf{m}) \tag{12.6a}$$

$$p_A(\mathbf{m}) = (2\pi)^{-M/2}(\det[\text{cov } \mathbf{m}]_A)^{-\frac{1}{2}}\exp(-\tfrac{1}{2}L(\mathbf{m}))$$
$$\text{with } L(\mathbf{m}) = \mathbf{1}^T[\text{cov } \mathbf{m}]_A^{-1}\mathbf{1} \text{ and } \mathbf{1} = \langle\mathbf{m}\rangle - \mathbf{m} \tag{12.6b}$$

Here, E is the weighted prediction error (measured with respect to a prediction $\mathbf{g}(\mathbf{m})$) and L is the weighted error in prior information (measured with respect to prior model $\langle\mathbf{m}\rangle$). Note that the factor of $(2\pi)^{-N/2}(\det[\text{cov } \mathbf{d}])^{-\frac{1}{2}}$ cancels from the Metropolis-Hastings formula, as neither the number of data nor their prior covariance vary with $\mathbf{m}$.

One should always keep in mind that the Metropolis-Hastings algorithm merely *samples* the posterior pdf $p(\mathbf{m}|\mathbf{d}^{obs})$; it does not *create* it. The posterior is specified by the practitioner through the process of choosing the likelihood and prior pdfs (as in Eq. 12.6), and once they are decided upon, the properties of the posterior are fixed, as are the properties of ensembles drawn from it (up to random variation). Consequently, attention should be given to choosing these pdfs so that they embody the noise properties of the data (in the case of the likelihood) and the prior information (in the case of the prior). However, in a complicated problem with a nonlinear model and/or non-Normal pdfs, the consequences of those choices may be hard to anticipate and may only be discovered by examining the properties of the ensemble.

An ensemble solution, such as one created using the MCMC method, can be put to better use than the inferring of a single estimated solution $\mathbf{m}^{est}$ from it. Nevertheless, $\mathbf{m}^{est}$ and its posterior covariance $[\text{cov } \mathbf{m}]$ can be inferred from an ensemble solution, simply by taking the sample mean and sample covariance of the ensemble. As described in more detail in Section 5.14, one possible use of these quantities is to explore the trade-off of model resolution and covariance, by applying the Backus-Gilbert method to the equation $\mathbf{m} = \mathbf{m}^{est}$. Viewed as an inverse problem of the form $\mathbf{Gm} = \mathbf{d}^{obs}$, with $\mathbf{G} = \mathbf{I}$, $\mathbf{d}^{obs} = \mathbf{m}^{est}$, and $[\text{cov } \mathbf{d}] = [\text{cov } \mathbf{m}]$, this equation defines the unique averages associated with the model resolution matrix $\mathbf{R}(a=1) = \mathbf{I}$, with covariance $\mathbf{C}_m(a=1) \equiv [\text{cov } \mathbf{m}]$ of the point $a=1$ on the trade-off curve. Here, α controls the relative weighting of spread of model resolution and size of covariance (see Section 5.10). The idea is to use the Backus-Gilbert method to attempt to identify an alternative value of α that corresponds to localized averages that have poorer (but still acceptably well-localized) model resolution and much better variance.

12.5 Examples of ensemble solutions

In Example 1, we compute a million-member ensemble solution to the same simple nonlinear inverse problem that we have been studying previously. Synthetic data are created by perturbing the true data with random noise with zero mean and variance $\sigma_d^2 = (0.3)^2$. Unlike previous examples, we add prior information of proximity of the solution to a prior value $\mathbf{m}_A = [1.43, 1.50]^T$ with prior variance $\sigma_m^2 = (0.5)^2$ (as in Eq. 12.6b). Transitions between $\mathbf{m}$ and $\mathbf{m}'$ are computed using the Normal pdf

$$q(m_i|m_i') = q(m_i'|m_i) = \frac{1}{\sqrt{2\pi}\sigma_q}\exp\left\{-\frac{(m_i - m_i')^2}{2\sigma_q^2}\right\} \tag{12.7}$$

We use $\sigma_q^2 = (0.05)^2$, so that the current and proposed solutions differ by a few percent of the likely range of the solution.

A million-member ensemble (not shown) is computed using the Metropolis-Hastings algorithm. Its members form a tight cluster in the (m_1, m_2)-plane, in the deepest part of minimum of the generalized error $\Psi = E + L$, with a center close to the point estimated by previous methods (Fig. 12.3A). Predicted data d_i^{obs} for each of the member models is computed and binned to produce an empirical pdf $p(d_i^{obs})$ (Fig. 12.3B). The scatter of these predicted models is narrower than the observed data and is narrowest near the peaks of the data, indicating that the most well-constrained property of the data is the peak positions.

In Example 2, we compute an ensemble solution to a Laplace transform-like problem (see Section 5.3), in which model parameters m_j are related to data d_i via

$$d_i = \sum_{j=1}^{M} m_j \exp\{-c_i z_j\} \text{ with } z_j = \frac{(j-1)}{M-1} \tag{12.8}$$

Here, $M=11$ and the cs are known constants in the range 0 to 0.6. The true model is the sine wave $m^{true}(z) = \sin(2\pi z)$. Synthetic data d^{obs} are created by adding random noise with variance σ_d^2 to the true data d^{true} (obtained by evaluating Eq. 12.8 with $m = m^{true}$). The prior $p_A(\mathbf{m})$ is chosen to be Normal, uncorrelated, with zero mean and uniform variance $\sigma_m^2 = (5)^2$ corresponding to weak information that the solution is most likely in the range ± 5. The pdfs $q(\mathbf{m}|\mathbf{m}')$ and $q(\mathbf{m}|\mathbf{m}')$ are chosen to be Normal, uncorrelated, with variance σ_q^2 (as in Eq. 12.7). We use $\sigma_q^2 = (0.05)^2$, so that the current and proposed solutions differ by a few percent of the likely range of the solution.

A histogram of the ensemble (Fig. 12.4) demonstrates that only the shallower (left hand) part of the solution is accurately recovered. As is typical of Laplace transform-type problems, the data have little sensitivity to the deeper part of the model, so the deeper part is strongly influenced mostly by the prior information (even though it is weak).

One could produce an estimated solution $\mathbf{m}^{est}$ by averaging the ensemble, but this would seem to defeat the purpose of generating the ensemble in the first place (as an estimate more easily could have been created using least squares). More in the spirit of ensembles is assessing the probability that the solution has some interesting property. For instance, the probability that the model has a peak near $z=0.2$ can be assessed by determining the percentage of models in the ensemble for which m_2 is the largest model parameter (63% in this case).

12.6 Trans-dimensional models

Suppose that you see something whiz by in the air above while taking a walk in a park. You're not sure what it was. Possibly, it was a ball. Or possibly, it was a bird. If it was a ball, then one model parameter suffices to describe it (say, its diameter m_1). Bur if it was a bird, then several are needed (say, its length m_1' and wingspan m_2').

The answer to the question, "What whizzed by?", would then be something like: there's a 25% probability that it was a ball, and if it was a ball, then its diameter is likely to be $m_1 = 10 \pm 2$ cm, and there's a 75% probability that it was a bird, and if it was a bird, then its length is likely to be $m_1' = 6 \pm 1$ cm and it wingspan is likely to be $m_2' = 21 \pm 2$ cm.

Welcome to the world of trans-dimensional models!

The idea is that the overall model is an amalgamation of D distinct model types, indexed by i, each having its own set of M_i model parameters. A full specification of the model is the probability of each type and the probabilities of the model parameters for that type, that is, $p(D, \mathbf{m})$.

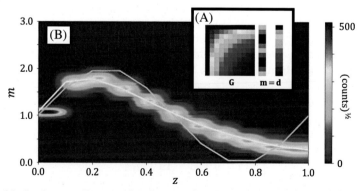

FIG. 12.4 Ensemble solution of the Laplace transform problem for model function $m(z)$ with depth z as introduced in Section 5.3. (A) Depiction of the equation $\mathbf{Gm} = \mathbf{d}^{obs}$ that is solved along with prior information $\mathbf{m} = \mathbf{m}_A = [0, \cdots, 0]^T$. (B) Histogram (colors) of a one-million-member ensemble solutions $\mathbf{m}^{(i)}(z)$ together with true solution (yellow curve) and ensemble-mean estimated solution (white curve) About 63% of the members have a peak in the vicinity of $z=0.2$. Scripts gdama12_04 and gdapy12_04.

In principle, the model types can be radically different, as in the ball-and-bird example mentioned before. However, in many cases, trans-dimensionality is used to implement a hierarchy of models that are all more or less the same, except for increasing spatial fidelity. For instance, a one-dimensional function $v(z)$ (where z is depth) can be considered to be *layered* (piecewise constant), with the model parameter m_i corresponding to the value of the field in a specific depth range (e.g., Bodin et al. 2012, 2014). Models with just a few layers are spatially coarse, and those with many are spatially fine. The trans-dimensional model simultaneously determines the probability of the number of layers and the probabilities of their values (and, possibly, their thicknesses, too).

The Metropolis-Hastings algorithm needs to be extended to the case where two states $\mathbf{m}$ and $\mathbf{m}'$ have different numbers, say M and M', of model parameters. Green and Hastie (2009) advocate that this be done by defining *slack* variables $\mathbf{u}$ and $\mathbf{u}'$ of dimensions, say P and P', so that $\mathbf{x} = [\mathbf{m}; \mathbf{u}]$ and $\mathbf{x}' = [\mathbf{m}'; \mathbf{u}']$ have the same dimension, that is, $M' + P' = M + P$. Furthermore, they advocate that the condition of detailed balance be derived by considering a *transformation of variables* that takes $\mathbf{x}$ into $\mathbf{x}'$. The usual rule for transforming probabilities (Eq. 3.15) is

$$p(\mathbf{m})g(\mathbf{u})A(\mathbf{m}',\mathbf{m}) = p(\mathbf{m}')g'(\mathbf{u}')\,A(\mathbf{m},\mathbf{m}')\,J \ \text{ where } \ J = \left|\det\frac{\partial(\mathbf{m}',\mathbf{u}')}{\partial(\mathbf{m},\mathbf{u})}\right| \tag{12.9}$$

Here, J is the Jacobian determinant. The slack variable $\mathbf{u}$ is drawn from a pdf $g(\mathbf{u})$, implying that the slack variable $\mathbf{u}'$ has pdf $g'(\mathbf{u}') = g[\mathbf{u}(\mathbf{m}',\mathbf{u}')]$. Then, following the same procedure as before, the As must satisfy

$$\frac{A(\mathbf{m}',\mathbf{m})}{A(\mathbf{m},\mathbf{m}')} = \alpha \ \text{ with } \ \alpha \equiv \frac{p(\mathbf{m}')g'(\mathbf{u}')\,J}{p(\mathbf{m})g(\mathbf{u})} \tag{12.10}$$

implying (as before)

$$A(\mathbf{m}',\mathbf{m}) = \min(1,\alpha) \ \text{ and } \ A(\mathbf{m},\mathbf{m}') = \min\left(1,\frac{1}{\alpha}\right) \tag{12.11}$$

Finally, trans-dimensionality is implemented by labeling the different possible moves between model types with an index, say m, and defining a probability $\varphi_m(\mathbf{m},\mathbf{u})$ that a particular move is attempted at state $(\mathbf{m},\mathbf{u})$. This alters the definition of α to

$$\alpha \equiv \frac{\varphi_m(\mathbf{m}',\mathbf{u}')\,p(\mathbf{m}')g'(\mathbf{u}')\,J}{\varphi_m(\mathbf{m},\mathbf{u})p(\mathbf{m})g(\mathbf{u})} \tag{12.12}$$

The α associated with $(\mathbf{m},\mathbf{u})$ transforming to $(\mathbf{m}',\mathbf{u}')$ is the reciprocal of α associated with $(\mathbf{m}',\mathbf{u}')$ transforming to $(\mathbf{m},\mathbf{u})$. Good practice when writing a trans-dimensional script is to check that this relationship holds.

In the examples we consider here, all moves are considered equally probable, so that the φs cancel from the equation. We also limit ourselves to linear transformations of the form $\mathbf{x}' = \mathbf{T}\mathbf{x}$ (for which $J = |\det(\mathbf{T})|$).

12.7 Examples of trans-dimensional solutions

In Example 1, the extended Metropolis-Hastings algorithm is used to sample a simple trans-dimensional pdf. The pdf consists of a one-dimensional model type, with probability β, model parameter m_1, and pdf $p(m_1)$, together with a two-dimensional model type, with probability $(1-\beta)$, model parameters m_1' and m_2', and pdf $p(m_1', m_2')$. The former pdf is Normal, with known mean and variance, and later pdf also is Normal, with known mean, uniform variance, and zero covariance. We use the linear transformations

$$\mathbf{x}' = \begin{bmatrix} m_1' \\ m_2' \\ u_1' \end{bmatrix} = \begin{bmatrix} 1 & 1 & 0 \\ 0 & 0 & 1 \\ 0 & 1 & 0 \end{bmatrix} \begin{bmatrix} m_1 \\ u_1 \\ u_2 \end{bmatrix} = \mathbf{T}\mathbf{x}$$

$$\mathbf{x}' = \begin{bmatrix} m_1' \\ u_1' \\ u_2' \end{bmatrix} = \begin{bmatrix} 1 & 0 & -1 \\ 0 & 0 & 1 \\ 0 & 1 & 0 \end{bmatrix} \begin{bmatrix} m_1 \\ m_2 \\ u_1 \end{bmatrix} = \mathbf{T}^{-1}\mathbf{x} \tag{12.13}$$

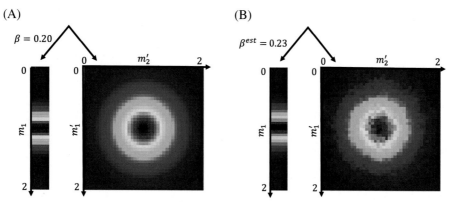

FIG. 12.5 (A) Trans-dimensional pdf consisting of a one-dimensional Normal pdf $p(m_1)$ with mean 1, variance ¼, and overall probability $\beta = 0.2$, and a two-dimensional Normal pdf $p(m_1', m_2')$ with mean (1,1), uniform variance 1, and overall probability $(1 - \beta)$. (B) Approximation to the pdf created by binning a one-million-member ensemble (not shown) sampled using the trans-dimensional Metropolis-Hastings algorithm. Scripts `gdama12_05` and `gdapy12_05`.

They are inverses of one another and both have $J = 1$. The forward transformation implies that, in a move from one dimension to two, m_1' is equal to m_1 plus random variation and that m_2' is randomly chosen. The pdfs $g(u_1)$, and $g(u_2)$ are chosen to be Normal, but with the variance of $g(u_2)$ much larger than that of $g(u_1)$, so that the value of m_1' is only slightly different than m_1, whereas the value of m_2' varies over a wide range. The parameter α is

$$\alpha = \frac{p(m_1', m_2')g'(u_1') J}{p(m_1)g(u_1)g(u_2)} = \frac{p(m_1', m_2')}{p(m_1)g(u_2)} \tag{12.14}$$

(since $J = 1$ and $g'(u_1') = g(u_1)$).

A one-million-member ensemble is generated and binned to produce an empirical histogram of the trans-dimensional pdf. It closely matches the true trans-dimensional pdf (Fig. 12.5).

In Example 2, the extended Metropolis-Hastings algorithm is used to sample the Bayesian pdf $p(\mathbf{m}) \propto p(\mathbf{d}^{obs}|\mathbf{m}) p_A(\mathbf{m})$, where the observed data $\mathbf{d}^{obs}$ scatter around a simple function $d(x)$ $(0 \leq x_i \leq 1)$. Two types of function are considered (Fig. 12.6A and C)

$$d(x_i) = m_1 \sin\left(\frac{\pi x_i}{L}\right) \text{ and } d(x_i) = m_1' \exp\left(-\frac{(x_i - \frac{1}{2})^2}{2m_2'}\right) \tag{12.15}$$

with the objectives of fitting the data to these two functions and assessing which function is more probable. The Bayesian pdfs are

$$p(m_1) \propto \exp(-\frac{1}{2}E(m_1)) \frac{1}{\sqrt{2\pi}\sigma_m} \exp(-\frac{1}{2}L(m_1))$$

$$p(m_1', m_2') \propto \exp\left(-\frac{1}{2}E(m_1', m_2')\right) \frac{1}{2\pi\sigma_m^2} \exp\left(-\frac{1}{2}L(m_1', m_2')\right) \tag{12.16}$$

Here, $E = \sigma_m^{-2}\|\mathbf{d}^{obs} - \mathbf{d}^{pre}\|_2^2$ is the weighted prediction error and $L = \sigma_m^{-2}\|\langle\mathbf{m}\rangle - \mathbf{m}\|_2^2$ is the weighted error in prior information (measured with respect to prior model $\langle\mathbf{m}\rangle$). Although it would be possible to include factors β and $(1 - \beta)$ representing the prior probability for model type, we have not done so here, but rather we have assumed equal probabilities, that is, $\beta = (1 - \beta) = \frac{1}{2}$.

Two cases are tested, one in which the true data scatter around the sinusoid, and the other in which they scatter around the Gaussian. In both cases, the true model type is found to be the most probable. And in both cases, the pdf associated with the correct model has most of its probability near the true model parameters (Fig. 12.6B and D).

In Example 3, the extended Metropolis-Hastings algorithm is applied to a hierarchy of layered models with increasing spatial fidelity, in order to find the most probable number of layers (and their values) that solve a Laplace transform-like problem (Fig. 12.7A). The field values m_j are related to data d_i via

$$d_i = \sum_{j=1}^{M_{max}} m_j \exp\{-c_i z_j\} \text{ with } z_j = (j - 1) \tag{12.17}$$

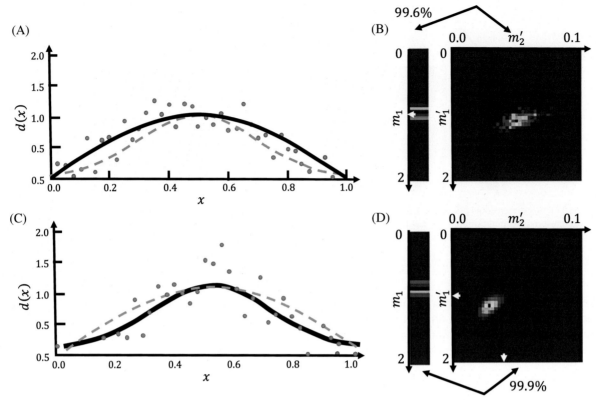

FIG. 12.6 Trans-dimensional curve-fitting example. Data are fit to a sinusoid of amplitude m_1 (one dimension) and a Gaussian of amplitude m_1' and variance m_2' (two dimensions). The solution is an empirical trans-dimensional pdf produced by binning a one-million-member ensemble. (Top row) True model is one-dimensional (A) True sinusoid (black curve), noisy data (circles), and Gaussian for comparison (green curve). (B) Empirical trans-dimensional pdf (colors) with m_1^{true} indicated with arrow (white). The one-dimensional part of the pdf has most (99%) of the probability. (Bottom row) True model is two-dimensional (A) True Gaussian (black curve), noisy data (circles), and sinusoid for comparison (green curve). (B) Trans-dimensional pdf (colors) recovered by inverting data together with weak prior information, with $\mathbf{m}'^{true}$ indicated with arrows (white). The two-dimensional part of the pdf has 99% of the probability. Scripts gdama12_06 and gdapy12_06.

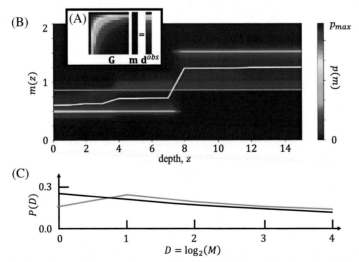

FIG. 12.7 Trans-dimensional solution of the Laplace transform problem for model function $m(z)$ with depth z. Models with between $2^0 = 1$ and $2^4 = 16$ layers are considered. (A) Synthetic data $\mathbf{d}^{obs} = \mathbf{Gm}^{true}$ are generated by adding random noise to the prediction of the true model $\mathbf{m}^{true} = [0.5, 1.5]^T$, which has $2^1 = 2$ layers. (B) Empirical pdf $p(m)$ (colors) of depth-interpolated models determined by binning the ensemble together with the mean model $\mathbf{m}^{est}$ (white curve). (C) Empirical probability $P(n)$ (green) for dimension n determined by binning a one-million-member ensemble (not shown) sampled using the trans-dimensional Metropolis-Hastings algorithm. The prior probability for dimension $P_A(n)$ (black) is shown for comparison. Scripts gdama12_07 and gdapy12_07.

Here, the cs are known constants. We assume that the number of layers M for a given model type is a power of two, that is, $M=2^D$, varying between $D=0$ ($M=1$) and $D=D_{max}$ ($M_{max}=2^{D_{max}}$), so that the layer thickness changes by a factor of two between model types. Consequently, the values v_i are represented by $M \leq M_{max}$ model parameters m_i, together with the rule that groups of M/M_{max} neighboring vs are equal to the same m. Thus the sequence $D=0$, 1, $\cdots D_{max}$ generates a sequence of models with increasing fidelity in depth z.

As in the previous example, we use Bayes theorem to construct $p(\mathbf{m})$, but modifying Eq. (12.6b) to include a prior distribution for model type $P_A(D)$, which monotonically declines with D. This factor represents the idea that low spatial-scale (low D) models are preferable because of their simplicity.

The transformation $\mathbf{T}^{(i,j)}$ implements moves, say from model type D_i to D_j (with M_i to M_j, respectively). We choose $M_i+P_i=2M_{max}$, for although this choice leads to some us being carried through unused on most of the moves, all the $\mathbf{T}s$ then are the same size, $2M_{max} \times 2M_{max}$. For the sake of brevity, we illustrate a few of the $\mathbf{T}s$ for $D_{max}=2$ (and $M_{max}=4$). For the $D_i=D_j=2$, we use the transformation

$$
\begin{bmatrix} m'_1 \\ m'_2 \\ m'_3 \\ m'_4 \\ u'_1 \\ u'_2 \\ u'_3 \\ u'_4 \end{bmatrix} = \begin{bmatrix} \begin{bmatrix} 1 & & & \\ & 1 & & \\ & & 1 & \\ & & & 1 \end{bmatrix} & \begin{bmatrix} 1 & & & \\ & 1 & & \\ & & 1 & \\ & & & 1 \end{bmatrix} \\ \begin{bmatrix} 0 & & & \\ & 0 & & \\ & & 0 & \\ & & & 0 \end{bmatrix} & \begin{bmatrix} 1 & & & \\ & 1 & & \\ & & 1 & \\ & & & 1 \end{bmatrix} \end{bmatrix} \begin{bmatrix} m_1 \\ m_2 \\ m_3 \\ m_4 \\ u_1 \\ u_2 \\ u_3 \\ u_4 \end{bmatrix}
\tag{12.18}
$$

which corresponds to adding random variation to the ms ($m'_i=m_i+u_i$) and carrying through the us ($u'_i=u_i$).

In order to increase D, say from $D_i=1$ to $D_j=2$, we must double the number of model parameters by duplication, that is, $m'_1=m'_2=m_1$ and $m'_3=m'_3=m_2$. The rest of $\mathbf{T}^{(1,2)}$ is determined by three conditions: assuring that the forward transformation $\mathbf{T}^{(1,2)}$ provides random variation to all the $m's$, assuring that the inverse transformation $\mathbf{T}^{(2,1)}=[\mathbf{T}^{(1,2)}]^{-1}$ provides random variation to all the ms, and assuring that the inverse transformation aggregated layers by nearest-neighbor averaging ($m_1'=\frac{1}{2}m_1+\frac{1}{2}m_2$ and $m_2'=\frac{1}{2}m_3+\frac{1}{2}m_4$). One possible set of transformations is

$$
\begin{bmatrix} m'_1 \\ m'_2 \\ m'_3 \\ m'_4 \\ u'_1 \\ u'_2 \\ u'_3 \\ u'_4 \end{bmatrix} = \begin{bmatrix} \begin{bmatrix} 1 & & & \\ 1 & & & \\ & 1 & & \\ & 1 & & \end{bmatrix} \begin{bmatrix} 1 & & & \\ -1 & 1 & & \\ & 1 & & \\ & -1 & & \end{bmatrix} \begin{bmatrix} 1 & & & \\ & 1 & & \\ & & 1 & \\ & & & 1 \end{bmatrix} \\ \begin{bmatrix} 0 & & & \\ & 0 & & \\ & & 0 & \\ & & & 0 \end{bmatrix} \begin{bmatrix} 1 & & & \\ & 1 & & \\ & & 1 & \\ & & & 1 \end{bmatrix} \end{bmatrix} \begin{bmatrix} m_1 \\ m_2 \\ u_1 \\ u_2 \\ u_3 \\ u_4 \\ u_5 \\ u_6 \end{bmatrix}
\tag{12.19}
$$

and

$$
\begin{bmatrix} m'_1 \\ m'_2 \\ u'_1 \\ u'_2 \\ u'_3 \\ u'_4 \\ u'_5 \\ u'_6 \end{bmatrix} = \begin{bmatrix} \begin{bmatrix} \frac{1}{2} & \frac{1}{2} & & \\ & & \frac{1}{2} & \frac{1}{2} \\ -\frac{1}{2} & \frac{1}{2} & & \\ & & -\frac{1}{2} & \frac{1}{2} \end{bmatrix} \begin{bmatrix} -\frac{1}{2} & -\frac{1}{2} & & \\ & & -\frac{1}{2} & -\frac{1}{2} \\ \frac{1}{2} & -\frac{1}{2} & & \\ & & \frac{1}{2} & -\frac{1}{2} \end{bmatrix} \\ \begin{bmatrix} 0 & & & \\ & 0 & & \\ & & 0 & \\ & & & 0 \end{bmatrix} \begin{bmatrix} 1 & & & \\ & 1 & & \\ & & 1 & \\ & & & 1 \end{bmatrix} \end{bmatrix} \begin{bmatrix} m_1 \\ m_2 \\ m_3 \\ m_4 \\ u_1 \\ u_2 \\ u_3 \\ u_4 \end{bmatrix}
\tag{12.20}
$$

The general algorithm for constructing such transformation matrices is stated in the scripts `gdama12_07` and `gdapy12_07`.

The true model has two layers, that is, $D^{true} = 1$, $M^{true} = 2$, and $\mathbf{m}^{true} = [0.5, 1.5]^T$. Synthetic data d^{obs} are created by adding random noise with variance $\sigma_d^2 = (0.01)^2$ to the true data d^{true} (obtained by evaluating Eq. 12.17 with $m = m^{true}$). The prior $p_A(\mathbf{m})$ is chosen to be Normal, uncorrelated, with zero mean and with uniform variance $\sigma_m^2 = (5)^2$, corresponding to weak information that the solution is most likely near a prior value of $\langle m_i \rangle = 1.0$. The pdfs $g(u_i)$ are all Normal with zero mean and variance $\sigma_u^2 = (0.01)^2$.

A one-million-member ensemble is generated and binned to produce empirical histograms. The empirical $P(D)$ peaks at the true value of 1 (that is, two layers), but not strongly so (only 28% probability), with both the one-layer model and the 16-layer model having significant probability (17% and 15%, respectively) (Fig. 12.7C). The modal model is indeed approximately two layered, with layer values very close to the true values of $[0.5, 1.5]^T$. However, the mean model, $\mathbf{m}^{est}$ (white curve in Fig. 12.7B), is far from the true model. This disparity is due to the high probability of the one-layer model, which has $m_1 = 0.85$, and which tends to reduce the extremes of the mean model. The moderately high probability of the one-layer model has the additional consequence of making $p[m(z_i)]$ bimodal.

Although the modal model is approximately two layered, it is composed of the superposition of two-, four-, eight-, and sixteen-layer models—these models are all *functionally* two-layer models (in the sense that $D = 2$ and $\mathbf{m} = [0.5, 0.5, 1.5, 1.5]^T$ correspond to the same vs as $D = 1$ $\mathbf{m} = [0.5, 1.5,]^T$). Consequently, the empirical distribution $P(D)$ of model type does not characterize the probability of "spatial complexity" in any meaningful way. Instead, we should scan through the ensemble and calculate the percentages that functionally have 2, 4, 8, $\cdots$ layers, based on a criterion that decides whether the values of two neighboring layers are sufficiently close that they can be considered as one layer.

This behavior brings out a conceptual problem with the concept of a hierarchy of spatial scales. A model with $M_2 > M_1$ model parameters is not necessarily more spatially complex than a model with M_1 model parameters. Spatial complexity depends, in part, on the values of the model parameters.

12.8   Problems

12.1 Modify your code for Problem 11.5 to substitute a Monte Carlo search in place of a grid search. If the modified code any more efficient?

12.2 This problem concerns the inverse problem shown in Figs. 11.8 and 12.3. (A) Modify the Newton's method solution script (gdama11_10 or gdapy12_10) to produce a linearized estimate of the posterior covariance $[\text{cov } \mathbf{m}^{(p+1)}] \approx \sigma_d^2 [\mathbf{G}_{(p)}^T \mathbf{G}_{(p)}]^{-1}$, where the posterior variance σ_d^2 is estimated from the prediction error. (B) Modify the MCMC method solution script (gdama11_08 or gdapy11_08) to estimate the posterior covariance as the sample covariance of the ensemble. Be sure that the prior data covariance matches between the code. Finesse the issue of the MCMC code having prior information by making it very weak, so that it has minimal effect on the solution. (C) The MCMC estimate is probably closer to the true covariance, so use it as the standard to judge the accuracy of the linearized version. How well does it do?

12.3 Modify the code for trans-dimensional problem shown in Fig. 12.5 (gdama12_05 or gdapy12_05) so that the one-dimensional pdf is a two-sided exponential pdf with the same mean and variance of the original Normal pdf. Does the method perform any better or worse?

References

Bodin, T., Sambridge, M., Tkalčić, H., Arroucau, P., Gallagher, K., Rawlinson, N., 2012. Transdimensional inversion of receiver functions and surface wave dispersion. J. Geophys. Res. Solid Earth 117. https://doi.org/10.1029/2011JB008560. JB008560.

Bodin, T., Yuan, H., Romanowicz, B., 2014. Inversion of receiver functions without deconvolution—application to the Indian craton. Geophys. J. Int. 196, 1025–1033. https://doi.org/10.1093/gji/ggt431.

Green, P.J., Hastie, D.I., 2009. Reversible jump MCMC. http://people.ee.duke.edu/~lcarin/rjmcmc_20090613.pdf.

Hastings, W.K., 1970. Monte Carlo sampling methods using Markov chains and their applications. Biometrika 57, 97–109. https://doi.org/10.1093/biomet/57.1.97. JSTOR 2334940. Zbl 0219.65008.

Kirkpatrick, S., Gelatt, C.D., Vecchi, M.P., 1983. Optimization by simulated annealing. Science 220 (4598), 671–680.

Metropolis, N., Rosenbluth, A.W., Rosenbluth, M.N., Teller, A.H., Teller, E., 1953. Equation of state calculations by fast computing machines. J. Chem. Phys. 21, 1087–1092. https://doi.org/10.1063/1.1699114.

13

Factor analysis

The data analysis techniques that we have discussed so far can be characterized as *directed*, in the sense that the practitioner made a conscious effort to identify model parameters **m** that have special significance, and that are linked to the data **d** through a theory, such as $\mathbf{d} = \mathbf{Gm}$, that specifically has been chosen as relevant to the problem. That type of data analysis focuses on estimating and interpreting **m**.

Sometimes, too little is known about the data to use a directed approach. Instead, the practitioner can only ask whether the data, themselves, reveal some kind of underlying pattern. Only after this pattern is discovered may the practitioner be in a position to think about its *meaning* and start to develop a theory capable of modeling it.

This chapter describes two interrelated *undirected* methods, *factor analysis* and *empirical orthogonal function analysis*, which aid in the detection of patterns within a data set about which little initially is known. Although described as "undirected," these methods do make some assumptions about the general kind of patterns contained in the data. Factor analysis assumes that the data consist of a *linear mixture* of a small set of simpler patterns. Empirical orthogonal function analysis extends this idea by allowing the patterns to vary with time and space.

13.1 The factor analysis problem

Consider an ocean whose sediment is derived from the simple mixing of continental source rocks A and B (Fig. 13.1). Suppose that the concentrations of three elements are determined for many samples of sediment and then plotted on a graph whose axes are percentages of those elements. Since all the sediments are derived from only two source rocks, the sample compositions lie on the triangular portion of a plane bounded by the compositions of A and B (Fig. 13.2).

The factor analysis problem is to deduce the number of the source rocks (called factors) and their composition from observations of the composition of the sediments (called samples) (Fig. 13.3). It is therefore a problem in inverse theory. We discuss it separately, because it provides an interesting example of the use of some of the vector space analysis techniques developed in Chapter 9.

The model proposes that the samples are simple mixtures (linear combinations) of the factors. If there are N samples containing M elements and if there are p factors, we can state this model algebraically with the equation

$$\mathbf{S} = \mathbf{CF} \tag{13.1}$$

where S_{ij} is the fraction of element j in sample i:

$$\mathbf{S} = \begin{bmatrix} \text{element 1 in sample 1} & \text{element 2 in sample 1} & \cdots & \text{element } M \text{ in sample 1} \\ \text{element 1 in sample 2} & \text{element 2 in sample 2} & \cdots & \text{element } M \text{ in sample 2} \\ \vdots & \vdots & \ddots & \vdots \\ \text{element 1 in sample } N & \text{element 2 in sample } N & \cdots & \text{element } M \text{ in sample } N \end{bmatrix} \tag{13.2}$$

The word *element* is used in the generic sense, as most factor analysis problems will not involve chemical elements. We will refer to individual samples as $\mathbf{s}^{(i)}$, with $\mathbf{s}^{(i)T}$ a row of **S**.

Similarly, F_{ij} is the fraction of element j in sample i:

$$\mathbf{F} = \begin{bmatrix} \text{element 1 in factor 1} & \text{element 2 in factor 1} & \cdots & \text{element } M \text{ in factor 1} \\ \text{element 1 in factor 2} & \text{element 2 in factor 2} & \cdots & \text{element } M \text{ in factor 2} \\ \vdots & \vdots & \ddots & \vdots \\ \text{element 1 in factor } p & \text{element 2 in factor } p & \cdots & \text{element } M \text{ in factor } p \end{bmatrix} \tag{13.3}$$

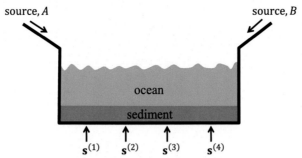

FIG. 13.1   Material from sources A and B is eroded into the ocean and deposited to form sediment. Samples $\mathbf{s}^{(i)}$ of the sediment are collected and their chemical composition is determined. The data are used to infer the composition of the sources.

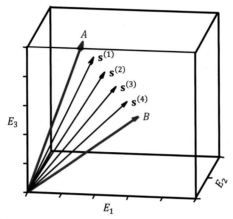

FIG. 13.2   The composition of the samples $\mathbf{s}^{(i)}$ (black arrows) lies on a triangular sector of a plane bounded by the composition of the sources A and B (red arrows). Scripts `gdama13_01` and `gdapy13_01`.

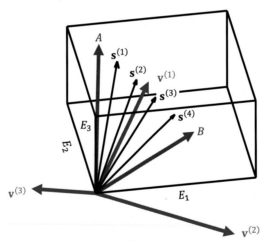

FIG. 13.3   Eigenvectors $\mathbf{v}^{(1)}$ and $\mathbf{v}^{(2)}$ lie in the plane of the samples ($\mathbf{v}^{(1)}$ is closest to the mean sample). Eigenvector $\mathbf{v}^{(3)}$ is Normal to the plane. Scripts `gdama13_02` and `gdapy13_02`.

We will refer to individual factors as $\mathbf{f}^{(i)}$, with $\mathbf{f}^{(i)T}$ a row of $\mathbf{F}$.

The matrix C_{ij} is the fraction of factor i in sample j:

$$\mathbf{C} = \begin{bmatrix} \text{factor 1 in sample 1} & \text{factor 2 in sample 1} & \cdots & \text{factor } p \text{ in sample 1} \\ \text{factor 1 in sample 2} & \text{factor 2 in sample 2} & \cdots & \text{factor } p \text{ in sample 2} \\ \vdots & \vdots & \ddots & \vdots \\ \text{factor 1 in sample } N & \text{factor 2 in sample } N & \cdots & \text{factor } p \text{ in sample } N \end{bmatrix} \tag{13.4}$$

The elements of the matrix $\mathbf{C}$ are referred to as the *factor loadings* (or sometimes just the *loadings*).

The inverse problem is to factor the matrix $\mathbf{S}$ into loadings $\mathbf{C}$ and factor $\mathbf{F}$. Each sample (each row of $\mathbf{S}$) is represented as a linear combination of factors (rows of $\mathbf{F}$), with the elements of $\mathbf{C}$ giving the coefficients of the combination. As long as we pick an $\mathbf{F}$ whose rows span the space spanned by the rows of $\mathbf{S}$, we can perform the factorization. For $p = M$, any linearly independent set of factors will do, so in this sense the factor analysis problem is completely nonunique. A much more interesting question is the minimum number of factors needed to represent the samples. Then the factor analysis problem is equivalent to examining the space spanned by $\mathbf{S}$ and determining its dimension. This problem can be solved by representing the sample matrix with its singular-value decomposition as

$$\mathbf{S} = \mathbf{U}_p \mathbf{\Lambda}_p \mathbf{V}_p^T = \mathbf{CF} \text{ with } \mathbf{C} \equiv (\mathbf{U}_p \mathbf{\Lambda}_p) \text{ and } \mathbf{F} \equiv \mathbf{V}_p^T \tag{13.5}$$

Only the eigenvectors with nonzero singular values appear in the decomposition. The number of factors is given by the number p of nonzero singular values. One possible set of factors is the corresponding set of p eigenvectors $\mathbf{V}_p$. This set of factors is not unique (Fig. 13.4), because any set of factors that spans the p-dimensional space will do. Mathematically, we transform the factors with any matrix $\mathbf{T}$ that possesses an inverse, in which case $\mathbf{S} = \mathbf{CF} = \mathbf{CIF} = (\mathbf{CT}^{-1})(\mathbf{TF}) = \mathbf{C}'\mathbf{F}'$ with the transformed factors being $\mathbf{F}' = \mathbf{TF}$ and the new loadings being $\mathbf{C}' = \mathbf{CT}^{-1}$.

If we write out Eq. (13.5), we find that the composition of the ith sample $\mathbf{s}^{(i)}$ is related to the eigenvectors $\mathbf{v}^{(i)}$ and singular values λ_i by

$$\mathbf{s}^{(1)} = \left[\mathbf{U}_p\right]_{11} \lambda_1 \mathbf{v}^{(1)} + \left[\mathbf{U}_p\right]_{12} \lambda_2 \mathbf{v}^{(2)} + \cdots + \left[\mathbf{U}_p\right]_{1p} \lambda_p \mathbf{v}^{(p)}$$
$$\vdots \tag{13.6}$$
$$\mathbf{s}^{(N)} = \left[\mathbf{U}_p\right]_{N1} \lambda_1 \mathbf{v}^{(1)} + \left[\mathbf{U}_p\right]_{N2} \lambda_2 \mathbf{v}^{(2)} + \cdots + \left[\mathbf{U}_p\right]_{Np} \lambda_p \mathbf{v}^{(p)}$$

When the singular values are arranged in descending order, most of each sample is composed of factor 1, with a smaller contribution from factor 2. Because $\mathbf{U}_p$ and $\mathbf{v}^{(i)}$ are composed of unit vectors, on average their elements are of equal size. We have identified the most "important" factors (Fig. 13.5). Even if $p = M$, it might be possible to neglect some of the smaller singular values and still achieve a reasonably good prediction of the sample compositions, that is, $\mathbf{S} \approx \mathbf{CF} = (\mathbf{U}_q \mathbf{\Lambda}_q)(\mathbf{V}_q^T)$ with $q < p$.

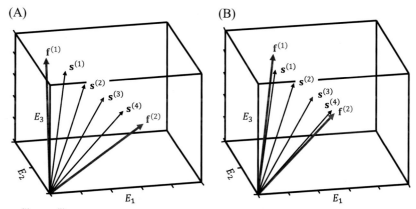

FIG. 13.4 Any two factors $\mathbf{f}^{(1)}$ and $\mathbf{f}^{(2)}$ (red arrows) that lie in the plane of the samples and that bound the range of sample compositions (black arrows) are acceptable, such as those shown in (A) and (B). Scripts gdama13_03 and gdapy13_03.

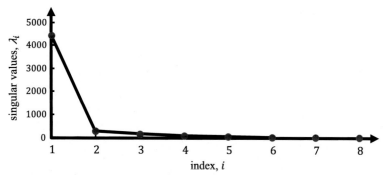

FIG. 13.5 Singular values λ_i of the Atlantic Ocean Rock data set. Scripts gdama13_04 and gdapy13_04.

The eigenvector with the largest singular value is near the mean of the sample vectors. It is easy to show that the sample mean $\langle \mathbf{s} \rangle$ maximizes the sum of dot products with the data $\sum_i^N \mathbf{s}^{(i)} \cdot \mathbf{v}$, whereas the eigenvector $\mathbf{v}$ with largest singular value maximizes the sum of squared dot products $\sum_i^N [\mathbf{s}^{(i)} \cdot \langle \mathbf{s} \rangle]^2$. (To show this, maximize the given functions using Lagrange multipliers, with the constraints that $\langle \mathbf{s} \rangle$ and $\mathbf{v}$ are unit vectors.) As long as most of the samples are in the same quadrant, these two functions have roughly the same maximum.

The code for computing the factors $\mathbf{F}$, loadings $\mathbf{C}$, and singular values λ_i via singular-value decomposition is

---

MATLAB®

```
[U, LAMBDA, V] = svd(S,0);
lambda = diag(LAMBDA);
F = V';
C = U*LAMBDA;
                                                [gdama13_04]
```

Python

```
U, lam, VT = la.svd(S,full_matrices=False);
lam = gda_cvec(lam);
V = VT.T;
Ns,i = np.shape(lam);
F = VT;
C = np.matmul(U,np.diag(lam.ravel()));
                                                [gdapy13_04]
```

Note that the MATLAB® version returns $\mathbf{U}$, $\mathbf{\Lambda}$, and $\mathbf{V}$ whereas the Python version returns $\mathbf{U}$, $\text{diag}(\mathbf{\Lambda})$, and $\mathbf{V}^T$. Here we use the "economy" version of the singular-value decomposition algorithm, in which the matrices $\mathbf{U}$, $\mathbf{\Lambda}$, and $\mathbf{V}^T$ have dimensions $N \times K$, $K \times K$, and $K \times M$, respectively, where $K = \min(N,M)$. The MATLAB® svd() function and the Python la.svd() method do not throw out any of the zero (or near-zero) eigenvalues; this is left to the user:

MATLAB®

```
P=5;
FP = F(1:P,:);
CP = C(:,1:P);
                                                [gdama13_04]
```

Python

```
P=5;
FP = np.copy(F[0:P,0:M]);
CP = np.copy(C[0:N,0:P]);
                                                [gdapy13_04]
```

We apply factor analysis to rock chemistry data taken from a petrologic database (PetDB at www.petdb.org). This database contains chemical information on igneous and metamorphic rocks collected from the floor of all the world's oceans, but we analyze here $N = 6356$ samples from the Atlantic Ocean that have the following chemical species: SiO_2, TiO_2, Al_2O_3, FeO_{total}, MgO, CaO, Na_2O, and K_2O in units of weight percent.

A plot of the singular values of the Atlantic Rock data set (Fig. 13.5) reveals that the first value is by far the largest, values 2 through 5 are intermediate in size, and values 6 through 8 are near-zero. The fact that the first singular value λ_1 is much larger than all the other reflects the composition of the rock samples having only a small range of variability. Thus all rock samples contain a large amount of the first factor $\mathbf{f}^{(1)}$—the typical sample. Only four additional factors $\mathbf{f}^{(2)}$, $\mathbf{f}^{(3)}$, $\mathbf{f}^{(4)}$, and $\mathbf{f}^{(5)}$ out of a total of eight are needed to describe the variability about the typical sample:

The role of each of the factors can be understood by examining its elements. Factor 2, for instance, increases the amount of MgO while decreasing mostly Al_2O_3 and CaO, with respect to the typical sample.

Element	$f^{(1)}$	$f^{(2)}$	$f^{(3)}$	$f^{(4)}$	$f^{(5)}$
SiO_2	+0.908	+0.007	-0.161	+0.209	+0.309
TiO_2	+0.024	-0.037	-0.126	+0.151	-0.100
Al_2O_3	+0.275	-0.301	+0.567	+0.176	-0.670
FeO_{total}	+0.177	-0.018	-0.659	-0.427	-0.585
MgO	+0.141	+0.923	+0.255	-0.118	-0.195
CaO	0.209	-0.226	+0.365	-0.780	+0.207
Na_2O	+0.044	-0.058	-0.042	+0.302	-0.145
K_2O	+0.003	-0.007	-0.006	+0.073	+0.015

The *factor analysis* has reduced the dimensions of variability of the rock data set from eight elements to four factors, improving the effectiveness of scatter plots. Three-dimensional plots are useful in this case, because any three of the four factors can be used as axes and the resulting three-dimensional scatter plot viewed from a variety of perspectives. The following code plots the coefficients of factors 2 through 4 for each sample:

```
                              MATLAB®
figure(2);
clf;
set(gca, 'LineWidth', 3);
set(gca, 'FontSize', 14);
hold on;
axis( [cmin, cmax, cmin, cmax, cmin, cmax] );
plot3( C(:,2), C(:,3), C(:,4), 'k.', 'LineWidth', 2 );
xlabel('C(2)');
ylabel('C(3)');
zlabel('C(4)');
view(3);
                                              [gdama13_04]
```

```
                              Python
# setup for 3D figure
fig2 = plt.figure(2,figsize=(12,8));   # smallish figure
ax2 = fig2.add_subplot(111, projection='3d');
ax2.axes.set_xlim3d(left=cmin, right=cmax);
ax2.axes.set_ylim3d(bottom=cmin, top=cmax);
ax2.axes.set_zlim3d(bottom=cmin, top=cmax);
myelev=15.0; myazim=125.0;
ax2.view_init(elev=myelev, azim=myazim);
ax2.set_xlabel("C1");
ax2.set_ylabel("C2");
ax2.set_zlabel("C3");
gdabox(ax2,cmin,cmax,cmin,cmax,cmin,cmax); # 3D box
ax2.plot3D( C[0:N,1], C[0:N,2], C[0:N,3], 'k.', lw=2 );
plt.show();
                                              [gdapy13_04]
```

In the Python version, the `gdabox()` function draws a three-dimensional bounding box. The plot can then be viewed from different perspectives (Fig. 13.6). Note that the samples appear to form two populations, one in which the variability is due to $\mathbf{f}^{(2)}$ and another due to $\mathbf{f}^{(3)}$.

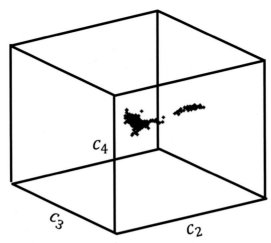

FIG. 13.6   Three-dimensional perspective view of the loadings C_i of factors 2, 3, and 4 in each of the rock samples (dots) of the Atlantic Ocean Rock data set. Scripts `gdama13_04` and `gdapy13_04`.

13.2 Normalization and physicality constraints

In many instances, an element can be important even though it occurs only in trace quantities. In such cases, one cannot neglect factors simply because they have small singular values. They may contain an important amount of the trace elements. It is therefore appropriate to normalize the matrix S so that there is a direct correspondence between singular-value size and importance. This is usually done by defining a diagonal matrix of weights W (usually proportional to the reciprocal of the standard deviations of measurement of each of the elements) and then forming a new weighted sample matrix $S' = SW$.

The singular-value decomposition enables one to determine a set of factors that span, or approximately span, the space of samples. These factors, however, are not unique in the sense that one can form linear combinations of factors that also span the space. Such transformations are useful because ordinarily the singular-value decomposition eigenvectors violate prior constraints on what "good" factors should be like. One such constraint is that the factors should have a unit L_1 norm, that is, their elements should sum to one. If the components of a factor represent fractions of chemical elements, for example, it is reasonable that the elements should sum to 100%. Another constraint is that the elements of both the factors and the factor loadings should be nonnegative. Ordinarily, a material is composed of a positive combination of components. Given an initial representation of the samples $S = CFW^{-1}$, we could imagine finding a new representation consisting of linear combinations of the old factors, defined by $F' = TF$, where T is an arbitrary $p \times p$ transformation matrix. The problem can then be stated.

Find T subject to the following constraints:

$$\sum_{j=1}^{M} \left[F'W^{-1} \right]_{ij} = 1 \text{ for all } i$$

(13.7)

$$\left[CT^{-1} \right]_{ij} \geq 0 \text{ and } \left[F'W^{-1} \right]_{ij} \geq 0 \text{ for all } i \text{ and } j$$

These conditions do not uniquely determine T, as can be seen from Fig. 13.4. Note that the second constraint is nonlinear in the elements of T. This is a very difficult constraint to implement and in practice is often ignored.

Prior information must be added to single out a unique solution. One possibility is to find a set of factors that maximize some measure of simplicity. One such measure is spikiness: the notion that a factor should have only a few large elements, with the other elements being near-zero. Minerals, for example, obey this principle. While a rock can contain upward of 20 chemical elements, typically it will be composed of minerals such as forsterite (Mg_2SiO_4), anorthite ($CaAl_2Si_2O_8$), rutile (TiO_2), each of which contains just a few elements. Spikiness is more or less equivalent to the idea that the elements of the factors should have high variance. The usual formula for estimated variance σ_d^2 of a data set d

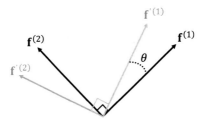

FIG. 13.7 Two mutually perpendicular factors $\mathbf{f}^{(1)}$ and $\mathbf{f}^{(2)}$ are rotated in their plane by an angle θ, creating two new mutually orthogonal vectors $\mathbf{f}'^{(1)}$ and $\mathbf{f}'^{(2)}$.

$$\sigma_d^2 = \frac{1}{N}\left(\sum_{i=1}^{N}(d_i - \bar{d})^2\right) = \frac{1}{N^2}\left(N\sum_{i=1}^{N}d_i^2 - \left(\sum_{i=1}^{N}d_i\right)^2\right) \tag{13.8}$$

(where $\bar{d}$ is the mean) can be generalized to a factor f_i

$$\sigma_f^2 = \frac{1}{M^2}\left(M\sum_{i=1}^{M}f_i^4 - \left(\sum_{i=1}^{N}f_i^2\right)^2\right) \tag{13.9}$$

Note that this is the variance of the squares of the elements of the factors. Thus a factor $\mathbf{f}$ has a large variance σ_f^2 when the absolute values of its elements have high variation. The signs of the elements are irrelevant.

The *varimax* procedure is a way of constructing a matrix $\mathbf{T}$ that increases the variance of the factors while preserving their orthogonality (Kaiser, 1958). It is an iterative procedure, with each iteration operating on only one pair of factors, with other pairs being operated upon in subsequent iterations. The idea is to view pairs of factors as vectors and to rotate them in their plane (Fig. 13.7) by an angle θ chosen to maximize the sum of their variances. The rotation changes only the two factors, leaving the other $p-2$ factors unchanged, as in the following example:

$$\begin{bmatrix} \mathbf{f}^{(1)T} \\ \mathbf{f}^{(2)T} \\ \cos(\theta)\mathbf{f}^{(3)T} + \sin(\theta)\mathbf{f}^{(5)T} \\ \mathbf{f}^{(4)T} \\ -\sin(\theta)\mathbf{f}^{(3)T} + \cos(\theta)\mathbf{f}^{(5)T} \\ \mathbf{f}^{(6)T} \end{bmatrix} = \begin{bmatrix} 1 & 0 & 0 & 0 & 0 & 0 \\ 0 & 1 & 0 & 0 & 0 & 0 \\ 0 & 0 & \cos(\theta) & 0 & \sin(\theta) & 0 \\ 0 & 0 & 0 & 1 & 0 & 0 \\ 0 & 0 & -\sin(\theta) & 0 & \cos(\theta) & 0 \\ 0 & 0 & 0 & 0 & 0 & 1 \end{bmatrix}\begin{bmatrix} \mathbf{f}^{(1)T} \\ \mathbf{f}^{(2)T} \\ \mathbf{f}^{(3)T} \\ \mathbf{f}^{(4)T} \\ \mathbf{f}^{(5)T} \\ \mathbf{f}^{(6)T} \end{bmatrix} \tag{13.10}$$

or $\mathbf{F}' = \mathbf{T}\mathbf{F}$. Here, only the pair $\mathbf{f}^{(3)}$ and $\mathbf{f}^{(5)}$ is changed.

In Eq. (13.10), the matrix $\mathbf{T}$ represents a rotation of one pair of vectors. The rotation matrix for many such rotations is just the product of a series of pairwise rotations. Note that the matrix $\mathbf{T}$ obeys the rule $\mathbf{T}^{-1} = \mathbf{T}^T$ (i.e., $\mathbf{T}$ is a unary matrix). For a given pair of factors $\mathbf{f}^A$ and $\mathbf{f}^B$, the rotation angle θ is determined by minimizing $\Phi \equiv M^2(\sigma_{f_A}^2 + \sigma_{f_B}^2)$ with respect to θ by solving $\partial\Phi/\partial\theta = 0$. The minimization requires a substantial amount of algebraic and trigonometric manipulation, so we omit it here. The result is (Kaiser, 1958)

$$\theta = \frac{1}{4}\tan^{-1}\left(\frac{A-B}{C-D}\right) \text{ with } A \equiv 2M\sum_i u_i v_i \text{ and } B \equiv \left(\sum_i u_i\right)\left(\sum_i v_i\right)$$

$$\text{and } C \equiv M\sum_i(u_i^2 - v_i^2) \text{ and } D \equiv \left(\left(\sum_i u_i\right)^2 - \left(\sum_i v_i\right)^2\right) \tag{13.11}$$

$$\text{and } u_i \equiv \left(f_i^A\right)^2 - \left(f_i^B\right)^2 \text{ and } v_i \equiv 2f_i^A f_i^B$$

By way of example, we note that the two orthogonal vectors

$$\mathbf{f}^A = \frac{1}{2}[1\ 1\ 1\ 1]^T \text{ and } \mathbf{f}^B = \frac{1}{2}[1\ -1\ 1\ -1]^T \tag{13.12}$$

are extreme examples of two non-spiky vectors because all their elements have the same absolute value. When applied to them, the varimax procedure returns

$$\mathbf{f}'^A = \frac{1}{\sqrt{2}}[1\ 0\ 1\ 0]^T \text{ and } \mathbf{f}'^B = \frac{1}{\sqrt{2}}[0\ -1\ 0\ -1]^T \tag{13.13}$$

which are significantly spikier than the original factors, because the absolute value of their elements range between zero and unity. Code that implements the transformation is

MATLAB®
| ```
u = fA.^2 - fB.^2;
v = 2* fA .* fB;
A = 2*M*u'*v;
B = sum(u)*sum(v);
top = A - B;
C = M*(u'*u-v'*v);
D = (sum(u)^2)-(sum(v)^2);
bot =  C - D;
q = 0.25 * atan2(top,bot);
cq = cos(q);
sq = sin(q);
fAp =  cq*fA + sq*fB;
fBp = -sq*fA + cq*fB;
``` |
| [gdama13_05] |

Python
| ```
u = np.power(fA,2) - np.power(fB,2);
v = 2.0* np.multiply(fA,fB);
A = 2.0*M*np.matmul(u.T,v);
B = np.sum(u)*np.sum(v);
top = A - B;
CC = M*(np.matmul(u.T,u)-np.matmul(v.T,v));
DD = (np.sum(u)**2)-(np.sum(v)**2);
bot =  CC - DD;
q = 0.25 * atan2(top,bot);
cq = cos(q);
sq = sin(q);
fAp =  cq*fA + sq*fB;
fBp = -sq*fA + cq*fB;
``` |
| [gdapy13_05] |

Here, the original pair of factors are fA and fB, and the rotated pair are fAp and fBp.

We now apply this procedure to factors 2 through 5 of the Atlantic Rock data set (i.e., the factors related to deviations about the typical rock). The varimax procedure is applied to all pairs of these factors and achieves convergence after several such iterations. The code for the loops is

---

MATLAB®

```
for iter = (1:3) % iterations
  for ii = (1:Nk)  % pairs of factors
    for jj = (ii+1:Nk)
        % spike factors i and j
        i=k(ii);
        j=k(jj);
        % copy factors from matrix to vectors
        fA = FP(i,:)';
        fB = FP(j,:)';
        % apply varimax rotation here
        . . .
        % put back into factor matrix, FP
        FP(i,:) = fAp';
        FP(j,:) = fBp';
        % accumulate rotation
        mA = MR(i,:)';
        mB = MR(j,:)';
        mAp =  cq*mA + sq*mB;
        mBp = -sq*mA + cq*mB;
        MR(i,:) = mAp';
        MR(j,:) = mBp';
    end
  end
end
```

[gdama13_05]

Python

```
for iter in range(3): # iterations
    for ii in range(Nk):  # pairs of factors
        for jj in range(ii+1,Nk):
            # spike factors i and j
            i=k[ii,0];
            j=k[jj,0];
            # copy factors from matrix to vectors
            fA = FP[i:i+1,0:M].T;
            fB = FP[j:j+1,0:M].T;
            # standard varimax procedure for rotation angle q
            ...
            # put back into factor matrix, FP
            FP[i:i+1,0:M] = fAp.T;
            FP[j:j+1,0:M] = fBp.T;
            # accumulate rotation
            mA = MR[i:i+1,0:M].T;
            mB = MR[j:j+1,0:M].T;
            mAp =  cq*mA + sq*mB;
            mBp = -sq*mA + cq*mB;
            MR[i:i+1,0:M] = mAp.T;
            MR[j:j+1,0:M] = mBp.T;
```

[gdapy13_05]

(A)   (B)

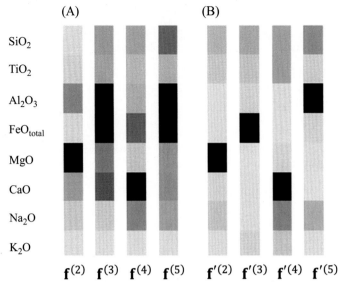

SiO$_2$

TiO$_2$

Al$_2$O$_3$

FeO$_\text{total}$

MgO

CaO

Na$_2$O

K$_2$O

$\mathbf{f}^{(2)}$ $\mathbf{f}^{(3)}$ $\mathbf{f}^{(4)}$ $\mathbf{f}^{(5)}$ $\mathbf{f}'^{(2)}$ $\mathbf{f}'^{(3)}$ $\mathbf{f}'^{(4)}$ $\mathbf{f}'^{(5)}$

FIG. 13.8 (A) Factors $\mathbf{f}^{(2)}$ through $\mathbf{f}^{(5)}$ of the Atlantic Rock data set, as calculated by singular-value decomposition. (B) Factors $\mathbf{f}'^{(2)}$ through $\mathbf{f}'^{(5)}$ after application of the varimax procedure. Scripts gdama13_05 and gdapy13_05.

Here, the rotated matrix of factors FP is initialized to the original matrix of factors F and then modified by the varimax procedure (omitted and replaced with a " . . . "), with each pass through the inner loop rotating one pair of factors. The procedure converges very rapidly, with three iterations of the outside loop being sufficient. The resulting factors (Fig. 13.8) are much spikier than the original ones. Each now involves mainly variations in one chemical element. For example, $\mathbf{f}'^{(2)}$ mostly represents variations in MgO, and $\mathbf{f}'^{(5)}$ mostly represents variations in Al$_2$O$_3$.

Another possible way of adding prior information is to find factors that are in some sense close to a set of prior factors. If closeness is measured by the L_1 or L_2 norm and if the constraint on the positivity of the factor loadings is omitted, then this problem can be solved using the techniques of Chapters 9 and 12. One advantage of this latter approach is that it permits one to test whether a particular set of prior factors can be factors of the problem (i.e., whether or not the distance between prior factors and actual factors can be reduced to an insignificant amount).

13.3 Q-mode and R-mode factor analysis

The eigenvectors $\mathbf{U}$ and $\mathbf{V}$ play completely symmetric roles in the singular-value decomposition of the sample matrix $\mathbf{S} = \mathbf{U}\mathbf{\Lambda}\mathbf{V}^T$. We introduced an asymmetry when we grouped them as $\mathbf{S} = (\mathbf{U}\mathbf{\Lambda})(\mathbf{V}^T) = \mathbf{C}\mathbf{F}$, to define the loadings $\mathbf{C}$ and factors $\mathbf{F}$.

This grouping is associated with the term, *R-mode factor analysis*. It is appropriate when the focus is on patterns among the elements, which is to say, reducing a large number of elements to a smaller number of factors. Thus, for example, we might note that the elements in the Atlantic Rock data set contain a pattern, associated with factor $\mathbf{f}^{(2)}$, in which Al$_2$O$_3$ and MgO are strongly and negatively correlated and another pattern, associated with factor $\mathbf{f}^{(3)}$, in which Al$_2$O$_3$ and FeO$_\text{total}$ are strongly and negatively correlated. The effect of these correlations is to reduce the effective number of elements, that is, to allow us to substitute a small number of factors for a large number of elements.

Alternately, we could have grouped the singular-value decomposition as $\mathbf{S} = (\mathbf{U})(\mathbf{\Lambda}\mathbf{V}^T)$, an approach associated with the term, *Q-mode factor analysis*. The equivalent transposed form $\mathbf{S}^T = (\mathbf{V}\mathbf{\Lambda})(\mathbf{U}^T)$ is more frequently encountered in the literature and is also more easily understood, since it can be interpreted as "normal factor analysis" applied to the matrix $\mathbf{S}^T$. The transposition has reversed the sense of samples and elements, so the factor matrix $\mathbf{U}^T$ quantifies patterns of variability among samples, in the same way that $\mathbf{V}^T$ quantifies patterns of variability among elements. To pursue this comparison further, consider an R-mode problem in which there is only one factor $\mathbf{v}^{(1)} = [1, 1, \cdots]^T$. This factor implies that all of the samples in the data table contain a 1:1 ratio of elements 1 and 2. Similarly, a Q-mode problem in which there is only one factor $\mathbf{u}^{(1)} = [1, 1, \cdots]^T$ implies that all of the elements in the transposed data table have a 1:1 ratio of samples 1 and 2. This approach is especially useful in detecting clustering among the samples; indeed, Q-mode factor analysis is often referred to as a form of *cluster analysis*.

13.4 Empirical orthogonal function analysis

Factor analysis need not be limited to data that contain actual mixtures of components. Given any set of vectors $\mathbf{s}^{(i)}$, one can perform the singular-value decomposition and represent $\mathbf{s}^{(i)}$ as a linear combination of a set of orthogonal factors. Even when the factors have no obvious physical interpretation, the decomposition can be useful as a tool for quantifying the similarities between the $\mathbf{s}^{(i)}$ vectors. This kind of factor analysis is often called *empirical orthogonal function (EOF)* analysis.

As an example of this application of factor analysis, consider the set of $N=14$ shapes shown in Fig. 13.9. These shapes might represent profiles of mountains or other subjects of interest. The problem we shall consider is how these profiles might be ordered to bring out the similarities and differences between the shapes. A geomorphologist might desire such an ordering because, when combined with other kinds of geological information, it might reveal the kinds of erosional processes that cause the shape of mountains to evolve with time.

We begin by discretizing each profile and representing it as a unit vector (in this case of length $M=11$). These unit vectors make up the matrix $\mathbf{S}$, on which we perform factor analysis. As the factors do not represent any particular physical object, there is no need to impose any positivity constraints on them, and we use the untransformed singular-value decomposition factors. The three most important EOFs (factors) (i.e., the ones with the three largest singular values) are shown in Fig. 13.10. The first EOF, as expected, appears to be simply an "average" mountain; the second seems to control the skewness, or degree of asymmetry, of the mountain; and the third, the sharpness of the mountain's summit. We emphasize, however, that this interpretation was made after the EOF analysis and was not based on any prior notions of how mountains might differ. We can then use the loadings as a measure of the similarities between the mountains. Since the amount of the first factor does not vary much between mountains, we use a two-dimensional ordering based on the relative amounts of the second and third factors in each of the mountain profiles (Fig. 13.11).

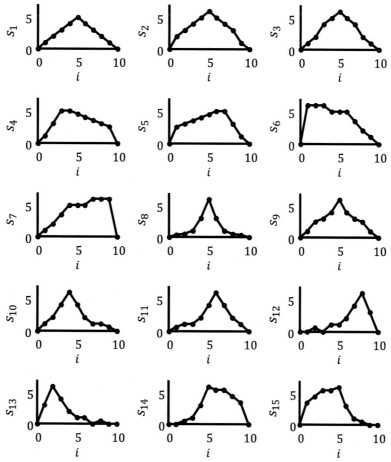

FIG. 13.9 A set of hypothetical mountain profiles. The variability of shape will be determined using factor analysis. Scripts gdama13_06 and gdapy13_06.

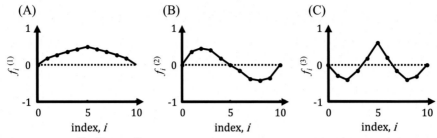

FIG. 13.10 The three largest factors $\mathbf{f}^{(1)}$, $\mathbf{f}^{(2)}$, and $\mathbf{f}^{(3)}$ in the representation of the mountain profiles in Fig. 13.9. (A) The factor with the largest singular value $\lambda_1 = 38.4$ has the shape of the average profile. (B) The factor with the second largest singular value $\lambda_2 = 12.7$ quantifies the asymmetry of the profiles. (C) The factor with the third largest singular value $\lambda_3 = 7.4$ quantifies the sharpness of the mountain summits. Scripts `gdama13_06` and `gdapy13_06`.

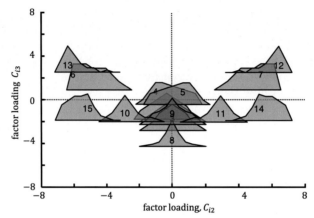

FIG. 13.11 The mountain profiles of Fig. 13.9, arranged according to the relative amounts of factors 2 and 3 contained in each profile's orthogonal decomposition, that is, by the size of the factor loadings C_{i2} and C_{i3}. Scripts `gdama13_06` and `gdapy13_06`.

EOF analysis is especially useful when the data have an ordering in space, time, or some other sequential variable. For instance, suppose that profiles in Fig. 13.12 are measured at sequential times. Then we can understand the model equation $\mathbf{S} = \mathbf{CF}$ to mean

$$S(t_i, x_j) = \sum_{k=1}^{p} C_k(t_i) F_k(x_j) \tag{13.14}$$

Here the quantity $S(t_i, x_j)$, which varies with both time and space, has been broken up into the sum of two sets of function, the EOFs $F_k(x_i)$, which vary only with space, and the corresponding loadings, $C_k(t_i)$, which vary only with time. A plot of the kth loading $C_k(t_i)$ as a function of time t_i reveals how the importance of the kth EOF varies with time.

This analysis can be extended to functions of two or more spatial dimensions by generalizing Eq. (13.14) to

$$S\left(t_i, \mathbf{x}^{(j)}\right) = \sum_{k=1}^{p} C_k(t_i) F_k\left(\mathbf{x}^{(j)}\right) \tag{13.15}$$

Here, the spatial observation points $\mathbf{x}^{(j)}$ are multidimensional, but they have been given a linear ordering through the index j. As an example, suppose that the data are a sequence of two-dimensional images, where each image represents a physical parameter observed on the (x, y) plane. This image can be unfolded (reordered) into a vector (see Section 5.12), which then becomes a row of the sample matrix $\mathbf{S}$. The resulting EOFs have this same ordering and must be folded back into two-dimensional images before being interpreted.

As an example, we consider a sequence of $N = 25$ images, each of which contains a grid of $20 \times 20 = 400$ pixels (Fig. 13.12). Each image represents the spatial variation of a physical parameter, such as pressure or temperature, at a fixed time, with the overall sequence being time sequential. These data are synthetic and are constructed by

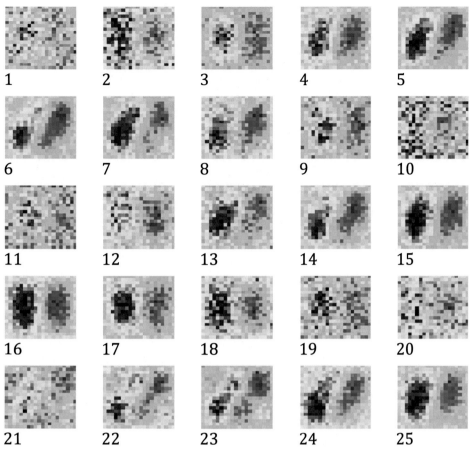

FIG. 13.12 Time sequence of $N=25$ images. Each image represents a parameter, such as pressure or temperature, that varies spatially in the (x,y) plane. Scripts gdama13_07 and gdapy13_07.

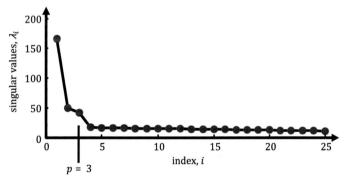

FIG. 13.13 Singular values λ_i of the image sequence shown in Fig. 13.12. Only $p=3$ singular values have significant amplitudes. Scripts gdama13_07 and gdapy13_07.

summing three spatial patterns (EOFs) with coefficients (loadings) that vary systematically with time, and then adding random noise. As expected, only $p=3$ singular values are found to be significant (Fig. 13.13). The corresponding EOFs and loadings are shown in Fig. 13.14. Had this analysis been based upon actual data, the time variation of each of the loadings, which have different periodicities, would be of special interest and might possibly provide insight into the physical processes associated with each of the EOFs. A similar example that uses actual ocean temperature data to examine the El Nino-Southern Oscillation climate instability is given by Menke and Menke (2011, their Section 8.5).

(A)

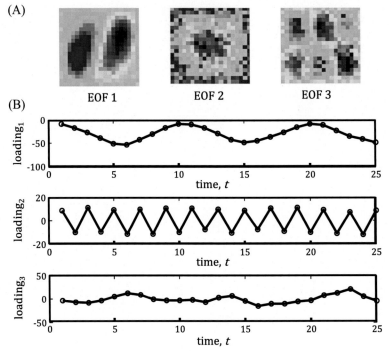

(B)

FIG. 13.14   (A) First three empirical orthogonal functions (EOFs) of the image sequence shown in Fig. 13.12. (B) Corresponding loadings as a function of time t. Scripts gdama13_07 and gdapy13_07.

13.5 Problems

13.1 Suppose that a set of $N > M$ samples are represented as $\mathbf{S} = \mathbf{C}_p \mathbf{F}_p$ where the matrix $\mathbf{F}_p$ contains $p < M$ factors whose values are prescribed (i.e., constitute prior information). (A) How can the loadings $\mathbf{C}_p$ be determined? (B) Write a script that implements your procedure for the case $M = 3, N = 10, p = 2$. (C) Make a three-dimensional plot of your results.

13.2 Write a script that verifies the varimax result given in Eq. (13.13). Use the following steps. (A) Compute the factors $\mathbf{f}'^A$ and $\mathbf{f}'^B$ for a complete suite of angles θ using the rotation

$$\mathbf{f}'^A = \cos(\theta)\mathbf{f}^A + \sin(\theta)\mathbf{f}^B \text{ and } \mathbf{f}'^B = -\sin(\theta)\mathbf{f}^A + \cos(\theta)\mathbf{f}^B \qquad (13.16)$$

(B) Compute and plot the variance $(\sigma_{f^A}^2 + \sigma_{f^B}^2)$ as a function of angle θ. (C) Note the angle of the minimum variance and verify that the angle is the one predicted by the varimax formula. (D) Verify that the factors corresponding to the angle are as stated in Eq. (13.13).

13.3 Suppose that a data set represents a function of three spatial dimensions, that is, with samples $S(t, x, y, z)$ on an evenly spaced three-dimensional grid. Write a script that unfolds these samples into a matrix $\mathbf{S}$.

References

Kaiser, H.F., 1958. The varimax criterion for analytic rotation in factor analysis. Psychometrika 23, 187–200.
Menke, W., Menke, J., 2011. Environmental Data Analysis with MATLAB. Academic Press, Elsevier Inc, Oxford. 263 pp.

14

Continuous inverse theory and tomography

In this chapter, we apply our experience of solving discrete inverse problems to continuous problems. As will be shown, almost all important concepts carry over. However, the mathematical level of our treatment is necessarily more advanced, for the techniques available for analyzing continuous functions are both more varied and more complicated than those associated with vectors and matrices. We have incorporated into our exposition short synopses of the most important techniques.

14.1   The Backus-Gilbert inverse problem

Discrete and continuous inverse problems differ in their assumptions about the model parameters. Whereas the model parameters are treated as a finite-length vector in discrete inverse theory, they are treated as a continuous function in continuous inverse theory. The standard form of the continuous linear inverse problem is

$$d_i^{obs} = \int\limits_V G_i(\mathbf{x})m(\mathbf{x})\mathrm{d}^L x \tag{14.1}$$

where $\mathrm{d}^L x$ is the volume element in the space of $\mathbf{x}$.

The "solution" of a discrete problem can be viewed as either an estimate of the model parameter vector $\mathbf{m}^{est}$ or a series of weighted averages of the model parameters $\mathbf{m}^{avg} = \mathbf{R}\mathbf{m}^{true}$, where $\mathbf{R}$ is the resolution matrix (see Section 4.3). If the discrete inverse problem is very underdetermined, then the interpretation of the solution in terms of weighted averages is most sensible, since a single model parameter is very poorly resolved. Continuous inverse problems can be viewed as the limit of discrete inverse problems as the number of model parameters becomes infinite, and they are inherently underdetermined. Attempts to estimate the model function $m(x)$ at a specific point $\mathbf{x} = \mathbf{x}'$ are futile. All determinations of the model function must be made in terms of local averages, which are simple generalizations of the discrete case, $m_i^{avg} = \sum_j G_{ij}^{-g} d_j^{obs} = \sum_j R_{ij} m_j^{true}$ where $R_{ij} = \sum_k G_{ik}^{-g} G_{kj}$

$$m^{avg}(\mathbf{x}') = \sum_{i=1}^N G_i^{-g}(\mathbf{x}')d_i^{obs} = \int\limits_V R(\mathbf{x}',\mathbf{x})m^{true}(\mathbf{x})\mathrm{d}^L x \tag{14.2}$$

where

$$R(\mathbf{x}',\mathbf{x}) = \sum_{i=1}^N G_i^{-g}(\mathbf{x}')G_i(\mathbf{x}) \tag{14.3}$$

Here, $G_i^{-g}(\mathbf{x}')$ is the continuous analog to the generalized inverse G_{ij}^{-g} and the averaging function $R(\mathbf{x}',\mathbf{x})$ (often called the *resolving kernel*) is the continuous analog to the model resolution matrix R_{ij}. The average is localized near the target point $\mathbf{x}'$ if the resolving kernel is peaked near $\mathbf{x}'$. The solution of the continuous inverse problem involves constructing the most peaked resolving kernel possible with a given set of measurements, that is, with a given set of data kernels $G_i(\mathbf{x})$. The spread of the resolution function is quantified by (compare with Eq. 5.22):

$$J(\mathbf{x}') = \int_V w(\mathbf{x}',\mathbf{x})R^2(\mathbf{x}',\mathbf{x})m(\mathbf{x})\mathrm{d}^L x \tag{14.4}$$

Here, $w(\mathbf{x}',\mathbf{x})$ is a nonnegative function that is zero at the point $\mathbf{x}'$ and that grows monotonically away from that point. One common choice is the quadratic function $w(\mathbf{x}',\mathbf{x}) = |\mathbf{x}' - \mathbf{x}|^2$. Other more complicated functions can be meaningful when the elements of $\mathbf{x}$ have an interpretation other than spatial position. After inserting the definition of the resolving kernel (Eq. 14.3) into the definition of the spread (Eq. 14.4), we find

$$\begin{aligned}
J(\mathbf{x}') &= \int_V w(\mathbf{x}',\mathbf{x})R^2(\mathbf{x}',\mathbf{x})\mathrm{d}^L x \\
&= \int_V w(\mathbf{x}',\mathbf{x}) \sum_{i=1}^N G_i^{-g}(\mathbf{x}')G_i(\mathbf{x}) \sum_{j=1}^N G_j^{-g}(\mathbf{x}')G_j(\mathbf{x})\mathrm{d}^L x \\
&= \sum_{i=1}^N \sum_{j=1}^N G_i^{-g}(\mathbf{x}')G_j^{-g}(\mathbf{x}') \int_V w(\mathbf{x}',\mathbf{x})G_i(\mathbf{x})G_j(\mathbf{x})\mathrm{d}^L x \\
&= \sum_{i=1}^N \sum_{j=1}^N G_i^{-g}(\mathbf{x}')G_j^{-g}(\mathbf{x}')[\mathbf{S}(\mathbf{x}')]_{ij}
\end{aligned} \tag{14.5}$$

where

$$[\mathbf{S}(\mathbf{x}')]_{ij} \equiv \int_V w(\mathbf{x}',\mathbf{x})G_i(\mathbf{x})G_j(\mathbf{x})\mathrm{d}^L x \tag{14.6}$$

Eq. (14.6) might be termed an *overlap integral*, because it is large only when the two data kernels overlap, that is, when they are simultaneously large in the same region of space. The continuous spread function has now been manipulated into a form completely analogous to the discrete spread function in Eq. (5.24). The generalized inverse that minimizes the spread of the resolution is the precise analogy of Eq. (5.34)

$$G_l^{-g}(\mathbf{x}') = \frac{\sum_{i=1}^N u_i [\mathbf{S}^{-1}(\mathbf{x}')]_{il}}{\sum_{i=1}^N \sum_{j=1}^N u_i [\mathbf{S}^{-1}(\mathbf{x}')]_{ij} u_j} \quad \text{where } u_i \equiv \int_V G_i(\mathbf{x})\mathrm{d}^L x \tag{14.7}$$

14.2 Trade-off of resolution and variance

As the data $\mathbf{d}$ are determined only up to some error quantified by the covariance matrix $[\text{cov } \mathbf{d}]$, the localized average $m^{avg}(\mathbf{x}')$ is determined up to some corresponding error

$$\text{var}(m^{avg}(\mathbf{x}')) = \sum_{i=1}^N \sum_{j=1}^N G_i^{-g}(\mathbf{x}')[\text{cov } \mathbf{d}]_{ij} G_j^{-g}(\mathbf{x}') \tag{14.8}$$

As in the discrete case, the generalized inverse that minimizes the spread of resolution may lead to a localized average with large error bounds. A slightly less localized average may be desirable because it may have much less error. This generalized inverse may be found by minimizing a weighted average of the spread of resolution and size of variance

$$\text{minimize } J'(\mathbf{x}') = \alpha \int_V w(\mathbf{x}',\mathbf{x})R^2(\mathbf{x}',\mathbf{x})\mathrm{d}^L x + (1-\alpha)\text{var}(m^{avg}(\mathbf{x}')) \tag{14.9}$$

The parameter α, which varies between 0 and 1, quantifies the relative weight given to spread of resolution and size of variance. As in the discrete case (see Section 5.11), the corresponding generalized inverse can be found by using Eq. (14.7), where all instances of $\mathbf{S}(\mathbf{x}')$ are replaced by

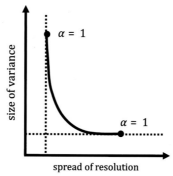

FIG. 14.1 Typical trade-off of resolution and variance for a linear continuous inverse problem. Note that the size(spread) function decreases monotonically with spread and that it is tangent to the two asymptotes at the endpoints, $\alpha=1$ and $\alpha=0$.

$$[\mathbf{S}'(\mathbf{x}')]_{ij} \equiv \alpha \int_V w(\mathbf{x}',\mathbf{x})G_i(\mathbf{x})G_j(\mathbf{x})\mathrm{d}^L x + (1-\alpha)[\mathrm{cov}\,\mathbf{d}]_{ij} \tag{14.10}$$

Backus and Gilbert (1968) prove a number of important properties of the trade-off curve (Fig. 14.1), which is a plot of size of variance against spread of resolution, including the fact that the variance decreases monotonically with spread.

14.3 Approximating a continuous inverse problem as a discrete problem

A continuous inverse problem can be converted into a discrete one with the assumption that the model function can be represented by a finite number M of coefficients, that is

$$m(\mathbf{x}) \approx \sum_{j=1}^{M} m_j f_j(\mathbf{x}) \tag{14.11}$$

Inserting Eq. (14.11) into the standard form of the continuous problem (Eq. 14.1) leads to a discrete problem $\mathbf{d}^{obs} = \mathbf{Gm}$

$$
\begin{aligned}
d_i^{obs} &= \int_V G_i(\mathbf{x})\left\{ \sum_{j=1}^{M} m_j f_j(\mathbf{x}) \right\}\mathrm{d}^L x = \\
&= \sum_{j=1}^{M}\left\{ \int_V G_i(\mathbf{x})f_j(\mathbf{x})\mathrm{d}^L x \right\}m_j = \sum_{j=1}^{M} G_{ij}m_j
\end{aligned}
\tag{14.12}
$$

$$\text{with } G_{ij} \equiv \int_V G_i(\mathbf{x})f_j(\mathbf{x})\mathrm{d}^L x$$

Many choices of $f_j(\mathbf{x})$ are possible, including those based on voxels, polynomial approximations, splines, and truncated Fourier series representations of $m(\mathbf{x})$. The particular choice of the functions $f_j(\mathbf{x})$ implies particular prior information about the behavior of $m(\mathbf{x})$, so the solution one obtains is sensitive to the choice.

Voxels are one commonly used set of functions. The model is assumed to be constant within certain subregions (voxels) V_j of the space of model parameters. In this case, $f_j(\mathbf{x})$ is unity inside the voxel and zero outside it, and m_j is the value of the model in the jth voxel. In one dimension, where the x-axis is divided into intervals of width Δx

$$m(x) \approx \sum_{j=1}^{M} m_j f_j(x) \text{ with } f_j(x) \equiv H(x - j\Delta x) - H(x - (j+1)\Delta x) \tag{14.13}$$

Here $H(x-\xi)$ is the Heaviside step function, which is zero for $x<\xi$ and unity for $x>\xi$.

The discrete data kernel involves an integral of the continuous version over the volume of the voxel. A common approximation assumes that the data kernel is approximately constant within the voxel, so that

$$G_{ij} \equiv \int_{V_j} G_i(\mathbf{x})\mathrm{d}^L x \text{ and } G_{ij} \approx G_i\left(\mathbf{x}_j^C\right)V_j \tag{14.14a,b}$$

Here, $\mathbf{x}_j^C$ and V_j are the center position and volume, respectively, of the jth voxel. Two different factors control the choice of the size of the voxels. The first is dictated by a prior assumption of the smoothness of the model. The second becomes important only when one uses Eq. (14.14b) in preference to Eq. (14.14a). Then the subregion must be small enough that the continuous data kernels $G_i(\mathbf{x})$ is approximately constant in the voxel. This second requirement often forces the voxels to be much smaller than dictated by the first requirement, so Eqs. (14.12) or (14.14a) should be used whenever the data kernels can be integrated analytically. Eq. (14.14b) fails completely whenever a data kernel has an integrable singularity within the subregion. This case commonly arises in problems involving the use of seismic rays to determine acoustic velocity structure.

14.4   Tomography and continuous inverse theory

The term "tomography" has come to be used in geophysics almost synonymously with the term "inverse theory." However, tomography is derived from the Greek word *tomos*, that is, slice, and denotes forming an image of an object from measurements made from slices (or rays) cutting through it. Consequently, we consider tomography to be a *subset* of inverse theory, distinguished by a special form of the data kernel that involves measurements made along rays. The model function in tomography is a function of two or more variables and is related to the data by

$$d_i = \int_{C_i} m(x(s), y(s))\, ds \tag{14.15}$$

Here, the model function $m(x,y)$ is integrated along a curved ray C_i having arc length s. This integral is equivalent to the one in a standard continuous problem (Eq. 14.2), in which the data kernel is $G_i(x,y) = \delta\{x(s) - x_i[y(s)]\}ds/dy$, where $\delta(x)$ is the Dirac delta function:

$$\iint_{-\infty}^{+\infty} G_i(x,y)m(x,y)dxdy = \iint_{-\infty}^{+\infty} \delta(x(s) - x_i[y(s)])\, m(x,y)\frac{ds}{dy}dxdy = \int_{C_i} m(x(s),y(s))ds \tag{14.16}$$

Here, x is supposed to vary with y along the curve C_i and y is supposed to vary with arc length s. The Dirac delta function $\delta(x - x_i)$ is further discussed in Section 14.7.

While the tomography problem is a special case of a continuous inverse problem, several factors limit the applicability of the formulas of the previous sections. First, the Dirac delta functions in the data kernel are not square integrable, so that the overlap integrals S_{ij} (Eq. 14.6) have nonintegrable singularities at points where rays intersect. Further, in three-dimensional cases, the rays may not intersect at all so that all the S_{ij} may be identically zero. Neither of these problems is insurmountable, and they can be overcome by replacing the rays with tubes of finite cross-sectional width. (Rays are often an idealization of a finite-width process anyway, as in acoustic wave propagation, where they are an infinitesimal wavelength approximation.) Since this approximation is equivalent to some statement about the smoothness of the model function $m(x,y)$, it often suffices to discretize the continuous problem by dividing m into constant-value subregions (voxels), where the subregions are large enough to guarantee a reasonable number containing more than one ray. The discrete inverse problem is then of the form $d_i = \sum_j^M G_{ij}m_j$, where the data kernel G_{ij} gives the arc length of the ith ray in the jth subregion. The concepts of resolution and variance, now interpreted in the discrete fashion of Chapter 5, are still applicable and of considerable importance.

14.5   The Radon transform

The simplest tomography problem involves straight-line rays and a two-dimensional model function $m(x,y)$ and is called *Radon's problem* (Radon, 1917). By historical convention, the straight-line rays C_i are parameterized by their perpendicular distance u from the origin and the angle θ that the perpendicular makes with the x-axis (Fig. 14.2). Position (x,y) and ray coordinates (u,s), where s is arc length, are related by

$$\begin{bmatrix} x \\ y \end{bmatrix} = \begin{bmatrix} \cos(\theta) & -\sin(\theta) \\ \sin(\theta) & \cos(\theta) \end{bmatrix} \begin{bmatrix} u \\ s \end{bmatrix} \text{ and } \begin{bmatrix} u \\ s \end{bmatrix} = \begin{bmatrix} \cos(\theta) & \sin(\theta) \\ -\sin(\theta) & \cos(\theta) \end{bmatrix} \begin{bmatrix} x \\ y \end{bmatrix} \tag{14.17}$$

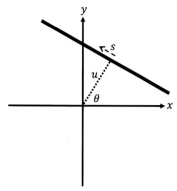

FIG. 14.2 The Radon transform is performed by integrating a function of (x, y) along straight lines (bold) parameterized by the arc length s, perpendicular distance u, and angle θ.

The forward problem is

$$d(u, \theta) = \int_{-\infty}^{+\infty} m(x = u\cos(\theta) - s\sin(\theta), y = u\sin(\theta) + s\cos(\theta))ds \qquad (14.18)$$

In realistic experiments, the data $d(u, \theta)$ is sampled only at discrete points $d_i \equiv d(u_i, \theta_i)$. Nevertheless, much insight can be gained into the behavior of Eq. (14.18) by regarding $d(u, \theta)$ as a continuous function. Eq. (14.18) is then an integral transform that transforms variables (x, y) into two new variables (u, θ) and is called a *Radon transform*. It can be symbolically abbreviated as

$$d(u, \theta) = \mathcal{R}_{xy \to u\theta}\, m(x, y) \qquad (14.19)$$

The inverse problem is to find the solution $m(x, y)$ that corresponds to the data $d(u, \theta)$. Or in other words, to perform the inverse Radon transform, symbolically abbreviated as

$$m(x, y) = \mathcal{R}_{u\theta \to xy}^{-1} d(u, \theta) \qquad (14.20)$$

14.6 The Fourier slice theorem

The inversion of the Radon transform relies on another integral transform, the Fourier transform. It transforms spatial position x into spatial *wave number* k_x

$$\widehat{f}(k_x) \equiv \int_{-\infty}^{+\infty} f(x)\exp(ik_x x)dx \equiv \mathcal{F}_{x \to k_x} f(x) \qquad (14.21)$$

Here, we use the hat to distinguish a function from its Fourier transform and abbreviate the transform with the symbols $\mathcal{F}_{x \to k_x}$. Important to this discussion is the fact that the inverse of the Fourier transform is known.

$$f(x) \equiv \frac{1}{2\pi}\int_{-\infty}^{+\infty} \widehat{f}(k_x)\exp(-ik_x x)dk_x \equiv \mathcal{F}_{k_x \to x}^{-1} \widehat{f}(k_x) \qquad (14.22)$$

It transforms wave number k_x back into position x and abbreviate it as $\mathcal{F}_{k_x \to x}^{-1}$. An important property of Fourier transforms is that they are unique. A given $f(x)$ implies a unique $\widehat{f}(k_x)$ and a given $\widehat{f}(k_x)$ implies a unique $f(x)$.

The following procedure inverts the Radon transform: First, the variable u is transformed to k_u via a Fourier transform

$$\widehat{d}(k_u, \theta) = \mathcal{F}_{u \to k_u} \mathcal{R}_{xy \to u\theta} m(x, y) =$$
$$= \int_{-\infty}^{+\infty} \int_{-\infty}^{+\infty} m(u\cos(\theta) - s\sin(\theta), u\sin(\theta) + s\cos(\theta))ds\ \exp(ik_u u)du \qquad (14.23)$$

Second, the variables of integration are changed from (s,u) to (x,y), using the fact that the Jacobian determinant $|\det \partial(x,y)/\partial(u,s)|$ is unity (see Eq. 14.17). The result is equivalent to two Fourier transforms in sequence, one that transforms x to $(k_u \cos\theta)$ and the other that transforms y to $(k_u \sin\theta)$

$$\widehat{d}(k_u,\theta) = \int_{-\infty}^{+\infty} \int_{-\infty}^{+\infty} m(x,y) \, \exp(ik_u x \cos\theta + ik_u y \sin\theta) dx dy$$

$$= \int_{-\infty}^{+\infty} \int_{-\infty}^{+\infty} m(x,y) \, \exp(ik_u x \cos\theta) dx \exp(ik_u y \sin\theta) dy \qquad (14.24)$$

$$= \mathcal{F}_{y \to k_u \sin\theta} \mathcal{F}_{x \to k_u \cos\theta} m(x,y) = \widehat{m}(k_u \cos\theta, k_u \sin\theta)$$

This result is called the *Fourier slice theorem*.

Third, one recognizes that the Fourier-transformed quantity $\widehat{d}(k_u,\theta)$ is simply the Fourier transform of the solution $\widehat{m}(k_x,k_y)$ evaluated along radial lines in the (k_x,k_y) plane (Fig. 14.3). If the Radon transform is known for all values of (u,θ), then the Fourier-transformed solution is known for all (k_x,k_y). As the Fourier transform and its inverse are unique, the Radon transform can be uniquely inverted if it is known for all values of (u,θ). Further, the method is practical, as the fast and widely available discrete Fourier transform algorithm can be used to approximate the integral Fourier transforms. However, u must be sampled sufficiently evenly that the $u \to k_u$ transform can be performed and θ must be sampled sufficiently finely that $\widehat{m}(k_x,k_y)$ can be sensibly interpolated onto a rectangular grid of (k_x,k_y) to allow the $k_x \to x$ and $k_y \to y$ transforms to be performed (see the discussion in scripts gdama14_01 and gdapy14_01). An example of the use of the Fourier slice theorem to invert tomography data is shown in Fig. 14.4.

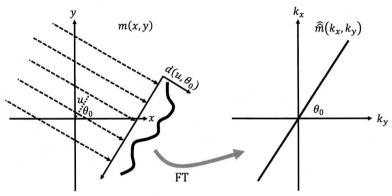

FIG. 14.3 (Left) The function $m(x,y)$ is integrated along a set of parallel lines (dashed) in a Radon transform to form the function $d(u,\theta_0)$. This function is called the *projection* of $m(x,y)$ at the angle θ_0. (Right) The Fourier slice theorem states that the Fourier transform (FT) of the projection is equal to the Fourier-transformed image evaluated along a line (bold) of angle θ_0 in the (k_x,k_y) plane.

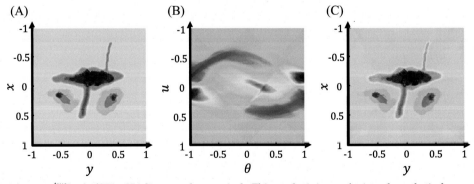

FIG. 14.4 (A) A test image $m^{true}(x,y)$ of 256×256 discrete values or pixels. This synthetic image depicts a hypothetical magma chamber beneath a volcano. (B) The Radon transform $d(u,\theta)$ of the image in (A), also evaluated at 256×256 points. (C) The image $m^{est}(x,y)$ reconstructed from its Radon transform by direct application of the Fourier slice theorem. Small errors in the reconstruction arise from the interpolation of the Fourier-transformed image onto a rectangular grid. Scripts gdama14_02 and gdapy14_02.

14.7 Linear operators

The continuous function $m(x)$ is the continuous analog to the vector $\mathbf{m}$ (Table 14.1). What is the continuous analog to the quantity $\mathbf{Lm}$, where $\mathbf{L}$ is a matrix? Let us call it $\mathcal{L}m$, that is, $\mathcal{L}$ operating on the function $m(x)$. In the discrete case, $\mathbf{Lm}$ is another vector, so by analogy $\mathcal{L}m$ is another function. Just as $\mathbf{L}$ is linear in the sense that $\mathbf{L}(\mathbf{m}^A + \mathbf{m}^B) = \mathbf{Lm}^A + \mathbf{Lm}^B$, we would like to choose $\mathcal{L}$ so that it is linear in the sense that $\mathcal{L}(m^A + m^B) = \mathcal{L}m^A + \mathcal{L}m^B$. A hint to the identity of $\mathcal{L}$ can be drawn from the fact that a matrix can be used to approximate derivatives and integrals, as well as multiplication by functions. Consider the examples,

$$\mathbf{L}^A = \frac{1}{\Delta x}\begin{bmatrix} -1 & 1 & 0 & \cdots & 0 \\ 0 & -1 & 1 & \cdots & 0 \\ \vdots & \vdots & \ddots & \ddots & \vdots \\ 0 & 0 & 0 & -1 & 1 \end{bmatrix} \text{ and } \mathbf{L}^B = \Delta x \begin{bmatrix} 1 & 0 & 0 & \cdots & 0 \\ 1 & 1 & 0 & \cdots & 0 \\ \vdots & \vdots & \ddots & \vdots & \vdots \\ 1 & 1 & 1 & 1 & 1 \end{bmatrix}$$

$$\text{and } \mathbf{L}^D = \begin{bmatrix} f_1 & 0 & 0 & \cdots & 0 \\ 0 & f_2 & 0 & \cdots & 0 \\ \vdots & \vdots & \ddots & \vdots & \vdots \\ 0 & 0 & 0 & 0 & f_M \end{bmatrix}$$

(14.25)

Here, $\mathbf{L}^A\mathbf{m}$ is the finite difference approximation to the first derivative $\mathcal{L}^A m \equiv dm/dx$, $\mathbf{L}^B\mathbf{m}$ is the Riemann sum approximation to the indefinite integral

$$\mathcal{L}^B m = \int_0^x m(x')dx'$$

(14.26)

and $\mathbf{L}^D\mathbf{m}$ approximates the product $\mathcal{L}^D m = f(x)m(x)$. Thus $\mathcal{L}$, which is called a linear operator, can be any linear combination of functions, integrals, and derivatives (Lanczos, 1961). In the general multidimensional case, where $m(\mathbf{x})$ is a

TABLE 14.1 Correspondences between discrete and continuous quantities.

Discrete	Continuous
i	x
j	ξ
m_i	$m(x)$
$\mathbf{L}$	$\mathcal{L}$
$\mathbf{L}^{-1}$	$\mathcal{L}^{-1}$
$\mathbf{I}$	$\mathcal{J}$
δ_{ij}	$\delta(x - \xi)$
$s = \mathbf{a}^T\mathbf{b}$	$s = (a, b)$
$s = (\mathbf{La})^T\mathbf{b}$	$s = (\mathcal{L}a, b)$
$\mathbf{L}^T$	$\mathcal{L}^\dagger$
$s = \mathbf{a}^T(\mathbf{L}^T\mathbf{b})$	$s = (a, \mathcal{L}^\dagger b)$
$\mathbf{b} = \mathbf{La}$	$b = \mathcal{L}a$
$b_i = \sum_j L_{ij}a_j$	$b(x) = (A(x, \xi), a(\xi))_\xi$ with $A(x, \xi) = \mathcal{L}_x\delta(x - \xi)$
$\mathbf{a} = \mathbf{L}^{-1}\mathbf{b}$	$a = \mathcal{L}^{-1}b$
$a_i = \sum_j [\mathbf{L}^{-1}]_{ij}b_j$	$a(x) = (B(x, \xi), b(\xi))_\xi$ with $\mathcal{L}_x B(x, \xi) = \delta(x - \xi)$

function of an N-dimensional position vector $\mathbf{x}$, $\mathcal{L}$ is built up of linear combinations of function, partial derivatives, and volume integrals. However, for simplicity, we restrict ourselves to the one-dimensional case here. Both the Radon and Fourier transforms (see Section 14.5) are linear operators and our abbreviation of them as $\mathcal{R}$ and $\mathcal{F}$ is a special case of the operator notation being developed here.

Now let us consider whether we can define an inverse operator $\mathcal{L}^{-1}$, that is, the continuous analog to the matrix inverse $\mathbf{L}^{-1}$. By analogy to $\mathbf{L}^{-1}\mathbf{Lm}=\mathbf{m}$, we want to choose $\mathcal{L}^{-1}$ so that $\mathcal{L}^{-1}\mathcal{L}m(x) = m(x)$. Actually, the derivative matrix $\mathbf{L}^A$ in Eq. (14.25) is problematical in this regard, for it is one row short of being square, and thus has no inverse. This corresponds to a function being determined only up to an integration constant when its derivative is known. This problem can be patched by adding to it a top row

$$\mathbf{L}^C = \frac{1}{\Delta x}\begin{bmatrix} 1 & 0 & 0 & \cdots & 0 \\ -1 & 1 & 0 & \cdots & 0 \\ 0 & -1 & 1 & \cdots & 0 \\ \vdots & \vdots & & \ddots & \vdots \\ 0 & 0 & 0 & -1 & 1 \end{bmatrix} \tag{14.27}$$

that fixes the value of m_1, that is, by specifying the integration constant. By analogy, $\mathcal{L}$ is not just a linear operator, but rather a linear operator plus one or more associated *boundary conditions*.

It is easy to verify that $\mathbf{L}^C\mathbf{L}^B = \mathbf{I}$. The analogous continuous result, that the derivative is the inverse operator of the integral, is the well-known *fundamental theorem of calculus*. It is denoted as $\mathcal{L}^C\mathcal{L}^B = \mathcal{J}$, where $\mathcal{J}$ is the identity operator. It obeys the rule

$$\mathcal{J}m(x) = m(x) \tag{14.28}$$

Although $\mathcal{J} \equiv 1$ is a mathematically acceptable identity operator, a far more useful version involves the *Dirac delta function* $\delta(x-\xi)$. It can be understood as a limited case of the Gaussian function $N(x-\xi,\sigma)$, as its variance σ approaches zero.

$$\delta(x-\xi) \equiv \lim_{\sigma\to 0} N(x-\xi,\sigma) \ \text{ with } \ N(x-\xi,\sigma) \equiv \frac{1}{\sqrt{2\pi}\sigma}\exp\left\{\frac{(x-\xi)^2}{2\sigma^2}\right\} \tag{14.29}$$

The Dirac function is zero everywhere except at the point $x=\xi$, where it has an integrable singularity. The integral of the product of a continuous function $f(x)$ and the finite-variance Gaussian function $N(x-\xi,\sigma)$ can be interpreted as a localized average of $f(x)$. The analogous integral involving the Dirac function is an indefinitely localized average, that is, the function, itself

$$f(x) = \int_{-\infty}^{+\infty} \delta(x-\xi)f(\xi)\,\mathrm{d}\xi \tag{14.30}$$

There is considerable advantage to defining the identity operator $\mathcal{J}$ as the integral of the product with the Dirac delta function, because it is a direct analog of a summation involving the identity matrix $f_i = \sum_j \delta_{ij}f_j$. The Dirac function is the continuous analog to the Kronecker delta δ_{ij} (the elements of the identity matrix).

Any linear differential equation can be abbreviated as $\mathcal{L}m(x) = f(x)$ and its solution can be abbreviated as $m(x) = \mathcal{L}^{-1}f(x)$. Because of linearity, solutions of such an equation superimpose; that is, if $m_A = \mathcal{L}^{-1}f_A$ and $m_B = \mathcal{L}^{-1}f_B$, then $(m_A + m_A) = \mathcal{L}^{-1}(f_A + f_B)$. This property is exploited in a useful method for solving a linear differential equation when its solution for a Dirac function source is known:

$$m(x) = \mathcal{L}^{-1}f(x) = \mathcal{L}^{-1}\int_{-\infty}^{+\infty}\delta(x-\xi)f(\xi)\,\mathrm{d}\xi$$
$$= \int_{-\infty}^{+\infty}\{\mathcal{L}^{-1}\delta(x-\xi)\}f(\xi)\,\mathrm{d}\xi = \int_{-\infty}^{+\infty}F(x,\xi)f(\xi)\,\mathrm{d}\xi \tag{14.31}$$
$$\text{where } \ \mathcal{L}F(x,\xi) = \delta(x-\xi)$$

The function $F(x,\xi)$ is known as the *Green function*. Eq. (14.31) indicates that inverse operator to a differential equation is a Green function integral.

Another important quantity ubiquitous in the formulas of inverse theory is the dot product between two vectors; written here generically as $s=\mathbf{a}^T\mathbf{b}=\sum_i a_i b_i$, where s is a scalar. The continuous analog is the integral

$$s = \int a(\mathbf{x})b(\mathbf{x})\mathrm{d}^L x \equiv (a,b) \tag{14.32}$$

where s is a scalar, $\mathrm{d}^L x$ is the volume element, and the integration is over the whole space of $\mathbf{x}$. This integral is called the *inner product* of the function $a(\mathbf{x})$ and $b(\mathbf{x})$ and is abbreviated as $s=(a,b)$. It is symmetric in its arguments, that is, $(a,b)=(b,a)$. As before, we restrict our discussion to the one-dimensional case:

$$s = \int a(x)b(x)\mathrm{d}x \equiv (a,b) \tag{14.33}$$

The linear inverse problem (Eq. 14.1) can be written in terms of an inner product $d_i^{obs}=(G_i(x),m(x))$. The identity operator and the Green function integral (Eq. 14.31) also can be written as one, as $m(x)=(\delta(x,\xi),m(\xi))_\xi$ and $m(x)=(F(x,\xi),f(\xi))_\xi$, respectively, where the subscript ξ is added to indicate the variable of integration. Similarly, the Fourier transform $\mathcal{F}$ and its inverse $\mathcal{F}^{-1}$ (see Section 14.6) can be written as inner products, that is, $\mathcal{F}m = (\exp(ik_x x), m(x))_x$ and $\mathcal{F}^{-1}\widehat{m} = \left((2\pi)^{-1}\exp(-ik_x x), \widehat{m}(k_x)\right)_{k_x}$. This latter result brings out that the analogy to the discrete transformations $\widehat{\mathbf{m}} = \mathbf{Tm}$ and $\mathbf{m} = \mathbf{T}^{-1}\widehat{\mathbf{m}}$, where in the Fourier case, $\mathbf{T}$ is a unary matrix whose elements are complex exponentials.

Many of the dot products that we encountered earlier in this book contained matrices, for example, $[\mathbf{La}]^T\mathbf{b}$, where $\mathbf{L}$ is a matrix and $\mathbf{a}$ and $\mathbf{b}$ are vectors. The continuous analog is an inner product containing a linear operator, that is, $(\mathcal{L}a,b)$. An extremely important property of the dot product $[\mathbf{La}]^T\mathbf{b}$ is that it can also be written as $\mathbf{a}^T[\mathbf{L}^T\mathbf{b}]$. We can propose the analogous relationship for linear operators:

$$(\mathcal{L}a, b) = \left(a, \mathcal{L}^\dagger b\right) \tag{14.34}$$

Here, the *dagger* symbol † is introduced as an analog to the matrix transpose symbol T. Eq. (14.34) serves as the definition of $\mathcal{L}^\dagger$; given a linear operator $\mathcal{L}$, then $\mathcal{L}^\dagger$ is whatever linear operator makes the equation true for all possible choices of a and b. The linear operator $\mathcal{L}^\dagger$ is also given a name, the *adjoint* of the linear operator $\mathcal{L}$.

Several approaches are available for determining the $\mathcal{L}^\dagger$ corresponding to a particular $\mathcal{L}$. The most straightforward is to start with the definition of the inner product. For instance, if $\mathcal{L} = c(x)$, where $c(x)$ is an ordinary function, then $\mathcal{L}^\dagger = c(x)$, too, as

$$(ca, b) = \int (ca)b\,\mathrm{d}x = \int a(cb)\,\mathrm{d}x = (a, cb) \tag{14.35}$$

When $\mathcal{L} = \mathrm{d}/\mathrm{d}x$, we can use integration by parts to find $\mathcal{L}^\dagger$

$$\int_{-\infty}^{+\infty} \frac{\mathrm{d}a}{\mathrm{d}x}b\,\mathrm{d}x = (ab)\big|_{-\infty}^{+\infty} - \int_{-\infty}^{+\infty} a\frac{\mathrm{d}b}{\mathrm{d}x}\mathrm{d}x \tag{14.36}$$

So, $\mathcal{L}^\dagger = -\mathrm{d}/\mathrm{d}x$, as long as the functions approach zero as x approaches $\pm\infty$. This same procedure, applied twice, can be used to show that $\mathcal{L} = -\mathrm{d}^2/\mathrm{d}x^2$ is *self-adjoint*, that is, it is its own adjoint. As another example, we derive the adjoint of the indefinite integral $\int_{-\infty}^x \mathrm{d}\xi$ by first writing it as

$$\int_{-\infty}^{x} a(\xi)\mathrm{d}\xi = \int_{-\infty}^{+\infty} H(x - \xi)a(\xi)\mathrm{d}\xi \tag{14.37}$$

where $H(x-\xi)$ is the Heaviside step function, which is unity if $x>\xi$ and zero if $x<\xi$. Then

$$(\mathcal{L}a, b) = \left(\int_{-\infty}^{x} a(\xi)\mathrm{d}\xi, b\right)_x = \int_{-\infty}^{+\infty}\int_{-\infty}^{x} a(\xi)\mathrm{d}\xi b(x)\mathrm{d}x =$$
$$\int_{-\infty}^{+\infty}\int_{-\infty}^{+\infty} H(x - \xi)a(\xi)\mathrm{d}\xi b(x)\mathrm{d}x = \int_{-\infty}^{+\infty} a(\xi)\int_{-\infty}^{+\infty} H(x - \xi)b(x)\mathrm{d}x\,\mathrm{d}\xi \tag{14.38}$$
$$= \int_{-\infty}^{+\infty}\int_{\xi}^{+\infty} b(x)\mathrm{d}x\, a(\xi)\mathrm{d}\xi = \left(a(\xi), \int_{\xi}^{+\infty} b(x)\mathrm{d}x\right)_\xi = (a, \mathcal{L}^\dagger b)$$

Hence, the adjoint of $\int_{-\infty}^x \mathrm{d}\xi$ is $\int_x^{+\infty}\mathrm{d}\xi$.

Another technique for computing an adjoint is to approximate the operator $\mathcal{L}$ as a matrix, transpose the matrix, and then to "read" the adjoint operator back by examining the matrix. In the cases of the integral and first derivative, the transposes are

$$\mathbf{L}^{BT} = \Delta x \begin{bmatrix} 1 & 1 & 1 & \cdots & 1 \\ 0 & 1 & 1 & \cdots & 1 \\ \vdots & \vdots & \ddots & \vdots & \vdots \\ 0 & 0 & 0 & 0 & 1 \end{bmatrix} \text{ and } \mathbf{L}^{CT} = \frac{1}{\Delta x}\begin{bmatrix} 1 & -1 & 0 & \cdots & 0 \\ 0 & 1 & -1 & \cdots & 0 \\ 0 & 0 & 1 & \cdots & 0 \\ \vdots & \vdots & \ddots & \vdots & \vdots \\ 0 & 0 & 0 & 0 & 1 \end{bmatrix} \tag{14.39}$$

The matrix $\mathbf{L}^{BT}$ represents the integral from progressively larger values of x to infinity and is equivalent to $\int_x^{+\infty} d\xi$. All but the last row of $\mathbf{L}^{CT}$ represents $-d/dx$, and the last row is the boundary condition. The boundary condition has moved from the top of $\mathbf{L}^C$ to the bottom of $\mathbf{L}^{CT}$. These results agree with those previously derived.

Adjoint operators have many of the properties of matrix transposes, including

$$
\left(\mathcal{L}^{\dagger}\right)^{\dagger} = \mathcal{L} \text{ and } \left(\mathcal{L}^{-1}\right)^{\dagger} = \left(\mathcal{L}^{\dagger}\right)^{-1}
$$

$$
\left(\mathcal{L}^A + \mathcal{L}^B\right)^{\dagger} = \left(\mathcal{L}^A\right)^{\dagger} + \left(\mathcal{L}^B\right)^{\dagger} \text{ and } \left(\mathcal{L}^A \mathcal{L}^B\right)^{\dagger} = \left(\mathcal{L}^B\right)^{\dagger} \left(\mathcal{L}^A\right)^{\dagger}
$$

(14.40)

14.8   The Fréchet derivative

Previously, we wrote the relationship between the model $m(x)$ and the data d_i as

$$
d_i = \int G_i(x)m(x)dx \equiv (G_i, m)
$$

(14.41)

But now, we redefine it in terms of perturbations around some reference model $m^{(0)}(x)$. We define $m(x) = m^{(0)}(x) + \delta m(x)$ where $m^{(0)}$ is a reference function and $\delta m(x)$ is a perturbation. If we write $d_i = d_i^{(0)} + \delta d_i$, where $d_i^{(0)} = (G_i, m^{(0)})$ is the data predicted by the reference model, then

$$
\delta d_i = \int G_i(x)\delta m(x)dx \equiv (G_i, \delta m)
$$

(14.42)

This equation says that a perturbation $\delta m(x)$ in the model causes a perturbation δd_i in the data. This formulation is especially useful in linearized problems, since then the data kernel can be approximate, that is, giving results valid only when δm is small.

Eq. (14.42) is reminiscent of the standard derivative formula that we have used in linearized discrete problems:

$$
\Delta d_i = \sum_{j=1}^{M} G_{ij}^{(0)} \Delta m_j \text{ with } G_{ij}^{(0)} \equiv \left.\frac{\partial d_i}{\partial m_j}\right|_{\mathbf{m}^{(0)}}
$$

(14.43)

The only difference is that $\Delta \mathbf{m}$ is a vector, whereas δm is a continuous function. Thus we can understand $G_i(x)$ in Eq. (14.42) as an analog to a derivative. We might use the notation

$$
G_i(x) \equiv \left.\frac{\delta d_i}{\delta m}\right|_{m^{(0)}}
$$

(14.44)

in which case it is called the *Fréchet derivative* of the datum d_i with respect to the model $m(x)$.

14.9   The Fréchet derivative of error

Fréchet derivatives of quantities other than the data are possible (Dahlen et al., 2000). One that is of particular usefulness is the Fréchet derivative of the error E with respect to the model, where the data $d(x)$ is taken to be a continuous variable.

$$
E \equiv \left(d^{obs} - d, d^{obs} - d\right) \text{ and } \delta E \equiv E - E^{(0)} = \left(\left.\frac{\delta E}{\delta m}\right|_{m^{(0)}}, \delta m\right)
$$

(14.45)

Here the error E is the continuous analog to the L_2 norm discrete error $E = (\mathbf{d}^{obs} - \mathbf{d})^T(\mathbf{d}^{obs} - \mathbf{d})$. Suppose now that the data are related to the model via a linear operator $d = \mathcal{L}m$. We compute the perturbation δE due to the perturbation δm as

$$\delta E = E - E^{(0)} = \left(d^{obs} - d, d^{obs} - d\right) - \left(d^{obs} - d^{(0)}, d^{obs} - d^{(0)}\right)$$

$$= -2\left(d, d^{obs}\right) + (d, d) + 2\left(d^{(0)}, d^{obs}\right) - \left(d^{(0)}, d^{(0)}\right)$$

$$= -2\left(d^{obs} - d^{(0)}, d - d^{(0)}\right) + \left(d - d^{(0)}, d - d^{(0)}\right) \tag{14.46}$$

$$= -2\left(d^{obs} - d^{(0)}, \delta d\right) + (\delta d, \delta d) \approx -2\left(d^{obs} - d^{(0)}, \delta d\right)$$

$$\approx -2\left(d^{obs} - d^{(0)}, \mathcal{L}\delta m\right)$$

Note that we have ignored a term involving the second-order quantity $(\delta d)^2$. Using the adjoint operator $\mathcal{L}^\dagger$, we find

$$\delta E = \left(-2\mathcal{L}^\dagger\left(d^{obs} - d^{(0)}\right), \delta m\right) \tag{14.47}$$

which implies that the Fréchet derivative of the error E is

$$\left.\frac{\delta E}{\delta m}\right|_{m^{(0)}} = -2\mathcal{L}^\dagger\left(d^{obs} - d^{(0)}\right) = -2\,\mathcal{L}^\dagger e^{(0)} \;\; \text{with} \;\; e^{(0)} \equiv d^{obs} - d^{(0)} \tag{14.48}$$

This result can be useful when solving the inverse problem using a gradient-descent method (see Section 11.9). As an example, consider the problem $d = \mathcal{L}m$, where

$$\mathcal{L}m = a\frac{d}{dx}m(x) + b\int_{-\infty}^{x} m(x')dx' \tag{14.49}$$

where a and b are constants. Using the adjoint relationships enumerated in Table 14.1, we find

$$\mathcal{L}^\dagger e^{(0)} = -a\frac{d}{dx}e^{(0)}(x) + b\int_{x}^{+\infty} e^{(0)}(x')dx' \tag{14.50}$$

and hence the Fréchet derivative of the error E is

$$\left.\frac{\delta E}{\delta m}\right|_{m^{(0)}} = -2\,\mathcal{L}^\dagger e^{(0)} = 2a\frac{d}{dx}e^{(0)}(x) - 2b\int_{x}^{+\infty} e^{(0)}(x')dx' \tag{14.51}$$

In order to use this result in a numerical scheme, one must discretize it; say by using voxels of width Δx and amplitude m for the model and d for the data, both of length $M = N$. Eq. (14.51) then yields the gradient $\partial E/\partial m_i$, which can be used to minimize E via the gradient-descent method (Fig. 14.5) (see also Section 11.9).

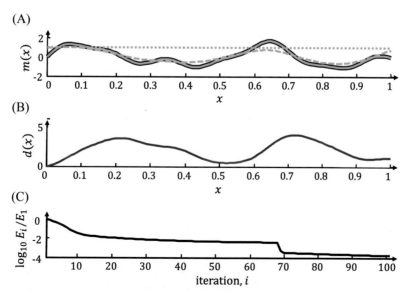

FIG. 14.5   Example of the solution of a continuous inverse problem using the gradient method to minimize the error E, where the adjoint method is used to compute ∇E. (A) An exemplary function $m^{true}(x)$ (green), trial function (dotted green), reconstructed function after 40 iterations (dashed green), and final reconstructed function $m^{est}(x)$ (green) after 15,890 iterations. (B) The data $d(t)$ satisfies $d(t) = \mathcal{L}m(t)$, where $\mathcal{L}$ is the linear operator discussed in the text. (C) Error E is a function of iteration number, for the first 100 iterations. Scripts gdama14_03 and gdapy14_03.

14.10 Back-projection

The formula for the Fréchet derivative of the error E (Eq. 14.48) is the continuous analog of the discrete gradient $\nabla E = -2\mathbf{G}^T(\mathbf{d}^{obs} - \mathbf{Gm})$. In Section 4.4, this expression for the gradient was set to zero, leading to the least-squares equation for the model parameters, $[\mathbf{G}^T\mathbf{G}]\mathbf{m} = \mathbf{G}^T\mathbf{d}^{obs}$. A similar result is achieved in the continuous case:

$$\left.\frac{\delta E}{\delta m}\right|_{m^{(0)}} = -2\,\mathcal{L}^\dagger\left(d^{obs} - d\right) = -2\,\mathcal{L}^\dagger\left(d^{obs} - \mathcal{L}m\right)$$

$$\text{or } \mathcal{L}^\dagger\mathcal{L}m = \mathcal{L}^\dagger d^{obs} \tag{14.52}$$

Inserting the identity operator into Eq. (14.52) yields

$$\left(\mathcal{L}^\dagger\mathcal{L} + \mathcal{J} - \mathcal{J}\right)m = \mathcal{L}^\dagger d^{obs} \text{ or}$$

$$m = \mathcal{J}m = \mathcal{L}^\dagger d^{obs} - \left(\mathcal{L}^\dagger\mathcal{L} - \mathcal{J}\right)m \tag{14.53}$$

In the special case that $\mathcal{L}^\dagger\mathcal{L} = \mathcal{J}$ (i.e., $\mathcal{L}^\dagger = \mathcal{L}^{-1}$), the solution is very simple, $m = \mathcal{L}^\dagger d^{obs}$. However, even in other cases the equation is still useful, since it can be viewed as a recursion relating an old estimate of the model parameters $m^{(i)}$ to a new one, $m^{(i+1)}$

$$m^{(i+1)} = \mathcal{L}^\dagger d^{obs} - \left(\mathcal{L}^\dagger\mathcal{L} - \mathcal{J}\right)m^{(i)} \tag{14.54}$$

If the recursion is started with $m^{(0)} = 0$, the new estimate is

$$m^{(1)} = \mathcal{L}^\dagger d^{obs} \tag{14.55}$$

As an example, suppose that $\mathcal{L}$ is the indefinite integral, so that

$$d^{obs}(x) = \mathcal{L}m(x) = \int_{-\infty}^x m(x')\mathrm{d}x' \text{ and } m^{(1)} = \mathcal{L}^\dagger d^{obs} = \int_x^{+\infty} d^{obs}(x')\mathrm{d}x' \tag{14.56a,b}$$

Eq. (14.56b) may seem crazy, as an indefinite integral is inverted by taking a derivative, not another integral. Yet this result, while approximate, is nevertheless quite good in some cases, at least up to an overall multiplicative factor (Fig. 14.6).

Eq. (14.56a) might be considered an ultrasimplified one-dimensional tomography problem, relating acoustic slowness $m(x)$ to travel time $d^{obs}(x)$. Note that the ray associated with travel time $d^{obs}(x)$ starts at $-\infty$ and ends at x; that is, the travel time at a point x depends only upon the slowness to the left of x. The formula for $m^{(1)}$ has a simple interpretation: the slowness at x is estimated by summing up all the travel times for rays ending at points $x' > x$; that is, summing up travel times only for those rays that sample the slowness at x. This process, which carries over to multidimensional tomography, is called *back-projection*. The inclusion, in a model parameter's own estimate, of just those data that are affected by it, is intuitively appealing.

FIG. 14.6   (A) True one-dimensional model $m^{true}(x)$. The data satisfy $d^{obs} = \mathcal{L}m^{true}$, where $\mathcal{L}$ is the indefinite integral. (B) Estimated model $m^{est} = \mathcal{L}^{-1}d^{obs}$, where $\mathcal{L}^{-1}$ is the first derivative. Note that $m^{est} = m^{true}$. (C) Back-projected model $m^{(1)} = \mathcal{L}^\dagger d^{obs}$, where $\mathcal{L}^\dagger$ is the adjoint of $\mathcal{L}$. Note that, up to an overall multiplicative factor $m^{(1)} \approx m^{true}$. Scripts gdama14_04 and gdapy14_04.

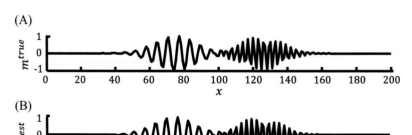

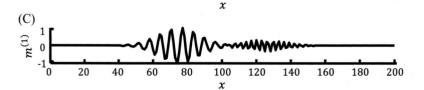

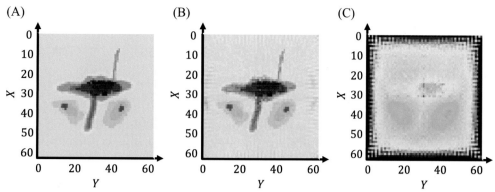

FIG. 14.7 Example of back-projection. (A) The true two-dimensional model, for which travel times associated with a dense and well-distributed set of rays are measured. (B) The estimated model, using damped least squares. (C) The estimated model, using back-projection. Scripts gdama14_05 and gdapy14_05.

Note that the accuracy of the back-projection depends upon the degree to which $\mathcal{L}^{\dagger}\mathcal{L} - \mathcal{J} = 0$ in Eq. (14.53). Some insight into this issue can be gained by examining the singular-value decomposition of a discrete data kernel $\mathbf{G} = \mathbf{U}_p\boldsymbol{\Lambda}_p\mathbf{V}_p^T$, which has transpose $\mathbf{G}^T = \mathbf{V}_p\boldsymbol{\Lambda}_p\mathbf{U}_p^T$ and generalized inverse $\mathbf{G}^{-g} = \mathbf{V}_p\boldsymbol{\Lambda}_p^{-1}\mathbf{U}_p^T$. Clearly, the accuracy of the approximation $\mathbf{G}^{-g} = \mathbf{G}^T$ will depend upon the degree to which $\boldsymbol{\Lambda}_p = \boldsymbol{\Lambda}_p^{-1}$, that is, to the degree to which the singular values have unit amplitude. The correspondence might be improved by scaling the rows of $\mathbf{G}$ and corresponding elements of $\mathbf{d}^{obs}$ by carefully chosen constants c_i; that is, G_{ij} is replaced with c_iG_{ij} and d_i^{obs} is replaced with $c_id_i^{obs}$, so that the singular values are of order unity. This analysis suggests a similar scaling of the continuous problem; that is, $\mathcal{L}$ is replaced by $c(x)\mathcal{L}$ and $d^{obs}(x)$ is replaced by $c(x)d^{obs}(x)$, where $c(x)$ is some function.

In tomography, a commonly used scaling is $c(x) = 1/L(x)$, where $L(x)$ is the length of ray x. The transformed data d^{obs}/L represents the average slowness along the ray. The back-projection process sums up the average slowness of all rays that interact with the model parameter at point x. This is a bit counterintuitive; one might expect that averaging the averages, as contrasted to summing the averages, would be more appropriate. Remarkably, the use of the summation introduces only long-wavelength errors into the image. An example of two-dimensional back projection is shown in Fig. 14.7.

14.11 Fréchet derivatives involving a differential equation

Eq. (14.1) links model parameters directly to the data. Some inverse problems are better analyzed when the link is indirect, through a field, $u(x)$. The model parameters are linked to the field, and the field to the data. Consider, for example, a problem in ocean circulation, where the model parameters $m(x)$ represent the force of the wind, the field $u(x)$ represents the velocity of the ocean water, and the data d_i represent the volume of water transported from one ocean to another. Wind forcing is linked to water velocity through a fluid-mechanical differential equation, and water transport is related to water velocity through an integral of the velocity across the ocean-ocean boundary. These relationships have the form

$$\mathcal{L}u^{(0)}(x) = m^{(0)}(x) \text{ and } \mathcal{L}\delta u(x) = \delta m(x) \tag{14.57a}$$

$$d_i^{(0)} = \left(h_i(x), u^{(0)}(x)\right) \text{ and } \delta d_i = (h_i(x), \delta u(x)) \tag{14.57b}$$

Here, Eq. (14.57a) is a differential equation with known boundary conditions, and Eq. (14.57b) is an inner product involving a known function $h_i(x)$. We have written the equations in terms of the unperturbed quantities $u^{(0)}$ and $d_i^{(0)}$ and the perturbations δm and δd_i. Symbolically, we can write the solution of the differential equation as

$$u^{(0)}(x) = \mathcal{L}^{-1}m^{(0)}(x) = \int F(x,\xi)m^{(0)}(x)d\xi \tag{14.58a}$$

$$\delta u(x) = \mathcal{L}^{-1}\delta m(x) = \int F(x,\xi)\delta m(\xi)d\xi \tag{14.58b}$$

where $F(x, \xi)$ is the Green function. Note that $\mathcal{L}^{-1}$ is a linear integral operator, whereas $\mathcal{L}$ is a linear differential operator. We now combine Eqs. (14.57b) and (14.58b):

$$\delta d_i = (h_i(x), \delta u(x)) = (h_i(x), \mathcal{L}^{-1}\delta m(x)) = (\mathcal{L}^{-1\dagger}h_i(x), \delta m(x)) = (\mathcal{L}^{\dagger-1}h_i(x), \delta m(x)) \tag{14.59}$$

Comparing to Eq. (14.42), we find

$$G_i(x) = \mathcal{L}^{\dagger-1}h_i(x) \quad \text{or} \quad \mathcal{L}G_i(x) = h_i(x) \tag{14.60}$$

The unperturbed field satisfies the equation $\mathcal{L}u^{(0)} = m^{(0)}$, and the data kernel satisfies the *adjunct equation* $\mathcal{L}^\dagger G_i = h_i$. In most applications, the scalar field $u^{(0)}$ must be computed by numerical solution of its differential equation. Thus the computational machinery is typically already in place to solve the adjoint differential equation and thus to construct the data kernel. Furthermore, in many cases, the operator $\mathcal{L}$ is self-adjoint, so one literally are solving differential equations that differs only in their right-hand sides.

An important special case is when the data is the field $u(x)$ itself, that is, $d_i = u(x_i)$. This choice implies that the weighting function in Eq. (14.60) is a Dirac delta function $h_i(x) = \delta(x - x_i)$, and that the data kernel is the Green function, say $Q(x, x_i)$, to the adjoint equation. Then

$$\delta d_i = (Q(x, x_i), \delta m(x)) \quad \text{with} \quad \mathcal{L}^\dagger Q(x, x_i) = \delta(x - x_i) \tag{14.61}$$

14.12   Case study: Heat source in problem with Newtonian cooling

As an example, we consider an object that is being heated by a flame, so we use time t instead of position x; the object is presumed to be of a spatially uniform temperature $u(t)$ that varies with time t. Suppose that temperature obeys the simple Newtonian heat flow equation

$$\mathcal{L}u(t) = \left\{\frac{\mathrm{d}}{\mathrm{d}t} + c\right\}u(t) = m(t) \tag{14.62}$$

where the function $m(t)$ represents the heating and c is a thermal constant. This equation implies that in the absence of heating, the temperature of the object decays away toward zero at a rate determined by c (a behavior called *Newtonian cooling*). We assume that the heating is restricted to some finite time interval, so that temperature satisfies the boundary condition $u(t \to \infty) = 0$.

The solution to the heat flow equation can be constructed from its Green function $F(t, \tau)$, which represents the temperature at time t due to an impulse of heat at time τ. It solves the equation

$$\left\{\frac{\mathrm{d}}{\mathrm{d}t} + c\right\}u(t) = \delta(t - \tau) \tag{14.63}$$

The solution to this equation can be shown to be

$$F(t, \tau) = H(t - \tau)\exp\{-c(t - \tau)\} \tag{14.64}$$

Here, $H(t - \tau)$ is the Heaviside step function, which is zero when $t < \tau$ and unity when $t > \tau$. This result can be verified by first noting that it satisfies the boundary condition and then by checking, through direct differentiation, that it solves the differential equation. The temperature of the object is zero before the heat pulse is applied, jumps to a maximum at the moment of application, and then exponentially decays back to zero at a rate determined by the constant c (Fig. 14.8A). A hypothetical heating function $m(t)$ and resulting temperature $u(t)$ are shown in Fig. 14.9B and C, respectively.

We now need to couple the temperature function $u(t)$ to observations, in order to build a complete forward problem. Suppose that a chemical reaction occurs within the object, at a rate that is proportional to temperature. We observe the amount of chemical product $d_i^{obs} = P(t_i)$, which is given by the integral

$$d_i^{obs} = P(t_i) = b\int_0^{t_i} u(t)\mathrm{d}t = b\int_0^\infty H(t_i - t)u(t)\mathrm{d}t = (bH(t_i - t), u) \tag{14.65}$$

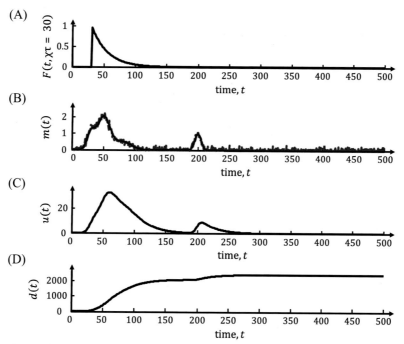

FIG. 14.8 Example of the solution of a continuous inverse problem involving a differential equation. (A) Green function $F(t, \tau)$ for $\tau = 30$. (B) True (black) and estimated (red) heat production function $m(t)$. (C) Temperature $u(t)$, which solves $\mathcal{L}u = m$. (D) Observed data $d(t)$, which is proportional to the integral of $u(t)$. Scripts `gdama14_06` and `gdapy14_06`.

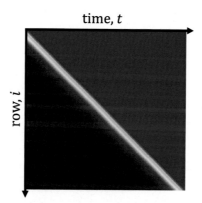

FIG. 14.9 Data kernel $G_i(t)$ for the continuous inverse problem involving a differential equation. See text for further discussion. Scripts `gdama14_06` and `gdapy14_06`.

where b is the proportionality factor. Thus $h_i = bH(t_i - t)$, where H is the Heaviside step function. The adjoint equation is

$$\mathcal{L}^\dagger G_i(t) = h_i \quad \text{or} \quad \left\{ -\frac{\mathrm{d}}{\mathrm{d}t} + c \right\} G_i(t) = bH(t_i - t) \tag{14.66}$$

because, as noted in Section 14.7, the adjoint of $\mathrm{d}/\mathrm{d}t$ is $-\mathrm{d}/\mathrm{d}t$ and the adjunct of the constant c is itself. The Green function of the adjoint equation can be shown to be

$$Q(t, \tau) = H(\tau - t) \exp\{ + c(t - \tau)\} \tag{14.67}$$

Note that $Q(t, \tau)$ is just a time-reversed version of $F(t, \tau)$, a relationship that arises because the two differential equations differ only by the sign of t. The data kernel $G_i(t)$ is given by the

Green function integral

$$G_i(t) = \int_0^\infty Q(t,\tau)h_i(\tau)\mathrm{d}\tau$$

$$= \int_0^\infty H(\tau - t)\exp\{ + c(t - \tau)\}bH(t_i - t)\mathrm{d}\tau$$

$$= b\int_0^\infty H(\tau - t)H(t_i - t)\exp\{ + c(t - \tau)\}\mathrm{d}\tau$$

$$= b\int_t^{t_i} \exp\{ + c(t - \tau)\}\mathrm{d}\tau$$

(14.68)

The integral is nonzero only in the case where $t_i > t$:

$$G_i(t) = \begin{cases} 0 & t_i \le t \\ -\dfrac{b}{c}\exp\{-c(t_i - t)\} & t_i > t \end{cases}$$

(14.69)

The data kernel $G_i(t)$ (Fig. 14.9) quantifies the effect of heat applied at time t on an observation made at time t_i. It being zero for times $t > t_i$ is a manifestation of causality; heat applied in the future of a given observation cannot affect it.

The problem that we have been discussing is completely linear. Furthermore, neither the differential equation relating $m(t)$ to $u(t)$ nor the equation relating $u(t)$ to d_i contains any approximations. The data kernel is therefore accurate for perturbations of any size and hence holds for $m(t)$ and d_i, as well as for the perturbations $\delta m(t)$ and δd_i:

$$d_i = (G_i(t), m(t))$$

(14.70)

We solve the inverse problem—determining the heating through observations of the chemical product—by first discretizing all quantities with a time increment Δt

$$d_i = (G_i(t), m(t)) \text{ becomes } \mathbf{d} = \mathbf{Gm}$$

$$\text{with } m_i \equiv m(i\Delta t) \text{ and } G_{ij} \equiv \Delta t\, G_i(j\Delta t)$$

(14.71)

Note that a factor of Δt has been included in the definition of $\mathbf{G}$ to account for the one that arises when the inner product is approximated by its Riemann sum. We assume that data is measured at all times between $t_1 = 0$ and t_N and then solve Eq. (14.71) by damped least squares (Fig. 14.8B). Some noise amplification occurs, owing to the very smooth data kernel. Rapid fluctuations in heating $m(t)$ have very little effect on the temperature $u(t)$, owing to the diffusive nature of heat, and thus are poorly constrained by the data. Adding smoothness constraints would lead to a better solution.

14.13   Derivative with respect to a parameter in a differential operator

The previous section was concerned with the effect on the data of a perturbation in forcing (meaning the function on the right-hand side of a differential equation). This section considers the effect on the data of a perturbation of a parameter in the differential operator $\mathcal{L}$ (Tromp et al., 2005).

As an example, suppose that the constant c in Eq. (14.63) is replaced with a function $c_0 + mc_1(t)$, where c_0 is a known constant, $c_1(t)$ is a known function, and m is a single model parameter (we will introduce multiple model parameters later):

$$\mathcal{L}u = \left\{\frac{\mathrm{d}}{\mathrm{d}t} + [c_0 + mc_1(t)]\right\}u(t) = f(t)$$

(14.72)

The forcing $f(t)$ is considered known. The partial derivative $\partial u/\partial m$ quantifies how a small change Δm in the model parameter m away from a reference value m_0 leads to a small perturbation δu of the solution away from its corresponding reference value $u^{(0)}$:

$$\delta u = \left.\frac{\partial u}{\partial m}\right|_{m_0}\Delta m$$

(14.73)

We derive an expression for this partial derivative, starting from the general case of the differential equation:

$$\mathcal{L}(m)u(t) = f(t)$$

(14.74)

and following these steps: First, the field is represented as $u = u^{(0)} + \delta u$, where the reference field $u^{(0)}$ solves the differential equation for $m = m_0$. Second, the operator $\mathcal{L}(m)$ is expanded in a Taylor series around the reference solution m_0 and terms higher than first order are discarded:

$$\mathcal{L}(m) \approx \mathcal{L}(m_0) + \left.\frac{\partial \mathcal{L}}{\partial m}\right|_{m_0} \Delta m \tag{14.75}$$

Third, these two approximations are inserted into the differential equation (Eq. 14.74) and only first-order terms are retained:

$$f \approx \left(\mathcal{L}(m_0) + \left.\frac{\partial \mathcal{L}}{\partial m}\right|_{m_0} \Delta m \right) \left(u^{(0)} + \delta u \right) \approx$$
$$\mathcal{L}(m_0) u^{(0)} + \left.\frac{\partial \mathcal{L}}{\partial m}\right|_{m_0} u^{(0)} \Delta m + \mathcal{L}(m_0) \delta u \tag{14.76}$$

Fourth, the reference equation $\mathcal{L}(m_0) u^{(0)} = f$ is subtracted out and the remaining terms are rearranged:

$$\mathcal{L}(m_0) \delta u = -\left.\frac{\partial \mathcal{L}}{\partial m}\right|_{m_0} u^{(0)} \Delta m \tag{14.77}$$

and finally, the differential equation is solved for δu

$$\delta u = \left(-\mathcal{L}^{-1}(m_0) \left.\frac{\partial \mathcal{L}}{\partial m}\right|_{m_0} u^{(0)} \right) \Delta m = \left.\frac{\partial u}{\partial m}\right|_{m_0} \Delta m$$
$$\text{with } \left.\frac{\partial u}{\partial m}\right|_{m_0} \equiv -\mathcal{L}^{-1}(m_0) \left.\frac{\partial \mathcal{L}}{\partial m}\right|_{m_0} u^{(0)} \tag{14.78}$$

This equation is known as the *Born approximation*. Here, the term in parentheses is identified as the derivative $\partial u / \partial m$. While the derivative $\partial \mathcal{L} / \partial m$ of an operator $\mathcal{L}$ might at first seem like a mysterious quantity, it is calculated simply by applying the $\partial / \partial m$ to functions of m and treating all other components of the operator as constants. For example, the operator in Eq. (14.72) has derivative:

$$\left.\frac{\partial \mathcal{L}}{\partial m}\right|_{m_0} = \left.\frac{\partial}{\partial m} \left\{ \frac{\mathrm{d}}{\mathrm{d}t} + [c_0 + m c_1(t)] \right\} \right|_{m_0} = 0 + [0 + c_1(t)] = c_1(t) \tag{14.79}$$

We now return to the case examined in the previous section where the data d_i are related to the field $u(t)$ through an inner product with a known functions $h_i(t)$, that is, $d_i = (h_i, u)$. Following Eq. (14.57), we have:

$$\delta d_i = (h_i, \delta u) = \left(h_i, \frac{\partial u}{\partial m} \Delta m \right) = \left(h_i, -\mathcal{L}^{-1}(m_0) \left.\frac{\partial \mathcal{L}}{\partial m}\right|_{m_0} u^{(0)} \right) \Delta m \tag{14.80}$$

The equation is now in the form $\delta d_i = (\partial d_i / \partial m) \Delta m$, the inner product part of the equation can be identified as the data kernel, $G_i = \partial d_i / \partial m$. Here, G_i lacks a second index because the problem has only one model parameter. We now use adjoint methods to manipulate the inner product:

$$G_i = \frac{\partial d_i}{\partial m} = \left(h_i, -\mathcal{L}_0^{-1} \frac{\partial \mathcal{L}_0}{\partial m} u^{(0)} \right) = -\left(\mathcal{L}_0^{\dagger -1} h_i, \frac{\partial \mathcal{L}_0}{\partial m} u^{(0)} \right) \equiv (\lambda_i, \xi)$$
$$\text{with } \mathcal{L}_0 \equiv \mathcal{L}(m_0) \text{ and } \frac{\partial \mathcal{L}_0}{\partial m} \equiv \left.\frac{\partial \mathcal{L}}{\partial m}\right|_{m_0} \tag{14.81}$$
$$\text{and } \lambda_i \equiv \mathcal{L}_0^{\dagger -1} h_i \text{ or } \mathcal{L}_0^{\dagger} \lambda_i \equiv h_i \text{ and } \xi \equiv -\frac{\partial \mathcal{L}_0}{\partial m} u^{(0)}$$

Here, we have used the fact that $\mathcal{L}^{\dagger -1} = \mathcal{L}^{-1\dagger}$ and have introduced the abbreviations $\mathcal{L}_0$, $\partial \mathcal{L}_0 / \partial m$, λ_i, and ξ. The quantities λ_i, $i = 1, \cdots, N$ are called the *adjoint fields*. There is one such field associated with each of the observations, and they each solve an *adjoint differential equation* $\mathcal{L}_0^{\dagger} \lambda_i = h_i$, the forcing for which is the data function h_i.

So far, our formulation is for one model parameter m only. The generalization to M model parameters m_j is very simple, as the only the quantity ξ depends upon it. The reference model parameters are now denoted $\mathbf{m}^{(0)}$ and the data kernel is

$$G_{ij}\left(\mathbf{m}^{(0)}\right) \equiv \left.\frac{\partial d_i}{\partial m_j}\right|_{\mathbf{m}^{(0)}} = \left(\lambda_i, \xi_j\right)$$

$$\text{with } \mathcal{L}_0 u^{(0)} = f(t) \text{ and } \mathcal{L}_0^\dagger \lambda_i = h_i(t) \text{ and } \xi_j \equiv -\frac{\partial \mathcal{L}_0}{\partial m_j} u^{(0)} \qquad (14.82)$$

$$\text{and } \mathcal{L}_0 \equiv \mathcal{L}\left(\mathbf{m}^{(0)}\right) \text{ and } \frac{\partial \mathcal{L}_0}{\partial m_j} \equiv \left.\frac{\partial \mathcal{L}}{\partial m_j}\right|_{\mathbf{m}^{(0)}}$$

The data kernel is calculated in the following four steps:

(1) Solve the unperturbed equation $\mathcal{L}(\mathbf{m}_0)u^{(0)}(t) = f(t)$ for the reference field $u^{(0)}(t)$.
(2) Solve the adjoint equations $\mathcal{L}_0^\dagger \lambda_i(t) = h_i(t)$ for the N adjoint fields $\lambda_i(t)$ (one for each data).
(3) Form the M quantities, $\xi_j(t) = -[\partial \mathcal{L}_0/\partial m_j]u^{(0)}(t)$ (one for each model parameter).
(4) Take the inner product of $\lambda_i(t)$ and $\xi_j(t)$, which yields the data kernel G_{ij}.

Most of the work is in solving the $M+1$ differential equations, but it is expedited by all the adjoint equations having the same left-hand side. The other two steps require minimal computational effort.

14.14 Case study: Thermal parameter in Newtonian cooling

As an example, consider the temperature equation (Eq. 14.72) when the forcing $f(t) = \delta(t)$ is a spike of heat at time zero and the perturbation in the material parameter $c_1(t) = \delta(t - t_0)$ is a spike at time t_0. We calculate the data kernel $G_i(m_0)$ when the data are integrals of the field (as in Eq. 14.80), so that $h_i = bH(t_i - t)$, and for the choice, $m_0 = 0$. The reference field is given by Eq. (14.64), with $\tau = 0$ and with c replaced by c_0:

$$u^{(0)}(t) = H(t)\exp(-c_0 t) \qquad (14.83)$$

where $H(t)$ is the Heaviside step function. The function $\xi(t)$ is then

$$\xi(t) = -c_0 H(t)\exp(-c_0 t)\delta(t - t_0) \qquad (14.84)$$

The Green function $Q(t - \tau)$ of the adjoint equation is given by Eq. (14.67), with c replaced by c_0. The adjoint field is given by the integral:

$$\begin{aligned}
\lambda_i(t) &= \int_{-\infty}^{+\infty} Q(t - \tau)h_i(\tau)\,d\tau \\
&= \int_{-\infty}^{+\infty} H(\tau - t)\exp\{c_0(t - \tau)\}bH(t_i - t)\,d\tau \\
&= b\exp(c_0 t)\int_t^{t_i} \exp(-c_0 \tau)\,d\tau \\
&= -\frac{b}{c_0}H(t_i - t)\exp(c_0 t)[\exp(-c_0 t_i) - \exp(-c_0 t)] \\
&= \frac{b}{c_0}H(t_i - t)[1 - \exp(-c_0(t_i - t))]
\end{aligned} \qquad (14.85)$$

As in the previous section, the adjoint field is "backward in time." The data kernel is then:

$$\begin{aligned}
G_i(m = 0) &= (\lambda_i, \xi) = \int_{-\infty}^{+\infty} \lambda_i(t)\xi(t)\,dt \\
&= -\frac{b}{c_0}\int_{-\infty}^{+\infty} H(t_i - t)[1 - \exp(-c_0(t_i - t))]H(t)\exp(-c_0 t)\delta(t - t_0)\,dt \\
&= -\frac{b}{c_0}H(t_i - t_0)H(t_0)[1 - \exp(-c_0(t_i - t_0))]\exp(-c_0 t_0)
\end{aligned} \qquad (14.86)$$

The data kernel's dependence on $H(t_0)$ implies it is zero when the spike in the material parameter occurs before the source time (at time zero). Its dependence on $H(t_i - t_0)$ implies that it is zero when the spike in the material parameter occurs after the time t_i that the data is measured. This data kernel is illustrated in Fig. 14.10.

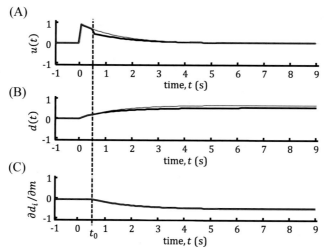

FIG. 14.10 Data kernel for a perturbation in the material parameter in the heat flow problem. (A) Reference field $u_0(t)$ (red curve) and perturbed field $u(t) \equiv u_0(t) + \delta u(t)$ (red curve) for a spike in the material parameter $c_1 = m\delta(t - t_0)$ at time t_0 (dashed line). (B) Corresponding unperturbed (black) and perturbed data (red). (C) Data kernel calculated by finite differences (black curve) and adjoint methods (red curve) match exactly. Scripts gdama14_07 and gdapy14_07.

One limitation of the data kernel derived before is that it is with respect to the very simple reference model $\mathbf{m}^{(0)} = 0$. In a realistic inverse problem, the reference model is updated during each iteration of Newton's method (see Section 14.5) and becomes increasingly complicated. Both differential equation and the adjoint differential equation must be solved numerically (as contrasted to analytically), because it becomes impractical to derive an analytic solution.

As an example, we estimate the thermal parameter $c(t)$. We use the same temperature equation $du/dt + c(t)u = \delta(t)$ as before, but we convert the impulsive forcing to the initial condition $u(t=0) = 1$. The correspondence is demonstrated by integrating the equation over a small time ε straddling the origin

$$\int_{-\varepsilon}^{+\varepsilon} \left(\frac{du}{dt} + c(t)u \right) dt = \int_{-\varepsilon}^{+\varepsilon} \delta(t) dt$$

$$u(\varepsilon) - u(-\varepsilon) + 2\varepsilon\overline{cu} = 1 \tag{14.87}$$

Here, $\overline{cu}$ is the average value of cu in the interval, which is finite. In the limit $\varepsilon \to 0$, and assuming $u(t < 0) = 0$, the initial value of the field is $u(0+) = 1$. As before, the data are given by $d_i = (h^{(i)}, u)$ with $h_i = bH(t_i - t)$. The true $c(t)$ is assumed to be the sum of a constant and a sinusoid. Noisy synthetic data then are used to estimate $c(t)$.

Key to implementing the inversion is the ability to solve both the original differential equation and the adjoint differential equation for variable $c(t)$ (in contrast to the example, before, where it was assumed constant). We use *Euler's method* because of its simplicity. The temperature is discretized as $u_i \equiv u(t_i)$ and the thermal parameter as $c_i \equiv c(t_i)$, with $t_i = (i-1)\Delta t$ and with Δt a time increment. The differential equation is approximated as

$$\frac{u_{k+1} - u_k}{\Delta t} + c_k = 0 \tag{14.88}$$

This equation can be rewritten as the forward recursion

$$u_k = u_{k-1} - \Delta t c_k u_{k-1} \quad \text{with} \quad u_1 = 1 \tag{14.89}$$

A simple generalization of the previous parameterization of the thermal function $c(t)$ is used

$$c(t) = c_0(t) + \sum_{i=1}^{M} m_i \delta(t - t_i) \tag{14.90}$$

so that the already-developed formula (Eq. 14.82) for the data kernel applies. The discrete version of this formula is $\mathbf{c} = \mathbf{c}_0 + \mathbf{m}/\Delta t$, where the factor of $1/\Delta t$ arises from smoothing out a delta function over one time increment. There is one adjoint field $\lambda^{(i)}$ and one data function $h^{(i)}$ for each datum d_i^{obs}. The backward recursion that arises when the adjoint field is discretized as

$$\lambda_k^{(i)} = \lambda_{k+1}^{(i)} - \Delta t c_{k+1} \lambda_{k+1}^{(i)} + \Delta t h_{k+1}^{(i)} \quad \text{with} \quad \lambda_N^{(i)} = 0 \tag{14.91}$$

FIG. 14.11 Inversion of noisy chemical concentration data for thermal parameter $c(t)$ appearing in the differential operator part of a differential equation (A) Linearized data kernel $G_{ij} = \partial d_i / \partial m_j$ for the first iteration of Newton's method (B). True thermal parameter (black), starting (green) and final (red). Scripts gdama14_09 and gdapy14_09.

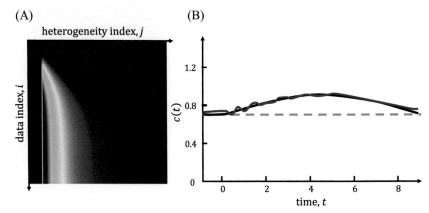

As previously, $\xi = -\delta(t - t_i)u(t)$ and $G_{ij} = (\lambda^{(i)}, \xi_j)$. The latter simplifies to $G_{ij} = -\lambda_j^{(i)}u_j$ because of the delta function. Consequently, $(N+1)$ differential equations need to be solved at each iteration (one for u and N for $\lambda^{(i)}$) during iteration of the linearized inversion. The data kernel (Fig. 14.11A) is very smooth and has its low amplitudes at large times, implying limited ability to estimate the latter part of $c(t)$. The data equation is $\mathbf{Gm} = \mathbf{d}^{obs} - \mathbf{d}$, where $\mathbf{d}$ is the data predicted by the current $\mathbf{u}$. Prior information of flatness (or smoothness) is needed to arrive at a well-behaved estimate.

The linearized inversion is started with $\mathbf{c}_0$ equal to a constant, and Generalized Least Squares is used to find an $\mathbf{m}$ that improves this starting value. In the example, the inversion includes prior information of smallness and flatness, with the strength of each hand-tuned to produce a well-behaved solution. The result of each iteration is an estimate of $\mathbf{m}$, which is used to update $\mathbf{c}_0$ according to the rule

$$\mathbf{c}_0 \text{ is replaced by } \mathbf{c}_0 + \mathbf{m}/\Delta t \tag{14.92}$$

The solution (Fig. 14.11B) is found to converge in about ten iterations. The sinusoid shape of $c(t)$ is well recovered but with some artifacts due to observational noise.

14.15   Application of the adjoint method to data assimilation

We return to the data assimilation problem introduced in Section 7.10 and discuss an adjoint method that is widely used in ocean and atmospheric circulation problems (Hall and Cacuci, 1983; Moore et al., 2004). The objective is to reconstruct a field $u(x, t)$ based on N noisy observations $u_i^{obs}(x_i, t_i)$, with the prior information that the field exactly satisfies a known differential equation. The only unknown is the boundary condition $u(x, t=0)$ or as is functionally equivalent (for first-order equations), an impulsive source $s(x, t) = m(x)\delta(t)$ acting at time $t = 0$ (with zero initial condition).

In order to facilitate our eventual discretization of the problem, we consider the source to be a line of M impulses of amplitude m_i and position x_i. Our intent that there will be one such source per pixel

$$m(x) = \sum_{i=1}^{M} m_i \delta(x - x_i) \tag{14.93}$$

The differential equation is then

$$\mathcal{L}u = s \text{ with } s(x, t) = \sum_{i=1}^{M} m_i \delta(x - x_i)\delta(t) \tag{14.94}$$

The initial conditions and boundary conditions (say at $x = 0$ and $x = x_L$) are zero. The individual observational errors e_i and total error E are

$$e_i \equiv u_i^{obs} - u\left(x_i^{obs}, t_i^{obs}\right) \text{ and } E \equiv \sum_{i=1}^{N} e_i^2 \tag{14.95}$$

Our solution strategy is to use the adjoint method to compute the derivative $\partial E/\partial m_i$ and then to use the gradient-descent method (Section 11.9) to find the m_i that minimizes the total error. This derivative is

$$\frac{\partial E}{\partial m_i} = -2 \sum_{j=1}^{N} e_j \frac{\partial u}{\partial m_i}\bigg|_{x_j^{obs},\,t_j^{obs}} \tag{14.96}$$

In order to apply the adjoint method, we must write this derivative as an inner product. This can be done by defining a continuous version of the error that contains delta functions

$$e(x,t) \equiv \sum_{i=1}^{N} e_i \delta\left(x - x_i^{obs}\right)\delta\left(t - t_i^{obs}\right) \tag{14.97}$$

and then noting that

$$\frac{\partial E}{\partial m_i} = -2\left(e, \frac{\partial u}{\partial m_i}\right) \tag{14.98}$$

Here, the inner product is over both space and time

$$(a,b) \equiv \int_0^{\infty} \int_0^{x_L} ab \, dt \, dx \tag{14.99}$$

The derivative $\partial u / \partial m_i$ is calculated by differentiating the solution of the differential equation:

$$u = \mathcal{L}^{-1}s \;\; \text{so} \;\; \frac{\partial u}{\partial m_i} = \mathcal{L}^{-1}\frac{\partial s}{\partial m_i} = \mathcal{L}^{-1}\delta(x - x_i)\delta(t) \tag{14.100}$$

This result is inserted into the inner product and the adjoint method is used to move the operator to its other side.

$$\frac{\partial E}{\partial s_i} = -2\left(e, \frac{\partial u}{\partial s_i}\right) = -2\left(e, \mathcal{L}^{-1}\delta(x - x_i)\delta(t)\right) = -2\left(\mathcal{L}^{-1\dagger}e, \delta(x - x_i)\delta(t)\right) \tag{14.101}$$

Using the rule that $\mathcal{L}^{-1\dagger} = \mathcal{L}^{\dagger-1}$, we introduce an adjoint field $\lambda \equiv \mathcal{L}^{\dagger-1}e$, so that

$$\frac{\partial E}{\partial m_i} = -2(\lambda, \delta(x - x_i)\delta(t)) \;\; \text{with} \;\; \mathcal{L}^{\dagger}\lambda = e \tag{14.102}$$

The adjoint field solves an adjoint differential equation whose source is the error. As is typical of problems involving first derivatives, the initial condition (at $t=0$) becomes a final condition (at $t=\infty$), implying that the adjoint equation must be solved backward in time.

Finally, the presence of the delta functions allows the inner product integral to be performed trivially

$$\frac{\partial E}{\partial m_i} = -2\lambda(x_i, t = 0) \tag{14.103}$$

Note that two differential equations must be solved to compute the derivative: the original equation, forward in time, and the adjoint equation, backward in time.

As an example, we return to the cooling rod problem that was introduced in Section 7.5. An insulated rod is being cooled from an initially warm temperature by holding its ends at zero temperature. Temperature in the rod satisfies the thermal diffusion equation, which has differential operator

$$\mathcal{L} = \frac{\partial}{\partial t} - \kappa \frac{\partial^2}{\partial x^2} \tag{14.104}$$

The differential equation $\mathcal{L}u = 0$ can be approximated with Euler's method as

$$\frac{u^{(m)} - u^{(m-1)}}{\Delta t} - \kappa \mathbf{D}^{(2)}\mathbf{u}^{(m-1)} = 0 \tag{14.105}$$

It is stepped forward in time from the initial condition $\mathbf{u}^{(1)} = \mathbf{m}$ via the recursion

$$\mathbf{u}^{(m)} = \mathbf{D}\mathbf{u}^{(m-1)} \;\; \text{with} \;\; \mathbf{D} \equiv \Delta t \kappa \mathbf{D}^{(2)} + \mathbf{I} \tag{14.106}$$

The adjoint differential operator is

$$\mathcal{L}^{\dagger} = -\frac{\partial}{\partial t} - \kappa \frac{\partial^2}{\partial x^2} \tag{14.107}$$

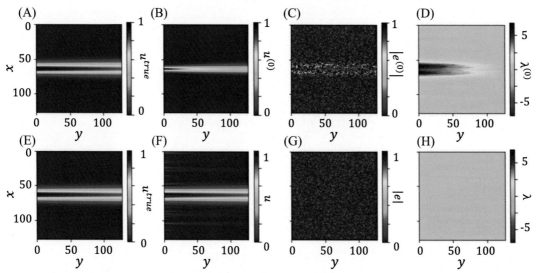

FIG. 14.12 Cooling rod data assimilation example. (Top row) Starting values. (A) True temperature u^{true} in a cooling rod. The initial condition is a Gaussian with half-width, $\sigma_x = 5$. (B) Starting temperature $u^{(0)}$ does not closely match the true temperature. (C) Absolute value of error $e^{(0)}$ of randomly chosen, noisy observations with respect to the starting temperature. (D) Adjoint field due to the source $e^{(0)}$. (Bottom row) Final values after 25 iterations of the gradient-descent method. (E) True temperature u^{true}. (F) Estimated temperature $u^{(25)}$ closely approximates the true temperature. (G) Absolute value of error $e^{(25)}$ is nearly zero. (H) Adjoint field due to source $e^{(25)}$ is nearly zero, too. Scripts gdama14_10 and gdapy14_10.

The adjoint equation $\mathcal{L}^\dagger \lambda = e$ can be approximated as

$$\frac{-\mathbf{u}^{(m)} + \mathbf{u}^{(m-1)}}{\Delta t} - \kappa \mathbf{D}^{(2)} \mathbf{u}^{(m)} = \mathbf{e}^{(m)} \tag{14.108}$$

and can be stepped backward in time from the final condition $\mathbf{u}^{(L_t)} = 0$ via the recursion

$$\mathbf{m}^{(m-1)} = \mathbf{D}\mathbf{m}^{(m)} + \Delta t \mathbf{e}^{(m)} \tag{14.109}$$

In the example, the true temperature field is computed by forward stepping the differential equation using a known impulsive source (corresponding to an initial condition) in the form of a Gaussian function with half-width, $\sigma_x = 5$ (Fig. 14.12A). Synthetic data are created by sampling the true temperature field at randomly selected positions and times and adding random noise. The starting guess for the initial condition is a Gaussian function with half-width, $\sigma_x = 2$ (Fig. 14.12B), which leads to an estimated temperature field with high error (Fig. 14.12C). The derivative $\partial E/\partial m_i$ is calculated and used to perturb the model parameters $\mathbf{m}$ in a direction that reduces the error. After 25 iterations, the gradient-descent method has reduced the error to negligible amount, resulting in a predicted temperature field (Fig. 14.12F) that closely matches the true one.

Although this data assimilation problem solves for the impulsive source $m(x)$ in many cases, it is the field predicted by this source, and not the source itself, that is of interest. That is, the process is used to fill in data gaps in a noisy data set, and possibly, to predict future values of the field, by stepping the differential equation forward additional increments of time.

14.16   Gradient of error for model parameter in the differential operator

The procedures developed in the previous section to handle model parameters in the source can easily be adapted to the case in which the model parameters are part of the differential operator. Eqs. (14.96) and (14.97), reproduced as follows, are still valid

$$\frac{\partial E}{\partial m_i} = -2\left(e, \frac{\partial u}{\partial m_i}\right) \text{ and } e(x,t) \equiv \sum_{i=1}^{N} e_i \delta\left(x - x_i^{obs}\right) \delta\left(t - t_i^{obs}\right) \tag{14.110}$$

In the previous case, where $s(x,t,\mathbf{m})$ is a source that depends on model parameter $\mathbf{m}$, we used the formula $\partial u/\partial m_i = \mathcal{L}^{-1} \partial s/\partial m_i$, which leads to

$$\frac{\partial E}{\partial m_i} = -2\left(\lambda, \frac{\partial s}{\partial m_i}\right) \text{ with } \mathcal{L}_0^\dagger \lambda = e \tag{14.111}$$

Now, we use the Born approximation (Eq. 14.78) for $\partial u / \partial m_i$, which leads to

$$\frac{\partial E}{\partial m_i} = 2\left(e, \mathcal{L}_0^{-1}\frac{\partial \mathcal{L}_0}{\partial m_i}u^{(0)}\right) = \left(\mathcal{L}_0^{-1\dagger}e, 2\frac{\partial \mathcal{L}_0}{\partial m_i}u^{(0)}\right) \tag{14.112}$$

Defining, as before, an adjoint field λ and functions ξ_i, we find

$$\frac{\partial E}{\partial m_i} = (\lambda, \xi_i) \text{ with } \mathcal{L}_0^\dagger \lambda = e \text{ and } \xi_i = 2\frac{\partial \mathcal{L}_0}{\partial m_i}u^{(0)} \tag{14.113}$$

An idiosyncrasy of the adjoint equation is that it needs to be solved for a source e that consists of a sum of Dirac delta functions (one per observation and each with the amplitude of the corresponding error), but this poses no special problem when it is solved numerically. The process is often referred to as *injecting* the error into the numerical solution. Furthermore, although a function ξ_i depends on the particular model parameter m_i, the adjoint field λ does not. Thus all M derivatives $\partial E / \partial m_i$ can be computed by solving just two differential equations, the original equation for $u^{(0)}$ and the adjoint equation for λ. The M derivatives do require a total of M inner products to be computed, but they typically require much less computation than do the two numerical solutions. Furthermore, in a voxel representation of the model parameters, the derivative $\partial \mathcal{L}_0 / \partial m_i$ contains Dirac delta functions in space, which allows the spatial part of inner product to be computed trivially, leaving just the time integration (a simplification discussed further in Section 16.4). Thus the inverse problem for estimating the model parameters can be very efficiently solved using the gradient-descent method (Section 11.9). However, as the whole process of computing the gradient must be repeated for every iteration of the method, a very substantial amount of computation is required to arrive at a solution.

14.17   Problems

14.1 This inverse problem is due to Robert Parker. Consider a radially stratified sphere of radius $R=1$ with unknown density $\rho(r)$. Its mass $M=4\pi \int \rho(r)\, r^2 dr$ and its moment of inertia $I=(8\pi/3) \int \rho(r)\, r^4 dr$ are measured. What is the best averaging kernel that can be designed that is centered about $\frac{1}{2}R$? As all the integrals involve only rational functions and all the matrices are 2×2, the problem can be solved analytically. Do as much of the problem as you can, analytically; do the parts that you cannot, numerically.

14.2 Suppose that a function $m(x)$ is discretized in each of two ways: (A) by dividing it into M rectangles, each of height m_i and width Δx; and (B) by representing it as a sum of M overlapping Gaussians, spaced at regular intervals Δx, prescribed variance σ^2, and unknown amplitude m_n'

$$m(x) = \sum_{n=1}^{M} m_n' \frac{\Delta x}{\sqrt{2\pi}\sigma} \exp\left\{-\frac{(x-n\Delta x)^2}{2\sigma^2}\right\}$$

Discuss the advantages and disadvantages of each representation. Include the respective effect of damping in the two cases.

14.3 How sensitive are the results of tomography to noise in the data $d(u,\theta)$? Run experiments with the Fourier slice script (gdama14_02 or gdapy14_02). Try two different types of noise: (A) uncorrelated random noise drawn from a Normal pdf and (B) a few large outliers. Comment upon the results.

14.4 Modify the example in Section 14.9 to use a truncated Fourier sine series representation of the model:

$$m^{(0)}(x) = \sum_{n=1}^{K} m_n^{(0)} \sin\left(\frac{n\pi x}{L}\right) \text{ and } \delta m^{(0)}(x) = \sum_{n=1}^{K} \delta m_n^{(0)} \sin\left(\frac{n\pi x}{L}\right)$$

where $m_n^{(0)}$ and $\delta m_n^{(0)}$ are scalar coefficients, $K=15$ and $0 < x < L$. You will need to use the chain rule

$$\left.\frac{\delta E}{\delta m_n}\right|_{\mathbf{m}^{(0)}} = \left(\left.\frac{\delta E}{\delta m}\right|_{\mathbf{m}^{(0)}}, \left.\frac{\delta m}{\delta m_n}\right|_{\mathbf{m}^{(0)}}\right)$$

14.5 Suppose that the Green function $F(t,\tau)$ of a particular differential equation $\mathcal{L}^F F(t,\tau) = \delta(t-\tau)$ depends only upon time differences, that is, $F(t,\tau)=F(t-\tau)$. Denote the Green function integral as the linear operator $\left(\mathcal{L}^F\right)^{-1}$. (A) Use the formula for the inner product to derive $\left(\mathcal{L}^F\right)^{-1\dagger}$. Call the function within this operator F'. (B) Suppose the differential equation $\mathcal{L}^{Q\dagger}Q(t,\tau) = \delta(t-\tau)$ has Green function $Q(t,\tau)$. Denote the Green function integral as the linear operator $\left(\mathcal{L}^{Q\dagger}\right)^{-1}$. By comparing $\left(\mathcal{L}^F\right)^{-1\dagger}$ and $\left(\mathcal{L}^{Q\dagger}\right)^{-1}$, establish the relationship between F' and Q. Explain your results in words.

14.6 The file, `../Data/prob1106_data.txt`, which can be generated by `gdama14_11` or `gdama14_11`, contains three columns, corresponding to $\mathbf{t}$, $\delta\mathbf{m}^{true}$, and $\mathbf{d}^{true}$ (all of length, N) in the temperature problem discussed in Section 14.13. The parameters associated with this problem include a source time of zero, $c_0 = 0.7$, $b = 0.4$, $m_0 = 0$, and

$$c_1(t) = \sum_{i=1}^{M} \delta m_i^{true}\delta(t - t_i)$$

where $M = N$. Construct a set of synthetic data $\mathbf{d}^{obs}$ by adding random noise to $\mathbf{d}^{true}$ and then use the data kernel in Eq. (14.86) to invert for $\delta\mathbf{m}^{est}$. Compare your result with $\delta\mathbf{m}^{true}$. You will need to add prior information of smallness or smoothness to the inversion, in order to prevent amplification of high-frequency noise.

References

Backus, G.E., Gilbert, J.F., 1968. The resolving power of gross earth data. Geophys. J. Roy. Astron. Soc. 16, 169–205.

Dahlen, F.A., Hung, S.-H., Nolet, G., 2000. Fréchet kernels for finite frequency travel times I. Theory. Geophys. J. Int. 141, 157–174.

Hall, M.C.G., Cacuci, D.G., 1983. Physical interpretation of the adjoint functions for sensitivity analysis of atmospheric models. J. Atmos. Sci. 40, 2537–2546.

Lanczos, C., 1961. Linear Differential Operators. Van Nostrand-Reinhold, Princeton, NJ.

Moore, A.M., Arango, H.G., Di Lorenzo, E., Cornuelle, B.D., Miller, A.J., Neilson, D.J., 2004. A comprehensive ocean prediction and analysis system based on the tangent linear and adjoint of a regional ocean model. Ocean Model 7, 227–258.

Radon, J., 1917. On the determination of functions from their integral values along certain manifolds, (in German). Reports on the Proceedings of the Royal Saxonian Academy of Sciences at Leipzig, Mathematical and Physical Section, 69, 262–277, Leipzig (Germany).

Tromp, J., Tape, C., Liu, Q., 2005. Seismic tomography, adjoint methods, time reversal and banana-doughnut kernels. Geophys. J. Int. 160, 195–216.

15

Sample inverse problems

15.1 An image enhancement problem

Suppose that a camera moves slightly during the exposure of an image, so that the picture is blurred. Also suppose that the amount and direction of motion are known. Can the image be "unblurred?"

The optical sensor in the camera consists of rows and columns of light-sensitive elements that measure the total amount of light received during the exposure. For simplicity, we shall consider that the camera moves parallel to a row of pixels. The data d_i are a set of numbers that represent the amount of light recorded at each of N pixels. Because the scene's brightness varies continuously, this is properly a problem in continuous inverse theory. We shall discretize it, however, by assuming that the scene can be adequately approximated by a row of small square elements, each with a constant brightness. These elements form M model parameters m_i. As the camera's motion is known, it is possible to calculate each scene element's relative contribution to the light recorded at a given camera pixel. For instance, if the camera moves through three scene elements during the exposure, then each camera element records the average of three neighboring scene brightness

$$d_i = \frac{1}{3}\left(m_{i-1} + m_i + m_{i+1}\right) \tag{15.1}$$

Note that this is a linear equation and can be written in the form $\mathbf{d} = \mathbf{Gm}$, where

$$\mathbf{G} = \frac{1}{3}\begin{bmatrix} 1 & 1 & 1 & 0 & 0 & \cdots & 0 \\ 0 & 1 & 1 & 1 & 0 & 0 & 0 \\ \vdots & \vdots & \vdots & \ddots & \ddots & \ddots & \vdots \\ 0 & 0 & 0 & 0 & 1 & 1 & 1 \end{bmatrix} \tag{15.2}$$

In general, there will be several more model parameters than data, so the problem will be underdetermined. In this case, $M = N + 2$, so there will be at least two null vectors. In fact, the problem is purely underdetermined, and there are only two null vectors, which can be identified by inspection as

$$\mathbf{m}^{(1)null} = \begin{bmatrix} 1 & 0 & -1 & 1 & 0 & -1 & \cdots & 1 & 0 & -1 \end{bmatrix}^T$$
$$\mathbf{m}^{(1)null} = \begin{bmatrix} 0 & 1 & -1 & 0 & 1 & -1 & \cdots & 0 & 1 & -1 \end{bmatrix}^T \tag{15.3}$$

These null vectors have rapidly fluctuating elements, which indicates that at best only the longer wavelength features can be recovered. To find a solution to the inverse problem, we must add prior information. We shall use a simplicity constraint and find the scene of shortest length. If the data are normalized so that zero represents gray (with white negative and black positive), then this procedure in effect finds the scene of least contrast that fits the data. We therefore estimate the solution with the damped minimum-length generalized inverse as

$$\mathbf{m}^{est} = \mathbf{G}^T\left[\mathbf{GG}^T + \varepsilon\mathbf{I}\right]^{-1}\mathbf{d}^{obs} \tag{15.4}$$

(where the damping is added in case $\mathbf{GG}^T$ is singular). Eq. (15.4) is solved in two steps: first, the linear equation $[\mathbf{GG}^T + \varepsilon\mathbf{I}]\mathbf{u} = \mathbf{d}^{obs}$, is solved for an intermediate vector $\mathbf{u}$; and second, the solution is calculated as $\mathbf{m}^{est} = \mathbf{G}^T\mathbf{u}$.

As an example, we shall solve the problem for the data kernel given earlier, for a blur width of 100 and with an image $M = 1500$ pixels wide. We first select a true scene (Fig. 15.1A) and blur it by multiplication with the data kernel to produce synthetic data (Fig. 15.1B). Note that the blurred image is much smoother than the true scene. We now try to

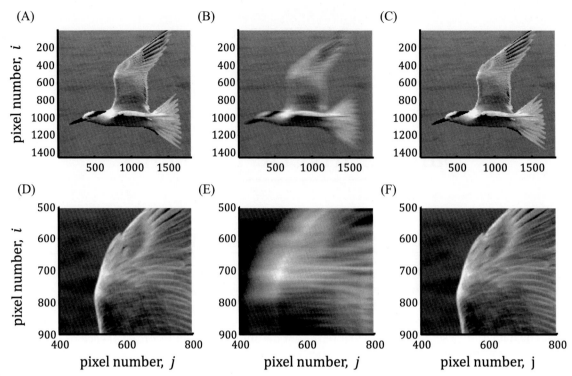

FIG. 15.1 Example of removing blur from image. (A) True image. (B) Image blurred with 100 pixel-wide boxcar filter. (C) Estimated image, unblurred using the minimum-length generalized inverse. (D–F) Enlargement of portion of images (A–C). Scripts gdama15_01 and gdapy15_01.

invert back for the true scene by premultiplying it by the generalized inverse. The result (Fig. 15.1C) has not only correctly captured some of the sharp features in the original scene but also contains a short wavelength oscillation not present in the true scene. This error results from creating a reconstructed scene containing an incorrect combination of null vectors.

In a script, the inverse problem for each row of the image is solved separately, with Eq. (15.4) evaluated in a loop over rows:

```
                              MATLAB®
gdaepsilon=1.0e-6; % damping, just in case GGT is singular
GGT = G*G' + gdaepsilon*eye(J);
GMG = G'/GGT;
for i = (1:I) % process image row-wise
    dobsrow = dobs(i,:)';
    mestrow = GMG*dobsrow;
    mest(i,:)=mestrow';
end
                                                   [gdama15_01]
```

```
                              Python
gdaepsilon=1.0e-6; % damping, just in case GGT is singular
GGT = np.matmul(G,G.T)+gdaepsilon*np.identity(J);
GMG = np.matmul(G.T,la.inv(GGT));
for i in range(I): # process image row-wise
    dobsrow = (dobs[i:i+1,0:J]).T;
    mestrow = np.matmul(GMG,dobsrow);
    mest[i:i+1,0:J]=mestrow.T;
                                                   [gdapy15_01]
```

(A)

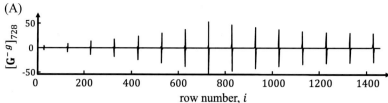

(B)

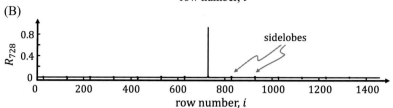

FIG. 15.2 (A) Central row (row 728) of the generalized inverse of the image deblurring problem. (B) Central row (row 728) of the corresponding resolution matrix. The sidelobes are about 6% of the amplitude of the central peak. Scripts gdama15_01 and gdapy15_01.

Here GGT is the damped minimum-length matrix product $[\mathbf{GG}^T + \varepsilon\mathbf{I}]$, where the small ε regularizes the problem in the event that $\mathbf{GG}^T$ is singular, and GMG is the generalized inverse $\mathbf{G}^{-g}$. We note that the matrix $\mathbf{G}$ is simple enough for the matrix product $[\mathbf{GG}^T]$ to be computed analytically. For example, for a blur of width 3

$$[\mathbf{GG}^T] = \frac{1}{9}\begin{bmatrix} 3 & 2 & 1 & 0 & 0 & 0 & \cdots & 0 \\ 2 & 3 & 2 & 1 & 0 & 0 & \cdots & 0 \\ 1 & 2 & 3 & 2 & 1 & 0 & \cdots & 0 \\ \vdots & \vdots & \vdots & \vdots & \vdots & \ddots & \ddots & \vdots \\ 0 & 0 & 0 & 0 & 0 & 1 & 2 & 3 \end{bmatrix} \tag{15.5}$$

In this problem, which deals with moderate-sized matrices, the effort saved in using the analytic version over a numerical computation is negligible. In larger inverse problems, however, considerable saving can result from a careful analytical treatment of the structures of the various matrices.

Each row of the generalized inverse states how a particular model parameter is constructed from the data. We might expect that this blurred image problem would have a localized generalized inverse, meaning that each model parameter would be constructed mainly from a few neighboring data. By examining the generalized inverse $\mathbf{G}^{-g}$, we see that this is not the case (Fig. 15.2A). We also note that the process of blurring is an integrating process; it sums neighboring data. The process of unblurring should be a sort of differentiating process, subtracting neighboring data. The generalized inverse is, in fact, just that.

The resolution matrix for this problem is seen to be very "spiky" (Fig. 15.2B); the diagonal elements are several orders of magnitude larger than the off-diagonal elements. On the other hand, the off-diagonal elements are all of uniform size, indicating that if the estimated model parameters are interpreted as localized averages, they are in fact not completely localized. It is interesting to note that the Backus-Gilbert inverse (not shown), which generally gives very localized resolution kernels, returns only the blurred image itself. The solution to this problem contains an unavoidable trade-off between the width of resolution and the presence of sidelobes.

15.2 Digital filter design

Suppose that two signals $d(t)$ and $g(t)$ are known to be related by convolution with a filter $m(t)$:

$$d(t) = g(t)*m(t) \equiv \int_{-\infty}^{+\infty} g(t-\tau)\, m(\tau)\, \mathrm{d}\tau \tag{15.6}$$

where τ is a dummy integration variable (see Section 2.11). Can $m(t)$ be found if $g(t)$ and $d(t)$ are known?

As the signals and filter are continuous functions, this is a problem in continuous inverse theory. We shall analyze it, however, by approximating the functions as time series. Each function will be represented by its value at a set of points spaced equally in time (with interval Δt). We shall assume that the signals are transient (have a definite beginning and end) so that $d(t)$ and $g(t)$ can be represented by time series of length N. Typically, the advantage of relating two signals by a filter is realized only when the filter length M is shorter than either signal, so $M < N$ is presumed. The convolution integral can then be approximated by the sum

$$d_i = \Delta t \sum_{j=1}^{M} g_{i-j+1} \, m_j \tag{15.7}$$

where $g_i \equiv 0$ if $i < 1$ or $i > N$. This equation is linear in the unknown filter coefficients and can be written in the form $\mathbf{Gm} = \mathbf{d}$, where

$$\mathbf{G} = \Delta t \begin{bmatrix} g_1 & 0 & 0 & 0 & \cdots & 0 \\ g_2 & g_1 & 0 & 0 & \cdots & 0 \\ g_3 & g_2 & g_1 & 0 & \cdots & 0 \\ \vdots & \vdots & \vdots & \vdots & \ddots & \vdots \\ g_N & g_{N-1} & g_{N-2} & g_{N-3} & \cdots & g_{N-M+1} \end{bmatrix} \tag{15.8}$$

The output time series d_i is identified with the data and the input time series m_i with the model parameters. The equation is an overdetermined linear system for $M < N$ filter coefficients. For this problem to be consistent with the tenets of probability theory, however, g_i must be known exactly, while d_i must contain uncorrelated Gaussian noise of uniform variance.

Many approaches are available for solving this inverse problem. The simplest is to use the least-squares equation $[\mathbf{G}^T\mathbf{G}]\mathbf{m}^{est} = \mathbf{G}^T\mathbf{d}^{obs}$. This formulation is especially attractive because the matrices $[\mathbf{G}^T\mathbf{G}]$ and $\mathbf{G}^T\mathbf{d}^{obs}$ can be computed analytically as

$$\mathbf{G}^T\mathbf{G} = (\Delta t)^2 \begin{bmatrix} \sum_{i=1}^{N} g_i^2 & \sum_{i=2}^{N} g_i g_{i-1} & \cdots \\ \sum_{i=2}^{N} g_i g_{i-1} & \sum_{i=1}^{N-1} g_i^2 & \cdots \\ \vdots & \vdots & \ddots \end{bmatrix} \text{ and } \mathbf{G}^T\mathbf{d}^{obs} = (\Delta t) \begin{bmatrix} \sum_{i=1}^{N} g_i d_i^{obs} \\ \sum_{i=2}^{N} g_{i-1} d_i^{obs} \\ \vdots \end{bmatrix} \tag{15.9}$$

Furthermore, the summations along each diagonal of $\mathbf{G}^T\mathbf{G}$ usually can be approximated adequately as all having the same limits, so that the resulting matrix is Toeplitz (meaning that it has constant diagonals). Then, the $\mathbf{G}^T\mathbf{G}$ matrix contains the autocorrelation of $\mathbf{g}$ (denoted $\mathbf{g} \star \mathbf{g}$) and $\mathbf{G}^T\mathbf{d}^{obs}$ the *cross-correlation* of $\mathbf{g}$ with $\mathbf{d}^{obs}$ (denoted $\mathbf{g} \star \mathbf{d}^{obs}$)

$$\left[\mathbf{G}^T\mathbf{G}\right]_{ij} = (\Delta t)^2 [\mathbf{g} \star \mathbf{g}]_{|i-j|+1} \text{ and } \left[\mathbf{G}^T\mathbf{d}^{obs}\right]_i = (\Delta t)\left[\mathbf{g} \star \mathbf{d}^{obs}\right]_i$$
$$\text{where } [\mathbf{a} \star \mathbf{b}]_i \equiv \sum_{j=1}^{M} a_j \, b_{i+j-1} \tag{15.10}$$

However, one often finds that the least-squares solution is very rough, because the high frequency components of $m(t)$ are poorly constrained (i.e., the problem is really mixed-determined).

An alternative approach is to incorporate prior information of smoothness using Generalized Least Squares $[\mathbf{F}^T\mathbf{F}]\mathbf{m}^{est} = \mathbf{F}^T\mathbf{d}^{obs}$, where the matrix $\mathbf{F}$ contains both $\mathbf{G}$ and a smoothness matrix $\mathbf{H}$ scaled by a damping parameter ε^2 (see Eq. 4.55). If the biconjugate gradient method is used to solve this equation (see Section 4.17), then the only quantity that need be computed is the product $\mathbf{F}^T(\mathbf{Fv}) = \mathbf{G}^T(\mathbf{Gv}) + \varepsilon^2 \mathbf{H}^T(\mathbf{Hv})$, where $\mathbf{v}$ is an arbitrary vector. The $\mathbf{G}^T(\mathbf{Gv})$ product can be computed extremely efficiently starting with $\mathbf{g}$ and using a standard cross-correlation function (see the comments accompanying gda_filtermul() function for details).

As an example of filter construction, we shall consider a time series $g(t)$, which represents a recording of the sound emitted by a seismic exploration airgun (Fig. 15.3A). Signals of this sort are used to detect layering at depth in the Earth through echo sounding. Ideally, a very spiky sound from the airgun is best because it allows echoes from layers at depth to be most easily detected. Engineering constraints, however, limit the airgun signal to a series of pulses. We shall attempt, therefore, to find a filter $m(t)$ that, when applied to the airgun pulse, produces a signal spike or delta function $\mathbf{d} = [0,0,0,\ldots,0,1,0,\ldots,0]^T$ centered on the largest pulse in the original signal. This filter can then be applied to the recorded echo soundings to remove the reverberation of the airgun and reveal the layering of the Earth. The least-squares estimate $\mathbf{m}^{est}$ of the filter, computed for this example using weighted damped least squares with both smoothness and length constraints and solved with the biconjugate gradient method, is shown in Fig. 15.3B and the resulting signal $\mathbf{d}^{pre} = \mathbf{g} * \mathbf{m}^{est}$ in Fig. 15.3C. Although the reverberations are reduced in amplitude, they are by no means completely removed.

The image deblurring problem in Section 15.1 has the form of a deconvolution, which implies that the methodology of this section can be applied to its solution.

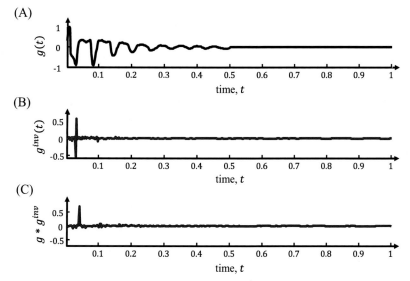

FIG. 15.3 (A) An airgun signal $g(t)$ after Smith (1975). Ideally, the inverse filter $g^{inv}(t)$ when convolved with $g(t)$ should produce the spike $\delta(t-t_0)$ centered at time t_0. (B) Estimate of the inverse filter $g^{inv}(t)$ for $t_0 = 0.04$ computed via generalized least squares with prior information on solution size and smoothness. (C) The convolution of $g(t)$ with the estimated $g^{inv}(t)$. While not a perfect spike, the result is significantly spikier than the airgun signal $g(t)$. Scripts `gdama15_02` and `gdapy15_02`.

15.3 Adjustment of crossover errors

Consider a set of radar altimetry data from a remote-sensing satellite. These data consist of measurements of the distance from the satellite to the surface of the Earth directly below the satellite. If the altitude of the satellite with respect to the Earth's center was known, then these data could be used to measure the elevation of the surface of the Earth. Unfortunately, while the height of the satellite during each orbit is approximately constant, its exact value is unknown. Since the orbits crisscross the Earth, one can try to solve for the satellite height in each orbit by minimizing the overall crossover error (Kaula, 1966).

Suppose that there are M orbits and that the unknown altitude of the satellite during the ith orbit is m_i. We shall divide these orbits into two groups (Fig. 15.4), the ascending orbits (when the satellite is traveling north) and the descending orbits (when the satellite is traveling south). The ascending and descending orbits intersect at, say, N points. At one such point ascending orbit number A_i intersects with descending orbit D_i (where the numbering refers to the ordering in **m**). At this point, the two orbits have measured a satellite-to-Earth distance of, say, S_{A_i} and S_{D_i}, respectively. The elevation of the ground is $m_{A_i} - S_{A_i}$, according to the data collected on the ascending orbit and $m_{D_i} - S_{D_i}$, according to the data from the descending orbit. The crossover error at the ith intersection is $e_i = (m_{A_i} - S_{A_i}) - (m_{D_i} - S_{D_i}) = (m_{A_i} - m_{D_i}) - (S_{A_i} - S_{D_i})$. The assertion that the crossover error should be zero leads to a linear equation of the form $\mathbf{Gm} = \mathbf{d}^{obs}$, where

$$G_{ij} \equiv \delta_{j,A_i} - \delta_{j,D_i} \quad \text{and} \quad d_i^{obs} \equiv S_{A_i} - S_{D_i} \tag{15.11}$$

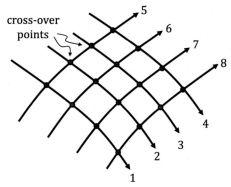

FIG. 15.4 Descending tracks 1–4 intersect ascending tracks 5–8 at 16 points. The height of the satellite along each track is determined by minimizing the crossover error at the intersections.

Each row of the data kernel contains one 1, one -1, and $(M-2)$ zeros. Note that the matrix **G** is extremely sparse; we would be well advised to define it as such in a script.

Initially, we might assume that we can use simple least squares to solve this problem. We note, however, that the solution is always to a degree underdetermined. Any constant can be added to all the ms without changing the cross-over error since the error depends only on the difference between the elevations of the satellite during the different orbits. This problem is therefore mixed-determined. We should therefore impose the prior constraint that $\sum_i^M m_i = 0$. Although this constraint is not physically realistic (implying as it does that the satellite has on average zero altitude), it serves to remove the underdeterminacy. Any desired constant can subsequently be added to the solution.

This constraint can be approximately implemented with damped least squares

$$\mathbf{m}^{est} = \left[\mathbf{G}^T\mathbf{G} + \varepsilon^2\mathbf{I}\right]^{-1}\mathbf{G}^T\mathbf{d}^{obs} \tag{15.12}$$

as long as damping parameter ε^2 is chosen carefully, the solution will very closely approximate the exact one. An example is shown in Fig. 15.5. As an alternative to damped least squares, the constraint can be implemented exactly using the Lagrange multiplier method (see Eq. 5.30)

$$\begin{bmatrix} [\mathbf{G}^T\mathbf{G}] & \mathbf{1} \\ \mathbf{1}^T & 0 \end{bmatrix}\begin{bmatrix} \mathbf{m} \\ \lambda \end{bmatrix} = \begin{bmatrix} [\mathbf{G}^T\mathbf{d}^{obs}] \\ 0 \end{bmatrix} \tag{15.13}$$

Here, **1** is a length-M column vector of ones, λ is a Lagrange multiplier, and the matrix on the left-hand side of the equation is $(M+1) \times (M+1)$. However, this approach is more complicated than damped least squares and offers little advantage.

A realistic problem may have thousands of orbits that intersect at millions of points. The data kernel will therefore be very large, with dimensions on the order of $1,000,000 \times 1000$. Nevertheless, because of it sparsity, the solution can be efficiently computed with an iterative method such as biconjugate gradients. An alternative is to compute analytically the Gram matrix $[\mathbf{G}^T\mathbf{G}]$ and vector $[\mathbf{G}^T\mathbf{d}^{obs}]$ and then to solve the linear system $[\mathbf{G}^T\mathbf{G} + \varepsilon^2\mathbf{I}]\mathbf{m}^{est} = \mathbf{G}^T\mathbf{d}^{obs}$ (which may be

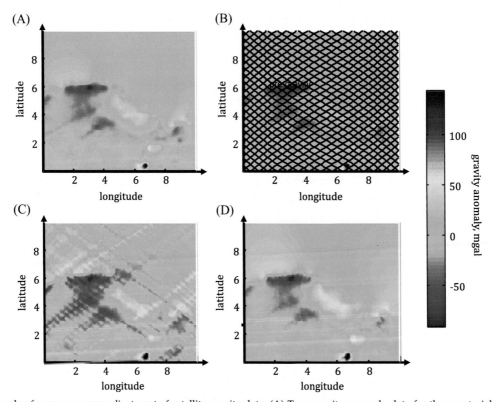

FIG. 15.5   Example of crossover error adjustment of satellite gravity data. (A) True gravity anomaly data for the equatorial Atlantic Ocean. It reflects variations in the depth of the seafloor and density variations within the oceanic crust. (B) Hypothetical satellite tracks, along which the gravity is measured. The measurements along each track have a constant offset reflecting errors in the assumed altitude of the satellite. (C) Reconstructed gravity anomaly without crossover correction. Artifacts parallel to the tracks are clearly visible. (D) Reconstructed gravity anomaly with crossover correction. The artifacts are eliminated. Scripts gdama15_03 and gdapy15_03. *Data courtesy of the late Bill Haxby, Lamont-Doherty Earth Observatory.*

"only" 1000×1000 in dimension) using a standard solver. In this case (as in many others that we have encountered previously), analytic formula for these quantities readily can be derived:

$$\left[\mathbf{G}^T\mathbf{G}\right]_{rs} = \sum_{i=1}^{N} G_{ir}G_{is} = \sum_{i=1}^{N}(\delta_{r,A_i} - \delta_{r,D_i})(\delta_{s,A_i} - \delta_{s,D_i})$$

$$= \sum_{i=1}^{N}(\delta_{r,A_i}\delta_{s,A_i} - \delta_{r,A_i}\delta_{s,D_i} - \delta_{r,D_i}\delta_{s,A_i} + \delta_{r,D_i}\delta_{s,D_i})$$

(15.14)

The diagonal elements of $[\mathbf{G}^T\mathbf{G}]$ are

$$\left[\mathbf{G}^T\mathbf{G}\right]_{rr} = \sum_{i=1}^{N}(\delta_{r,A_i}\delta_{r,A_i} - 2\delta_{r,A_i}\delta_{r,D_i} + \delta_{r,D_i}\delta_{r,D_i})$$

(15.15)

The first term contributes to the sum whenever the ascending orbit is r, and the third term contributes whenever the descending orbit is r. The second term is zero since an orbit never intersects itself. The rth element of the diagonal is the number of times the rth orbit is intersected by other orbits.

Only the two middle terms of the sum in the expression for $[\mathbf{G}^T\mathbf{G}]_{rs}$ contribute to the off-diagonal elements. The second term contributes whenever $A_i = r$ and $D_i = s$, and the third when $A_i = s$ and $D_i = r$. The (r,s) off-diagonal element is the number of times the rth and sth orbits intersect, multiplied by -1.

The vector product is

$$\left[\mathbf{G}^T\mathbf{d}^{obs}\right]_r = \sum_{i=1}^{N} G_{ir}d_i^{obs} = \sum_{i=1}^{N}(\delta_{r,A_i} - \delta_{r,D_i})d_i^{obs}$$

(15.16)

The delta functions can never both equal 1, as an orbit can never intersect itself. Therefore $[\mathbf{G}^T\mathbf{d}^{obs}]_r$ is the sum of all the ds that have ascending orbit number $A_i = r$ minus the sum of all the ds that have descending orbit number $D_i = r$.

We can then compute the Gram matrix and vector. We first prepare a table that gives the ascending orbit number A_i, descending orbit number D_i, and elevation difference d_i^{obs} for each of the N orbital intersections. We then start with $[\mathbf{G}^T\mathbf{G}]$ and $[\mathbf{G}^T\mathbf{d}^{obs}]$ initialized to zero and, for each row i of the table, execute the following steps:

1. Add 1 to the $r=A_i$, $s=A_i$ element of $[\mathbf{G}^T\mathbf{G}]_{rs}$.
2. Add 1 to the $r=D_i$, $s=D_i$ element of $[\mathbf{G}^T\mathbf{G}]_{rs}$.
3. Subtract 1 from the $r=A_i$, $s=D_i$ element of $[\mathbf{G}^T\mathbf{G}]_{rs}$.
4. Subtract 1 from the $r=D_i$, $s=A_i$ element of $[\mathbf{G}^T\mathbf{G}]_{rs}$.
5. Add d_i^{obs} to the $r=A_i$ element of $[\mathbf{G}^T\mathbf{d}^{obs}]_r$.
6. Subtract d_i^{obs} from the $r=D_i$ element of $[\mathbf{G}^T\mathbf{d}^{obs}]_r$.

The final form for $[\mathbf{G}^T\mathbf{G}]$ is relatively simple. If two orbits intersect at most once, then it will contain only zeros and ones on its off-diagonal elements.

15.4 An acoustic tomography problem

An acoustic tomography problem was discussed previously in Section 2.7. In that simple case, travel times d_i of sound rays are measured through the rows and columns of a square 4×4 grid of bricks, each having height and width h and acoustic slowness m_i. The data kernel G_{ij} represents the length of the ith ray in the jth brick. As the sound ray paths are either exactly horizontal or exactly vertical, each of the lengths is either h (if the ray intersects the pixel) or zero (if it doesn't). The data kernel is

$$\mathbf{G} = h\begin{bmatrix} 1 & 1 & 1 & 1 & 0 & 0 & 0 & 0 & 0 & 0 & 0 & 0 & 0 & 0 & 0 & 0 \\ 0 & 0 & 0 & 0 & 1 & 1 & 1 & 1 & 0 & 0 & 0 & 0 & 0 & 0 & 0 & 0 \\ \vdots & \vdots & \vdots & \vdots & \vdots & \vdots & \vdots & \vdots & \vdots & \vdots & \vdots & \vdots & \vdots & \vdots & \vdots & \vdots \\ 0 & 0 & 0 & 1 & 0 & 0 & 0 & 1 & 0 & 0 & 0 & 1 & 0 & 0 & 0 & 1 \end{bmatrix}$$

(15.17)

The matrix $\mathbf{G}$ is sparse, since the typical ray crosses just a small fraction of the total number of bricks. In the general case, where the image is divided up into rectangular pixels and where the ray paths are slanted and/or curved, the

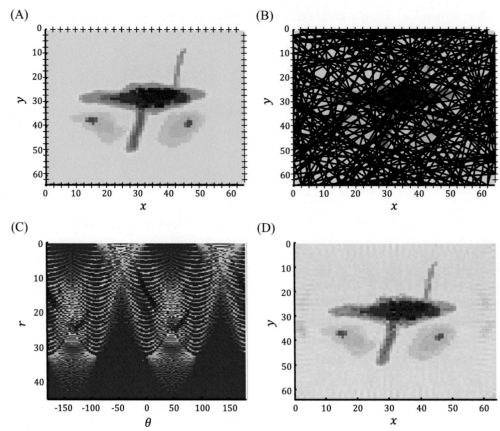

FIG. 15.6   Acoustic tomography problem. (A) True image. (B) Ray paths (only a few percent of rays are shown, else the image would be black). (C) Travel time data, organized by the distance r and angle θ of the ray from the midpoint of the image. (D) Reconstructed image. See text for further discussion. Scripts `gdama15_04` and `gdapy15_04`.

data kernel G_{ij} still represents the length of the ith ray in the jth pixel, but now that length is no longer constant. These variable lengths may be difficult to calculate analytically; instead, one must resort to numerical approximations.

One commonly used technique begins by initializing the data kernel to zero and then stepping along each ray i in arc length increments Δs, chosen to be much smaller than the size of the pixels. The pixel index j of the center of each increment is determined, and the whole increment is added to the corresponding element of the data kernel; that is, G_{ij} is replaced with $G_{ij} + \Delta s$.

We examine a test case where a square object is divided into a 256×256 grid of pixels (Fig. 15.6A), with the model parameter m_i representing the acoustic slowness within the pixels. A set of evenly distributed source and receiver points are placed on the four edges of the square and connected with straight-line rays (Fig. 15.6A). The data kernel is constructed using the approximate ray-stepping method described earlier, and a data set of synthetic travel times is constructed via $\mathbf{d} = \mathbf{Gm}^{true}$. Having an effective way of plotting travel time data is important, for instance, in detecting outliers. We parameterize each ray by its perpendicular distance r from the center of the object and by its angle θ from the horizontal and form an (r, θ) image of the data (Fig. 15.6C) for this purpose. The estimated image (Fig. 15.6D), computed using damped least squares solved with the biconjugate gradient method, recovers most of the long-wavelength features of the image, but contains faint streaks along ray paths. These streaks can be eliminated by increasing the number of rays (not shown).

The success of tomography is critically dependent upon the ray coverage, which must not only be spatially dense but must—at every point—cover a 90 degree suite of angles (from horizontal to vertical). Tomographic reconstructions based on incomplete ray coverage (Fig. 15.7) generally have poor resolution.

15.5   One-dimensional temperature distribution

As a hot slab within a uniform whole space cools, heat flows from the slab to the surrounding material, slowly warming it. The width of the warm zone slowly grows with time and the boundary between the slab and the whole space, initially distinct, slowly fades away. When several hot slabs are present, the temperatures from each blend

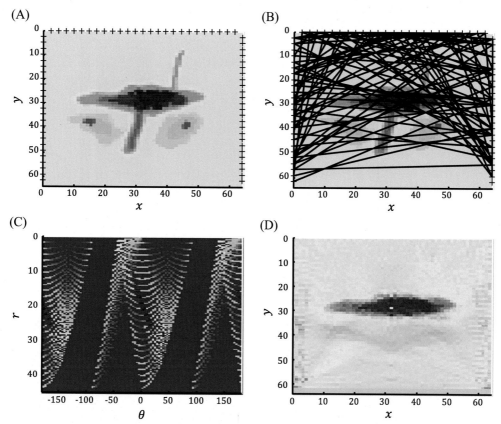

FIG. 15.7 Acoustic tomography problem with deficient distribution of rays. (A) True image. (B) Ray paths (only a few percent of rays are shown, else the image would be black). (C) Travel time data, organized by the distance r and angle θ of the ray from the midpoint of the image. (D) Reconstructed image. See text for further discussion. Scripts `gdama15_05` and `gdapy15_05`.

together as time increases, making them hard to distinguish. The question that we pose is how well the initial pattern of temperatures can be reconstructed, given the temperature profile measured at some later time.

For a single slab, the temperature $T(x,t)$ at position x and time t can be shown to be (Menke and Abbott, 1990, their Section 6.3.3)

$$T(x,t) = \frac{T_0}{2}\left\{\mathrm{erf}\left[\frac{x - (\xi - \tfrac{1}{2}h)}{\sqrt{t}}\right] - \mathrm{erf}\left[\frac{x - (\xi + \tfrac{1}{2}h)}{\sqrt{t}}\right]\right\} \equiv T_0\, g(x,t,\xi) \tag{15.18}$$

Here, x is the distance measured perpendicular to the face of the slab, h is the thickness of the slab, and ξ is its position of its center (Fig. 15.8A). The slab has initial temperature T_0, with its surroundings at zero temperature. The thermal diffusivity of both materials is taken, for simplicity, to be unity. The special function erf(.) is called the *error function*; MATLAB®'s and Python's implementations of it have the same name.

When several slabs of different initial temperature m_i are placed face-to-face within the whole space, the temperature at position x_i and time t is

$$T(x_i, t) = \sum_{j=1}^{M} g\left(x_i, t, \xi_j\right) m_j \tag{15.19}$$

The inverse problem that we consider is how well the initial temperature distribution m_i can be reconstructed by making measurements of the temperature at all positions, but at a fixed time t. We might anticipate that the reconstruction will be well resolved at short times, as heat will not have much time to flow and the initial pattern of temperature will still be partly preserved. On the other hand, it will be poorly resolved at long times, since the initial pattern of temperature will have faded away.

Our test scenario has $M = 100$ slabs located in the distance interval $|x| \leq 20$, which have a sinusoidal temperature pattern with five oscillations, overall (Fig. 15.8B). The details of this pattern fade with time, so that after about $t \approx 20$ only a broad warm zone is present. The width of this warm zone slowly grows with time, roughly doubling by $t \approx 250$.

(A) (B)

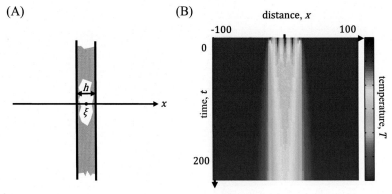

FIG. 15.8   (A) Single hot slab of thickness h located at position $x = \xi$. (B) Temporal evolution of the temperature $T(x,t)$ of 100 adjacent slabs. The initial temperature distribution of the slabs $T(x, t=0)$ is taken to be the model parameter vector $\mathbf{m}$. It is nonzero only for slabs near $|x| \leq 20$. The temperature $T(x,t)$ at subsequent times t can be computed from the initial temperature distribution, since the data kernel can be calculated from the physics of heat transport. Note that the band of hot temperatures widens with increasing time, and that fine scale temperature fluctuations are preferentially attenuated. Scripts gdama15_06 and gdapy15_06.

(A) (B) (C)

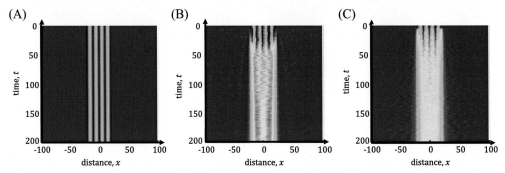

FIG. 15.9   Model for the temperature distribution problem. (A) The true model is the initial temperature distribution $T(x, t=0)$. Although this function is not a function of time, it is displayed on the (x,t) image for comparison purposes. (B) Minimum-length (ML) estimate of the model $T(x, t=0)$ for a data set consisting of observations at all distances at a single time $t > 0$. (C) Corresponding Backus-Gilbert (BG) estimate. In both the ML and BG cases, the ability of the data to resolve fine details declines with time, with the BG case declining fastest. Scripts gdama15_06 and gdapy15_06.

We reconstruct the initial temperature using $N = 100$ observations of temperature $d_i^{obs} = T(x_i, t)$ from equally spaced between $-100 < x < 100$ all made at a fixed time t. Two different inversion methods are used, minimum length and Backus-Gilbert (Fig. 15.9). As hypothesized, their quality falls off rapidly with observation time t with little detail being present after $t \approx 50$. The minimum-length inversion does better at recovering the sinusoidal pattern, but it also contains artifacts, especially at long time, where two instead of five oscillations appear to be present. In contrast, the Backus-Gilbert inversion provides a poorer reconstruction, but one that is artifact free. These differences can be understood by examining the model resolution matrices of the two methods (Fig. 15.10). The minimum-length inversion has the narrower resolution but also the stronger sidelobes.

15.6 L_1, L_2, and L_∞ fitting of a straight line

The L_1, L_2, and L_∞ problem is to fit the straight line $d_i = m_1 + m_2 z_i$ to a set of (z, d) pairs by minimizing the prediction error under a variety of norms. This is a linear problem with an $N \times 2$ data kernel (see Section 4.3)

$$\mathbf{G} = \begin{bmatrix} 1 & z_1 \\ 1 & z_2 \\ \vdots & \vdots \\ 1 & z_N \end{bmatrix} \tag{15.20}$$

The L_2 norm is the simplest to implement. It implies that the error follows a Normal probability density function. The simple least-squares solution $\mathbf{m}^{est} = [\mathbf{G}^T\mathbf{G}]^{-1}\mathbf{G}^T\mathbf{d}^{obs}$ is appropriate when the measurement error is uncorrelated and

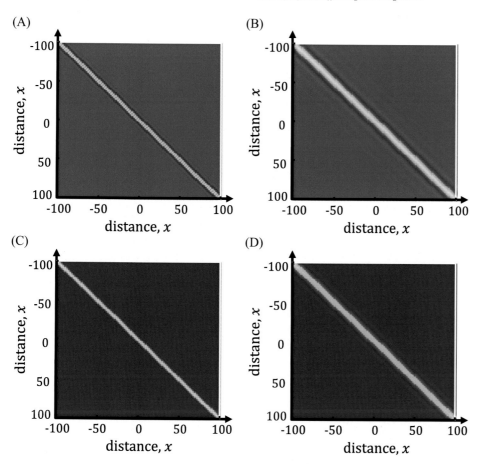

FIG. 15.10 Model resolution matrices for the temperature distribution problem. (A) Minimum-length (ML) solution for data at time $t=10$. (B) ML solution for data at time $t=40$. (C) Backus-Gilbert (BG) solution for data at time $t=10$. (D) BG solution for data at time $t=40$. Note that the BG resolution matrix has smaller sidelobes than the corresponding ML case. Scripts `gdama15_06` and `gdapy15_06`.

with uniform variance σ_d^2, However, for a more complicated covariance matrix, the weighted least-squares solution $\mathbf{m}^{est}=[\mathbf{G}^T[\text{cov } \mathbf{d}]^{-1}\mathbf{G}]^{-1}\mathbf{G}^T[\text{cov } \mathbf{d}]^{-1}\mathbf{d}^{obs}$ must be used. As the L_2 problem has been discussed previously (see Section 4.5), we shall not treat it in detail here. However, it is interesting to compute the data resolution matrix $\mathbf{N}=\mathbf{G}\mathbf{G}^{-g}$

$$\mathbf{N} = \begin{bmatrix} 1 & z_1 \\ 1 & z_2 \\ \vdots & \vdots \\ 1 & z_N \end{bmatrix} \frac{1}{N\sum_k^N z_k^2 - \left(\sum_k^N z_k\right)^2} \begin{bmatrix} \sum_k^N z_k^2 & -\sum_k^N z_k \\ -\sum_k^N z_k & N \end{bmatrix} \begin{bmatrix} 1 & 1 & \cdots & 1 \\ z_1 & z_2 & \cdots & z_N \end{bmatrix}$$

$$N_{ij} = \frac{\sum_k^N z_k^2 - (z_i + z_j)\sum_k^N z_k + z_i z_j N}{N\sum_k^N z_k^2 - \left(\sum_k^N z_k\right)^2} \equiv A_i + B_i z_j$$

(15.21)

Each row of the data resolution matrix N_{ij} is a linear function of z_j, so that the elements with the largest absolute value are at an edge of the matrix and not along its main diagonal. The resolution is not at all localized; instead, the points with most extreme z_j control the fit of the straight line.

The L_1 and L_∞ estimates can be determined by using the transformation to a linear programming problem described in Chapter 10. Although more efficient algorithms exist, we shall set up the problems so that they can be solved with a standard linear programming algorithm.

This algorithm determines a vector $\mathbf{y}$ that minimizes the quantity $\mathbf{c}^T\mathbf{y}$ subject to the inequality constraints $\mathbf{A}\mathbf{y} \geq \mathbf{b}$ and $\mathbf{y} \geq 0$. The first step is to define two new variables $\mathbf{m}'$ and $\mathbf{m}''$, such that $\mathbf{m}=\mathbf{m}'-\mathbf{m}''$. This definition relaxes the positivity constraints on the model parameters. For the L_1 problem, we define three additional vectors $\boldsymbol{\alpha}$, $\mathbf{x}$, and $\mathbf{x}'$ and then arrange them in the form of a linear programming problem in $(2M+3N)$ variables and $2N$ constraints as

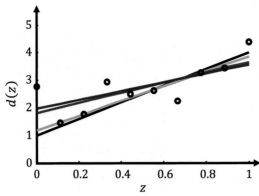

FIG. 15.11 Fitting a straight line to data $d(z)$. The true model (black line) is the straight line $d^{true}(z) = 1 + 3z$. The $N = 10$ observations d_i^{obs} are the true data perturbed with exponentially distributed noise with variance $\sigma_2^2 = (0.4)^2$. Three different fits have been computed by minimizing the L_1, L_2, and L_∞ norms of the error. The corresponding predicted data are shown in green, blue, and red, respectively. Scripts `gdama15_07` and `gdapy15_07`.

$$\mathbf{y} = \left[\left[m'_1, \cdots, m'_M\right], \left[m''_1, \cdots, m''_M\right], [\alpha_1, \cdots, \alpha_N], [x_1, \cdots, x_N], [x'_1, \cdots, x'_N]\right]^T$$

$$\mathbf{c} = \left[[0, \cdots, 0], [0, \cdots, 0], [1, \cdots, 1], [0, \cdots, 0], [0, \cdots, 0]\right]^T$$

$$\mathbf{A} = \begin{bmatrix} \mathbf{G}_{N\times M} & -\mathbf{G}_{N\times M} & -\mathbf{I}_{N\times N} & \mathbf{I}_{N\times N} & \mathbf{0}_{N\times N} \\ \mathbf{G}_{N\times M} & -\mathbf{G}_{N\times M} & \mathbf{I}_{N\times N} & \mathbf{0}_{N\times N} & -\mathbf{I}_{N\times N} \end{bmatrix} \qquad (15.22)$$

$$\mathbf{b} = \left[[d_1, \cdots, d_N], \ [d_1, \cdots, d_N]\right]^T$$

The L_∞ problem is transformed into a linear programming problem with additional variables x and x' and a scalar parameter α. The transformed problem in $(2M + 2N + 1)$ unknowns and $2N$ constraints is

$$\mathbf{y}^T = \left[\left[m'_1, \cdots, m'_M\right], \left[m''_1, \cdots, m''_M\right], [\alpha], [x_1, \cdots, x_N], [x'_1, \cdots, x'_N]\right]^T$$

$$\mathbf{c} = \left[[0, \cdots, 0], [0, \cdots, 0], [1], [0, \cdots, 0], [0, \cdots, 0]\right]^T$$

$$\mathbf{A} = \begin{bmatrix} \mathbf{G}_{N\times M} & -\mathbf{G}_{N\times M} & -\mathbf{1}_{N\times 1} & \mathbf{I}_{N\times N} & \mathbf{0}_{N\times N} \\ \mathbf{G}_{N\times M} & -\mathbf{G}_{N\times M} & \mathbf{1}_{N\times 1} & \mathbf{0}_{N\times N} & -\mathbf{I}_{N\times N} \end{bmatrix} \qquad (15.23)$$

$$\mathbf{b} = \left[[d_1, \cdots, d_N], \ [d_1, \cdots, d_N]\right]^T$$

We illustrate the results of using these three different norms to data with exponentially distributed error (Fig. 15.11). Note that the L_1 line is by far the best; it is designed to give little weight to outliers, which are common in data with exponentially distributed noise. Both the L_1 and L_∞ fits may be nonunique. Most versions of the linear programming algorithm will find only one solution, so the process of identifying the complete range of minimum solutions may be difficult.

15.7   Finding the mean of a set of unit vectors

Suppose that a set of measurements of direction (defined by unit vectors in a three-dimensional Cartesian space) are thought to scatter randomly about a mean direction (Fig. 15.12). How can the mean vector be determined?

This problem is similar to that of determining the mean of a group of scalar quantities (Eq. 3.5) and is solved by direct application of the principle of maximum likelihood. In the scalar mean problem of Section 6.1, we assume that the data are Normally distributed and then apply the principle of maximum likelihood to estimate two unknowns, the mean and the variance. The Normal pdf is not applicable to directional data because they are defined on a different interval $(0, \pi)$ and not $(-\infty, +\infty)$. A better choice is the *Fisher probability density function* (Fisher, 1953). Its vectors are clumped near the mean direction with no preferred azimuthal direction. It proposes that the probability of finding a vector in an increment of solid angle $d\Omega = \sin(\theta)d\theta d\varphi = d(\cos(\theta))d\varphi$ (where $\cos(\theta)$ is being treated as an independent variable), located at an angle with inclination θ and azimuth φ from the mean direction (Fig. 15.13), is

$$p(\theta, \varphi) = \frac{\kappa}{4\pi \sinh(\kappa)} \exp\{\kappa \cos(\theta)\} \qquad (15.24)$$

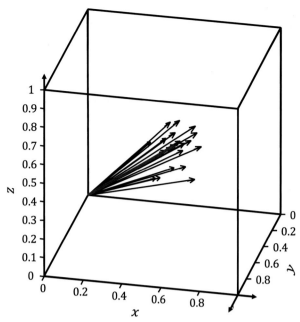

FIG. 15.12 Several unit vectors (black) scattering about a central direction (red). Scripts `gdama15_08` and `gdapy15_08`.

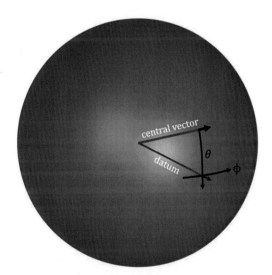

FIG. 15.13 Unit sphere, showing coordinate system used in Fisher distribution. Scripts `gdama15_09` and `gdapy15_09`.

This pdf is normalized so that the integral over the unit sphere ($0 \leq \theta \leq \pi$, $0 \leq \varphi < 2\pi$) is unity. It is peaked at mean direction $\theta = 0$ and has a width that depends on the value of the *precision parameter* κ (Fig. 15.14). The reciprocal of this parameter serves a role similar to the variance in the Normal pdf. When $\kappa = 0$, the distribution is uniform on the sphere, but when $\kappa \gg 1$ it becomes very peaked near the mean direction, with a shape that is very similar to that of a Normal pdf.

The data are specified by N Cartesian unit vectors $\mathbf{x}^{obs} \equiv (x_i^{obs}, y_i^{obs}, z_i^{obs})$, or equivalently, by N pairs of polar angles θ_i^{obs} and φ_i^{obs}. The mean direction is described by its Cartesian unit vector $\mathbf{m} \equiv [m_1, m_2, m_3]^T$, then the cosine of the inclination for any datum is just the dot product $\cos(\theta_i) = \mathbf{m}^T \mathbf{x}^{obs} = m_1 x_i + m_2 y_i + m_3 z_i$. The likelihood $p(\mathbf{x}^{obs}|\mathbf{m}, \kappa)$ of the data set as a whole being observed is the product of the N univariate pdfs

$$p(\mathbf{x}^{obs}|\mathbf{m}, \kappa) = \left[\frac{\kappa}{4\pi \sinh(\kappa)} \right]^N \exp\left[\kappa \sum_{i=1}^{N} \cos(\theta_i) \right] \qquad (15.25)$$

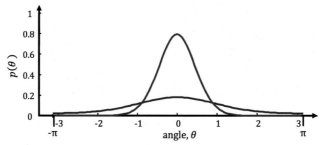

FIG. 15.14 Fisher probability density function $p(\theta, \varphi)$ for a large precision parameter ($\kappa = 5$, blue curve) and a small precision parameter ($\kappa = 1$, red curve). Scripts gdama15_10 and gdapy15_10.

The log-likelihood function is

$$
L = \log(p) = N \log(\kappa) - N \log(4\pi) - N \log(\sinh(\kappa))
$$
$$
+ \kappa \sum_{i=1}^{N} \left[x_i m_1 + y_i m_2 + z_i m_3 \right] \tag{15.26}
$$

The best estimate of model parameters $\mathbf{m}$ and precision parameter κ are those that maximize the log-likelihood. However, as the solution is assumed to be a unit vector, L must be maximized with respect to the ms and κ under the constraint that $\mathbf{m}$ has unit length, that is, $\sum_{i=1}^{3} m_i^2 = 1$. This maximization is best achieved using the method of Lagrange multipliers (see Section 17.1), a technique that allows a constrained minimization problem (e.g., minimize $L(\mathbf{m}, \kappa)$ with respect to $\mathbf{m}$ and κ subject to the constraint $C(\mathbf{m}) \equiv \mathbf{m}^T \mathbf{m} - 1 = 0$) to be converted to an unconstrained one (e.g., minimize $L(\mathbf{m}) + \lambda C(\mathbf{m})$ with respect to $\mathbf{m}$ and κ). Here λ is a new unknown, the Lagrange multiplier. The Lagrange multiplier equations for the problem are then

$$
\kappa \sum_i x_i - 2\lambda m_1 = 0
$$
$$
\kappa \sum_i y_i - 2\lambda m_2 = 0
$$
$$
\kappa \sum_i z_i - 2\lambda m_3 = 0 \tag{15.27}
$$
$$
\frac{N}{\kappa} - N \frac{\cosh(\kappa)}{\sinh(\kappa)} + \sum_{i=1}^{N} \left[x_i m_1 + y_i m_2 + z_i m_3 \right] = 0
$$

where λ is the Lagrange multiplier. The first three equations can be solved simultaneously along with the constraint equation for the model parameters as

$$
[m_1, m_2, m_3]^T = \frac{\left[\sum_i^N x_i, \sum_i^N y_i, \sum_i^N z_i \right]^T}{\left(\left(\sum_i^N x_i \right)^2 + \left(\sum_i^N y_i \right)^2 + \left(\sum_i^N z_i \right)^2 \right)^{1/2}} \tag{15.28}
$$

Note that the mean vector is the normalized vector sum of the individual observed unit vectors. The fourth equation is an implicit transcendental equation for the precision parameter κ. If it can be assumed that $\kappa > 5$, we can use the approximation $\cosh(\kappa) / \sinh(\kappa) \approx 1$, and the fourth equation yields

$$
\kappa \approx \frac{N}{N - \sum_i \cos(\theta_i)} \tag{15.29}
$$

An example is shown in Fig. 15.15.

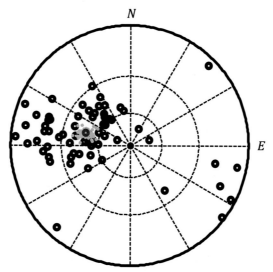

FIG. 15.15 Lower hemisphere stereonet showing P-axes of deep (300–600 km) earthquakes in the Kurile-Kamchatka subduction zone. The axes mostly dip to the west, parallel to the direction of subduction, indicating that the subducting slab is in down-dip compression. (Black circles) Axes of individual earthquakes. (Red circle) Central axis, computed by maximum likelihood technique. (Blue dots) Scatter of central axis determined by bootstrapping. Scripts `gdama15_11` and `gdapy15_11`. *Data courtesy of the Global CMT Project.*

15.8 Gaussian and Lorentzian curve fitting

Many types of spectral data consist of several overlapping peaks, each of which has either *Gaussian* or *Lorentzian* shape. (Both the Gaussian and the Lorentzian have a single maximum, but the Lorentzian is much longer tailed.) The problem is to determine the location, area, and width of each peak through least-squares curve fitting.

Suppose that the data consist of N pairs of (z_i, d_i) where the auxiliary variable z_i represents spectral frequency. Each of the, say, q peaks is parameterized by its area A_i, center frequency f_i, and width c_i. There are then $M = 3q$ model parameters $m = [A_1, f_1, c_i, \cdots, A_q, f_q, c_q]^T$. The model is nonlinear and of the form $d^{obs} = g(m)$

$$\text{Gaussian} : g_i = \sum_{j=1}^{q} \frac{A_j}{\sqrt{2\pi}c_j} \exp\left\{ -\frac{\left(z_i - f_j\right)^2}{2c_j^2} \right\}$$

$$\text{Lorentzian} : g_i = \sum_{j=1}^{q} \frac{A_j c_j^2}{\left(z_i - f_j\right)^2 + c_j^2}$$

(15.30)

In the case of the Gaussian, the quantity c_j^2 has the interpretation of variance, but in the case of the Lorentzian (which has infinite variance), it does not. If the data have Normally distributed error, Newton's method can be used to solve this problem. Furthermore, the problem will typically be overdetermined, at least if $N > M$ and if the peaks do not overlap completely. The equation is linearized around trial solution $\mathbf{m}^{(k)}$, using Taylor's theorem, as in Section 11.5. This linearization involves computing a matrix $\mathbf{G}^{(k)}$ of derivatives

$$\mathbf{G}^{(k)} = \begin{bmatrix} \frac{\partial g_1}{\partial A_1} & \frac{\partial g_1}{\partial f_1} & \frac{\partial g_1}{\partial c_1} & \cdots & \frac{\partial g_1}{\partial A_p} & \frac{\partial g_1}{\partial f_p} & \frac{\partial g_1}{\partial c_p} \\ \frac{\partial g_2}{\partial A_1} & \frac{\partial g_2}{\partial f_1} & \frac{\partial g_2}{\partial c_1} & \cdots & \frac{\partial g_2}{\partial A_p} & \frac{\partial g_2}{\partial f_p} & \frac{\partial g_2}{\partial c_p} \\ \vdots & \vdots & \vdots & \ddots & \vdots & \vdots & \vdots \\ \frac{\partial g_N}{\partial A_1} & \frac{\partial g_N}{\partial f_1} & \frac{\partial g_N}{\partial c_1} & \cdots & \frac{\partial g_N}{\partial A_{1p}} & \frac{\partial g_N}{\partial f_p} & \frac{\partial g_N}{\partial c_p} \end{bmatrix}_{\mathbf{m}^{(k)}}$$

(15.31)

In this problem the Gaussian and Lorentzian models are simple enough for the derivatives to be computed analytically as

Gaussian :

$$\frac{\partial g_i}{\partial A_j} = \frac{f_{ij}}{\sqrt{2\pi}c_j} \quad \text{with } f_{ij} \equiv \exp\left\{-\frac{\left(z_i - f_j\right)^2}{2c_j^2}\right\}$$

$$\frac{\partial g_i}{\partial f_j} = \frac{A_j}{\sqrt{2\pi}c_j}\frac{\left(z_i - f_j\right)}{c_j^2}f_{ij}$$

$$\frac{\partial g_i}{\partial c_j} = \frac{A_j}{\sqrt{2\pi}c_j^2}\left[\frac{\left(z_i - f_j\right)^2}{c_j^2} - 1\right]f_{ij} \tag{15.32}$$

Lorentzian :

$$\frac{\partial g_i}{\partial A_j} = c_j^2 h_{ij} \quad \text{with } h_{ij} \equiv \frac{1}{\left(z_i - f_j\right)^2 + c_j^2}$$

$$\frac{\partial g_i}{\partial f_j} = 2A_j c_j^2 \left(z_i - f_j\right)h_{ij}^2$$

$$\frac{\partial g_i}{\partial c_j} = 2A_j c_j h_{ij} - 2A_j c_j^3 h_{ij}^2$$

These derivatives are evaluated at the trial solution $\mathbf{m}^{(k)}$ with the starting value (when $k=1$) chosen from visual inspection of the data. Improved estimates of the model parameters are found using the recursion $\mathbf{m}^{(k+1)} = \mathbf{m}^{(k)} + \Delta\mathbf{m}$, where the solution update $\Delta\mathbf{m}$ satisfies the equation $\mathbf{G}^{(k)}\Delta\mathbf{m} = \mathbf{d}^{obs} - \mathbf{g}(\mathbf{m}^{(k)})$ and is solved with simple least squares. The iterations are terminated when $\Delta\mathbf{m}$ becomes negligibly small (for instance, when the absolute value of each element becomes less than some given tolerance). An example is shown in Fig. 15.16.

Occasionally, prior information requires that the separation between two peaks be equal to a known value $\langle s_{ij}\rangle \equiv f_i - f_j$. This constraint is of the form $\mathbf{Hm} = \mathbf{h}$, where each row of the equation represents one pair of peaks. Each row of $\mathbf{h}$ specifies their separation and the corresponding row of $\mathbf{H}$ with the form $[0,0,0,\cdots,0,1,0,\cdots 0,-1,0,\cdots 0,0,0]$ specifies the frequencies being differenced. The problem can be solved with the linearized version of Generalized Least Squares (see Method Summary 3 in Chapter 17):

$$\begin{bmatrix}\mathbf{G}^{(k)} \\ \varepsilon\mathbf{H}\end{bmatrix}\Delta\mathbf{m} = \begin{bmatrix}\mathbf{d}^{obs} - \mathbf{g}(\mathbf{m}^{(k)}) \\ \varepsilon(\mathbf{h} - \mathbf{Hm}^{(k)})\end{bmatrix} \tag{15.33}$$

Here, ε is set to a very large number.

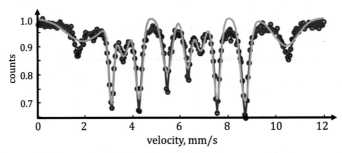

FIG. 15.16   Example of fitting the sum of 10 Lorentzian (blue) or Gaussian (green) curves to Mossbauer spectroscopic data (red). The Lorentzian curves are better able to fit the shape of the curve, with the ratio of estimated variances being about 4. An F-test indicates that a null hypothesis that any difference between the two fits can be ascribed to random variation can be rejected to better than 99.99%. Scripts gdama15_12 and gdapy15_12. *Data courtesy of NASA and the University of Mainz.*

One of the drawbacks to these iterative methods is that the solution may oscillate wildly or diverge from one iteration to the next if the initial guess is too far off. There are several possible remedies for this difficulty. One is to force the perturbation $\Delta\mathbf{m}$ to be less than a specified maximum length. This result can be achieved by examining the perturbation on each iteration and, if it is longer than the limit, decreasing its length (but not changing its direction). This procedure will prevent the method from wildly "overshooting" the true minimum but will also slow the convergence. Another possibility is to constrain some of the model parameters to equal prior values for the first few iterations, thus allowing only some of the model parameters to vary. The constraints are relaxed after convergence, and the iteration is continued until the unconstrained solution converges.

15.9   Fourier analysis

The term *Fourier analysis* refers to the process of fitting sines and cosines to data

$$d_p = \sum_{q=1}^{N/2+1} A_q \cos(\omega_q t_p) + \sum_{q=2}^{N/2} B_q \sin(\omega_q t_p) \tag{15.34}$$

Here, the observations are presumed to be uniformly spaced in time, that is, $d_p = d(t_p)$, with $t_p = (p-1)\Delta t$. The As and Bs are model parameters, and the angular frequencies ω_q are auxiliary variables. Angular frequencies are measured in radians/s and relate to frequency in Hertz by $f_q = \omega_q/(2\pi)$. Because of *Nyquist's principle*, which states that frequencies greater than $f_{max} = 1/(2\Delta t)$ cannot be resolved, the frequencies are usually taken to be evenly spaced between zero and ω_{max}, that is, $\omega_q = (q-1)\Delta\omega$, with $\Delta\omega = 2\pi f_{max}/(N/2)$. The problem is then even-determined, as the $\sin(\omega_1 t_q)$ and $\sin(\omega_{N/2+1} t_q)$ terms can be shown to be identically zero (which is why they are omitted from the sum in Eq. 15.34).

Fourier transforms have a variety of uses, among which is the detection of periodicities. As both $A_q \cos(\omega_q t_p)$ and $B_q \sin(\omega_q t_p)$ oscillate with the same frequency ω_q, the quantity $s_q = [A_q^2 + B_q^2]^{1/2}$ quantifies the overall amount of amplitude at that frequency. A plot of s_q against angular frequency ω_q (or frequency f_q or period $p_q = 1/f_q$) will have peaks at frequencies at which especially prominent periodicities occur. In practice, the sines and cosines version of the model (Eq. 15.34) is not used and an equivalent version involving complex exponentials is used instead.

$$d(t_p) = \sum_{q=1}^{N/2+1} C_q^+ \exp(i\omega_q t_p) + \sum_{q=2}^{N/2} C_q^- \exp(-\omega_q t_p) \tag{15.35}$$

Here, we assume that N is even, so that $(N/2+1)$ is an integer. In order for the summation to represent a real time series, the coefficients C_q^+ and C_q^- must be complex conjugates of one another. *Euler's formula* $\exp(ix) = \cos(x) + i\sin(x)$ can then be used to show that $A_q = 2\,\mathrm{real}(C_q^+)$ and $B_q = -2\,\mathrm{imag}(C_q^+)$, so $s_q = 2|C_q^+|$. Eq (15.35) can be viewed as a matrix equation of the form

$$\mathbf{d} \equiv \begin{bmatrix} d_1 \\ d_2 \\ \vdots \\ \vdots \\ d_{N-2} \\ d_N \end{bmatrix} = \mathbf{G} \begin{bmatrix} C_1^+ \\ \vdots \\ C_{N/2+1}^+ \\ C_{N/2}^- \\ \vdots \\ C_2^- \end{bmatrix} \equiv \mathbf{Gm} \tag{15.36}$$

Here, the coefficients C_q^+ and C_q^- have been grouped into a single model parameter vector $\mathbf{m}$, which omits C_1^- and $C_{N/2+1}^-$, as they are redundant. The matrix $\mathbf{G}$ contains the complex exponentials. The least-squares solution $\mathbf{m}^{est} = [\mathbf{G}^{*T}\mathbf{G}]^{-1}\mathbf{G}^{*T}\mathbf{d}$ (see Section 17.2) is especially easy to compute, as it can be shown that $[\mathbf{G}^{*T}\mathbf{G}] = N\mathbf{I}$, from which it follows that $\mathbf{m}^{est} = N^{-1}\mathbf{G}^{*T}\mathbf{d}$. Furthermore, a very efficient algorithm, called the *Fast Fourier Transform (FFT)*, is available for computing the product $\mathbf{G}^{*T}\mathbf{d}$

MATLAB®

```
N = 2*floor(N/2);  % truncate to an even number of points
d = d(1:N);

% frequency axis, cycles per day
fmax = 1.0/(2.0*Dt);
Df = fmax/(N/2);
M = N/2+1;  % non-negative frequencies
f = Df*[0:M-1]';

% hamming taper, a bell-shaped curve
a0 = 0.53836;
a1 = 0.46164;
ham = a0 - a1 * cos(2*pi*[0:N-1]'/(N-1));

% amplitude spectral density
dp = ham.*(d-mean(d)); % remove mean and hamming taper
m = fft(dp); Fourier transform
s = sqrt(2/N)*abs(m(1:M)); # amplitude spectral density
% Note normalization of sqrt(2/N) is appropriate for
% stationary timeseries
```

[gdama15_13]

Python

```
N = 2*int(N/2);    #  truncate to an even number of points
d = D[0:N,1:2];    #  sea surface in feet sampple every hour

# frequency axis, cycles per day
fmax = 1.0/(2.0*Dt);
Df = fmax/(N/2);
M = int(N/2+1);  # non-negative frequencies
f = gda_cvec(Df*np.linspace(0,M-1,M));

# hamming taper, a bell-shaped curve
a0 = 0.53836;
a1 = 0.46164;
ha= gda_cvec(a0-a1*np.cos(2.0*pi*np.linspace(0,N-1,N)/(N-1)));

# amplitude power spectral density
# remove mean and hamming taper

dp = np.multiply( ham, d-np.mean(d));
# fourier transform. Note axis=0 necessary
m = np.fft.fft(dp, axis=0); # compute a.s.d.
# Note normalization of sqrt(2/N) is appropriate for
# stationary timeseries
s = sqrt(2.0/N)*np.abs(m[0:M,0:1]);
```

[gdapy15_13]

(A)

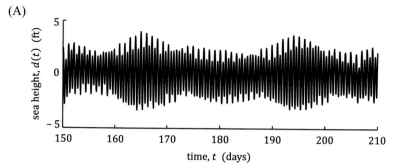

(B)

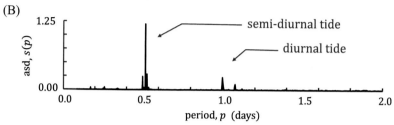

FIG. 15.17 Fourier analysis example. (A) Sea surface height $d(t)$ as a function of time t for a tide gauge at The Battery (New York, United States). Two months of the 12-month long data set are shown. (B) Amplitude spectral density (asd) $s(p)$ computed via the FFT algorithm and plotted as a function of period $p = 1/f$. Note peaks at the semidiurnal and diurnal periods of 0.5 and 1.0 days, respectively. Scripts gdama15_13 and gdapy15_13. *Data courtesy of the US National Oceanic and Atmospheric Administration.*

In these scripts, the time series is truncated to an even number of points, so that the formulas given before for frequency can be applied. The amplitude spectrum is formed from the first $(N/2+1)$ elements of $\mathbf{m}^{est}$, that is, from the C_q^+s. Note that the data has been *demeaned* (by subtracting its mean) and *tapered* (multiplied by a bell-shaped window function $h(t)$) before being Fourier transformed. This standard procedure reduces edge effects arising from the abrupt beginning and end of the data. In this case, we used the Hamming taper

$$h(t_n) = a_0 - a_1 \cos\left(\frac{2\pi(n-1)}{N}\right)$$

$$\text{with } a_0 = 0.53836 \text{ and } a_1 = 0.46164$$

(15.37)

The size of the quantity s_q in part depends on the length N of the time series. For this reason, it is preferable to use the normalized version $(2/N)^{1/2}s_q$, which is independent of it. It is called the *amplitude spectral density*.

We compute the amplitude spectral density of sea surface height measured by a tide gauge at The Battery (New York, United States) using the FFT algorithm. The data set is one year long, with hourly measurements of height. We compute the amplitude spectral density $s(f)$ using the FFT algorithm. However, to aid our interpretation, we plot it against period $p = 1/f$, as contrasted to frequency f. We expect that the amplitude spectrum will be dominated by ocean tides, and this notion is borne out by the amplitude spectrum, which has strong peaks at periods of a little about a half-day and about a day. These peaks represent the semidiurnal and diurnal tides. Each peak is a *doublet*, which is due to the slightly different forces exerted on the Earth by the Sun and Moon (Fig. 15.17).

15.10 Earthquake location

When a fault ruptures within the Earth, seismic compressional P and shear S waves are emitted. These waves propagate through the Earth and are recorded by seismometers on the Earth's surface. The earthquake location problem is to determine the *hypocenter* (location $\mathbf{x}^{(0)} = [x_0, y_0, z_0]^T$) and *origin time* (time of occurrence t_0) of the rupture on the basis of measurements of the arrival of the P and S waves at the seismometers. Although actual geologic faults have finite size, and their rupture takes finite time, they are being approximated as point-like and impulsive.

The data are the *arrival times* t_i^P of P waves and the t_i^S of S waves at a set of $i = 1 \cdots K$ receivers (seismometers) located at $\mathbf{x}^{(i)} = [x_i, y_i, z_i]^T$. The model is the arrival time equals the origin time plus the travel time of the wave from the source (geologic fault) to the receiver. In a constant velocity medium, the waves travel along straight-line rays connecting source and receiver, so the travel time is the length of the line divided by the P or S wave velocity. In heterogeneous media, the rays are curved (Fig. 15.18) and the ray paths and travel times are much more difficult to calculate. We shall not discuss the process of *ray tracing* here, but merely assume that the travel times T_i^P and T_i^S can be calculated for arbitrary source and receiver locations. Then

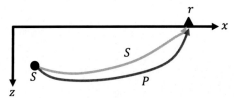

FIG. 15.18   Compressional P and shear S waves travel along rays from earthquake source s (circle) to receiver r (triangle).

$$t_i^P = T_i^P\left(\mathbf{x}^{(i)}, \mathbf{x}^{(0)}\right) + t_0$$
$$t_i^S = T_i^S\left(\mathbf{x}^{(i)}, \mathbf{x}^{(0)}\right) + t_0 \tag{15.38}$$

These are nonlinear equations of the form $\mathbf{d} = \mathbf{g}(\mathbf{m})$, with $N = 2K$ data, $\mathbf{d} \equiv [t_1^P, \cdots, t_N^P, t_1^S, \cdots, t_N^S]^T$, model parameters $\mathbf{m} \equiv [x_0, y_0, z_0, t_0]^T$. When many observations are made, the equations are overdetermined and an iterative least-squares approach based on Newton's method may be used. This method requires that the gradient of P wave travel time $[\ \partial T_i^P / \partial x_0, \partial T_i^P / \partial y_0, \partial T_i^P / \partial z_0]^T$, and the analogous gradient of S wave travel time, be computed for arbitrary values of the hypocentral parameters. Unfortunately, no simple, differentiable analytic formula for travel time is known. One possible solution is to calculate derivatives numerically using the finite difference formula. For the P wave, we find

$$\frac{\partial T_i^P}{\partial x_0} \approx \frac{T_i^P\left(\mathbf{x}^{(i)}, x_0 + \Delta x, y_0, z_0\right) - T_i^P\left(\mathbf{x}^{(i)}, x_0, y_0, z_0\right)}{\Delta x}$$

$$\frac{\partial T_i^P}{\partial y_0} \approx \frac{T_i^P\left(\mathbf{x}^{(i)}, x_0, y_0 + \Delta y, z_0\right) - T_i^P\left(\mathbf{x}^{(i)}, x_0, y_0, z_0\right)}{\Delta y} \tag{15.39}$$

$$\frac{\partial T_i^P}{\partial z_0} \approx \frac{T_i^P\left(\mathbf{x}^{(i)}, x_0, y_0, z_0 + \Delta z\right) - T_i^P\left(\mathbf{x}^{(i)}, x_0, y_0, z_0\right)}{\Delta z}$$

and similarly for the S wave. Here, Δx, Δy, and Δz are small distance increments. These equations represent moving the location of the earthquake a small distance along a coordinate direction and then computing the change in travel time. This approach has two disadvantages. First, if the increments are made very small so that the finite difference approximates a derivative very closely, the terms in the numerator become nearly equal and computer round-off error can become very significant. Second, this method requires that the travel time be computed for three additional earthquake locations and, therefore, is four times as expensive as computing travel time alone.

In some inverse problems, there is no alternative to finite element derivatives. Fortunately, it is possible in this problem to deduce the gradient of travel time by examining the geometry of a ray as it leaves the source (Fig. 15.19). If the earthquake is moved a small distance s parallel to the ray in the direction of the receiver, then the travel time is simply decreased by an amount s/v_0, where $v_0 = v(\mathbf{x}^{(0)})$ is the P or S wave velocity at the source. If it is moved a small distance perpendicular to the ray, then the change in travel time is negligible, as the new ray path will have nearly the same length as the old. The gradient is therefore $\mathbf{s}/v_0$, where $\mathbf{s}$ is a unit vector tangent to the ray at the hypocenter that points

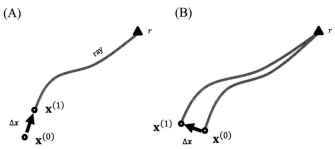

FIG. 15.19   (A) Moving the source a distance Δx from $\mathbf{x}^{(0)}$ to $\mathbf{x}^{(1)}$ parallel to the ray path leads to a large change in travel time. (B) Moving the source perpendicular to the ray path leads to no (first order) change in travel time. The partial derivative of travel time with respect to distance can therefore be determined with minimal extra effort.

toward the receiver. This insight is due to Geiger (1912), and linearized earthquake location is often referred to as *Geiger's method*. Since we must calculate the ray path to find the travel time, no extra computational effort is required to find the gradient. If $\mathbf{s}^{(i)P}$ is the direction of the P wave ray to the ith receiver (of a total of K receivers), and similarly $\mathbf{s}^{(i)S}$ is for the S wave, the linearized problem is

$$
\begin{bmatrix} t_1^P - T_1^P \\ \vdots \\ t_K^P - T_K^P \\ t_1^S - T_1^S \\ \vdots \\ t_K^S - T_K^S \end{bmatrix} = \begin{bmatrix} -s_1^{(1)P}/v_0^p & -s_2^{(1)P}/v_0^p & -s_3^{(1)P}/v_0^p & 1 \\ \vdots & \vdots & \vdots & \vdots \\ -s_1^{(K)P}/v_0^p & -s_2^{(K)P}/v_0^p & -s_3^{(K)P}/v_0^p & 1 \\ -s_1^{(1)S}/v_0^S & -s_2^{(1)S}/v_0^S & -s_3^{(1)S}/v_0^S & 1 \\ \vdots & \vdots & \vdots & \vdots \\ -s_1^{(K)S}/v_0^S & -s_1^{(K)S}/v_0^S & -s_3^{(K)S}/v_0^S & 1 \end{bmatrix} \begin{bmatrix} x_0 \\ y_0 \\ z_0 \\ t_0 \end{bmatrix}
\tag{15.40}
$$

This equation is then solved iteratively using Newton's method. An example is shown in Fig. 15.20.

There are some instances in which the matrix can become underdetermined and the least-squares method will fail. This possibility is especially likely if the problem contains only P wave arrival times. The matrix equation consists of only the top half of Eq. (15.21). If all the rays leave the source in such a manner that one or more components of their unit vectors are all equal, then the corresponding column of the matrix will be proportional to the constant vector $[1,1,1,\cdots,1]^T$ and will therefore be linearly dependent on the fourth column of the matrix. The earthquake location is then nonunique and can be traded off with origin time (Fig. 15.21). This problem occurs when the earthquake is far from all of the stations. The addition of S wave arrival times resolves the underdeterminacy, as the columns are then proportional to $[1,1,1,\cdots,1,v_0^P/v_0^S,v_0^P/v_0^S,\cdots,v_0^P/v_0^S]^T$ and are not linearly dependent on the fourth column. It is therefore wise to use singular-value decomposition and the natural inverse to solve earthquake location problems, permitting easy identification of underdetermined cases.

Another problem can occur if all the stations are on the horizontal plane. Then the solution is nonunique, with the solution $[x_0,y_0,z_0,t_0]^T$ and its mirror image $[x_0,y_0,-z_0,t_0]^T$ having exactly the same error. This problem can be avoided by including prior information that the earthquake occurs beneath the ground and not in the air.

The earthquake location problem can also be extended to locate the earthquake and determine the velocity structure of the medium simultaneously. It then becomes similar to the tomography problem (Section 15.4) except that the

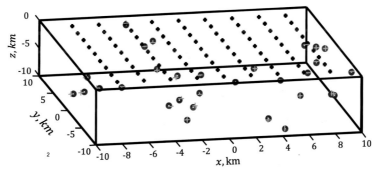

FIG. 15.20 Earthquake location example. Arrival times of P and S waves from earthquakes (red circles) are recorded on an array of 81 stations (black crosses). The observed arrival times include random noise with variance $\sigma_d^2 = (0.1)^2\,\mathrm{s}^2$. The estimated locations (green crosses) are computed using Geiger's method. Scripts gdama15_14 and gdapy15_14.

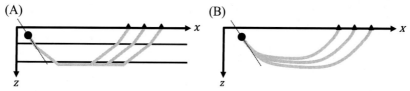

FIG. 15.21 (A) In layered media, rays follow "refracted" paths, such that the rays to all receivers (triangles) leave the source (circle) at the same angle. The position and travel time of the source on the dashed line therefore trade-off. (B) This phenomenon also occurs in nonlayered media if the source is far from the receivers.

orientations of the ray paths are unknown. To ensure that the medium is crossed by a sufficient number of rays to resolve the velocity structure, one must locate simultaneously a large set of earthquakes—on the order of 100 when the velocity structure is assumed to be one-dimensional (Crosson, 1976) but thousands to millions when it is fully three-dimensional.

15.11   Vibrational problems

There are many inverse problems that involve determining the structure of an object from measurements of its characteristic frequencies of vibration (*eigenfrequencies*). For instance, the solar and terrestrial vibrational frequencies can be inverted for the density and elastic structure of those two bodies.

The forward problem of calculating the *modes* (spatial patterns) and eigenfrequencies of vibration of a body of a given structure is quite formidable. A commonly used approximate method is based on perturbation theory. The modes and eigenfrequencies are first computed for a simple unperturbed structure and then they are perturbed to match a more complicated structure. Consider, for instance, an acoustic problem in which the pressure modes $p_n(x)$ and corresponding eigenfrequencies ω satisfy the differential equation

$$-\omega_n^2 \, p_n(x) = v^2(\mathbf{x})\nabla^2 p_n(x) \tag{15.41}$$

with homogeneous boundary conditions. Many properties of the solutions to this equation can be deduced without actually solving it. For instance, two modes $p_n(x)$ and $p_m(x)$ can be shown to obey the orthogonality relationship

$$\int p_n(\mathbf{x})p_m(\mathbf{x})v^{-2}(\mathbf{x})\mathrm{d}^3 x = \delta_{nm} \tag{15.42}$$

(where δ_{nm} is the Kronecker delta) as long as their eigenfrequencies are different, that is, $\omega_n \neq \omega_m$. Actually determining the modes and characteristic frequencies is a difficult problem for an arbitrary complicated velocity $v(\mathbf{x})$. Suppose, however, that velocity can be written as

$$v(\mathbf{x}) = v_0(\mathbf{x}) + \varepsilon v_1(\mathbf{x}) \tag{15.43}$$

where ε is a small number. Furthermore, suppose that the modes $p_n^{(0)}(\mathbf{x})$ and eigenfrequencies $\omega_n^{(0)}$ of the unperturbed problem (i.e., with $\varepsilon = 0$) are known. The idea of *perturbation theory* is to determine approximate versions of the perturbed modes $p_n(\mathbf{x})$ and corresponding perturbed eigenfrequencies ω_n that are valid for small ε.

The special case where all the unperturbed eigenfrequencies are distinct (i.e., with numerically different values) is easiest (and the only one we solve here). We first assume that the eigenfrequencies and perturbed modes can be written as a series in powers in ε

$$\begin{aligned}
\omega_n &= \omega_n^{(0)} + \varepsilon\omega_n^{(1)} + \varepsilon^2\omega_n^{(2)}\cdots \\
p_n(\mathbf{x}) &= p_n^{(0)}(\mathbf{x}) + \varepsilon p_n^{(1)}(\mathbf{x}) + \varepsilon^2 p_n^{(2)}(\mathbf{x})\cdots
\end{aligned} \tag{15.44}$$

Except for $i=0$, the functions $p_n^{(i)}(\mathbf{x})$ and constants $\omega_n^{(i)}$ are unknown. When ε is small, it suffices to solve merely for $p_n^{(1)}(\mathbf{x})$ and $\omega_n^{(1)}$. We first represent $p_n^{(1)}(\mathbf{x})$ as a sum of unperturbed modes, but omitting the $p_n^{(0)}(\mathbf{x})$ term as it already appears in the equation for $p_n(\mathbf{x})$

$$p_n^{(1)}(\mathbf{x}) = \sum_{\substack{m=1 \\ \omega_m^{(0)} \neq \omega_n^{(0)}}}^{\infty} b_{nm}p_m^{(0)}(\mathbf{x}) \tag{15.45}$$

where b_{nm} are unknown coefficients. After substituting Eqs. (15.45) and (15.44) into Eq. (15.41), equating terms of equal power in ε, and applying the orthogonality relationship, the first-order quantities can be shown to be (see, e.g., Menke and Abbott, 1990, their Section 8.6.7)

$$\omega_n^{(1)} = \omega_n^{(0)} \int \left[p_n^{(0)}(\mathbf{x})\right]^2 v_0^{-3}(\mathbf{x})v_1(\mathbf{x})\mathrm{d}^3 x$$

$$b_{nm} = \frac{2\left(\omega_m^{(0)}\right)^2}{\left(\omega_m^{(0)}\right)^2 - \left(\omega_n^{(0)}\right)^2} \int p_n^{(0)}(\mathbf{x})p_m^{(0)}(\mathbf{x})v_0^{-3}(\mathbf{x})v_1(\mathbf{x})\mathrm{d}^3 x \tag{15.46}$$

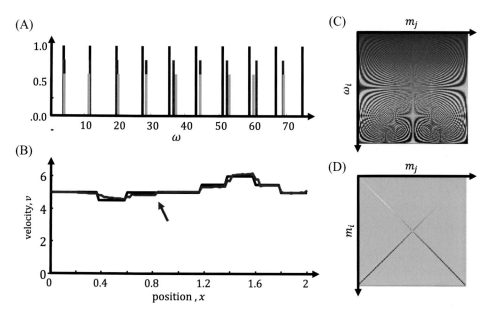

FIG. 15.22 Organ pipe example. (A) Ladder diagram for unperturbed (black), true (red), and observed (green) eigenfrequencies of the organ pipe. (B) True (black) and estimated (red) velocity structure. Arrow points to an artifact in the estimated solution. (C) Data kernel $\mathbf{G}$. (D) Model resolution matrix $\mathbf{R}$. Scripts gdama15_15 and gdapy15_15.

The equation for $\omega_n^{(1)}$ is the relevant one to this discussion, for it constitutes a linear inverse problem for the velocity perturbation $v_1(\mathbf{x})$

$$\omega_n^{(1)} = \int G_n(\mathbf{x})v_1(\mathbf{x})\mathrm{d}^3x \quad \text{with} \quad G_n(\mathbf{x}) \equiv \omega_n^{(0)}\left[p_n^{(0)}(\mathbf{x})\right]^2 v_0^{-3}(\mathbf{x}) \tag{15.47}$$

As $v_1(\mathbf{x})$ is a continuous function, this is properly a problem in continuous inverse theory. However, our approach to its solution will involve discretizing it.

As an example, we will consider a one-dimensional organ pipe of length h open at one end and closed at the other (corresponding to the boundary conditions $p_n|_{x=0} = \mathrm{d}p_n/\mathrm{d}x_{x=0} = 0$). We consider the unperturbed problem of a constant velocity structure, which has eigenfrequencies and modes given by

$$p_n^{(0)}(x) = \frac{2v_0^2}{h}\sin\left(\frac{(n-\frac{1}{2})\pi}{h}x\right)$$

$$\omega_n^{(0)} = \frac{(n-\frac{1}{2})\pi v_0}{h} \quad \text{with} \quad n = 1, 2, 3, \cdots \tag{15.48}$$

The eigenfrequencies are evenly spaced in frequency (Fig. 15.22A). We now imagine that we measure the first N eigenfrequencies ω_n of an organ pipe whose velocity structure $v(x)$ is unknown (Fig. 15.22A). In order to avoid carrying around factors of ε, we arbitrarily set that parameter to unity. The data are the deviations of these frequencies from those of the unperturbed problem $d_n \equiv \omega_n^{(1)} = \omega_n - \omega_n^{(0)}$. The model parameters are the continuous function $m(x) \equiv v_1(x) = v(x) - v_0(x)$, which we will discretize into a model parameter vector $\mathbf{m}$ by sampling it at M evenly spaced values, with spacing Δx. The data kernel in the discrete equation $\mathbf{d} = \mathbf{Gm}$ is the discrete version of Eq. (15.47) (Fig. 15.22C)

$$G_{nm} = G_n(x_m)\Delta x = \omega_n^{(0)}\left[p_n^{(0)}(x_m)\right]^2 v_0^{-3}\Delta x \tag{15.49}$$

The equation can be solved using damped least squares. Superficially, the estimated solution (red curve in Fig. 15.22B) looks reasonably good, except that it seems to have small steps (arrow in Fig. 15.22B) in locations where the true solution is smooth. However, a close inspection of the model resolution matrix (Fig. 15.22D), which is X-shaped, indicates a serious problem: two widely separated model parameters are trading off. The steps in the estimated solution are artifacts—steps from elsewhere in the model that are mapped into the wrong position. This is an inherent nonuniqueness of this and many other eigenfrequency-type problems. In seismology, it is addressed by combining eigenfrequency data with other data types (e.g., travel time measurements along rays) that do not suffer from this nonuniqueness.

15.12 Problems

15.1 Modify the MATLAB® or Python script for the airgun problem (gdama15_02 or gdapy15_02) to compute a shorter filter $m(t)$. How short can it be and still do a reasonable job at spiking the airgun pulse?

15.2 Modify the MATLAB® or Python script for the tomography problem (gdama15_04 or gdapy15_04) so that it retains receivers on the top and bottom of the model but omits them on the sides. Interpret the results.

15.3 Modify the MATLAB® or Python script for the L_1 straight-line problem (gdama15_07 or gdapy15_07) to allow for data of different accuracy. Test it against synthetic data, where the first $N/2$ data have variance σ_L^2 and the last $N/2$ have variance σ_R^2, both for the case where $\sigma_L = 10\sigma_R$ and $\sigma_R = 10\sigma_L$.

15.4 Modify the MATLAB® or Python script for the earthquake location problem (gdama15_14 or gdapy15_14) to include prior information that all the earthquakes occur near depth z_A. Adjust the true locations of the earthquakes to reflect this information. [Hint: See Method Summary 3 in Chapter 17 for instructions on how to implement the prior information. The problem can be solved using biconjucate gradients.]

15.5 Modify the MATLAB® or Python script for the earthquake location problem (gdama15_14 or gdapy15_14) to use differential P wave travel time data, but no absolute travel time data. Each of these new data is the time difference between the arrival times of two earthquakes observed at a common station. Compare the locations to the results of the absolute-arrival time script, in the case where the variance of the differential data is four orders of magnitude smaller than the variance of the absolute data. [Hint: You may have to use a better starting guess than was used in the absolute-arrival time script.]

References

Crosson, R.S., 1976. Crustal structure modeling of earthquake data: 1. Simultaneous least squares estimation of hypocenter and velocity parameters. J. Geophys. Res. 81, 3036–3046.

Fisher, R.A., 1953. Dispersion on a sphere. Philos. Trans. R. Soc. Lond. A 217, 295–305.

Geiger, L., 1912. Probability method for the determination of earthquake epicenters from the arrival time only (translated from Geiger's 1910 German article). Bull. St. Louis Univ. 8 (1), 56–71.

Kaula, W.M., 1966. Theory of Satellite Geodesy. Ginn (Blaisdell), Boston, MA.

Menke, W., Abbott, D., 1990. Geophysical Theory. Columbia University Press, New York. 457 pp.

Smith, S.G., 1975. Measurement of airgun waveforms. Geophys. J. R. Astron. Soc. 42, 273–280.

16

Applications of inverse theory to solid earth geophysics

16.1 Earthquake location and determination of the velocity structure of the earth from travel time data

The problem of determining the *hypocentral parameters* of an earthquake (i.e., its location and its origin time) and the problem of determining the velocity structure of the Earth from travel time data are very closely coupled in geophysics, because earthquakes, which occur naturally at initially unknown locations, are a primary source of structural information.

When the wavelength of the seismic waves is smaller than the scale of the velocity heterogeneities, the propagation of seismic waves can be described with ray theory, an approximation in which energy is assumed to propagate from source to observer along a curved path called a *ray*. In this approximation, the wave field is completely determined by the pattern of rays and the travel time along them.

In areas where the velocity structure is already known, earthquakes can be located using Geiger's (1912) method (see Section 15.10). The travel time $T(\mathbf{x})$ of an earthquake with *hypocenter* (location) $\mathbf{x}$ and its ray tangent $\mathbf{t}(\mathbf{x})$ (pointing from earthquake to receiver) is calculated using ray theory. The arrival time is $t = \tau + T(\mathbf{x})$, where τ is the origin time (starting time) of the earthquake. The perturbation in arrival time t due to a change in the earthquake location from $\mathbf{x}^{(0)}$ to $\mathbf{x}^{(0)} + \delta\mathbf{x}$ and a change in the origin time from τ_0 to $\tau_0 + \delta\tau$ is

$$\delta t = t - t_0 = t - \left[\tau_0 + T\left(\mathbf{x}^{(0)}\right) \right] \approx \delta\tau - s\left(\mathbf{x}^{(0)}\right) \mathbf{t}\left(\mathbf{x}^{(0)}\right) \cdot \delta\mathbf{x} \tag{16.1}$$

(see Section 15.10) where s is the slowness (reciprocal velocity) of the Earth. Note that this linearized equation has four model parameters τ and the three components of $\delta\mathbf{x}$.

The forward problem, that of finding the ray path connecting earthquake and station and its travel time, is computationally intensive. Two alternative strategies have been put forward, one based on first finding the ray path using a shooting or bending strategy and then calculating the travel time by an integral of slowness along the ray (Cerveny, 2001), or conversely, by first finding the travel time by solving its partial differential equation (the Eikonal equation) and then calculating the ray path by taking the gradient of the travel time (Vidale, 1990).

The velocity structure can be represented with varying degrees of complexity. The simplest is a vertically stratified structure consisting of a stack of homogeneous layers or, alternatively, a continuously varying function of depth. Analytic formula for the travel time and its derivative with source parameters can be derived in some cases (Aki and Richards, 2002, their Section 9.3). The most complicated cases are fully three-dimensional velocity models represented with voxels or splines. Compromises between these extremes are also popular and include two-dimensional models (in cases where the structure is assumed to be uniform along the strike of a linear tectonic feature) and regionalized models (which assign a different vertically stratified structure to each tectonically distinct region but average them when computing travel times of rays that cross from one region to another).

In the simplest formulation, each earthquake is located separately. The data kernel of the linearized problem is an $N \times 4$ matrix, where N is the number of arrival time observations, and is typically solved by singular-value decomposition. A very commonly used computer program that uses a layered velocity model is HYPOINVERSE (Klein, 1985). However, locations can often be improved by simultaneously solving for at least some aspect of the velocity model, which requires the earthquakes to be located simultaneously. The simplest modification is to solve for a station correction for each station; that is, a time increment that can be added to predicted travel times to account for near-station velocity heterogeneity not captured by the velocity model. More complicated formulations solve for a vertically stratified structure (Crosson, 1976) and for fully three-dimensional models (Menke, 2005; Zelt, 1998; Zelt and Barton, 1998). In the three-dimensional case, the inverse problem is properly one of tomographic inversion, as the perturbation in arrival time δt includes a line integral

$$\delta t = \delta\tau - s\left(\mathbf{x}^{(0)}\right) \mathbf{t}\left(\mathbf{x}^{(0)}\right) \cdot \delta\mathbf{x} + \int_{\substack{unperturbed \\ ray}} \delta s \, dl \tag{16.2}$$

Here, δs is the perturbation in slowness with respect to the reference model and dl is an increment of arc-length along the unperturbed ray. This problem is $N \times (4K + L)$, where N is the number of arrival time observations, K is the number of earthquakes, and L is the number of unknown parameters in the velocity model. One of the limitations of earthquake data is that earthquakes tend to be spatially clustered. Thus the ray paths often poorly sample the model volume, and prior information, usually in the form of smoothness information, is required to achieve useful solutions. Even so, earthquake location (and especially earthquake depth) will often trade off strongly with velocity structure.

A very useful algorithm due to Pavlis and Booker (1980) simplifies the inversion of the very large matrices that result from simultaneously inverting for the source parameter of many thousands of earthquakes and an even larger number of velocity model parameters. Suppose that the linearized problem has been converted into a standard discrete linear problem $\mathbf{d} = \mathbf{Gm}$, where the model parameters can be arranged into two groups $\mathbf{m} = [\mathbf{m}^{(1)}, \mathbf{m}^{(2)}]^T$, where $\mathbf{m}^{(1)}$ is a vector of the earthquake source parameters and $\mathbf{m}^{(2)}$ is a vector of the velocity parameters. Then the inverse problem can be written

$$\mathbf{d} = \mathbf{Gm} = \begin{bmatrix} \mathbf{G}^{(1)} & \mathbf{G}^{(2)} \end{bmatrix} \begin{bmatrix} \mathbf{m}^{(1)} \\ \mathbf{m}^{(2)} \end{bmatrix} = \mathbf{G}^{(1)}\mathbf{m}^{(1)} + \mathbf{G}^{(2)}\mathbf{m}^{(2)} \tag{16.3}$$

Now suppose that $\mathbf{G}^{(1)}$ has singular-value decomposition $\mathbf{G}^{(1)} = \mathbf{U}_p^{(1)}\mathbf{\Lambda}_p^{(1)}\mathbf{V}_p^{(1)T}$, with p nonzero singular values, so that $\mathbf{U}_p^{(1)} = \begin{bmatrix} \mathbf{U}_p^{(1)} & \mathbf{U}_0^{(1)} \end{bmatrix}$ where $\mathbf{U}_p^{(1)T}\mathbf{U}_0^{(1)} = 0$. Premultiplying the inverse problem (Eq. 16.3) by $\mathbf{U}_0^{(1)T}$ yields

$$\mathbf{U}_0^{(1)T}\mathbf{d} = \mathbf{U}_0^{(1)T}\mathbf{G}^{(1)}\mathbf{m}^{(1)} + \mathbf{U}_0^{(1)T}\mathbf{G}^{(2)}\mathbf{m}^{(2)} = \mathbf{U}_0^{(1)T}\mathbf{G}^{(2)}\mathbf{m}^{(2)}$$

$$\text{or } \mathbf{d}' = \mathbf{G}'\mathbf{m}^{(2)} \text{ with } \mathbf{d}' \equiv \mathbf{U}_0^{(1)T}\mathbf{d} \text{ and } \mathbf{G}' = \mathbf{U}_0^{(1)T}\mathbf{G}^{(2)} \tag{16.4}$$

Here, we have used the fact that $\mathbf{U}_0^{(1)T}\mathbf{G}^{(1)}\mathbf{m}^{(1)} = \mathbf{U}_0^{(1)T}\mathbf{U}_p^{(1)}\mathbf{\Lambda}_p^{(1)}\mathbf{V}_p^{(1)T} = 0$. This problem, which does not depend upon $\mathbf{m}^{(1)}$, can now be solved for $\mathbf{m}^{(2)}$. Premultiplying by the inverse problem (Eq. 16.3) by $\mathbf{U}_p^{(1)T}$ yields

$$\mathbf{U}_p^{(1)T}\mathbf{d} = \mathbf{U}_p^{(1)T}\mathbf{G}^{(1)}\mathbf{m}^{(1)} + \mathbf{U}_p^{(1)T}\mathbf{G}^{(2)}\mathbf{m}^{(2)} = \mathbf{\Lambda}_p^{(1)}\mathbf{V}_p^{(1)T}\mathbf{m}^{(1)} + \mathbf{U}_p^{(1)T}\mathbf{G}^{(2)}\mathbf{m}^{(2)}$$

$$\text{or } \mathbf{d}'' = \mathbf{G}''\mathbf{m}^{(1)} \text{ with } \mathbf{d}'' \equiv \mathbf{U}_p^{(1)T}\mathbf{d} - \mathbf{U}_p^{(1)T}\mathbf{G}^{(2)}\mathbf{m}^{(2)} \text{ and } \mathbf{G}'' \equiv \mathbf{\Lambda}_p^{(1)}\mathbf{V}_p^{(1)T} \tag{16.5}$$

This problem can be solved for $\mathbf{m}^{(1)}$, as $\mathbf{m}^{(2)}$ is known. The inverse problem has been partitioned into two equations, an equation for $\mathbf{m}^{(2)}$ (the velocity parameters) that can be solved first, and an equation for $\mathbf{m}^{(1)}$ (the source parameters) that can be solved once $\mathbf{m}^{(2)}$ is known (see scripts gdama16_01 and gdapy16_01). When the data consist of only absolute travel times (as contrasted to differential travel times, described later), the source parameter data kernel $\mathbf{G}^{(1)}$ is block diagonal, as the source parameters of a given earthquake affect only the travel times associated with that earthquake. The problem of computing the singular-value decomposition of $\mathbf{G}^{(1)}$ can be broken down into the problem of computing the singular-value decomposition of the submatrices.

Signal processing techniques based on waveform cross-correlation are able to determine the difference $\Delta t^{A,B} \equiv t^A - t^B$ in arrival times of two earthquakes (say, labeled A and B), observed at a common station, to several orders of magnitude better precision than the absolute arrival times t^A and t^B can be determined (e.g., Menke and Menke, 2015). Thus earthquake locations can be vastly improved, either by supplementing absolute arrival time data Δt with differential arrival time data Δt or by relying on differential data alone. The perturbation in differential arrival time for two earthquakes A and B is formed by differencing two versions of Eq. (15.40)

$$\delta\Delta t^{A,B} = \delta\tau^A - s\left(\mathbf{x}^{A(0)}\right)\mathbf{t}\left(\mathbf{x}^{A(0)}\right) \cdot \delta\mathbf{x}^A - \delta\tau^B + s\left(\mathbf{x}^{B(0)}\right)\mathbf{t}\left(\mathbf{x}^{B(0)}\right) \cdot \delta\mathbf{x}^B \tag{16.6}$$

Each equation involves eight unknowns. Early versions of this method focused on using differential data to improve the relative location between earthquakes (e.g., Spence and Alexander, 1968). The modern formulation, known as the *double-difference method*, acknowledges that the relative locations are better determined than the absolute locations but solves for both (e.g., Slunga et al., 1995; Waldhauser and Ellsworth, 2000; Menke and Schaff, 2004). Double-difference methods have also been extended to include the tomographic estimation of velocity structure (Zhang and Thurber, 2003).

At the other extreme is the case where the origin times and locations of the seismic sources are known and the travel time T of elastic waves through the Earth is used to determine the slowness $s(\mathbf{x})$ in the Earth

$$\delta t = \int_{\substack{perturbed \\ ray}} \delta s \, dl \approx \int_{\substack{unperturbed \\ ray}} \delta s \, dl \tag{16.7}$$

The approximation relies on *Fermat's Principle*, which states that perturbations in ray path cause only second-order changes in travel time. (See Menke (2020) for a derivation that follows the notation used in this book.) Tomographic inversion based on Eq. (16.7) is now a standard tool in seismology and has been used to image the Earth at many scales, from kilometer-scale geologic structures to the whole Earth. Model estimation is typically performed using generalized least squares to solve a discrete version of Eq. (16.7), often with Newton iteration to correct the ray pattern as the slowness model is updated.

The gradient method can also be used to solve ray-based travel time tomography problems, as adjoint methods provide an efficient mean of computing the derivative of the total travel time error with respect to a slowness model parameter, that is, $\partial E / \partial m_i$ (Bin Waheed et al., 2016; Menke, 2020).

An alternative to model estimation is presented by Vasco (1986) who uses extremal inversion to determine quantities such as the maximum difference between the velocities in two different parts of the model, assuming that the velocity perturbation is everywhere within a given set of bounds (see Section 8.7).

Tomographic inversions are often performed separately from earthquake location but then are degraded by whatever error is present in the earthquake locations and origin times. In the case where the earthquakes are all very distant from the volume being imaged, the largest error arises from the uncertainty in origin time, since errors in location do not change the part of the ray path in the imaged volume by very much. Suppose that the slowness in the imaged volume is represented as $s(x,y,z) = s_1(z) + s_2(x,y,z)$, with (x,y) horizontal and z vertical position, and with the mean of $s_2(x,y,z)$ zero at every depth, so that it represents horizontal variations in velocity, only. Poor knowledge of origin times affects only the estimates of $s_1(z)$ and not $s_2(x,y,z)$ (Aki et al., 1976; Menke et al., 2006). Estimates of the horizontal variations in velocity contain important information about the geologic structures that are being imaged, such as their (x,y) location and dip direction. Consequently, the method, called teleseismic tomography, has been applied in many areas of the world.

16.2   Moment tensors of earthquakes

Seismometers measure the motion or *displacement* of the ground, a three-component vector $\mathbf{u}(\mathbf{x},t)$ that is a function of position $\mathbf{x}$ and time t. At wavelengths longer than the spatial scale of a seismic source, such as a geologic fault, an earthquake can be approximated by a system of nine force-couples acting at a point $\mathbf{x}^{(0)}$, the *centroid* of the source. The amplitude of these force-couples is described by a 3×3 symmetric matrix $\mathbf{M}(t)$, called the *moment tensor*, which is a function of time t (Aki and Richards, 2002, their Section 3.3). Ground displacement depends on the time derivative $d\mathbf{M}/dt$, called the *moment-rate tensor*. Its six independent elements constitute six continuous model parameters $m_i(t)$. The ground displacement at $(\mathbf{x},t)$ due to a set of force-couples at $\mathbf{x}^{(0)}$ is linearly proportional to the elements of the moment-rate tensor. The forward problem of predicting the displacement given a known moment tensor is

$$u_i^{pre}(\mathbf{x},t) = \sum_{j=1}^{6} \int G_{ij}\left(\mathbf{x}, \mathbf{x}^{(0)}, t, t_0\right) m_j(t_0) \, dt_0$$

$$\text{with } \mathbf{m} = \frac{d}{dt}[M_{11} \; M_{22} \; M_{33} \; M_{12} \; M_{13} \; M_{23}]^T \tag{16.8}$$

Here, $G_{ij}(\mathbf{x},\mathbf{x}^{(0)},t,t_0)$ are data kernels that describe the ith component displacement at $(\mathbf{x},t)$ due to the jth force-couple at $(\mathbf{x}^{(0)},t_0)$. These kernels depend upon Earth structure and, though challenging to calculate, are usually assumed to be known. If the location $\mathbf{x}^{(0)}$ of the source is also known, then the problem of estimating the moment-rate tensor from observations of ground displacement is linear (Dziewonski and Woodhouse, 1983; Dziewonski et al., 1981). The time variability of the moment-rate tensor is often parameterized, with a triangular function of fixed width (or several overlapping triangles) being popular, so that the total number of unknowns is $M = 6L$, where L is the number of triangles. The error $e_i^{(j)}(t_k)$ associated with component i, observation location j, and observation time k is

$$e_i^{(j)}(t_k) \equiv u_i^{obs}\left(\mathbf{x}^{(j)}, t_k\right) - u_i^{pre}\left(\mathbf{x}^{(j)}, t_k\right) \tag{16.9}$$

When the total error E is defined as the usual L_2 prediction error

$$E = \sum_i \sum_j \sum_k \left[e_i^{(j)}(t_k) \right]^2$$

(16.10)

the problem can be solved with simple least squares. Most natural sources, such as earthquakes, are thought to occur without causing volume changes, so some authors add the linear constraint $m_1 + m_2 + m_3 = 0$ (at all times). Sometimes, the location $\mathbf{x}^{(0)} \equiv [x_0, y_0, z_0]^T$ of the source (or just its depth z_0) is also assumed to be unknown. The inverse problem is then nonlinear and can be solved using a grid search over source location, that is, by solving the linear inverse problem for a grid of fixed source locations and then choosing the one with smallest total error E. The Fréchet derivatives of E with respect to hypocentral parameters are known; so alternately, the gradient-descent method (Section 11.9) can be used. Adjoint methods based on Eq. (14.111) are also applicable to this problem (Kim et al., 2011).

Earthquake locations are routinely calculated by the U.S. Geological Survey (http://earthquake.usgs.gov/earthquakes/recenteqsww) and seismic moment tensors by the Global Centroid Moment Tensor (CMT) Project (http://www.globalcmt.org). They are standard data sets used in earthquake and tectonic research.

16.3   Adjoint methods in seismic imaging

The fields most important to seismology depend upon both position $\mathbf{x}$ and time t. They include the displacement $\mathbf{u}(\mathbf{x}, t)$, which quantifies the motion of particles in the Earth, the corresponding particle velocity $\dot{\mathbf{u}}(\mathbf{x}, t)$, particle acceleration $\ddot{\mathbf{u}}(\mathbf{x}, t)$, and the body force $\mathbf{f}(\mathbf{x}, t)$, which describes seismic sources such as earthquakes.

The displacement field $\mathbf{u}(\mathbf{x}, t)$ satisfies a partial differential equation of the form:

$$\mathcal{L}(\mathbf{m})\mathbf{u} = \mathbf{f}$$

(16.11)

Here, $\mathcal{L}$ is is a 3×3 second-order matrix partial differential operator that depends on material parameters, such as density $\rho(\mathbf{x})$, Lamè parameter $\lambda(\mathbf{x})$, and shear modulus $\mu(\mathbf{x})$, which we discretize and lump together into a model parameter vector $\mathbf{m}$. The seismic imaging problem is to estimate these model parameters using observations of the displacement $\mathbf{u}(\mathbf{x}, t)$, itself, or of parameters derived from it (such as finite-frequency travel time, discussed in Section 16.6). In some applications, the force $\mathbf{f}$ is known, but in others it is an additional model parameter that must be determined as the inverse problem is solved.

The most general form of $\mathcal{L}(\mathbf{m})$ is very complicated. The simplest case is for an isotropic, homogeneous elastic medium, where it has the form

$$\mathcal{L} = \rho \frac{\partial^2}{\partial t^2} \mathbf{I} - \begin{bmatrix} (\lambda + 2\mu)\frac{\partial^2}{\partial x^2} + \mu\frac{\partial^2}{\partial y^2} + \mu\frac{\partial^2}{\partial z^2} & (\lambda + 2\mu)\frac{\partial^2}{\partial x \partial y} & (\lambda + 2\mu)\frac{\partial^2}{\partial x \partial z} \\ (\lambda + 2\mu)\frac{\partial^2}{\partial x \partial y} & \mu\frac{\partial^2}{\partial x^2} + (\lambda + 2\mu)\frac{\partial^2}{\partial y^2} + \mu\frac{\partial^2}{\partial z^2} & (\lambda + 2\mu)\frac{\partial^2}{\partial y \partial z} \\ (\lambda + 2\mu)\frac{\partial^2}{\partial x \partial z} & (\lambda + 2\mu)\frac{\partial^2}{\partial y \partial z} & \mu\frac{\partial^2}{\partial x^2} + \mu\frac{\partial^2}{\partial y^2} + (\lambda + 2\mu)\frac{\partial^2}{\partial z^2} \end{bmatrix}$$

(16.12)

Here, $\mathbf{I}$ is the 3×3 identity matrix. In this case, the partial derivative of the operator with respect to density is also very simple. Suppose the density is $\rho = \rho_0 + m\delta(\mathbf{x} - \mathbf{x}_H)$, which contains a point density heterogeneity of unknown amplitude m and known position $\mathbf{x}_H$. After inserting this formula into Eq. (16.12) and differentiating, we find:

$$\frac{\partial \mathcal{L}}{\partial m} = \delta(\mathbf{x} - \mathbf{x}_H)\frac{\partial^2}{\partial t^2}\mathbf{I}$$

(16.13)

As seismic fields vary in both space and time, the inner product, previously defined as a spatial integral in Eq. (14.32), must be modified to include a time integral. The inner product between two vectors, $\mathbf{a}(\mathbf{x}, t)$ and $\mathbf{b}(\mathbf{x}, t)$, is defined as:

$$(\mathbf{a}, \mathbf{b}) \equiv \int_{-\infty}^{+\infty} \iiint_V \mathbf{a}^T \mathbf{b} \, d^3x \, dt = \sum_{i=1}^{3} \int_{-\infty}^{+\infty} \iiint_V a_i b_i \, d^3x \, dt$$

(16.14)

Here, $\mathbf{a}^T\mathbf{b} \equiv \sum_i a_i b_i$ is the dot product and V is the volume of the Earth.

Adjoints can be used to manipulate inner products, using the rule $(\mathbf{a}, \mathcal{L}\mathbf{b}) = (\mathcal{L}^\dagger\mathbf{a}, \mathbf{b})$. It can be shown that the adjoint of a matrix operator is the matrix transpose of the adjoint of its elements, that is

$$\left[\mathcal{L}^\dagger\right]_{ij} = \left[\mathcal{L}_{ji}\right]^\dagger \tag{16.15}$$

The wave equation operator in Eq. (16.12) is self-adjoint, that is, $\mathcal{L}^\dagger = \mathcal{L}$. This property is shared by many other second-order wave equations, including more complicated forms of the seismic wave equation.

A characteristic of seismic observations is that sampling in space is often irregular and sparse (because each observation point requires a separate geophone), whereas sampling in time is always regular and closely spaced (because seismic recording systems use very high-speed electronics). Consequently, seismic data are often treated as continuous functions of time made at discrete points in space. For example, $u_i^{(k)} \equiv u_i(\mathbf{x}^{(k)}, t)$ is the ith component of displacement measured by the kth receiver. Here, $i = 1, 2, 3$ and $k = 1, \ldots, K$ when three-component observations are made at K receivers.

A datum, similar to the one defined in Eq. (14.57b), can be derived from a single component i of displacement measured at a single receiver k using the rule:

$$d_i^{(k)} = \int\limits_{-\infty}^{+\infty} h(t) u_i\left(\mathbf{x}^{(k)}, t\right) dt \tag{16.16}$$

Here, $h(t)$ is a known function. This equation is not in the form of an inner product, as defined in Eq. (16.14), because it omits integration over space. It can be manipulated into that form by inserting a Dirac impulse function and by defining a vector function $\mathbf{h}^{(i)}$ that is zero, except for its ith component $h_j^{(i)} \equiv h(t)\delta_{ij}$. Then,

$$\left(\mathbf{h}^{(i)}\delta\left(\mathbf{x} - \mathbf{x}^{(k)}\right), \mathbf{u}(\mathbf{x}, t)\right) = \sum_{j=1}^{3} \int\limits_{-\infty}^{+\infty} \iiint_V h(t)\delta_{ij}\delta\left(\mathbf{x} - \mathbf{x}^{(k)}\right) u_j(\mathbf{x}, t)\, d^3x\, dt$$

$$= \int\limits_{-\infty}^{+\infty} h(t) u_i\left(\mathbf{x}^{(k)}, t\right) dt = d_i^{(k)} \tag{16.17}$$

The data kernel evaluated for a reference model $\mathbf{m}_0$ can then be found using adjoint methods:

$$\frac{\partial}{\partial m} d_i^{(k)}\bigg|_{\mathbf{m}_0} = \left(\mathbf{h}^{(i)}\delta\left(\mathbf{x} - \mathbf{x}^{(k)}\right), \frac{\partial \mathbf{u}}{\partial m}\bigg|_{\mathbf{m}_0}\right)$$

$$= \left(\mathbf{h}^{(i)}\delta\left(\mathbf{x} - \mathbf{x}^{(k)}\right), -\mathcal{L}_0^{-1}\frac{\partial \mathcal{L}_0}{\partial m}\mathbf{u}^{(0)}(\mathbf{x}, t)\right) \tag{16.18}$$

$$= -\left(\mathcal{L}_0^{-1\dagger}\mathbf{h}^{(i)}\delta\left(\mathbf{x} - \mathbf{x}^{(k)}\right), \frac{\partial \mathcal{L}_0}{\partial m}\mathbf{u}^{(0)}(\mathbf{x}, t)\right)$$

Here, we have used the vector form of the Born approximation, which is analogous to Eq. (14.78) (and is derived in a similar manner). The expression can be simplified by defining an adjoint field λ and a quantity ξ

$$\frac{\partial d_i^{(k)}}{\partial m}\bigg|_{\mathbf{m}_0} = (\lambda, \xi)$$

$$\mathcal{L}_0\mathbf{u}^{(0)} = \mathbf{f} \quad \text{and} \quad \mathcal{L}_0 \equiv \mathcal{L}(\mathbf{m}_0) \quad \text{and} \quad \frac{\partial \mathcal{L}_0}{\partial m} \equiv \frac{\partial \mathcal{L}}{\partial m}\bigg|_{\mathbf{m}_0} \tag{16.19}$$

$$\text{and} \quad \mathcal{L}_0^\dagger\lambda \equiv \mathbf{h}^{(i)}\delta\left(\mathbf{x} - \mathbf{x}^{(k)}\right) \quad \text{and} \quad \xi \equiv -\frac{\partial \mathcal{L}_0}{\partial m}\mathbf{u}^{(0)}(\mathbf{x}, t)$$

The reference field $\mathbf{u}^{(0)}$ emanates from the source and propagates forward in time. The adjoint field λ emanates from the receiver and propagates backward in time. When the version of $\partial\mathcal{L}_0/\partial m$ corresponding to a point density heterogeneity (Eq. 16.13) is substituted into Eq. (16.19), the spatial part of the inner product can be performed trivially:

$$\frac{\partial d_i^{(k)}}{\partial m}\bigg|_{\mathbf{m}_0} = -\left(\lambda, \frac{\partial \mathcal{L}_0}{\partial m}\mathbf{u}^{(0)}\right) = -\left(\lambda(\mathbf{x}, t), \delta(\mathbf{x} - \mathbf{x}_H)\ddot{\mathbf{u}}^{(0)}(\mathbf{x}, t)\right)$$

$$= -\int\limits_{-\infty}^{+\infty} \lambda^T(\mathbf{x}_H, t)\ddot{\mathbf{u}}^{(0)}(\mathbf{x}_H, t) \tag{16.20}$$

The data kernel is constructed by taking the dot product of the adjoint field and the reference acceleration, both evaluated at the heterogeneity, and then time integrating the result. This process is often referred to as *correlating* λ and $\ddot{\mathbf{u}}^{(0)}$ (as it corresponds to the zero-lag cross-correlation of the two fields).

A similar process can be used to construct the derivative of the total wave field error E. When three-component, individual wave field errors $\mathbf{e}^{(k)}$ (with $\mathbf{e} = \mathbf{u}^{obs} - \mathbf{u}$) are measured at K stations, it is defined as

$$E \equiv \sum_{k=1}^{K} \int_{-\infty}^{+\infty} \left[\mathbf{e}^{(k)} \right]^{T} \mathbf{e}^{(k)} dt \tag{16.21}$$

In order to put this expression into the form of an inner product over both space and time, we must introduce Dirac delta functions at each of the receiver locations $\mathbf{x}^{(k)}$. Then, an expression equivalent to Eq. (16.21) is

$$E = \sum_{k=1}^{N} \left(\mathbf{e}\, \delta\!\left(\mathbf{x} - \mathbf{x}^{(k)} \right), \mathbf{e} \right) \tag{16.22}$$

As only the Dirac delta function involves the receiver index k, the summation can be moved inside of the inner product

$$E = (s\mathbf{e}, \mathbf{e}) \quad \text{with} \quad s \equiv \sum_{k=1}^{N} \delta\!\left(\mathbf{x} - \mathbf{x}^{(k)} \right) \tag{16.23}$$

Differentiating with respect to the model yields:

$$\left.\frac{\partial E}{\partial m}\right|_{\mathbf{m}_0} = \left.\frac{\partial}{\partial m}(s\mathbf{e}, \mathbf{e})\right|_{\mathbf{m}_0} = 2\left(s\mathbf{e}, \left.\frac{\partial \mathbf{e}}{\partial m}\right|_{\mathbf{m}_0} \right) = -2\left(s\mathbf{e}, \left.\frac{\partial \mathbf{u}}{\partial m}\right|_{\mathbf{m}_0} \right) \tag{16.24}$$

Here we have used $\partial \mathbf{e}^{(k)}/\partial m = -\partial \mathbf{u}^{(k)}/\partial m$, together with the fact that s is not a function of m. The rest of the derivation parallels the data kernel case. We insert the Born approximation, manipulate the inner product with adjoint methods, and define an adjoint field λ and a quantity ξ, so that

$$\left.\frac{\partial E}{\partial m}\right|_{\mathbf{m}_0} = (\lambda, \xi)$$

$$\text{with} \quad \mathcal{L}_0 \mathbf{u}^{(0)} = \mathbf{f} \quad \text{and} \quad \mathcal{L}_0^{\dagger} \lambda \equiv s\mathbf{e} \quad \text{and} \quad \xi \equiv 2\frac{\partial \mathcal{L}_0}{\partial m} \mathbf{u}^{(0)}(\mathbf{x}, t) \tag{16.25}$$

The *adjoint source* $s\mathbf{e}$ can be further simplified by recognizing that, because of the Dirac delta functions in s, it is only dependent on the value of the error at the receiver locations. Consequently, the adjoint equation becomes:

$$\mathcal{L}_0^{\dagger} \lambda \equiv \sum_{k=1}^{N} \mathbf{e}^{(k)} \delta\!\left(\mathbf{x} - \mathbf{x}^{(k)} \right) \tag{16.26}$$

Both the calculation of the data kernel (Eq. 16.20) and the error derivative (Eq. 16.22) require the numerical solution of differential equations—a very computationally intensive process. The reference field must be propagated from its source, forward in time, throughout the entire Earth model (or at least to a volume large enough to encompass all the heterogeneities). Similarly, Eq. (16.26) indicates that the adjoint field must be propagated backwards in time, from it source, throughout the entire Earth model (or at least to a volume large enough to encompass all heterogeneities). The source of the reference field is usually a spatially localized earthquake. In contrast, the source of the adjoint field is spatially distributed, but in a distinctive way, with each receiver acting as a point source. However, this behavior poses no special problems when the adjoint equation is solved numerically. The process of solving the adjoint equation is often referred to *as injecting the error into it* (as the adjoint source depends on the wave field error).

In this formulation, the model parameter m appears only in the function ξ through its dependence on $\partial \mathcal{L}_0/\partial m$. Consequently, the same adjoint field can be used to calculate any number of derivatives $\partial E/\partial m_i$ (although a different correlation must be performed for each of them). Consequently, in a problem with M model parameter, two partial differential equations (forward and adjoint) must be solved, and M correlations must be performed.

16.4 Wavefield tomography

In order to study the essential character of the error derivative, we step away from the complications of vector wave fields and consider the simple case where a "displacement" $u(x,t)$, which satisfies the scalar wave equation:

$$\mathcal{L}u \equiv \left(s^2 \frac{\partial^2}{\partial t^2} - \nabla^2\right)u = f \tag{16.27}$$

The wave equation has one material parameter, the slowness $s(\mathbf{x})$. We consider a point source at $\mathbf{x}_S$ with source time function $w(t)$ so that $f(\mathbf{x},t) = w(t)\delta(\mathbf{x} - \mathbf{x}_S)$. We limit our analysis to the special case of a constant background slowness s_0 and a small spatially variable perturbation $\delta s(\mathbf{x})$, so that $s = s_0 + \delta s$ and $s^2 = s_0^2 + 2s_0\delta s$. Furthermore, we assume a point heterogeneity of the form $2s_0\delta s = m\delta(\mathbf{x} - \mathbf{x}_H)$, with position $\mathbf{x}_H$ and unknown amplitude m. The reference solution with $m = 0$ can be shown to be:

$$u_0(\mathbf{x},t) = \frac{w(t - T_{SX})}{4\pi R_{SX}} \quad \text{with} \quad R_{SX} = |\mathbf{x} - \mathbf{x}_S| \quad \text{and} \quad T_{SX} = s_0 R_{SX} \tag{16.28}$$

Here, R_{SX} is the distance from the source to $\mathbf{x}$ and T_{SX} is the corresponding travel time. The solution is a spherical wave diverging from the source, with an amplitude proportional to $1/R_{SX}$ and with a pulse shape that is a delayed version of the source time function $w(t)$.

For the point heterogeneity stated before, the derivative of the operator in Eq. (16.27) is:

$$\frac{\partial \mathcal{L}_0}{\partial m} = \delta(\mathbf{x} - \mathbf{x}_H)\frac{\partial^2}{\partial t^2} \tag{16.29}$$

Consequently,

$$\xi(\mathbf{x},t) = 2\frac{\partial \mathcal{L}_0}{\partial m}u_0(\mathbf{x},t) = 2\delta(\mathbf{x} - \mathbf{x}_H)\ddot{u}_0(\mathbf{x},t) = 2\delta(\mathbf{x} - \mathbf{x}_H)\frac{\ddot{w}(t - T_{SX})}{4\pi R_{SX}} \tag{16.30}$$

The operator in the adjoint equation is the same as in the wave equation, since the latter is self-adjoint. The adjoint field emanates from the receiver and has a time function given by the wave field error, that is, $f(\mathbf{x},t) = e_0(\mathbf{x},t)\delta(\mathbf{x} - \mathbf{x}_R)$. The solution to the adjoint equation is analogous to Eq. (16.28), except that it is backward in time:

$$\lambda(\mathbf{x},t) = \frac{e_0(\mathbf{x}_R, t + T_{RX})}{4\pi R_{RX}} \tag{16.31}$$

The error derivative for a single receiver R is then:

$$\frac{\partial E_R}{\partial m} = (\lambda(\mathbf{x},t), \xi(\mathbf{x},t)) = 2\frac{1}{4\pi R_{SH}}\frac{1}{4\pi R_{RH}}\int_{-\infty}^{+\infty} e_0(t + T_{RH})\ddot{w}(t - T_{SH})\mathrm{d}t \tag{16.32}$$

The spatial pattern of the derivative is axially symmetric about a line drawn from source to receiver, since R_{SH}, R_{RH}, T_{RH}, and T_{SH} depend only on the perpendicular distance of the heterogeneity from the $\overline{SR}$ line. When $\ddot{w}$ is oscillatory, the time integral is oscillatory and the derivative consists of a series of concentric ellipses of alternating sign, with foci at the source and receiver (Fig. 16.1). The ellipses represent surfaces of equal travel time from source to heterogeneity to receiver. The scattered fields from heterogeneities on a given ellipse all arrive at the receiver at the same time and all contribute to changing the error at that time, only; they have no effect on the error at other times. The amplitude of the derivative varies across the surface of an ellipse, because it depends upon the product of the source-to-heterogeneity and heterogeneity-to-receiver distances, rather than their sum.

16.5 Seismic migration

Seismic migration is a technique for imaging the heterogeneities that cause reflected (or *scattered*) arrivals on seismic reflection profiles. In a seismic reflection experiment, a down-going incident wave of known waveform is sent into the Earth and up-going reflected waves (or scattered waves), caused by the interaction of the incident wave with heterogeneities at depth, are recorded by an array of receivers on the Earth's surface (Fig. 16.2). Seismic migration is based on the *imaging principle* that at the moment and location that the incident wave u_0 interacts

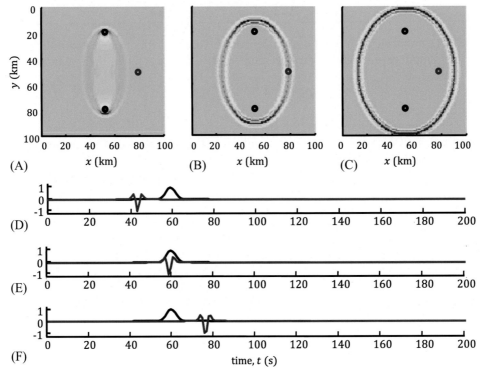

FIG. 16.1   Partial derivative $\partial E/\partial m$ of the wave field error $E(\mathbf{x}_R)$ for an acoustic wave propagation problem. The source $\mathbf{x}_S$ and receiver $\mathbf{x}_R$ (black circles) are separated by a distance of $R = 60$ km. The perturbation in acoustic velocity is $\delta v = (m/2v_0)\delta(\mathbf{x} - \mathbf{x}_h)$, where $v_0 = 1$ km/s is a constant reference velocity and $\mathbf{x}_h$ is the heterogeneity's location. The wave field error $e(t)$ is a Gaussian pulse at time $t = R/v_0 + \tau$. (A–C) Partial derivative (colors) for a suite of $\mathbf{x}_h$s covering the (x, y) plane, for $\tau = 25$, 40, and 57 s, respectively. Case (B) corresponds to the situation when the reference field emanating from source and the adjoint field emanating from the receiver arrive at a particular heterogeneity, $\mathbf{x}_H$ (red circle), at the same time, leading to a large value of the derivative. (D–F) Reference field $u_0(\mathbf{x}_H, t)$ (black) and adjoint field $\lambda(\mathbf{x}_H, t)$ (red), for $\tau = 25$, 40, and 57 s, respectively. Scripts `gdama16_02` and `gdapy16_02`.

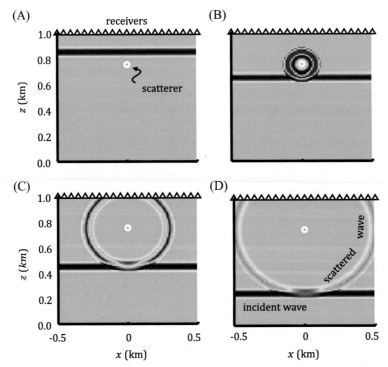

FIG. 16.2   (A–D) Sequence of four snapshots of a down-going incident plane wave (red band) interacting with a point-like scatterer (white circle) and generating a scattered spherical wave. The wave field is observed by an array of receivers (triangles) on the Earth's surface. Scripts `gdama16_03` and `gdapy16_03`.

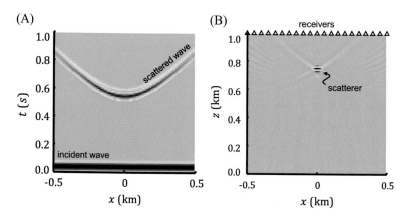

FIG. 16.3 (A) Seismic reflection data recorded by the array of receivers in Fig. 16.4, containing incident and scattered waves. (B) The data from (A) is migrated to produce the correlation $C(x,z)$, which is a proxy for the estimated heterogeneity $m^{est}(x,z)$. The correlation is at the location of the true heterogeneity (compare with Fig. 16.2). Scripts `gdama16_03` and `gdapy16_03`.

with the heterogeneity to produce the scattered wave δu, the waveform of the scattered waves exactly matches $\ddot{u}_0$, where the d^2/dt^2 arises from the scattering interaction. This behavior suggests that the heterogeneity can be imaged by (1) propagating the incident wave forward in time from its starting point to all possible heterogeneity locations $\mathbf{x}_H$, (2) propagating the scattered wave field backward in time from the receivers to these same locations, and (3) correlating (time integrating) the scattered wave field with the second derivative of the incident wave field:

$$C(\mathbf{x}_H) \equiv \int_{-\infty}^{+\infty} \ddot{u}_0(\mathbf{x}_H, t) \, \delta u^{obs}(\mathbf{x}_H, t) \, dt \qquad (16.33)$$

The correlation $C(\mathbf{x}_H)$ will be peaked at the location of the heterogeneity (or, if several heterogeneities are present, at all their locations) and can be used as a proxy for the heterogeneous structure of the Earth (Fig. 16.3).

If this process of propagating some fields forward in time, others backward in time, and then correlating sounds familiar—it should! It is exactly the same as the process used in Eq. (16.32) to compute the error derivative $\partial E/\partial m$ with the observed scattered wave field $\delta u^{obs}(\mathbf{x}_H, t)$ being used as a proxy for the wave field error $e_0(t + T_{RH})$. The imaging principle of seismic migration is equivalent to the idea that the error derivative is a proxy for the estimated heterogeneities $m^{est}(\mathbf{x})$. This equivalence is explored further by Zhu et al. (2009).

16.6   Finite-frequency travel time tomography

Originally, seismic tomography was based on arrival times *picks*; that is, measurements of the onset of motion of a seismic phase on a seismogram, either determined "by eye" or by an algorithm that detects a sudden increase in the amplitude of the ground motion. Such travel times are easy to measure on short-period seismograms but problematical at longer periods, owing to the emergent onset of the waveforms. A more suitable measurement technique for these data involves cross-correlating the observed seismic phase with a synthetic reference seismogram, because cross-correlation can accurately determine the small time difference, say τ, between two smooth pulses. However, the results of cross-correlation are dependent upon the frequency band of measurement; a phase that is observed to arrive earlier than the reference phase for one frequency band may well arrive later than it for another. Consequently, *finite-frequency travel times* must be interpreted in the context of the frequency band at which they are measured. Finite-frequency travel time tomography is based upon a derivative $\partial \tau/\partial m$ (where m is a model parameter) that incorporates the frequency-dependent behavior of cross-correlations (Dahlen et al., 2000).

At the long wavelengths used in finite-frequency tomography, seismic energy does not propagate along an infinitesimally thin ray connecting source and receiver, as it does in the high-frequency limits, so that $\partial \tau/\partial m$ is not simply an integral of heterogeneities along the ray path (as we have assumed previously, see Sections 2.7, 2.8, and 15.4). Instead, heterogeneities in entire Earth volume contribute to it. As we demonstrate later, the finite-frequency kernel has a diffuse shape that narrows with frequency, and which takes on the appearance of a ray only in the infinite frequency limit. The effect of a small heterogeneity on τ is less at low frequency than at high (giving low-frequency measurements a disadvantage). On the other hand, it has some effect at low frequencies even if, at high frequencies, the corresponding ray misses the heterogeneity entirely (giving low-frequency measurements an advantage).

For clarity, the derivative $\partial \tau / \partial m$ is first derived for the scalar wave field case; the result will be extended later to measurements made on one of the components of a vector field. The differential travel time τ_0 between an observed field $u^{obs}(\mathbf{x}, t)$ and a reference field $u_0(\mathbf{x}, t)$ is defined as the one that maximizes the cross-correlation:

$$C(\mathbf{x}, \tau; u_0) \equiv \int_{-\infty}^{+\infty} u^{obs}(\mathbf{x}, t - \tau)\, u_0(\mathbf{x}, t)\; dt \tag{16.34}$$

The first derivative of the cross-correlation C is zero at τ_0, as by definition, C has a maximum at this point:

$$\left. \frac{d}{d\tau} C(\mathbf{x}, \tau; u_0) \right|_{\tau_0} = 0 \tag{16.35}$$

Now let the field be perturbed from u_0 to $u = u_0 + \delta u$. The cross-correlation experiences a corresponding perturbation:

$$C(\mathbf{x}, \tau; u) = C(\mathbf{x}, \tau_0; u) + \delta C(\mathbf{x}, \tau) \;\; \text{with} \;\; \delta C(\mathbf{x}, \tau) \equiv \int_{-\infty}^{+\infty} u^{obs}(\mathbf{x}, t - \tau)\, \delta u(\mathbf{x}, t)\, dt \tag{16.36}$$

The perturbed cross-correlation has a maximum at, say, $\tau = \tau_0 + \delta\tau$. Expanding $C(\mathbf{x}, \tau; u)$ in a Taylor series up to second order in small quantities yields:

$$C(\mathbf{x}, \tau; u) \approx C(\mathbf{x}, \tau_0; u_0) + 0 + \frac{1}{2} \left. \left(\frac{d^2}{d\tau^2} C(\mathbf{x}, \tau, u_0) \right) \right|_{\tau_0} (\delta\tau)^2 + \delta C(\mathbf{x}, \tau_0) + \left. \left(\frac{d}{d\tau} \delta C(\mathbf{x}, \tau) \right) \right|_{\tau_0} \delta\tau \tag{16.37}$$

As is shown in Eq. (16.35), the second term on the right-hand side is zero. The maximum occurs where the derivative of Eq. (16.37) with respect to $\delta\tau$ is zero:

$$\frac{d}{d\delta\tau} C(\mathbf{x}, \tau; u) = 0 \approx \left. \left(\frac{d^2}{d\tau^2} C(\mathbf{x}, \tau; u_0) \right) \right|_{\tau_0} \delta\tau + \left. \left(\frac{d}{d\tau} \delta C(\mathbf{x}, \tau) \right) \right|_{\tau_0}$$

$$\text{so} \;\; \delta\tau = -\frac{1}{D} \left. \left(\frac{d}{d\tau} \delta C(\mathbf{x}, \tau; u_0) \right) \right|_{\tau_0} \;\; \text{with} \;\; D \equiv \left. \left(\frac{d^2}{d\tau^2} C(\mathbf{x}, \tau; u_0) \right) \right|_{\tau_0} \tag{16.38}$$

This expression can be further simplified by differentiating the cross-correlation:

$$\left. \frac{d\delta C}{d\tau} \right|_{\mathbf{x}, \tau_0} = \frac{d}{d\tau} \int_{-\infty}^{+\infty} u^{obs}(\mathbf{x}, t - \tau)\, \delta u(\mathbf{x}, t)\, dt = - \int_{-\infty}^{+\infty} \dot{u}^{obs}(\mathbf{x}, t - \tau)\, \delta u(\mathbf{x}, t)\, dt$$

$$D(\mathbf{x}, \tau_0) = \left. \left(\frac{d^2}{d\tau^2} C(\mathbf{x}, \tau, u_0) \right) \right|_{\tau_0} = \left. \frac{d^2}{d\tau^2} \int_{-\infty}^{+\infty} u^{obs}(\mathbf{x}, t - \tau)\, u_0(\mathbf{x}, t)\, dt \right|_{\tau_0} \tag{16.39}$$

$$= \int_{-\infty}^{+\infty} \ddot{u}^{obs}(\mathbf{x}, t - \tau_0)\, u_0(\mathbf{x}, t)\, dt$$

The scalar field $u(\mathbf{x}, t)$ is now equated with the ith component of a vector field $\mathbf{u}(\mathbf{x}, t)$. Eq. (16.38) can be manipulated into the form of an inner product:

$$\delta\tau^{(i)}(\mathbf{x}_R) = \left(\mathbf{h}^{(i)} \delta(\mathbf{x} - \mathbf{x}_R), \delta\mathbf{u}(\mathbf{x}, t) \right) \;\; \text{with}$$

$$\left[\mathbf{h}^{(i)} \right]_j \equiv -\frac{\dot{u}_j^{obs} \left(\mathbf{x}_R, t - \tau_0^{(j)} \right)}{D_j \left(\mathbf{x}_R, \tau_0^{(j)} \right)} \delta_{ij} \;\; \text{and} \;\; D_j(\mathbf{x}, \tau_0) \equiv \int_{-\infty}^{+\infty} \ddot{u}_j^{obs} \left(\mathbf{x}_R, t - \tau_0^{(j)} \right) u_j^{(0)}(\mathbf{x}_R, t)\, dt \tag{16.40}$$

Here, $\mathbf{h}^{(i)}(t)$ is a vector with only one nonzero component, corresponding to the component of displacement for which travel time is being computed. The total travel time is $\tau^{(i)} = \tau_0^{(i)} + \delta\tau^{(i)}$ and its derivative with respect to the model is:

$$\left. \frac{\partial \tau^{(i)}}{\partial m} \right|_{m_0} = \left. \frac{\partial \delta\tau^{(i)}}{\partial m} \right|_{m_0} = \left(\mathbf{h}^{(i)} \delta(\mathbf{x} - \mathbf{x}_R), \left. \frac{\partial \mathbf{u}(\mathbf{x}, t)}{\partial m} \right|_{m_0} \right) \tag{16.41}$$

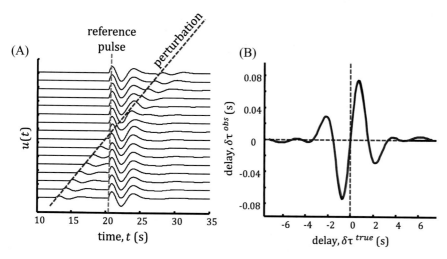

FIG. 16.4 Finite-frequency travel times. (A) A suite of signals $u(t)$ (black curves) are created by adding a small perturbation (red) to a reference pulse u_0 (blue), where the signals differ only by the delay $\delta\tau^{true}$ of the perturbation. (B) Cross-correlation (black curve) and the first-order approximation to it (Eq. 16.41) (red curve) are used to estimate the travel time perturbation $\delta\tau^{obs}$. Both yield similar results. Owing to wave interference, $\delta\tau^{obs}$ behaves like $\delta\tau^{obs} \approx \delta\tau^{true}$ only for very small delays straddling the origin and behaves in an oscillatory manner at larger delays. Scripts `gdama16_04` and `gdapy16_04`.

This equation, due to Marquering et al. (1999), is exactly the same form as the one for the data kernel in Eq. (16.17) so it can be calculated using the adjoint method of Eq. (16.19). Numerical tests (Fig. 16.4) show that the formula, though based on a first-order approximation, nevertheless is very accurate.

Sensitivity kernels (derivatives) have been derived for many other kinds of seismological quantities, including the differential travel times (Yuan et al., 2016), the envelope of the wave field (Yuan et al., 2015), and the cross-convolution measure (Menke, 2017b).

16.7 Banana-doughnut kernels

Once again, we step away from the complications of vector wave fields and use the scalar wave equation (Eq. 16.27) to study the behavior finite-frequency travel times. In keeping with the idea that the wave field is narrow band, we assume a band-limited source time function $w(t)$:

$$w(t) \equiv \text{sinc}(\omega_1 t) - \text{sinc}(\omega_2 t) \tag{16.42}$$

This function has a spectrum that is nonzero only at angular frequencies between ω_1 and ω_2.

We consider the case where the observed field equals the reference field, that is, $u^{obs}(\mathbf{x}_R, t) = u_0(\mathbf{x}_R, t)$. The "observations" align perfectly with the reference field, so $\tau_0 = 0$. The reference field $u_0(\mathbf{x}, t)$ and the related quantity $\xi(\mathbf{x}, t)$ are the same as given by Eqs. (16.28) and (16.30), respectively. The adjoint equation and its solution are

$$\mathcal{L}^\dagger \lambda = -\frac{\dot{u}^{obs}(\mathbf{x}_R, t)}{D(\mathbf{x}_R)} \delta(\mathbf{x} - \mathbf{x}_R) \quad \text{and} \quad \lambda(\mathbf{x}, t) = -\frac{1}{D(\mathbf{x}_R)} \frac{\dot{u}^{obs}(\mathbf{x}_R, t + T_{XR})}{4\pi R_{XR}} \tag{16.43}$$

Assuming $m_0 = 0$ for the reference value of the model parameters, the data kernel is

$$\left.\frac{\partial\tau}{\partial m}\right|_{\mathbf{m}_0} = (\lambda, \xi) = \frac{-2}{D(\mathbf{x}_R)} \frac{1}{4\pi R_{SH}} \frac{1}{4\pi R_{HR}} \int_{-\infty}^{+\infty} \dot{u}^{obs}(\mathbf{x}_R, t + T_{RH}) \, \ddot{w}(t - T_{SH}) \, dt \tag{16.44}$$

The effect of a small heterogeneity is to perturb the travel time by an amount:

$$\left.\frac{\partial\tau}{\partial m}\right|_{\mathbf{m}_0} = \frac{-2}{D(\mathbf{x}_R)} \frac{1}{4\pi R_{SH}} \frac{1}{4\pi R_{HR}} \frac{1}{4\pi R_{SR}} \int_{-\infty}^{+\infty} \dot{w}(t - T_{SR} + T_{RH}) \, \ddot{w}(t - T_{SH}) \, dt$$

$$= \frac{-2}{D(\mathbf{x}_R)} \frac{1}{4\pi R_{SH}} \frac{1}{4\pi R_{HR}} \frac{1}{4\pi R_{SR}} \int_{-\infty}^{+\infty} \dot{w}(t' - \Delta T) \, \ddot{w}(t') \, dt' \tag{16.45}$$

$$\text{where} \quad \Delta T \equiv T_{SR} - (T_{RH} + T_{SH})$$

The last step uses the transformation of variables $t' = t - T_{SH}$. The quantity ΔT represents the difference in travel time between the direct $\overline{SR}$ path and the scattered $\widehat{SHR}$ path. The quantity $D(\mathbf{x}_R)$ is given by

$$D(\mathbf{x}_R) = \int\limits_{-\infty}^{+\infty} \ddot{u}^{obs}(\mathbf{x}_R, t)\, u_0(\mathbf{x}_R, t)\, dt = \left(\frac{1}{4\pi R_{SR}}\right)^2 \int\limits_{-\infty}^{+\infty} \ddot{w}(t')\, w(t')\, dt' \tag{16.46}$$

Here, we have used the transformation of variables $t' = t - T_{SR}$. The derivative $\partial\tau/\partial m$ is axially symmetric about the $\overline{SR}$ line, as R_{SH} and R_{HR} depend only on the perpendicular distance r of $\mathbf{x}_H$ from the line. Sliced perpendicular to the line, the derivative $\partial\tau/\partial m$ is doughnut shaped. The derivative $\partial\tau/\partial m$ is zero whenever $\Delta T = 0$. This behavior follows from d/dt being an antiself-adjoint operator, as any quantity equal to its negative is zero:

$$\int\limits_{-\infty}^{+\infty} \dot{w}(t)\, \ddot{w}(t)\, dt = -\int\limits_{-\infty}^{+\infty} \ddot{w}(t)\, \dot{w}(t)\, dt = -\int\limits_{-\infty}^{+\infty} \dot{w}(t)\, \ddot{w}(t)\, dt = 0 \tag{16.47}$$

The time difference ΔT is zero when the heterogeneity is between S and R and on the $\overline{SR}$ line, so $\partial\tau/\partial m = 0$ in this case. This zero makes the hole in the center of the doughnut.

At finite frequency, the travel time kernel is zero along the ray path. This result is paradoxical, for the ray theoretical kernel, which is valid in the high-frequency limit, is zero everywhere *except* on the ray path. This counterintuitive behavior arises because the model perturbations have been parameterized as Dirac impulse functions, which are functions of infinitesimal size. Because it lacks a scale length, the Dirac function cannot advance or delay the scattered wave field. Advances and delays result from differences in the scattered and reference path lengths, only, and for a heterogeneity on the ray path, this difference is zero.

Now consider an oscillatory, band-limited source time function with a characteristic period P. Suppose we construct the elliptical volume surrounding the points $\mathbf{x}_S$ and $\mathbf{x}_R$, for which $\Delta T < \frac{1}{2}P$. The time integral in Eq. (16.45) will have the same sign everywhere in this volume, as will $\partial\tau/\partial m$. This is a banana-shaped region enclosing the source and receiver (Fig. 16.5). The banana is thinner for short periods than for long periods. Moving away from the $\overline{SR}$ line along its perpendicular, the time integral, and hence the derivative, oscillates in sign, as the $\dot{w}(t - \Delta t)$ and $\ddot{w}(t)$ factors beat against one another. The derivative also decreases in amplitude because the factors R_{RH} and R_{SH} grow with distance. Consequently,

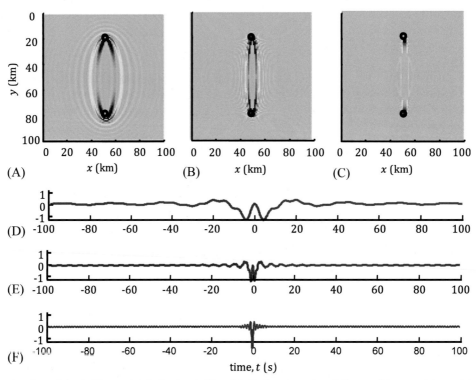

FIG. 16.5 Banana-doughnut kernels (the partial derivative $\partial\tau/\partial m$ of finite-frequency travel time τ with respect to model parameter m) for an acoustic wave propagation problem. The source x_S and receiver x_R (black circles) are separated by a distance of $R = 60$ km. The perturbation in acoustic velocity is $\delta v = (m/2v_0)\delta(\mathbf{x} - \mathbf{x}_h)$, where $v_0 = 1$ km/s is a constant reference velocity and $\mathbf{x}_h$ is the heterogeneity's location. (A–C) Kernels (colors) for a suite of $\mathbf{x}_h$s covering the (x, y) plane, for frequency bands centered at 0.025, 0.125, and 0.500 Hz, respectively. Note that the "banana" shape of the kernel narrows with frequency, becoming increasingly ray-like. (D–F) Corresponding band-limited source time functions used in the calculation. Scripts gdama16_05 and gdapy16_05.

the central banana is surrounded by a series of larger, but less intense, bananas of alternating sign. On account of these behaviors, the finite-frequency travel time kernel is colloquially called a *banana-doughnut kernel* (Tromp et al., 2005).

16.8 Velocity structure from free oscillations and seismic surface waves

The Earth, being a finite body, can oscillate only with certain characteristic patterns of motion (eigenfunctions) at certain corresponding discrete frequencies, ω_{klm} (eigenfrequencies), where the indices (k,l,m) increase with the complexity of the motion. Most commonly, the frequencies of vibration are measured from the amplitude spectral density (see Section 15.9) of seismograms of very large earthquakes, and the corresponding indices are estimated by comparing the observed frequencies of vibration to those predicted using a reference model of the Earth's structure. The inverse problem is to use small differences between the measured and predicted frequencies to improve the reference model. The most fundamental part of this inverse problem is the Fréchet derivatives relating small changes in the Earth's material properties to the resultant changes in the frequencies of vibration. These derivatives are computed through the perturbative techniques described in Section 15.11 and are three-dimensional analogs of Eq. (15.49) (Takeuchi and Saito, 1972). All involve integrals of the perturbation in Earth structure with the eigenfunctions of the reference model, so the problem is one in linearized continuous inverse theory. The complexity of this problem is determined by the assumptions made about the Earth model—whether it is isotropic or anisotropic, radially stratified or laterally heterogeneous—and how it is parameterized. A radially stratified inversion that incorporates both body wave travel times and eigenfrequencies and that has become a standard in the seismological community is the Preliminary Reference Earth Model of Dziewonski and Anderson (1981).

Seismic surface waves, such as the Love wave and the Rayleigh wave, occur when seismic energy is trapped near the Earth's surface by the rapid increase in seismic velocities with depth. The two waves differ by the Love wave being horizontally polarized in the direction perpendicular to propagation, and the Rayleigh wave being polarized in the vertical plane containing the direction of propagation. The propagation velocity of these waves is very strongly frequency dependent, since the lower-frequency waves extend deeper into the Earth than the higher-frequency waves and are affected by the higher velocities that occur there. The surface wave caused by an impulsive source, such as an earthquake, is not itself impulsive, but rather is dispersed, since its component frequencies propagate at different phase velocities.

The horizontal phase velocity c of the surface wave is defined as $c = \omega(k)/k$, where ω is angular frequency and k is wavenumber (wavenumber is 2π divided by wavelength). The function $c(\omega)$ (or alternatively $c(k)$) is called the *dispersion function* (or, sometimes, *dispersion curve*) of the surface wave. It is easily measured from observations of earthquake wave fields and easily predicted for vertically stratified Earth models. The inverse problem is to infer the Earth's material properties from observations of the dispersion function.

As in the case of free oscillations, Fréchet derivatives can be calculated using a perturbative approach. In an Earth model in which the material properties, shear modulus $\mu(z)$, Lamé parameter $\lambda(z)$, and density $\rho(z)$, depend only upon depth z, the perturbation in phase velocity c can be calculated using perturbation theory (Aki and Richards, 2002, their Eq. 7.71). The Love wave perturbation is

$$\left(\frac{\delta c}{c}\right)_\omega = \int_0^\infty G_\mu^L(z)\delta\mu(z)\,dz + \int_0^\infty G_\rho^L(z)\delta\rho(z)\,dz$$

$$\text{with } G_\mu^L(z) \equiv C^{-1}\left[k^2 l_1^2 + \left(\frac{dl_1}{dz}\right)^2\right] \text{ and } G_\rho^L(z) = -C^{-1}\omega^2 l_1^2 \tag{16.48}$$

$$\text{and } C \equiv 2k^2 \int_0^\infty \mu_0(z)\, l_1^2\, dz$$

Here, shear modulus $\mu(z) \equiv \mu_0(z) + \delta\mu(z)$ and density $\rho(z) \equiv \rho(z) + \delta\rho(z)$ are vertically stratified in depth z and the horizontal displacement of the Love wave is $u_y \equiv l_1(z)\exp(ikx - i\omega t)$. The corresponding formula for the Rayleigh wave is similar in form but involves perturbations in three material parameters, Lamè parameter $\lambda(z)$, shear modulus $\mu(z)$, and density $\rho(z)$.

$$\left(\frac{\delta c}{c}\right)_\omega = \int_0^\infty G_\lambda^R(z)\delta\lambda(z)\,dz + \int_0^\infty G_\mu^R(z)\delta\mu(z)\,dz + \int_0^\infty G_\rho^R(z)\delta\rho(z)\,dz \tag{16.49}$$

The reader is referred to Aki and Richards (2002, their Eq. 7.78) for the specific form of the kernels $G_\lambda^R(z)$, $G_\mu^R(z)$, and $G_\rho^R(z)$.

When the Earth model contains three-dimensional heterogeneities, the inverse problem of determining Earth structure from surface waves dispersion is just a type of wavefield tomography (when seismograms are being matched) or finite-frequency tomography (when dispersion functions are being matched). The dispersion function is now understood to encode the frequency-dependent time shift between an observed and predicted surface wave field. Adjoint methods can be used to compute its Fréchet derivatives (Sieminski et al., 2007).

Ray-theoretical ideas can be applied to surface waves, which at sufficiently short periods can be approximated as propagating along a horizontal ray path connecting source and receiver (Wang and Dahlen, 1995). Each point on the Earth's surface is presumed to have its own dispersion function $c(\omega, x, y)$. As in travel time tomography (Eq. 16.7), the problem is most efficiently formulated in terms of phase slowness $s(\omega, x, y) \equiv 1/c(\omega, x, y)$. The phase slowness function $s^{(i)}(\omega)$ observed for a particular source-observer path i is then calculated using the *pure path approximation*

$$s^{(i)}(\omega) = \frac{1}{L_i} \int_{\text{path } i} s(\omega, x, y) \, dl \tag{16.50}$$

Here, L_i is the length of path i and dl is an increment of arc length along the path. The path is often approximated as the great circle connecting source and receiver. This is a two-dimensional tomography problem relating line integrals of $s(\omega, x, y)$ to observations of $s^{(i)}(\omega)$—hence the term *surface wave tomography*. The $s(\omega, x, y)$ then can be inverted for estimates of the vertical structure using the vertically stratified method described earlier (e.g., Boschi and Ekström, 2002). This two-step process is not as accurate as full waveform tomography but is computationally much faster.

Eqs. (16.48) and (16.49) are inverse problems relating several model functions to a single data function. However, a single data function is insufficient to uniquely determine several model functions. Consider, for example, the discrete version of the problem

$$\mathbf{d} = \mathbf{G}^{(A)}\mathbf{m}^{(A)} + \mathbf{G}^{(B)}\mathbf{m}^{(B)} \tag{16.51}$$

where data vector $\mathbf{d}$ is of length N and model vectors $\mathbf{m}^{(A)}$ and $\mathbf{m}^{(B)}$ each are all of length $M = N$. One can arbitrarily choose $\mathbf{m}^{(A)}$ and then solve for $\mathbf{m}^{(B)}$ via

$$\mathbf{d} - \mathbf{G}^{(A)}\mathbf{m}^{(A)} = \mathbf{G}^{(B)}\mathbf{m}^{(B)} \tag{16.52}$$

The model parameter vectors $\mathbf{m}^{(A)}$ and $\mathbf{m}^{(B)}$ completely trade off. However, a special case can arise in which some linear combinations of $\mathbf{d}$ are controlled by $\mathbf{m}^{(A)}$, only, and others by $\mathbf{m}^{(B)}$, only. This is the case where the null space of $\mathbf{m}^{(A)}$ exactly matches the p-space of $\mathbf{m}^{(B)}$ and the null space of $\mathbf{m}^{(B)}$ exactly matches the p-space of $\mathbf{m}^{(A)}$, that is, $S_0(\mathbf{m}^{(A)}) = S_p(\mathbf{m}^{(B)})$ and $S_0(\mathbf{m}^{(B)}) = S_p(\mathbf{m}^{(A)})$. Then, $\mathbf{G}^{(A)}$ and $\mathbf{G}^{(B)}$ have singular-value decompositions that share the same eigenvector matrices and have singular value matrices that are orthogonal (Menke, 2017a)

$$\mathbf{G}^{(A)} = \mathbf{U}\boldsymbol{\Lambda}^{(A)}\mathbf{V}^{(T)} \text{ and } \mathbf{G}^{(B)} = \mathbf{U}\boldsymbol{\Lambda}^{(B)}\mathbf{V}^{(T)}$$
$$\text{with } \mathbf{U} = \begin{bmatrix} \mathbf{U}^{(A)} & \mathbf{U}^{(B)} \end{bmatrix} \text{ and } \mathbf{V} = \begin{bmatrix} \mathbf{V}^{(A)} & \mathbf{V}^{(B)} \end{bmatrix}$$
$$\text{and } \boldsymbol{\Lambda}^{(A)} = \begin{bmatrix} \boldsymbol{\Lambda}_p^{(A)} & \mathbf{0} \\ \mathbf{0} & \mathbf{0} \end{bmatrix} \text{ and } \boldsymbol{\Lambda}^{(B)} = \begin{bmatrix} \mathbf{0} & \mathbf{0} \\ \mathbf{0} & \boldsymbol{\Lambda}_p^{(B)} \end{bmatrix} \text{ so } \boldsymbol{\Lambda}^{(A)}\boldsymbol{\Lambda}^{(B)} = 0 \tag{16.53}$$

Premultiplying Eq. (16.51) by $\mathbf{U}^T$ yields

$$\begin{bmatrix} \mathbf{U}_A^T\mathbf{d} \\ \mathbf{U}_B^T\mathbf{d} \end{bmatrix} = \begin{bmatrix} \boldsymbol{\Lambda}_p^{(A)}\mathbf{V}_A^T\mathbf{m}^{(A)} \\ 0\mathbf{V}_B^T\mathbf{m}^{(A)} \end{bmatrix} + \begin{bmatrix} 0\mathbf{V}_A^T\mathbf{m}^{(B)} \\ \boldsymbol{\Lambda}_p^{(B)}\mathbf{V}_B^T\mathbf{m}^{(B)} \end{bmatrix} \tag{16.54}$$

This equation constrains the values of $\mathbf{V}_A^T\mathbf{m}^{(A)}$ and $\mathbf{V}_B^T\mathbf{m}^{(B)}$ but not the values of $\mathbf{V}_B^T\mathbf{m}^{(A)}$ and $\mathbf{V}_A^T\mathbf{m}^{(B)}$ and has the solution

$$\mathbf{m}^{(A)} = \mathbf{V}_A\left[\boldsymbol{\Lambda}_p^{(A)}\right]^{-1}\mathbf{U}_A^T\mathbf{d} + \mathbf{V}_B\boldsymbol{\alpha}_A \text{ and } \mathbf{m}^{(B)} = \mathbf{V}_B\left[\boldsymbol{\Lambda}_p^{(B)}\right]^{-1}\mathbf{U}_B^T\mathbf{d} + \mathbf{V}_A\boldsymbol{\alpha}_B \tag{16.55}$$

where $\boldsymbol{\alpha}_A$ and $\boldsymbol{\alpha}_B$ are arbitrary. Although $\mathbf{m}^{(A)}$ and $\mathbf{m}^{(B)}$ are determined independently of one another, neither is uniquely determined by the data alone; prior information (in the form of $\boldsymbol{\alpha}_A$ and $\boldsymbol{\alpha}_B$) is required. In this case, it is possible to find unique (and possibly localized) averages $\langle \mathbf{m}^{(A)} \rangle \equiv \mathbf{a}^T\mathbf{m}^{(A)}$ and $\langle \mathbf{m}^{(B)} \rangle \equiv \mathbf{b}^T\mathbf{m}^{(B)}$ by choosing averaging vectors $\mathbf{a}$ and $\mathbf{b}$, such that $\mathbf{a}^T\mathbf{V}_B\boldsymbol{\alpha}_A = 0$ and $\mathbf{b}^T\mathbf{V}_A\boldsymbol{\alpha}_B = 0$.

This analysis carries over to the continuous case, where the functions $\mathbf{m}^{(A)}$ and $\mathbf{m}^{(B)}$ become the functions $m^{(A)}$ and $m^{(B)}$, the matrices $\mathbf{G}^{(A)}$ and $\mathbf{G}^{(B)}$ become the linear operators $\mathcal{G}^{(A)}$ and $\mathcal{G}^{(A)}$, and the matrices $\mathbf{U}$ and $\mathbf{V}$ become forward and inverse integral transforms, respectively. Menke (2017a) has examined Eq. (16.49) and determined that it conforms to the general case in which $m^{(A)}$ and $m^{(B)}$ completely trade off (Eq. 16.52), and not to the special case described before. No localized averages dependent upon only one of $\delta\lambda$, $\delta\mu$, and $\delta\rho$ can be formed.

The incompleteness of the surface wave dispersion problem can be addressed by supplementing the dispersion data with other seismological data, such as receiver functions (Juliá et al., 2000; Bodin et al., 2012), or by reducing the number of model function to one by imposing a scaling rule between $\delta\lambda$, $\delta\mu$, and $\delta\rho$.

16.9 Seismic attenuation

Internal friction within the Earth causes seismic waves to lose amplitude as they propagate. The amplitude loss can be characterized by the attenuation factor $\alpha(\omega, \mathbf{x})$, which varies with angular frequency ω and position $\mathbf{x}$. In a regime where ray theory is accurate, the fractional loss of amplitude A of the seismic wave is

$$\frac{A}{A_0} = \exp\left\{-\int_{ray} \alpha(\omega, \mathbf{x})\, \mathrm{d}l\right\} \tag{16.56}$$

Here, A_0 is the initial amplitude. After linearization through taking its logarithm, this equation is a standard tomography problem (identical in form to the X-ray problem described in Section 2.8)—hence the term *attenuation tomography*. Jacobson et al. (1981) discuss its solution when the attenuation is assumed to vary only with depth. Dalton and Ekström (2006) perform a global two-dimensional tomographic inversion using surface waves. Many authors have performed three-dimensional inversions (see review by Romanowicz, 1998).

An alternative parameterization of attenuation, called the *quality factor*, is often encountered in the seismological literature. A simple behavior that was observed in early experiments is that the decrease in amplitude depends only on the number n of wavelengths propagated, that is $A/A_0 = \exp(-cn)$, where c is a constant. Historically, this constant was denoted as $c = \pi/Q$, where Q is the quality factor. The wavelength λ, velocity v, and angular frequency ω are related by $\omega\lambda/2\pi = v$, so the number of wavelengths is $n = x/\lambda = \omega x/2\pi v$, and amplitude decreases as

$$A/A_0 = \exp\left(-\tfrac{1}{2}\omega \int_{ray} \frac{1}{Qv}\, \mathrm{d}l\right) \text{ with } \alpha = \frac{\omega}{2Qv} \text{ and } Q = \frac{\omega}{2\alpha v} \tag{16.57}$$

The original intention of this parameterization was to have Q vary with position but not frequency, allowing observations at several frequencies to be combined in an inversion for $Q(\mathbf{x})$, or more often, its reciprocal $1/Q(\mathbf{x})$. However, a growing body of evidence has indicated that quality factor is frequency dependent, leading some authors to parameterize it as $Q(\omega, \mathbf{x}) = Q_0(\omega/\omega_0)^\eta$, where $Q_0(\mathbf{x})$ is the quality factor at a reference frequency ω_0 and the exponent $\eta(\mathbf{x})$ is in the interval $0 < \eta < 1$. The inversion then solves for both $Q_0(\mathbf{x})$ and $\eta(\mathbf{x})$ (although most authors further assume that η has no spatial dependence).

Attenuation measurements sometimes are reported as t^* values, pronounced *tee-star*, measured in seconds. Tee-star is defined as

$$t^* \equiv 2\omega^{-1} \log(A_0/A) \text{ so that } A/A_0 = \exp\left(-\tfrac{1}{2}\omega t^*\right) \tag{16.58}$$

Tee-star increases with distance (or travel time t) along the ray as $t^* = \int_{ray}(Qv)^{-1}\mathrm{d}l = \int_{ray}Q^{-1}\mathrm{d}t$. As $Q \gg 1$ in most parts of the Earth, tee-star is typically a few percent of the travel time of the seismic wave.

A common problem in attenuation tomography is that the initial amplitude A_0 (or equivalently, the centroid moment tensor (CMT) of the seismic source) is unknown (or poorly known). Menke et al. (2006) show that including A_0 as an unknown parameter in the inversion introduces a nonuniqueness that is mathematically identical to the one encountered when an unknown origin time is introduced into the velocity tomography problem.

Attenuation also has an effect on the free oscillations of the Earth, causing a widening in the spectral peaks associated with the eigenfrequencies ω_{klm}. The amount of widening depends on the relationship between the spatial variation of the attenuation and the spatial dependence of the eigenfunction and can be characterized by a quality factor Q_{klm} (i.e., a fractional loss of amplitude per oscillation) for that mode. Pertubative methods are used to calculate the Fréchet derivatives of Q_{klm} with the attenuation factor. An early example of such an inversion is that of Masters and Gilbert (1983).

16.10   Signal correlation

Geologic processes record signals only imperfectly. For example, while variations in oxygen isotopic ratios $r(t)$ through geologic time t are recorded in oxygen-bearing sediments as they are deposited, the sedimentation rate is itself a function of time. Measurements of isotopic ratio $r(z)$ as a function of depth z cannot be converted to the variation of $r(t)$ without knowledge of the sedimentation function $z(t)$ or equivalently $t(z)$.

Under certain circumstances, the function $r(t)$ is known prior information (for instance, oxygen isotopic ratio correlates with temperature, which can be estimated independently). In these instances, it is possible to use the observations $r^{obs}(z)$ and the predictions $r^{pre}(t)$ to invert for the function $t(z)$. This is essentially a problem in signal correlation: distinctive features that can be correlated between $r^{obs}(z)$ and $r^{pre}(t)$ establish the function $t(z)$. The inverse problem

$$r^{obs}(z) = r^{pre}[t(z)] \tag{16.59}$$

is therefore a problem in nonlinear continuous inverse theory. The unknown *mapping function* $t(z)$ must increase monotonically with z. The solution of this problem is discussed by Martinson et al. (1982) and Shure and Chave (1984).

16.11   Tectonic plate motions

The motion of the Earth's rigid tectonic plates can be described by an *Euler vector* $\boldsymbol{\omega}$, whose orientation gives the pole of rotation and whose magnitude gives its rate. Euler vectors can be used to represent relative rotations, that is, the rotation of one plate relative to another, or absolute motion, that is, motion relative to the Earth's mantle. If we denote the relative rotation of plate A with respect to plate B as $\boldsymbol{\omega}_{AB}$, then the Euler vectors of three plates A, B and C satisfy the relationship $\boldsymbol{\omega}_{AB}+\boldsymbol{\omega}_{BC}+\boldsymbol{\omega}_{CA}=0$. Once the Euler vectors for plate motion are known, the relative velocity between two plates at any point on their boundary can easily be calculated from trigonometric formulas.

Several geologic features provide information on the relative rotation between plates, including the faulting directions of earthquakes at plate boundaries and the orientation of transform faults, which constrain the direction of the relative velocity vectors, and spreading rate estimates based on magnetic lineations at ridges, which constrain the magnitude of the relative velocity vectors.

These data can be used in an inverse problem to determine the Euler vectors (Chase and Stuart, 1972; DeMets et al., 2010; Minster et al., 1974). The main differences between various authors' approaches are in the manner in which the Euler vectors are parameterized and the types of data that are used. Some authors use Cartesian components of the Euler vectors; others use magnitude, azimuth, and inclination. The two inversions behave somewhat differently in the presence of noise; Cartesian parameterizations seem to be more stable. Data include seafloor spreading rates, azimuths of strike-slip plate boundaries, earthquake moment tensors, and Global Positioning System strain measurements.

16.12   Gravity and geomagnetism

Inverse theory plays an important role in creating representations of the Earth's gravity and magnetic fields. Field measurements made at many points about the Earth need to be combined into a smooth representation of the field, a problem which is mainly one of interpolation in the presence of noise and incomplete data (see Section 7.4). Both spherical harmonic expansions and various kinds of spline functions (Sandwell, 1987; Shure et al., 1982, 1985; Wessel and Becker, 2008) have been used in the representations. In either case, the trade-off of resolution and variance is very important.

Studies of the Earth's core and geo-dynamo require that measurements of the magnetic field at the Earth's surface be extrapolated to the core-mantle boundary. This is an inverse problem of considerable complexity, since the short-wavelength components of the field that are most important at the core-mantle boundary are very poorly measured at the Earth's surface. Furthermore, the rate of change of the magnetic field with time (called *secular variation*) is of critical interest. This quantity must be determined from fragmentary historical measurements, including measurements of compass deviation recorded in old ship logs. Consequently, these inversions introduce substantial prior constraints on the behavior of the field near the core, including the assumption that the root-mean-square time rate of change of the field is minimized and the total energy dissipation in the core is minimized (Bloxam, 1987). Once the magnetic field and its time derivative at the core-mantle boundary have been determined, they can be used in inversions for the fluid velocity near the surface of the outer core (Bloxam, 1992).

The Earth's gravity field $\mathbf{g}(\mathbf{x})$ is determined by its density structure $\rho(\mathbf{x})$. In parts of the Earth where electric currents and magnetic induction are unimportant, the magnetic field vector $\mathbf{H}(\mathbf{x})$ is caused by the magnetization vector $\mathbf{M}(\mathbf{x})$ of the rocks. These quantities are related by

$$g_i(\mathbf{x}) = \iiint_V G_i^g\left(\mathbf{x}, \mathbf{x}^{(0)}\right) \rho\left(\mathbf{x}^{(0)}\right) dV_0 \text{ and } H_i(\mathbf{x}) = \iiint_V \sum_{j=1}^{3} G_{ij}^H\left(\mathbf{x}, \mathbf{x}^{(0)}\right) M_j\left(\mathbf{x}^{(0)}\right) dV_0$$

$$\text{with } G_i^g\left(\mathbf{x}, \mathbf{x}^{(0)}\right) = -\frac{\gamma\left(x_i - x_i^{(0)}\right)}{|\mathbf{x} - \mathbf{x}^{(0)}|^3} \tag{16.60}$$

$$\text{and } G_{ij}^H\left(\mathbf{x}, \mathbf{x}^{(0)}\right) = \frac{1}{4\pi\mu_0}\left[\frac{\delta_{ij}}{|\mathbf{x} - \mathbf{x}^{(0)}|^3} - \frac{3\left(x_i - x_i^{(0)}\right)\left(x_j - x_j^{(0)}\right)}{|\mathbf{x} - \mathbf{x}^{(0)}|^5}\right]$$

Here, γ is the universal gravitational constant and μ_0 is the magnetic permeability of the vacuum. In both cases, the gravity and magnetic fields are linearly related to their sources, density and magnetization, respectively. Consequently, these are linear, continuous inverse problems. Nevertheless, inverse theory has proved to have little application to these problems, owing to their inherent underdetermined nature. In both cases, it is possible to show (e.g., Menke and Abbott, 1989, their Section 5.14) that the field outside a finite body can be generated by an infinitesimally thick spherical shell of mass or magnetization surrounding the body and below the level of the measurements. The null space of these problems is so large that it is generally impossible to formulate any useful solution (as we encountered in Fig. 9.08), except when an enormous amount of prior information is added to the problem.

Some progress has been made in special cases where prior constraints can be sensibly stated. For instance, the magnetization of sea mounts can be computed from their magnetic anomaly, assuming that their magnetization vector is everywhere parallel (or nearly parallel) to some as yet undetermined direction and that the shape of the magnetized region closely corresponds to the observed bathymetric expression of the sea mount (Grossling, 1970; Parker, 1988; Parker et al., 1987).

16.13   Electromagnetic induction and the magnetotelluric method

Electric currents are induced within a conducting medium when it is illuminated by a plane electromagnetic wave. In a homogeneous medium, the field decays exponentially with depth, as energy is dissipated by the currents. In a heterogeneous medium, the behavior of the field is more complicated and depends on the details of the electrical conductivity $\sigma(\mathbf{x})$.

Consider the special case of a vertically stratified Earth illuminated by a vertically incident electromagnetic wave (e.g., from space). The plane wave consists of a horizontal electric field E_x and a horizontal magnetic field H_y, with their ratio on the Earth's surface, $z = 0$, being determined by the conductivity $\sigma(z)$, which is assued to vary only with depth z. The quantity

$$Z(\omega) \equiv \frac{E_x(\omega, z = 0)}{i\omega H_y(\omega, z = 0)} \tag{16.61}$$

that relates the two fields is called the *impedance*. It is frequency dependent, because the depth of penetration of an electromagnetic wave into a conductive medium depends upon its frequency, causing lower frequencies to sense deeper parts of the conductivity structure. The *magnetotelluric problem* is to determine the conductivity $\sigma(z)$ from measurements of the impedance $Z(\omega)$ at a suite of angular frequencies ω. This one-dimensional nonlinear inverse problem is relatively well understood. For instance, Fréchet derivatives are relatively straightforward to compute (Parker, 1970, 1977; McGillivray et al., 1994). Oldenburg (1983) discusses the discretization of the problem and the application of extremal inversion methods (see Section 8.6). Key (2016) presents a two-dimensional inversion method, based on adjoint method and numerical solution of the electromagnetic wave equation.

In three-dimensional problems, all the components of magnetic $\mathbf{H}$ and electric $\mathbf{E}$ fields are now relevant, and the impedance matrix $\mathbf{Z}(\omega, x, y)$ that connects them via

$$\mathbf{E}(\omega, x, y, z = 0) = \mathbf{Z}(\omega, x, y)\mathbf{H}(\omega, x, y, z = 0) \tag{16.62}$$

constitutes the data. The inverse problem of determining $\sigma(\mathbf{x})$ from observations of $\mathbf{Z}(\omega, x, y)$ at a variety of frequencies and positions is much more difficult than the one-dimensional version (Chave et al., 2012), especially in the presence of large heterogeneities in electoral resistivity caused, for instance, by topography (e.g., Baba and Chave, 2005).

16.14   Problems

16.1 Show that the differential operator in Eq. (16.12) is self-adjoint.

16.2 Modify the seismic migration example (gdama16_03 or gdama16_03) to include $N_H = 5$ heterogeneities at locations of your choosing.

References

Aki, K., Richards, P.G., 2002. Quantitative Seismology, second ed. University Science Books, Sausalito, CA, 700 pp.

Aki, K., Christoffersson, A., Husebye, E., 1976. Three-dimensional seismic structure under the Montana LASA. Bull. Seismol. Soc. Am. 66, 501–524.

Baba, K., Chave, A., 2005. Correction of seafloor magnetotelluric data for topographic effects during inversion. J. Geophys. Res. 110, B12105. https://doi.org/10.1029/2004JB003463.

Bin Waheed, U., Flagg, G., Yarman, C.E., 2016. First-arrival traveltime tomography for anisotropic media using the adjoint-state method. Geophysics 81, R147–R155. https://doi.org/10.1190/geo2015-0463.1.

Bloxam, J., 1987. Simultaneous inversion for geomagnetic main field and secular variation, 1. A large scale inversion problem. J. Geophys. Res. 92, 11597–11608.

Bloxam, J., 1992. The steady part of the secular variation of the Earth's magnetic field. J. Geophys. Res. 97, 19565–19579.

Bodin, T., Sambridge, M., Tkalčić, H., Arroucau, P., Gallagher, K., Rawlinson, N., 2012. Transdimensional inversion of receiver functions and surface wave dispersion. J. Geophys. Res. Solid Earth 117, JB008560.

Boschi, J., Ekström, G., 2002. New images of the Earth's upper mantle from measurements of surface wave phase velocity anomalies. J. Geophys. Res. 107, 2059.

Cerveny, V., 2001. Seismic Ray Theory. Cambridge University Press, New York. 607 pp.

Chase, E.P., Stuart, G.S., 1972. The N plate problem of plate tectonics. Geophys. J. R. Astron. Soc. 29, 117–122.

Chave, A., Jones, A., Mackie, R., Rodi, W., 2012. The Magnetotelluric Method, Theory and Practice. Cambridge University Press, New York. 590 pp.

Crosson, R.S., 1976. Crustal structure modeling of earthquake data: 1. Simultaneous least squares estimation of hypocenter and velocity parameters. J. Geophys. Res. 81, 3036–3046.

Dahlen, F.A., Hung, S.-H., Nolet, G., 2000. Frechet kernels for finite frequency traveltimes—I. Theory. Geophys. J. Int. 141, 157–174.

Dalton, C.A., Ekström, G., 2006. Global models of surface wave attenuation. J. Geophys. Res. 111, B05317.

DeMets, C., Gordon, R.G., Argus, D.F., 2010. Geologically current plate motions. Geophys. J. Int. 181, 1–80.

Dziewonski, A.M., Anderson, D.L., 1981. Preliminary reference Earth model. Phys. Earth Planet. Inter. 25, 297–358.

Dziewonski, A.M., Woodhouse, J.H., 1983. An experiment in systematic study of global seismicity: centroid-moment tensor solutions for 201 moderate and large earthquakes of 1981. J. Geophys. Res. 88, 3247–3271.

Dziewonski, A.M., Chou, T.-A., Woodhouse, J.H., 1981. Determination of earthquake source parameters from waveform data for studies of global and regional seismicity. J. Geophys. Res. 86, 2825–2852.

Geiger, L., 1912. Probability method for the determination of earthquake epicenters from the arrival time only (translated from Geiger's 1910 German article). Bull. St. Louis Univ. 8, 56–71.

Grossling, B.F., 1970. Seamount magnetism. In: Maxwell, A.E. (Ed.), The Sea. vol. 4. Wiley, New York, pp. 129–156.

Jacobson, R.S., Shor, G.G., Dorman, L.M., 1981. Linear inversion of body wave data—part 2: attenuation vs. depth using spectral ratios. Geophysics 46, 152–162.

Julià, J., Ammon, C.J., Herrmann, R.B., Correig, A.M., 2000. Joint inversion of receiver function and surface wave dispersion observations. Geophys. J. Int., 143, 99–112.

Key, K., 2016. MARE2DEM: a 2-D inversion code for controlled-source electromagnetic and magnetotelluric data. Geophys. J. Int. 207, 571–588.

Kim, Y., Liu, Q., Tromp, J., 2011. Adjoint centroid-moment tensor inversions. Geophys. J. Int. 186, 264–278.

Klein, F.W., 1985. User's Guide to HYPOINVERSE, a Program for VAX and PC350 Computers to Solve for Earthquake Locations. U.S. Geologic Survey Open File Report 85–515, 24 pp.

Marquering, H., Dahlen, F.A., Nolet, G., 1999. Three-dimensional sensitivity kernels for finite frequency traveltimes: the banana-doughnut paradox. Geophys. J. Int. 137, 805–815.

Martinson, D.G., Menke, W., Stoffa, P., 1982. An inverse approach to signal correlation. J. Geophys. Res. 87, 4807–4818.

Masters, G., Gilbert, F., 1983. Attenuation in the Earth at low frequencies. Philos. Trans. R. Soc. Lond. Ser. A 388, 479–522.

McGillivray, P.R., Oldenburg, D.W., Ellis, R.G., Habashy, T.M., 1994. Calculation of sensitivities for the frequency-domain electromagnetic problem. Geophys. J. Int. 116, 1–4.

Menke, W., 2005. Case studies of seismic tomography and earthquake location in a regional context. In: Levander, A., Nolet, G. (Eds.), Seismic Earth: Array Analysis of Broadband Seismograms. Geophysical Monograph Series, vol. 157. American Geophysical Union, Washington, DC, pp. 7–36.

Menke, W., 2017a. The uniqueness of single data function, multiple model functions, inverse problems including the Rayleigh wave dispersion problem. Pure Appl. Geophys. 174, 1699–1710.

Menke, W., 2017b. Sensitivity kernels for the cross-convolution measure. Bull. Seismol. Soc. Am. 107, 2213–2224. https://doi.org/10.1785/0120170045.

Menke, W., 2020. A connection between geometrical spreading and the adjoint field in travel time tomography. Appl. Math. 11, 84–96. https://doi.org/10.4236/am.2020.112009.

Menke, W., Abbott, D., 1989. Geophysical Theory (Textbook). Columbia University Press, New York. 458 pp.

Menke, W., Schaff, D., 2004. Absolute earthquake location with differential data. Bull. Seismol. Soc. Am. 94, 2254–2264.

Menke, W., Holmes, R.C., Xie, J., 2006. On the nonuniqueness of the coupled origin time—velocity tomography problem. Bull. Seismol. Soc. Am. 96, 1131–1139.

Menke, W., Menke, H., 2015. Improved precision of delay times determined through cross correlation achieved by out-member averaging. Jökull 64, 15–22.

Minster, J.B., Jordan, T.H., Molnar, P., Haines, E., 1974. Numerical modeling of instantaneous plate tectonics. Geophys. J. R. Astron. Soc. 36, 541–576.

Oldenburg, D.W., 1983. Funnel functions in linear and nonlinear appraisal. J. Geophys. Res. 88, 7387–7398.

Parker, R.L., 1970. The inverse problem of the electrical conductivity of the mantle. Geophys. J. R. Astron. Soc. 22, 121–138.

Parker, R.L., 1977. The Fréchet derivative for the one dimensional electromagnetic induction problem. Geophys. J. R. Astron. Soc. 39, 543–547.

Parker, R.L., 1988. A statistical theory of seamount magnetism. J. Geophys. Res. 93, 3105–3115.

Parker, R.L., Shure, L., Hildebrand, J., 1987. An application of inverse theory to seamount magnetism. Rev. Geophys. 25, 17–40.

Pavlis, G.L., Booker, J.R., 1980. The mixed discrete-continuous inverse problem: application to the simultaneous determination of earthquake hypocenters and velocity structure. J. Geophys. Res. 85, 4801–4809.

Romanowicz, B., 1998. Attenuation tomography of the Earth's mantle: a review of current status. Pure Appl. Geophys. 153, 257–272.

Sandwell, D.T., 1987. Biharmonic spline interpolation of GEOS-3 and SEASAT altimeter data. Geophys. Res. Lett. 14, 139–142.

Shure, L., Chave, A.D., 1984. Comments on "An inverse approach to signal correlation". J. Geophys. Res. 89, 2497–2500.

Shure, L., Parker, R.L., Backus, G.E., 1982. Harmonic splines for geomagnetic modeling. Phys. Earth Planet. Inter. 28, 215–229.

Shure, L., Parker, R.L., Langel, R.A., 1985. A preliminary harmonic spline model for MAGSAT data. J. Geophys. Res. 90, 11505–11512.

Sieminski, A., Liu, Q., Trampert, J., Tromp, J., 2007. Finite-frequency sensitivity of surface waves to anisotropy based upon adjoint methods. Geophys. J. Int. 168, 1153–1174.

Slunga, R., Rognvaldsson, S.T., Bodvarsson, B., 1995. Absolute and relative locations of similar events with application to microearthquakes in southern Iceland. Geophys. J. Int. 123, 409–419.

Spence, W., Alexander, S., 1968. A method of determining the relative location of seismic events. Earthq. Notes 39, 13.

Takeuchi, H., Saito, M., 1972. Seismic surface waves. In: Bolt, B.A. (Ed.), Methods of Computational Physics. vol. 11. Academic Press, New York, pp. 217–295.

Tromp, J., Tape, C., Liu, Q., 2005. Seismic tomography, adjoint methods, time reversal and banana-doughnut kernels. Geophys. J. Int. 160, 195–216.

Vasco, D.W., 1986. Extremal inversion of travel time residuals. Bull. Seismol. Soc. Am. 76, 1323–1345.

Vidale, J., 1990. Finite difference calculation of travel times in 3-D. Geophysics 55, 521–526.

Waldhauser, F., Ellsworth, W., 2000. A double-difference earthquake location algorithm; method and application to the northern Hayward Fault, California. Bull. Seismol. Soc. Am. 90, 1353–1368.

Wang, Z., Dahlen, F.A., 1995. Validity of surface-wave ray theory on a laterally heterogeneous Earth. Geophys. J. Int. 123, 757–773.

Wessel, P., Becker, J., 2008. Gridding of spherical data using a Green's function for splines in tension. Geophys. J. Int. 174, 21–28.

Yuan, Y.O., Simons, F.J., Bozdağ, E., 2015. Multiscale adjoint waveform tomography for surface and body waves. Geophysics 80, R281–R302.

Yuan, Simons, Tromp, 2016. Double-difference adjoint seismic tomography. Geophys. J. Int. 206, 1599–1618.

Zelt, C.A., 1998. FAST: 3-D First Arrival Seismic Tomography Programs. http://terra.rice.edu/department/faculty/zelt/fast.html.

Zelt, C.A., Barton, P.J., 1998. 3D seismic refraction tomography: a comparison of two methods applied to data from the Faeroe Basin. J. Geophys. Res. 103, 7187–7210.

Zhang, H., Thurber, C.H., 2003. Double-difference tomography; the method and its application to the Hayward fault, California. Bull. Seismol. Soc. Am. 93, 1875–1889.

Zhu, H., Luo, Y., Nissen-Meyer, T., Morency, C., Tromp, J., 2009. Elastic imaging and time-lapse migration based on adjoint methods. Geophysics 74, WCA167–WCA177.

Important algorithms and method summaries

17.1 Implementing constraints with Lagrange multipliers

Consider the problem of minimizing a function of two variables, say $E(x,y)$, with respect to x and y, subject to the constraint that $C(x,y)=0$. One way to solve this problem is to first use $C(x,y)=0$ to write y as a function of x and then substitute this function into $E(x,y)$. The resulting function $E(x,y(x))$ of a single variable x can now be minimized by setting $dE/dx=0$. The constraint equation is used to explicitly reduce the number of independent variables.

One problem with this method is that it is rarely possible to solve $C(x,y)=0$ explicitly for either $y(x)$ or $x(y)$. The method of Lagrange multipliers provides a method of dealing with the constraints in their implicit form.

The constraint equation $C(x,y)=0$ defines a curve in (x,y) on which the solution must lie (Fig. 17.1). As a point is moved along this curve, the value of E changes, achieving a minimum at (x_0,y_0). This point is not the global minimum of E, which lies, in general, off the curve. However, it must be at a point at which ∇E is perpendicular to the tangent $\hat{t}$ to the curve, or else the point could be moved along the curve to further minimize E. The tangent is perpendicular to ∇C, so the condition that $\nabla E \perp \hat{t}$ is equivalent to $\nabla E \propto \nabla C$. Writing the proportionality factor as $-\lambda$, we have $\nabla E + \lambda \nabla C = 0$ or $\nabla(E+\lambda C)=0$. Thus the constrained minimization of E subject to $C=0$ is equivalent to the unconstrained minimization of $\Phi = E + \lambda C$, except that a new variable λ is introduced that needs to be determined as part of the problem.

The two equations $\nabla\Phi=0$ and $C=0$ must be solved simultaneously to yield (x_0,y_0) and λ. In N-dimensions and with $L<N$ constraint equations, each constraint $C_i=0$ describes an $N-1$ dimensional surface. Their intersection is an $N-L$ dimensional surface. We treat the $L=N-1$ case here, where the intersection is a curve (the argument for the other cases being similar). As in the two-dimensional case, the condition that a point (x_0,y_0) is at a minimum of E on this curve is that ∇E is perpendicular to the tangent $\hat{t}$ to the curve. The curve necessarily lies within all of the intersecting surfaces, and so $\hat{t}$ is perpendicular to any surface normal ∇C_i. Thus the condition that $\nabla E \propto \hat{t}$ is equivalent to the requirement that ∇E be constructed from a linear combination of surface normals ∇C_i. Calling the coefficients of the linear combination $-\lambda_i$, we have

$$\nabla E = - \sum_{i=1}^{L} \lambda_i \nabla C_i \ \text{ or } \ \nabla E + \sum_{i=1}^{L} \lambda_i \nabla C_i = 0 \ \text{ or } \ \nabla\Phi=0 \ \text{ with } \ \Phi \equiv E + \sum_{i=1}^{L} \lambda_i C_i \tag{17.1}$$

As stated earlier, the constrained minimization of E subject to $C_i=0$ is equivalent to the unconstrained minimization of Φ, but now Φ contains one Lagrange multiplier λ_i for each of the L constraints.

17.2 L_2 inverse theory with complex quantities

Some inverse problems (especially those that deal with data that have been operated on by Fourier or other transforms) involve complex quantities that are considered random variables. A complex random variable, say $z=x+iy$, has real and imaginary parts, x and y, that are themselves real-valued random variables. The probability density function $p(z)$ gives the probability that the real part of z is between x and $x+\Delta x$ and an imaginary part is in between y and $y+\Delta y$. The pdf is normalized so that its integral over the complex plane is unity:

$$\int_{-\infty}^{+\infty} \int_{-\infty}^{+\infty} p(z) \, dxdy = 1 \tag{17.2}$$

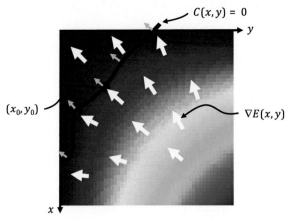

FIG. 17.1   Graphical interpretation of the method of Lagrange multipliers, in which the function, $E(x, y)$, is minimized subject to the constraint that $C(x, y) = 0$. The solution (bold dot) occurs at the point, (x_0, y_0), on the curve, $C(x, y) = 0$, where the perpendicular direction (gray arrows) is parallel to the gradient, $\nabla E(x, y)$ (white arrows). At this point, E can only be further minimized by moving the point, (x_0, y_0), off of the curve, which is disallowed by the constraint. Scripts `gdama17_01` and `gdapy17_01`.

The mean or expected value of z is then the obvious generalization of the real-valued case

$$\langle z \rangle = \int_{-\infty}^{+\infty} \int_{-\infty}^{+\infty} z\, p(z)\, \mathrm{d}x \mathrm{d}y \tag{17.3}$$

In general, the real and imaginary parts of z can be of unequal variance and be correlated. However, the uncorrelated, equal variance case, which is called a *circular random variable*, suffices to describe the behavior of many systems. Its variance is defined as

$$\sigma = \int_{-\infty}^{+\infty} \int_{-\infty}^{+\infty} \left(z - \langle z \rangle\right)^* \left(z - \langle z \rangle\right) p(z)\, \mathrm{d}x \mathrm{d}y = \int_{-\infty}^{+\infty} \int_{-\infty}^{+\infty} |z - \langle z \rangle|^2\, p(z)\, \mathrm{d}x \mathrm{d}y \tag{17.4}$$

Here, the asterisk signifies complex conjugation. A vector $\mathbf{z}$ of N circular random variables, with probability density function $p(\mathbf{z})$ has mean and covariance given by

$$\langle z \rangle_i = \int_{-\infty}^{+\infty} \int_{-\infty}^{+\infty} z_i\, p(\mathbf{z})\, \mathrm{d}^N x \mathrm{d}^N y$$
$$[\mathrm{cov}\, \mathbf{z}]_{ij} = \int_{-\infty}^{+\infty} \int_{-\infty}^{+\infty} \left(z_i - \langle z \rangle_i\right)^* \left(z_j - \langle z \rangle_j\right) p(\mathbf{z})\, \mathrm{d}^N x \mathrm{d}^N y \tag{17.5}$$

Note that $[\mathrm{cov}\, \mathbf{z}]$ is a *Hermitian matrix*, that is, a matrix whose transform is its complex conjugate.

The definition of the L_2 norm must be changed to accommodate the complex nature of quantities. The appropriate change is to define the squared length of a vector $\mathbf{v}$ to be

$$\|\mathbf{v}\|_2^2 \equiv \mathbf{v}^H \mathbf{v} \tag{17.6}$$

where $\mathbf{v}^H$ is the *Hermitian transpose*, that is, the transpose of the complex conjugate of the vector $\mathbf{v}$, that is, $\mathbf{v}^H \equiv \mathbf{v}^{*T}$. This choice ensures that the norm is a nonnegative real number. When the results of L_2 inverse theory are rederived for circular complex vectors, they are very similar to those derived previously; the only difference is that all the ordinary transposes are replaced by Hermitian transposes. For instance, the least-squares solution is

$$\mathbf{m}^{est} = \left[\mathbf{G}^H \mathbf{G}\right]^{-1} \mathbf{G}^H \mathbf{d}^{obs} \tag{17.7}$$

Note that all the square symmetric matrices of real inverse theory now become square Hermitian matrices:

$$[\mathrm{cov}\, \mathbf{d}], [\mathrm{cov}\, \mathbf{m}], \mathbf{G}^H \mathbf{G}, \mathbf{G} \mathbf{G}^H, \text{etc.} \tag{17.8}$$

An example of complex least squares is given in scripts `gdama17_02` and `gdapy17_02`.

Hermitian matrices have real eigenvalues, so no special problems arise when deriving eigenvalues or singular values. The matrices $\mathbf{U}$ and $\mathbf{V}$ in the singular-value decomposition are complex unary matrices, obeying $\mathbf{U}^H = \mathbf{U}^{-1}$ and $\mathbf{V}^H = \mathbf{V}^{-1}$. The modification made to Householder transformations is also very simple. The requirement that the Hermitian length of a vector be invariant under transformation implies that a unitary transformation must satisfy

$\mathbf{T}^H\mathbf{T} = \mathbf{I}$. The most general unitary transformation is therefore $\mathbf{T} = \mathbf{I} - 2\mathbf{v}\mathbf{v}^H/\mathbf{v}^H\mathbf{v}$, where $\mathbf{v}$ is any complex vector. The Householder transformation that annihilates the jth column of $\mathbf{G}$ below its main diagonal then becomes

$$\mathbf{T}_j = \mathbf{I} - \frac{1}{|\alpha_j|(|\alpha_j| + G_{j,j})} \begin{bmatrix} 0 \\ \vdots \\ 0 \\ (G_{j,j} - \alpha_j) \\ G_{j+1,j} \\ \vdots \\ G_{N,j} \end{bmatrix} \begin{bmatrix} 0 & \cdots & 0 & (G_{j,j} - \alpha_j)^* & G_{j+1,j}^* & \cdots & G_{N,j}^* \end{bmatrix} \tag{17.9}$$

$$\text{where } |\alpha_j|^2 = \sum_{i=j}^{N} |G_{ij}|^2$$

The phase of α is chosen to be π away from the phase of $G_{j,j}$. This choice guarantees that the transformation is in fact a unitary transformation and that the denominator is not zero.

17.3   Inverse of a "resized" matrix

Occasions arise in which an inverse problem needs to be repeatedly re-solved, as old data is deemed obsolete and new data is obtained. Many such problems involve a symmetric matrix, say $\mathbf{A}$ (such as the Gram matrix), which is repeatedly modified by removing some of its rows and columns (the ones associated with the obsolete data) and by adding new one (the ones arising from the new data). The process of solving the inverse problem requires that the matrix inverse $\mathbf{A}^{-1}$ to be recalculated. When the total number of modified rows and columns is just a few percent of the total, this matrix inverse can be obtained through a modification process that is more efficient than merely recalculating $\mathbf{A}^{-1}$ from scratch (Menke, 2022).

We begin with the process of removing rows and columns from $\mathbf{A}$ to obtain a smaller matrix $\mathbf{Y}$. At the start of the process, the $N_A \times N_A$ matrix $\mathbf{A}$ and its inverse $\mathbf{A}^{-1}$ are known. These matrices are partitioned into submatrices:

$$\mathbf{A} \equiv \begin{bmatrix} \mathbf{X} & \mathbf{Z}^T \\ \mathbf{Z} & \mathbf{Y} \end{bmatrix} \text{ and } \mathbf{A}^{-1} \equiv \begin{bmatrix} \mathbf{P} & \mathbf{R}^T \\ \mathbf{R} & \mathbf{Q} \end{bmatrix} \tag{17.10}$$

Here, $\mathbf{X}$ and $\mathbf{P}$ are $N_1 \times N_1$, $\mathbf{Y}$ and $\mathbf{Q}$ are $N_2 \times N_2$, and $\mathbf{Z}$ and $\mathbf{R}$ are $N_2 \times N_1$, with $N_1 + N_2 = N_A$. We imagine that the rows and columns associated with $\mathbf{X}$ and $\mathbf{Z}$ are discarded, leaving the smaller matrix $\mathbf{Y}$. Our goal is to use our knowledge of $\mathbf{P}$, $\mathbf{Q}$, and $\mathbf{R}$, to compute $\mathbf{Y}^{-1}$.

The following matrix identity involving $\mathbf{Z}$, $\mathbf{R}$, $\mathbf{P}$, and $\mathbf{Q}$ will be used later in the process:

$$\mathbf{Z}^T \mathbf{R} \mathbf{P}^{-1} [\mathbf{I} - \mathbf{R}^T \mathbf{Z}] + \mathbf{Z}^T \mathbf{Q} \mathbf{Z} = \mathbf{Z}^T \mathbf{R} \mathbf{P}^{-1} - \mathbf{Z}^T \mathbf{R} \mathbf{P}^{-1} \mathbf{R}^T \mathbf{Z} + \mathbf{Z}^T \mathbf{Q} \mathbf{Z} = 0 \tag{17.11}$$

It can be derived by multiplying out the block matrix form of $\mathbf{A}^{-1}\mathbf{A} = \mathbf{I}$, solving the diagonal elements for $\mathbf{X}$ and $\mathbf{Y}$, and substituting the result into the off-diagonal elements. The process of removing the first N_1 data corresponds to replacing $\mathbf{A}$ with $\mathbf{Y}$, and $\mathbf{A}^{-1}$ with $\mathbf{Y}^{-1}$.

The method for computing $\mathbf{Y}^{-1}$ relies on the *Woodbury identity* (Woodbury, 1950):

$$(\mathbf{M} + \mathbf{U}\mathbf{W}\mathbf{V})^{-1} = \mathbf{M}^{-1} - \mathbf{M}^{-1}\mathbf{U}(\mathbf{W}^{-1} + \mathbf{V}\mathbf{M}^{-1}\mathbf{U})^{-1}\mathbf{V}\mathbf{M}^{-1} \tag{17.12}$$

Here, $\mathbf{M}$, $\mathbf{W}$, $\mathbf{U}$, and $\mathbf{V}$ are conformable matrices. The reader may confirm by direct substitution that

$$\mathbf{W} = \mathbf{I} \text{ and } \mathbf{M} = \mathbf{A} \text{ and } (\mathbf{M} + \mathbf{U}\mathbf{W}\mathbf{V})^{-1} = \begin{bmatrix} \mathbf{X}^{-1} & \mathbf{0} \\ \mathbf{0} & \mathbf{Y}^{-1} \end{bmatrix}$$

$$\mathbf{U}_{N,N_1} = \begin{bmatrix} \mathbf{I}_{N_1,N_1} & \mathbf{0}_{N_1,N_1} \\ \mathbf{0}_{N_2,N_1} & -\mathbf{Z}_{N_2,N_1} \end{bmatrix} \text{ and } \mathbf{V}_{N_1,N} = \begin{bmatrix} \mathbf{0}_{N_1,N_1} & -[\mathbf{Z}^T]_{N_1,N_2} \\ \mathbf{I}_{N_1,N_1} & \mathbf{0}_{N_1,N_2} \end{bmatrix} \tag{17.13}$$

are consistent choices of the several matrices (with the subscripts indicating the sizes of selected matrices). Thus $\mathbf{Y}^{-1}$ can be read from the lower right-hand block of $\mathbf{A}^{-1} - \mathbf{A}^{-1}\mathbf{U}(\mathbf{I} + \mathbf{V}\mathbf{A}^{-1}\mathbf{U})^{-1}\mathbf{V}\mathbf{A}^{-1}$. Note that the matrix $(\mathbf{I} + \mathbf{V}\mathbf{A}^{-1}\mathbf{U})$ is $(2N_1) \times (2N_1)$, so that its inverse requires less computational effort than does the $N_2 \times N_2$ matrix $\mathbf{Y}$ as long as a relatively few data are being removed. (Following, we will develop an analytic formula for $(\mathbf{I} + \mathbf{V}\mathbf{A}^{-1}\mathbf{U})^{-1}$ that further reduces the size of the necessary matrices to $N_1 \times N_1$.) An explicit formula for $\mathbf{Y}^{-1}$ is obtained by manipulating the block matrix form of the Woodbury formula, starting with:

$$[\mathbf{I} + \mathbf{V}\mathbf{A}^{-1}\mathbf{U}] = \mathbf{I} + \begin{bmatrix} 0 & -\mathbf{Z}^T \\ \mathbf{I} & 0 \end{bmatrix} \begin{bmatrix} \mathbf{P} & \mathbf{R}^T \\ \mathbf{R} & \mathbf{Q} \end{bmatrix} \begin{bmatrix} \mathbf{I} & 0 \\ 0 & -\mathbf{Z} \end{bmatrix} = \begin{bmatrix} [\mathbf{I}_{N_1,N_1} - \mathbf{Z}^T\mathbf{R}] & \mathbf{Z}^T\mathbf{Q}\mathbf{Z} \\ \mathbf{P} & [\mathbf{I}_{N_1,N_1} - \mathbf{R}^T\mathbf{Z}] \end{bmatrix} \qquad (17.14)$$

Note that the off-diagonal elements of this matrix are symmetric, and the off-diagonal elements are transposes of one another. Its inverse has the form:

$$[\mathbf{I} + \mathbf{V}\mathbf{A}^{-1}\mathbf{U}]^{-1} \equiv \begin{bmatrix} \mathbf{I}_{N_1,N_1} & \mathbf{C}_{N_1,N_1} \\ \mathbf{D}_{N_1,N_1} & \mathbf{I}_{N_1,N_1} \end{bmatrix}$$

$$\text{with } \mathbf{C} = \mathbf{Z}^T\mathbf{R}\mathbf{P}^{-1} = \mathbf{Z}^T\mathbf{R}\mathbf{P}^{-1}\mathbf{R}^T\mathbf{Z} - \mathbf{Z}^T\mathbf{Q}\mathbf{Z} \qquad (17.15)$$

$$\text{and } \mathbf{D} = \mathbf{R}^T\mathbf{Z}[\mathbf{Z}^T\mathbf{Q}\mathbf{Z}]^{-1} = -[\mathbf{P}^{-1} - \mathbf{C}]^{-1}$$

The formulas for $\mathbf{C}$ and $\mathbf{D}$ were derived by multiplying out the block diagonal form of $[\mathbf{I} + \mathbf{V}\mathbf{A}^{-1}\mathbf{U}]^{-1}[\mathbf{I} + \mathbf{V}\mathbf{A}^{-1}\mathbf{U}] = \mathbf{I}$, solving the diagonal elements for $\mathbf{C}$ and $\mathbf{D}$, and applying the identity from Eq. (17.11). Note that both $\mathbf{C}$ and $\mathbf{D}$ are symmetric matrices. The products $\mathbf{A}^{-1}\mathbf{U}$ and $\mathbf{V}\mathbf{A}^{-1}$ have the forms:

$$\mathbf{A}^{-1}\mathbf{U} = \begin{bmatrix} \mathbf{P} & \mathbf{R}^T \\ \mathbf{R} & \mathbf{Q} \end{bmatrix} \begin{bmatrix} \mathbf{I} & 0 \\ 0 & -\mathbf{Z} \end{bmatrix} = \begin{bmatrix} \mathbf{P} & -\mathbf{R}^T\mathbf{Z} \\ \mathbf{R} & -\mathbf{Q}\mathbf{Z} \end{bmatrix} \equiv \begin{bmatrix} \mathbf{P} & \mathbf{S} \\ \mathbf{R} & \mathbf{T} \end{bmatrix}$$

$$\mathbf{V}\mathbf{A}^{-1} = \begin{bmatrix} 0 & -\mathbf{Z}^T \\ \mathbf{I} & 0 \end{bmatrix} \begin{bmatrix} \mathbf{P} & \mathbf{R}^T \\ \mathbf{R} & \mathbf{Q} \end{bmatrix} = \begin{bmatrix} -\mathbf{Z}^T\mathbf{R} & -\mathbf{Z}^T\mathbf{Q} \\ \mathbf{P} & \mathbf{R}^T \end{bmatrix} \equiv \begin{bmatrix} \mathbf{S}^T & \mathbf{T}^T \\ \mathbf{P} & \mathbf{R}^T \end{bmatrix} \qquad (17.16)$$

$$\text{with } \mathbf{S} \equiv -\mathbf{R}^T\mathbf{Z} \text{ and } \mathbf{T} = -\mathbf{Q}\mathbf{Z}$$

Thus the expression $\mathbf{A}^{-1}\mathbf{U}(\mathbf{I} + \mathbf{V}\mathbf{A}^{-1}\mathbf{U})^{-1}\mathbf{V}\mathbf{A}^{-1}$ can be simplified to:

$$\mathbf{A}^{-1}\mathbf{U}(\mathbf{I} + \mathbf{V}\mathbf{A}^{-1}\mathbf{U})^{-1}\mathbf{V}\mathbf{A}^{-1} = \begin{bmatrix} \mathbf{P} & \mathbf{S} \\ \mathbf{R} & \mathbf{T} \end{bmatrix} \begin{bmatrix} \mathbf{I} & \mathbf{C} \\ \mathbf{D} & \mathbf{I} \end{bmatrix} \begin{bmatrix} \mathbf{S}^T & \mathbf{T}^T \\ \mathbf{P} & \mathbf{R}^T \end{bmatrix} =$$

$$\begin{bmatrix} \{\mathbf{P}(\mathbf{S}^T + \mathbf{C}\mathbf{P}) + \mathbf{S}(\mathbf{D}\mathbf{S}^T + \mathbf{P})\} & \{\mathbf{P}(\mathbf{T}^T + \mathbf{C}\mathbf{R}^T) + \mathbf{S}(\mathbf{D}\mathbf{T}^T + \mathbf{R}^T)\} \\ \{\mathbf{R}(\mathbf{S}^T + \mathbf{C}\mathbf{P}) + \mathbf{T}(\mathbf{D}\mathbf{S}^T + \mathbf{P})\} & \{\mathbf{R}(\mathbf{T}^T + \mathbf{C}\mathbf{R}^T) + \mathbf{T}(\mathbf{D}\mathbf{T}^T + \mathbf{R}^T)\} \end{bmatrix} \qquad (17.17)$$

The lower right-hand block of (11) becomes the new $\mathbf{A}^{-1}$:

$$\mathbf{A} \text{ is replaced with } \mathbf{Y}$$

$$\mathbf{A}^{-1} \text{ is replaced with } \mathbf{Y}^{-1} = \mathbf{Q} - ((\mathbf{R}\mathbf{T}^T + \mathbf{T}\mathbf{R}^T) + \mathbf{R}\mathbf{C}\mathbf{R}^T + \mathbf{T}\mathbf{D}\mathbf{T}^T) \qquad (17.18)$$

We now describe the process, based on the well-known *Bordering method* (Householder, 1975), of adding rows and columns to a symmetric matric $\mathbf{X}$ to obtain a larger matrix $\mathbf{A}$. Both $\mathbf{X}$ and its inverse $\mathbf{X}^{-1}$ are known $N_2 \times N_2$ matrices. We seek to add the blocks $\mathbf{Y}$ and $\mathbf{Z}$ associated with the N_3 new data, creating an augmented $N_B \times N_B$ matrix $\mathbf{A}$ with $N_B = N_2 + N_3$. The matrix $\mathbf{A}$ and its inverse satisfy:

$$\mathbf{A}^{-1}\mathbf{A} \equiv \begin{bmatrix} \mathbf{P} & \mathbf{R}^T \\ \mathbf{R} & \mathbf{Q} \end{bmatrix} \begin{bmatrix} \mathbf{X} & \mathbf{Z}^T \\ \mathbf{Z} & \mathbf{Y} \end{bmatrix} = \begin{bmatrix} \mathbf{I}_{N_2N_2} & 0_{N_2N_3} \\ 0_{N_3N_2} & \mathbf{I}_{N_3N_3} \end{bmatrix} \qquad (17.19)$$

Multiplying out the equation and solving for $\mathbf{P}$, $\mathbf{Q}$, and $\mathbf{R}$ yields:

$$\mathbf{P}\mathbf{X} + \mathbf{R}^T\mathbf{Z} = \mathbf{I} \text{ so } \mathbf{P} = [\mathbf{I} - \mathbf{R}^T\mathbf{Z}]\mathbf{X}^{-1}$$

$$\mathbf{P}\mathbf{Z}^T + \mathbf{R}^T\mathbf{Y} = 0 \text{ so } \mathbf{R}^T = -\mathbf{P}\mathbf{Z}^T\mathbf{Y}^{-1}$$

$$\mathbf{R}\mathbf{X} + \mathbf{Q}\mathbf{Z} = 0 \text{ so } \mathbf{R} = -\mathbf{Q}\mathbf{Z}\mathbf{X}^{-1} \qquad (17.20)$$

$$\mathbf{R}\mathbf{Z}^T + \mathbf{Q}\mathbf{Y} = \mathbf{I} \text{ so } -\mathbf{Q}\mathbf{Z}\mathbf{X}^{-1}\mathbf{Z}^T + \mathbf{Q}\mathbf{Y} = \mathbf{I} \text{ so } \mathbf{Q} = [\mathbf{Y} - \mathbf{Z}\mathbf{X}^{-1}\mathbf{Z}^T]^{-1}$$

The matrices then become:

$$A \text{ is replaced with } \begin{bmatrix} X & Z^T \\ Z & Y \end{bmatrix}$$

$$A^{-1} \text{ is replaced with } \begin{bmatrix} P & R^T \\ R & Q \end{bmatrix} \tag{17.21}$$

Although the delete/append operation can be repeated many (e.g., hundreds) of times without significant loss of precision—at least for the class of matrices encountered in inverse theory. However, as a precaution, we recommend that the inverse be corrected every few hundred iterations, either though the direct calculation of A^{-1} from A or through one application of the *Schultz method* (Li, 2014):

$$A^{-1} \text{ is replaced by } A^{-1}\left(2I - AA^{-1}\right) \tag{17.22}$$

These formulas have been tested numerically and are correct to within machine precision (scripts gdama17_03 and gdapy17_03). Accuracy and speed are improved thorough the use of coding techniques that utilize and enforce the symmetry of all the symmetric matrices.

17.4 Method summaries

Method summary 1, generalized least squares

Step 1: State the problem in words
How are the data related to the model?

Step 2: Organize the problem in standard form
Identify the data d (length N) and the model parameters m (length M).
Define the data kernel G so that $Gm = d^{obs}$.

Step 3: Examine the data
Make plots of the data and look for outliers and other problems.

Step 4: Establish the accuracy of the data
State a prior covariance matrix [cov d] based on accuracy of the measurement technique.
For uncorrelated data with uniform variance [cov d]$= \sigma_d^2 I$.

Step 5: State the prior information in words, for example:
A linear function of model parameters is close to a known value h^{pri}.
The mean of the model parameters is close to a known value.
The model parameters vary smoothly with space and/or time.

Step 6: Organize the prior information in standard form:
$Hm = h^{pri}$ where h^{pri} is of length K

Step 7: Establish the accuracy of the prior information
State a prior covariance matrix [cov h] based on accuracy of the prior information.
For uncorrelated information of uniform variance [cov h]$=\sigma_h^2 I$.

Step 8: Estimate model parameter m^{est} and their covariance [cov m]

$$m^{est} = \left[F^T F\right]^{-1} F^T f^{obs} \text{ and } [\text{cov } m] = \left[F^T F\right]^{-1}$$

$$\text{with } F = \begin{bmatrix} [\text{cov } d]^{-\frac{1}{2}} G \\ [\text{cov } h]^{-\frac{1}{2}} H \end{bmatrix} \text{ and } f^{obs} = \begin{bmatrix} [\text{cov } d]^{-\frac{1}{2}} d^{obs} \\ [\text{cov } h]^{-\frac{1}{2}} h^{pri} \end{bmatrix}$$

For the uniform, uncorrelated case:

$$F = \begin{bmatrix} \sigma_d^{-1} G \\ \sigma_h^{-1} H \end{bmatrix} \text{ and } f^{obs} = \begin{bmatrix} \sigma_d^{-1} d^{obs} \\ \sigma_h^{-1} h^{pri} \end{bmatrix}$$

Step 9: Examine the model resolution

$$\mathbf{R}^G = \mathbf{G}^{-g}\mathbf{G} \text{ with } \mathbf{G}^{-g} = \left[\mathbf{F}^T\mathbf{F}\right]^{-1}\mathbf{G}^T[\text{cov } \mathbf{d}]^{-1}$$

Any departure of $\mathbf{R}^G$ from a diagonal matrix indicates that only averages of the model parameters can be resolved.

Are these averages localized?

Step 10: State estimates and their 95% confidence intervals

$$m_i^{true} = m_i^{est} \pm 2\sigma_{mi} \text{ (95\%) with } \sigma_{mi} = \sqrt{[\text{cov } \mathbf{m}]_{ii}}$$

Step 11: Examine the individual errors

$$\mathbf{d}^{pre} = \mathbf{G}\mathbf{m}^{est} \text{ and } \mathbf{e} = [\text{cov } \mathbf{d}]^{-\frac{1}{2}}\left(\mathbf{d}^{obs} - \mathbf{d}^{pre}\right)$$

$$\mathbf{h}^{pre} = \mathbf{H}\mathbf{m}^{est} \text{ and } \mathbf{l} = [\text{cov } \mathbf{h}]^{-\frac{1}{2}}\left(\mathbf{h}^{pri} - \mathbf{h}^{pre}\right)$$

Make plots of e_i vs. i and l_i vs. i.

Make scatter plots of d_i^{pre} vs. d_i^{obs} and scatter plot of h_i^{pre} vs. h_i^{pre}.

Any unusually large errors?

Step 12: Examine the total error Φ^{est}

$$\Phi^{est} = E^{est} + L^{est} \text{ with } E^{est} = \mathbf{e}^T\mathbf{e} \text{ and } L^{est} = \mathbf{l}^T\mathbf{l}$$

Φ^{est} is chi-squared distributed with $\nu = N + K - M$ degrees of freedom.

Use a chi-squared test to assess the likelihood of the null hypothesis that Φ^{est} is different than the expected vale of ν only due to random variation.

The 95% confidence interval is

$$\nu - 2(2\nu)^{\frac{1}{2}} \leq \Phi \leq \nu + 2(2\nu)^{\frac{1}{2}}$$

Step 13: Two different models, A and B?

Use an F-test on

$$F = \left(\Phi_A^{est}/\nu_A\right)/\left(\Phi_B^{est}/\nu_B\right)$$

to assess the likelihood of the null hypothesis that F is different from unity only due to random variation.

(See scripts `gdama17_04` and `gdapy17_04`)

Method summary 2, the grid search

Step 1: Which model parameters are to be searched?

Grid searches over more than a few parameters are impractical.

Dividing model parameters into two groups, one found through a search and the other by least squares, may be possible.

Step 2: Identify the range and sampling of the search

The range must bracket the solution.

The sampling of grid nodes $\mathbf{m}^{(k)}$ must not be so coarse that the point of minimum error is missed.

Step 3: Determine the error $E(\mathbf{m}^{(k)})$ at each grid point $\mathbf{m}^{(k)}$

Predict the data $\mathbf{d}^{pre} = \mathbf{g}(\mathbf{m}^{(k)})$.

Compute the least squared error $E(\mathbf{m}^{(k)}) = \mathbf{e}^T\mathbf{e}$ with $\mathbf{e} = \mathbf{d}^{obs} - \mathbf{d}^{pre}$.

Step 4: Find the grid node that minimizes the error

$\mathbf{m}^{min}$ is the $\mathbf{m}^{(k)}$ that minimizes $E(\mathbf{m}^{(k)})$.

$\mathbf{m}^{min}$ is a preliminary estimate of $\mathbf{m}^{est}$.

Step 5: Refine the solution using a quadratic approximation

Determine E_0, $\mathbf{b}$, and $\mathbf{B}$ in

$E(\mathbf{m}) = E_0 + \mathbf{b}^T\Delta\mathbf{m} + \frac{1}{2}(\Delta\mathbf{m})^T\mathbf{B}\Delta\mathbf{m}$ with $\Delta\mathbf{m} = \mathbf{m} - \mathbf{m}^{min}$

using least square on a few grid nodes surrounding $\mathbf{m}^{min}$.

then $\mathbf{m}^{est} = \mathbf{m}^{min} + \Delta\mathbf{m}$ with $\Delta\mathbf{m} = -\mathbf{B}^{-1}\mathbf{b}$

Step 6: Estimate the covariance of the solution

$[\text{cov } \mathbf{m}] = \sigma_d^2 [\tfrac{1}{2}\mathbf{B}]^{-1}$ with $\sigma_d^2 = E(\mathbf{m}^{est})/(N-M)$

(See scripts gdama17_05 and gdapy17_05)

Method summary 3, nonlinear least squares

Step 1: State the problem in words

How are the data related to the model parameters?

What prior information is applicable?

Step 2: Organize the problem in standard form

N data, M model parameters, K pieces of information.

Data equation $\mathbf{g}(\mathbf{m}) = \mathbf{d}^{obs}$ and prior information equation $\mathbf{Hm} = \mathbf{h}^{pri}$.

Step 3: Decide up a reasonable trial solution

Trial solution $\mathbf{m}^{(k)}$ for $k=1$.

Step 4: Linearize the data equation

$$\mathbf{G}^{(k)}\Delta\mathbf{m}^{(k)} = \Delta\mathbf{d}^{(k)} \text{ and } \mathbf{H}\Delta\mathbf{m}^{(k)} = \Delta\mathbf{h}^{(k)}$$

$$\text{with } G_{ij}^{(k)} = \left.\frac{\partial g_i}{\partial m_j}\right|_{\mathbf{m}^{(k)}}$$

$$\text{and } \Delta\mathbf{d}^{(k)} = \mathbf{d}^{obs} - \mathbf{g}\left(\mathbf{m}^{(k)}\right) \text{ and } \Delta\mathbf{h}^{(k)} = \mathbf{h}^{pri} - \mathbf{Hm}^{(k)}$$

Step 5: Form the combined equation

$$\mathbf{F}^{(k)}\Delta\mathbf{m}^{(k)} = \mathbf{f}^{(k)} \text{ with}$$

$$\mathbf{F}^{(k)} = \begin{bmatrix} [\text{cov } \mathbf{d}]^{-\frac{1}{2}}\mathbf{G}^{(k)} \\ [\text{cov } \mathbf{h}]^{-\frac{1}{2}}\mathbf{H} \end{bmatrix} \text{ and } \mathbf{f}^{(k)} = \begin{bmatrix} [\text{cov } \mathbf{d}]^{-\frac{1}{2}}\Delta\mathbf{d}^{(k)} \\ [\text{cov } \mathbf{h}]^{-\frac{1}{2}}\Delta\mathbf{h}^{(k)} \end{bmatrix}$$

Step 6: Iteratively improve the solution

Solution: $\Delta\mathbf{m}^{(k)} = [\mathbf{F}^{(k)T}\mathbf{F}^{(k)}]^{-1}\mathbf{F}^{(k)T}\mathbf{f}^{(k)}$.

Update rule: $\mathbf{m}^{(k+1)} = \mathbf{m}^{(k)} + \Delta\mathbf{m}^{(k)}$.

Recompute $\mathbf{G}^{(k)}$ and $\mathbf{F}^{(k)}$ between each iteration.

Step 7: Stop iterating when

$\Delta\mathbf{m}^{(k)} \approx 0$ or $\Phi^{est} \approx N+K-M$.

Solution $\mathbf{m}^{est}$ is $\mathbf{m}^{(k)}$ of final iteration.

Covariance $[\text{cov } \mathbf{m}]$ is $[\mathbf{F}^{(k)T}\mathbf{F}^{(k)}]^{-1}$ of final iteration.

Steps 8–12: Examine the results

Use the $\mathbf{G}^{(k)}$ and $\mathbf{F}^{(k)}$ from the last iteration to examine the model resolution.

State 95% confidence intervals for $\mathbf{m}^{est}$.

Examine the individual errors. $\mathbf{e} = \mathbf{d}^{obs} - \mathbf{g}(\mathbf{m}^{est})$.

Apply chi-squared test to the total error, Φ^{est}.

Test the difference between two different models by following the procedures outlined in Method Summary 1 (but recognizing that these results are only approximate).

(See scripts gdama17_06 and gdapy17_06)

Method summary 4, MCMC inversion

Step 1: Organize the problem in standard form

Identify the data $\mathbf{d}$ (length N) and the model parameters $\mathbf{m}$ (length M)

Define the relationship $\mathbf{g}(\mathbf{m}) = \mathbf{d}^{obs}$ that relates model to data

Step 2: Decide whether MCMC is appropriate

Intermediate number M of model parameters.

Theory $\mathbf{g}(\mathbf{m})$ that is fast to evaluate.

Properties of $p(\mathbf{d}^{obs}|\mathbf{m})$ unsuited for Newton's method (e.g., multimodal).

A need to assess complicated properties of ensemble solution.

Step 3: Examine the data

Make plots of the data and look for outliers and other problems.

Step 4: Establish the accuracy of the data
State a prior covariance matrix [cov **d**] based on accuracy of the measurement technique.
For uncorrelated data with uniform variance [cov **d**] $= \sigma_d^2 \mathbf{I}$.

Step 5: Choose a likelihood pdf describing the behavior of the data
A Normal pdf is often appropriate

$$p\left(\mathbf{d}^{obs}|\mathbf{m}\right) \propto \exp\left\{-\frac{1}{2}\left[\mathbf{d}^{obs} - \mathbf{g}(\mathbf{m})\right]^{T}[\text{cov } \mathbf{d}]^{-1}\left[\mathbf{d}^{obs} - \mathbf{g}(\mathbf{m})\right]\right\}$$

Step 6: State the prior information
The model parameters are close to a known value $\langle \mathbf{m} \rangle$.

Step 7: State the accuracy of the prior information
The prior knowledge has an accuracy of $[\text{cov } \mathbf{m}]_A$.
For uncorrelated prior information with uniform variance $[\text{cov } \mathbf{m}]_A = \sigma_m^2 \mathbf{I}$.

Step 8: Choose a prior pdf describing the behavior of the prior model
A Normal pdf is often appropriate

$$p_A(\mathbf{m}) \propto \exp\left\{-\frac{1}{2}[\mathbf{m} - \langle \mathbf{m} \rangle]^{T}[\text{cov } \mathbf{m}]_A^{-1}[\mathbf{m} - \langle \mathbf{m} \rangle]\right\}$$

Step 9: Form the logarithm of the posterior pdf

$$L(\mathbf{m}) = \log p\left(\mathbf{m}|\mathbf{d}^{obs}\right) = \log p\left(\mathbf{d}^{obs}|\mathbf{m}\right) + \log p_A(\mathbf{m})$$

Step 10: Choose a pdf for finding a proposed successor
Successor $\mathbf{m}'$ should be in the neighborhood of $\mathbf{m}^{(i)}$.
The pdf $q(\mathbf{m}'|\mathbf{m}^{(i)})$ must be easy to sample.
A symmetrical pdf obeying $q(\mathbf{m}'|\mathbf{m}^{(i)}) = q(\mathbf{m}^{(i)}|\mathbf{m}')$ simplifies the math.
A Normal $q(\mathbf{m}'|\mathbf{m}^{(i)})$ with mean $\mathbf{m}^{(i)}$ and variance σ_q^q controlling size of neighborhood is often appropriate:

$$q\left(\mathbf{m}'|\mathbf{m}^{(i)}\right) = q\left(\mathbf{m}^{(i)}|\mathbf{m}'\right) \propto \exp\left(-\frac{1}{2}\sigma_q^{-2}\left(\mathbf{m}' - \mathbf{m}^{(i)}\right)^{T}\left(\mathbf{m}' - \mathbf{m}^{(i)}\right)\right)$$

Step 11: Choose length and first value of the Markov chain of models
Decide upon the length K of the chain needed to adequately sample $p(\mathbf{m}|\mathbf{d}^{obs})$ $\mathbf{m}^{(1)}$ should have a reasonably large $L(\mathbf{m}^{(1)})$.

Step 12: Build the Markov chain using the Metropolis-Hastings algorithm:
Create a proposed successor $\mathbf{m}'$ to the current link in the chain $\mathbf{m}^{(i)}$ by sampling $q(\mathbf{m}'|\mathbf{m}^{(i)})$.
Form test parameter $\gamma = L(\mathbf{m}') - L(\mathbf{m}^{(i)}) + \log q(\mathbf{m}^{(i)}|\mathbf{m}') - \log q(\mathbf{m}'|\mathbf{m}^{(i)})$.
Note $\log q$ terms cancel when q is symmetric.
If $\gamma \geq 0$ accept successor, so $\mathbf{m}^{(i+1)} = \mathbf{m}'$.
if $\gamma < 0$ generate a uniformly distributed random number $0 \leq r \leq 1$ and accept successor if $\exp(\gamma) \geq r$, $\mathbf{m}^{(i+1)} = \mathbf{m}'$ else set $\mathbf{m}^{(i+1)} = \mathbf{m}^{(i)}$.
Repeat $K-1$ times to build chain.
Discard beginning of chain to allow for burn-in.
Consider concatenating several chains with different $\mathbf{m}^{(1)}$s.

Step 13: Assess the ensemble solution
Compute histograms of interesting scalar function $s_i = f(\mathbf{m}^{(i)})$ of model parameters.
Sample mean, sample covariance proxies for $\mathbf{m}^{est}$ and the posterior [cov **m**]
(but they are typically similar to those estimated using Newton's method).
(See scripts gdama17_07 and gdapy17_07)

Method summary 5, bootstrap confidence intervals

Step 1: Identify the quantity q for which confidence intervals are needed
The quantity q might be a model parameter in an inverse problem or a function of several model parameters.
Is a quicker method available for determining the confidence of q?
If so, consider using it in preference to bootstrapping.

Step 2: Identify the data and a method of estimating q from the data
The data $\mathbf{d}^{obs}$ are of length N.
The method usually involves solving an inverse problem.
Step 3: For reference, estimate q^{est} from the observed data $\mathbf{d}^{obs}$
Step 4: Resample the data and recompute q
Let $\mathbf{j}$ be a list of N random integers, each in the range 1 to N.
The resampled data is $d_i^{rs} = d_{j_i}^{obs}$.
Determine q^{rs} from $\mathbf{d}^{rs}$ using the method in Step 2.
Step 5: Repeat Step 4 many times
Repeat many, say $N_r \approx 1000$, times.
Tabulate the results in a vector $\mathbf{q}^{rs}$ of length N_r.
Step 6: Construct an empirical pdf for q
Decide upon a reasonable number N_q and range (q^{min}, q^{max}) of histogram bins by examining the range of values in $\mathbf{q}^{rs}$.
Define a vector $\mathbf{q}$ of histogram bin locations

$$q_i = q^{min} + (i-1)\Delta q \ \text{ with } \ \Delta q = (q^{max} - q^{min})/(N_q - 1)$$

Compute a histogram $\mathbf{h}$ of the values in $\mathbf{q}^{rs}$ using this binning.
Normalize the histogram to unit area so it is an empirical pdf.

$$p(q_i) = h_i/A \text{ with } A = \Delta q \sum_{j=1}^{N_q} h_j$$

Integrate $p(q)$ into the cumulative probability $P(q)$.

$$P(q_i) = \Delta q \sum_{j=1}^{N_q} p(q_j)$$

Step 7: Determine 95% confidence limits for q from $P(q)$
The left bound q_L is the q for which $P(q) \approx 0.025$.
The right bound q_R is the q for which $P(q) \approx 0.975$.
Plot $p(q)$, q_L, q_R, and q^{est} and interpret results.
The 95% confidence interval for q^{true} is approximately $q_L < q^{true} < q_R$.
(See scripts gdama17_08 and gdapy17_08)

Method summary 6, factor analysis

Step 1: State the problem in words
What are samples, elements, and factors in this problem?
Is the idea that samples are a linear mixture of factors appropriate?
Are there natural orderings in space, time, etc.?
Step 2: Organize the data as a sample matrix S

$$\mathbf{S}^{obs} = \begin{bmatrix} \text{element 1 in sample 1} & \cdots & \text{element } M \text{ in sample 1} \\ \cdots & \cdots & \cdots \\ \text{element 1 in sample } N & \cdots & \text{element } M \text{ in sample } N \end{bmatrix}$$

Step 3: Establish weights that reflect the importance of the elements
State w_i for $i = 1, \cdots, M$
One possibility: $w_i = 1/\sigma_i$ with σ_i^2 the prior variance of element i.
Step 4: Perform singular-value decomposition
$\mathbf{U}\mathbf{\Lambda}\mathbf{V}^T = \mathbf{S}^{obs}\mathbf{W}$ with $W_{ij} = \delta_{ij} w_i$
Step 5: Determine the number P of important factors
Plot the diagonal of $\mathbf{\Lambda}$ as a function of row index i and choose P to include all rows with "large" Λ_{ii}.

Step 6: Reduce the number of factors from M to P and form the factor matrix $\mathbf{F}_P$ and loading matrix $\mathbf{C}_P$

$$\mathbf{F}_P = \mathbf{V}_P^T \mathbf{W}_P^{-1} \text{ and } \mathbf{C}_P = \mathbf{U}_P \mathbf{\Sigma}_P$$

$$\mathbf{F}_P = \begin{bmatrix} \text{element 1 in factor 1} & \cdots & \text{element } M \text{ in factor 1} \\ \cdots & \cdots & \cdots \\ \text{element 1 in factor } P & \cdots & \text{element } M \text{ in factor } P \end{bmatrix}$$

$$\mathbf{C}_P = \begin{bmatrix} \text{factor 1 in sample 1} & \cdots & \text{factor } P \text{ in sample 1} \\ \cdots & \cdots & \cdots \\ \text{factor 1 in sample } N & \cdots & \text{factor } P \text{ in sample } N \end{bmatrix}$$

Step 7: Predict the data and examine the error

$\mathbf{S}^{pre} = \mathbf{C}_P \, \mathbf{F}_P$ and $E = \mathbf{S}^{obs} - \mathbf{S}^{pre}$

Step 8: Interpret the factors and their loadings

Pattern of elements in the factors?

Spatial/temporal pattern of the loading?

(See scripts gdama17_09 and gdapy17_09)

References

Householder, A.S., 1975. The Theory of Matrices in Numerical Analysis. Dover, New York. ISBN-13: 978-0486449722.

Li, R.C., 2014. Matrix Perturbation Theory, in Hogben, Handbook of Linear Algebra, second ed. CRC Press, Boca Raton, FL, ISBN: 978-1-138199-89-7.

Menke, W., 2022. Rolling Gaussian process regression with application to regime shifts. Appl. Math. 13, 859–868.

Woodbury, M., 1950. Inverting modified matrices. Memorandum Report Stat. Res. Group Princeton 42, Princeton Universit, Princeton, NJ.

Index

Note: Page numbers followed by *f* indicate figures and *t* indicate tables.